ANIMAL SCIENCE, ISSUES AND PROFESSIONS

POST-CAMBRIAN TESTATE FORAMINIFERA AS A SYSTEM IN ITS EVOLUTION

Animal Science, Issues and Professions

Additional books in this series can be found on Nova's website under the Series tab.

Additional e-books in this series can be found on Nova's website under the e-book tab.

ANIMAL SCIENCE, ISSUES AND PROFESSIONS

POST-CAMBRIAN TESTATE FORAMINIFERA AS A SYSTEM IN ITS EVOLUTION

V. I. MIKHALEVICH

Zoological Institute, Russian Academy of Sciences,
Laboratory of Protozoology, Russia

nova publishers

New York

For permission to use material from this book please contact us:
Telephone 631-231-7269; Fax 631-231-8175
Web Site: http://www.novapublishers.com

Library of Congress Cataloging-in-Publication Data
Mikhalevich, V. I. (Valeriia Iosifovna)
Post-Cambrian testate foraminifera as a system in its evolution / Alla-Valeria Mikhalevich.
pages cm
Includes bibliographical references and index.
ISBN: 978-1-62618-772-6 (hardcover)
1. Foraminifera. 2. Foraminifera--Evolution. 3. Foraminifera, Fossil. I. Title.
QL368.F6M584 2013
561'.994--dc23

2013028054

Published by Nova Science Publishers, Inc. † New York

To the memory of Dagmara Rauser-Chernousova,
Alexander Fursenko, Alfred Richard Loeblich Jr.,
and Helen Tappan

CONTENTS

PREFACE

Foraminifera is a big group of unicellular eukaryotic and mostly sea animals, having a hard shell of complexity striking for unicellular organisms. It includes over 4,500 genera and about 40,000 species, mostly now extinct. Their usual size is about 0.5 mm but some giant forms may reach 17 to 20 cm. The multi-chamberedness of their advanced forms with their complex inner integrative systems represents the evolutionary attempt to overcome a unicellular level of organization. The role of the ramified system of the thinnest canals in their multi-chambered shells may be compared with the role of the blood system in Metazoans. The beauty of their shells attracted Leonardo da Vinci to describe them in his ornaments. The highest morphological variability of their shells along with the small number of their chromosomes may permit their usage in the future as classical objects of genetic-morphological studies. Their species are used as bio-indicators of recent and paleo environments, as well as markers of the carbohydrates bearing strata.

This book summarizes the author's and literary data on the ultrastructure of foraminiferal living cells and skeleton along with some biological data and represents fundamentally new concepts of the evolution and classification of the phylum Foraminifera elaborated by the author since the beginning of 1980s and later supported by the molecular data of the Western school. Foraminifera is the group having the most full and representative geological record providing a way to see and study their evolution in the most demonstrative way. This permits the author on the basis of comparative morphological analysis of this and some other related unicellular eukaryotic groups to study in detail the main evolutionary regularities at the unicellular level of life which previously have been studied mostly in multicellular organisms. Research is concentrated about the evolutionary regularities representing the mainstream of evolutionary development and permitting the structures to achieve a new and higher level of organization, opening pathways to further evolutionary changes. The general character of these regularities which can be applied to the social level of life as well is examined.

This book may be of interest to specialists in protozoology, micropaleontology, oceanography, ecology, stratigraphy, and evolution, and to university courses and students, and everybody else for whom the question of how life is organized and how it develops is of interest.

ACKNOWLEDGMENTS

I am thankful to my Alma Mater – the Zoological Institute of the Russian Academy of Sciences and its Laboratory of Protozoology (founded by V.A. Dogiel, later headed by A.A. Strelkov, and from 1975 on by M.V. Krylov) for the opportunity given to me to work in my chosen path. I am also deeply indebted to the outstanding Russian biologists-evolutionists Ja. I. Starobogatov, O.G. Kusakin, V.R. Dolnik, V.V. Khlebovich, Ju. A. Labas, E.S. Sugonjaev, and to the protozoologists M.V. Krylov, S.S. Shulman, L.N. Seravin, I.B. Raykov. They have all influenced my formation as a scientist. The support of foreign colleagues, especially that of H. Tappan and A.R. Loeblich became a source of strength when I took my first steps in the new direction of foraminiferal macrotaxonomy. Many thanks go to J.-P. Debenay for the opportunity to study foraminifera in living cultures in his laboratory at the University of Anger in France and to the colleagues of the laboratory V. Stauff, F. Redois, F. Sylvestre et al. for their attention and every-day help. I am grateful to V. Yanko-Hombach who invited me to my first conference abroad after the fall of the iron curtain and to M. McGann and P. Hallock for their active support of my studies. My special thanks go to my colleagues Mary McGann and Sev Kender who generously agreed to edit the English of the present book and also to Andrew Gooday and Pamela Hallock who edited the book scientifically. Their comments proved highly valuable and helped me in improving the manuscript.

And I am most grateful to M. Kaminski for his constant support and for organizing in 2005 an International Working Group on the Foraminiferal Classification (IWGFC) which permitted me to discuss at an international level my ideas and the results of my previous work published in Russian. I gratefully acknowledge the Grzybowski Foundation for according me a Honorable Fellowship and the Grzybowski Award 2006.

Chapter I

PHYLUM FORAMINIFERA D'ORBIGNY, 1826

Unicellular eukaryotic organisms whose body, in all but the most primitive forms, is protected by a shell. In primitive forms the shell is organic and is composed of proteins and mucopolysacharids. In the more advanced forms shell may be agglutinated or agglutinated with partly secreted cement. The most advanced have shells secreted throughout, occurring as either monofontinal (=monolamellar) (microgranular, porcellaneous, hyaline-radial) with pseudopores or bifontinal (= biamellar) with complex pore canaliculi. The shell may be unilocular, pseudocolonial, pseudochambered and multi-chambered, with chamber number usually 6 - 15, 30, or more rarely, up to several hundreds (super-multi-chambered). In the latter case, the chambers may be strongly differentiated. Chambers are connected with each other by foramina and in the advanced groups by more complex integrative systems: stolons, tunnels, inner apertural systems and canals. Shells have one, two or several external openings referred to as apertures, although in the most primitive forms the aperture may be absent. In advanced groups the inner apertural structures (teeth, ribs, and tooth-plates of differering complexity) and additional apertures arise. The aperture serves for protrusion of the granuloreticulopodia, which usually are branching and anastomosing, forming a reticulum around the shell. Life cycles are characterized by an alternation of sexual and asexual generations (with the intermediate reduction). Gametes are mostly biflagellate and heterocontous, although more rarely three flagellate or amoeboid. One of the generations (more often the sexual one) may be absent. Nuclear apparatus with the complex changes during the life cycle, only in the lower forms part of the life cycle is mononuclear, in the rest ones it is multinucliate during the whole life cycle (except zygote), in the most advanced forms nuclear dimorphism arises. Meiosis is two-stepped, with the nucleus dividing in the process of closed pleuromitoses. Mitochondrial cristae are tubular. Planktonic forms have cryptosomes and microvilli. Foraminifera are mostly sea bottom free-living organisms, but a planktonic group also exists. Some primitive fresh-water and terrestrial forms are also known. Parasitic forms are very rare, epibiotic forms more abundant. Geological record gives stable data on the occurrence of their skeletal forms since the Cambrian to Holocene, whereas naked ones are first recorded from the Prae-Cambrian (the origin of the whole group according to molecular timescale - about 950 and 1350 mil. years ago (Groussin et al. 2011).

Chapter II

GENERAL CHARACTERISTICS

Foraminifera represent one of the biggest, most peculiar, and most advanced groups among the unicellular eukaryotes. They now include over 4500 genera and about 40 000 species, most of which are extinct, as the number of recent species ranges from 5000 to 10 000 according to various authors. There are nearly 200 000 first descriptions in the Foraminiferal Catalogue (USA) (Hottinger, 1982) and more than 275 thousands foraminiferal species mentioned, obviously including synonyms, in the popular literature. Foraminifera represent one third of all the known protists and 2.5% of all the known Earth organisms (Loeblich & Tappan, 1964). Their position among the other unicellular eukaryotes is, according to Keeling, 2001 and Nikolaev et al., 2004, within supergroup Rhizaria, close to the heterogenous assemblage Cercozoa and to Haplosporidia and Gromiidae.

One of the most characteristic features of the group in the majority of its representatives is the presence of a shell (also referred to as a test), which may be organic in the simplest cases (soft-walled forms), later agglutinated (foreign particles, mostly sand grains, bound on the cell surface with organic or calcareous cement), and then fully calcareous secreted in their most advanced forms.This particular study focuses on the Post-Cambrian testate forms. Foraminiferal shell form is extremely variable (Plates **1, 1a, 2**). The shell structure is different in the different foraminiferal classes which represent special phylogenetic developmental lineages. The shell ultrastructure differs only in the more advanced subclasses with the secreted calcareous shell wall, hence it can be used as a designative feature of rather high (subclass level) taxonomic rank, nevertheless as a character of subordinate meaning comparing the shell morphology. The foraminiferal shell is usually covered by the cytoplasmic layer, hence, in a strict sense, their skeleton can be considered being the inner rather than the outer layer (Plate **16**, figure-s 2,3, Plate **18**, figure 5).

Even the more important designative feature is the structure of their shell serving as the main base in the construction of the foraminiferal system. In the simplest case the shells are unilocular or bilocular. In the majority of the multi-chambered shells, chamber number does not exceed from five to thirty, but there are also highly organized forms with the chamber number up to several hundreds and more (Plate **1 a, 2, 5, 35, 37**). In such cases, the chambers may be differentiated in their form and sometimes even in their function, showing a high organizational level among the unicellular eukaryotes (see Part V, 1.1). The multiformity of the foraminiferal shells and the variability of their types of symmetry is so large that these organisms are an exceptional model in the investigation of the fundamental principles of morphogenesis. In the multi-chambered forms, different types of the integrative systems arose (Plates **5**, figure - s, **4** c - e, **40**; see below, Part V. 1. 3), they permit the organism to overcome the disconnection of their separate sub-units elements. These systems are represented by stolons, inner apertural integrative systems, tunnels, interlocular cavities and canals (Plates **5**, figure – s, **4** c - e, **40**; see below, Part V. 1. 3). The canals have their own calcareous walls and are characterized with a special complexity, the ectoplasm is circulating in them. The canal function may be compared with that of the blood system in vertebrates but with the vessels having hard calcareous walls. The canal system is a unique organismal system representing a high level of organization which occurs among unicellular eukaryotes only in Foraminifera and is of complexity striking for unicellular organisms. The development of a high degree of polymerization and differentiation of the organismic structures, as well as a rise of complex integrative systems, represent a unique evolutionary development comparable in some aspects with the development of the multicellular organisms (see below, Part VIII). Not only the complexity of their organism but their life span also resembles that of multicellular animals: their life cycle usually takes a year, and in some large multi-chambered forms the agamonts (see Part VI. 5) need several years to reach reproductive size (Krueger et al., 1996).

The foraminiferal cytoplasmic body is contained mainly inside the shell, coming in contact with the outside by means of a net of fine anastomosing granuloreticulopodia, through special openings called apertures (see Part VI. 1, Plate **16**, figure 4 A, Plate **23**, figure – s 4 - 8). Their main function is for movement and capturing food. Granuloreticulopodia are morhphologically distinct from the locomotive structures of other unicellular organisms such as the wide amoebian lobopodia and even more so from cilia and flagella, thus serving as a diagnostic feature of the phylum rank and defining Foraminifera among the other protists.

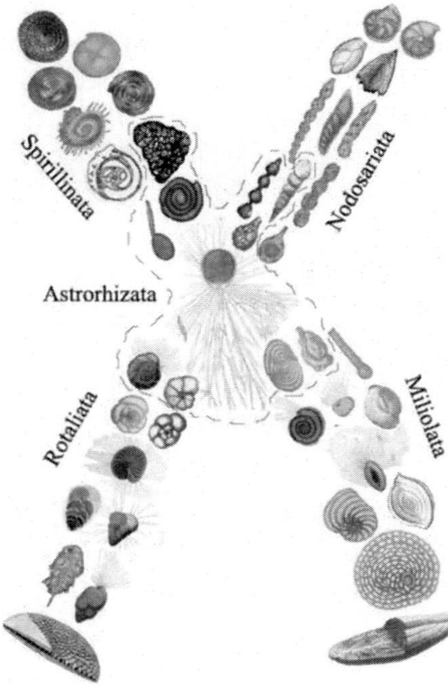

Plate 1. The main phyletic foraminiferal lines (classes) according to the classification of Mikhalevich (1992) (scheme, not to scale) (the dotted line outlines the previous heterogenous taxon Textulariina of Loeblich & Tappan, 1987).

Note the unilocular shells at the beginning of Nodosariata and Miliolata, pseudo-two-chambered shells with the long tubular second chamber at the beginning of Spirillinata and Miliolata lines, and super-multi-chambered shells at the end of Miliolata and Rotaliata lines.

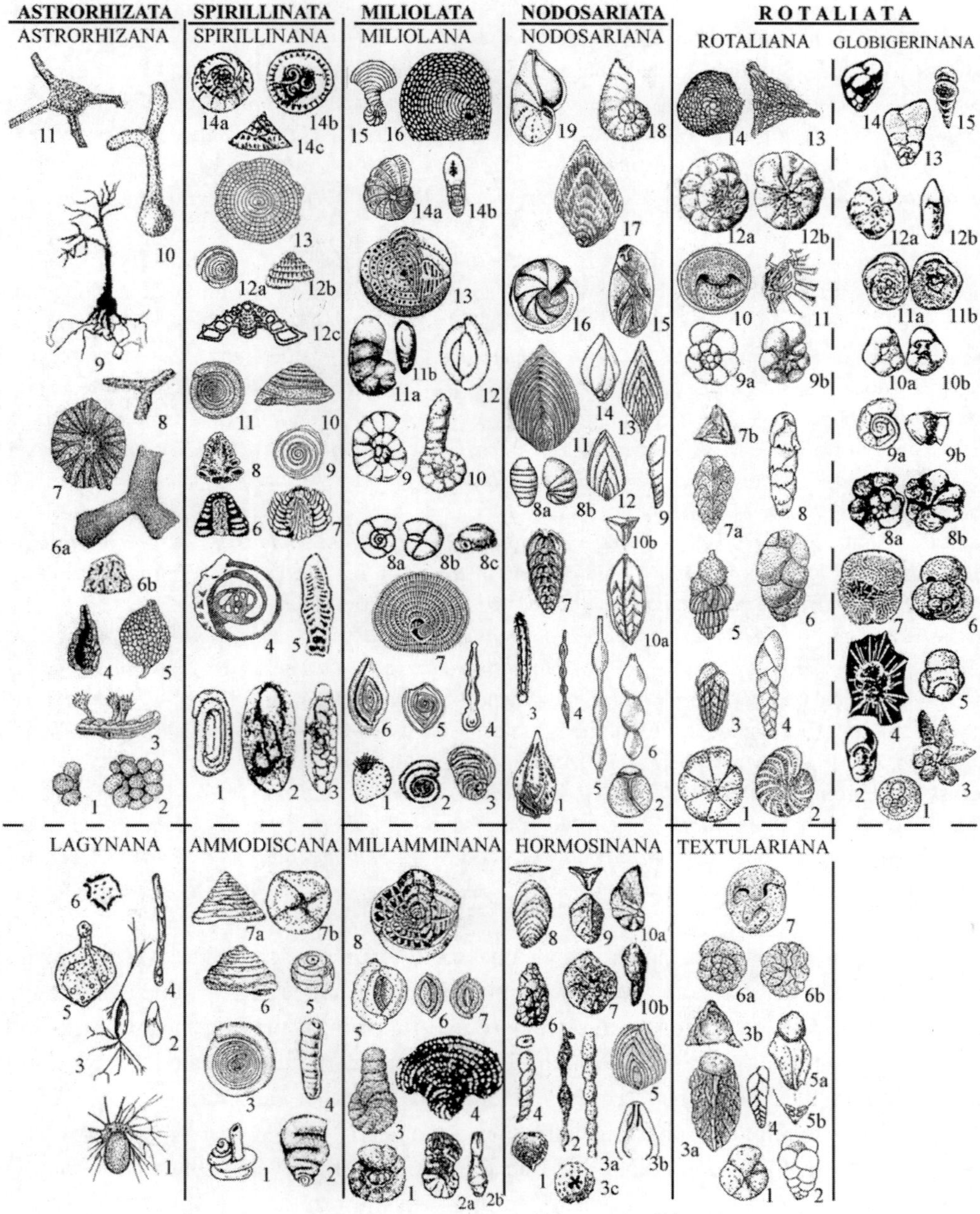

Plate 1a. The main phyletic Foraminiferal classes and subclasses (unbroken line – boundary between the classes, dotted line – between the subclasses) (after Mikhalevich, 2000, modified).

I. Class Astrorhizata. Subclass Lagynana: 1 – *Diplogromia*, 2 – *Lagynis*, 3 – *Phainogulmia*, 4 - *Xenothekella*, 5 – *Heterogromia*, 6 - *Microcometes*. Subclass Astrorhizana: 1 – *Sorostomasphaera*, 2 – *Thuramminopsis*, 3 – *Halyphysema*, 4 – *Lagenammina*, 5 – *Astrammina*, 6 a, b - *Schizammina* (a - the shell, b – inner structure of the labyrintic wall), 7 – *Masonella* (with canaliculi in the thick wall), 8 – *Rhabdammina*, 9 – *Notodendrodes*, 10 – *Astrorhiza*, 11 – *Saccorhiza*.

II. Class Spirillinata. Subclass Ammodiscana: 1 – *Ammovertellina*, 2 *Ammovertella*, 3 – *Ammodiscus*, 4 – *Turritelella*, 5 - *Repmanina*, 6 – *Arenoturrispirillina,* 7a,b - *Tetrataxis*.
Subclass Spirillinana: 1 – *Miliospirella*, 2 – *Glomodiscus*, 3 – *Archaediscus dubitabilis*, 4 – *A. karreri*, 5 - *Cylindrotrocholina*, 6 - *Howchinia*, 7 – *Lasiotrochus*, 8 – *Babelispirrillina*, 9 – *Coronipora*, 10 – *Trocholina*, 11 – *Spirillina*, 12 – *Spirotrocholina* (a,b - views of the shell, c – scheme of the canal, axial section), 13 – *Annulopatellina*, 14 – *Paleopatellina*.
III. Class Miliolata. Subclass Rzehakrinana: 1 – *Recurvoides*, 2a,b – *Charentia*, 3 – *Lituola*, 4 – *Alzonella*, 5 – *Dentostomina*, 6 - *Sigmoilopsis*, 7 – *Ammomassilina,* 8 - *Reticulinella*. Subclass Miliolana: 1 – *Squamulina*, 2 – *Cornuspira*, 3 – *Cornuspiroides*, 4 – *Gheorgianina*, 5 – *Cornuloculina*, 6 – *Spirophthalmidium*, 7 – *Discospirina*, 8 - *Fischerinella*, 9 – *Zoyaella*, 10 – *Spirolina*, 11a,b – *Danubiella*, 12 - *Quinqueloculina*, 13 – *Neoalveolina*, 14a,b -*Dendritina*, 15 – *Laevipeneroplis*, 16 - *Parasorites*.
IV. Class Nodosariata. Subclass Hormosinana: 1 – *Saccammina*, 2 - *Hormosinella*, 3 a, b, c - *Nodosinum* (a – view of the shell, b – view of the radial aperture in section with the inner longitudinal ribs, c – view from the apertural side), 4 - *Adelungia*, 5 – *Pseudopalmula*, 6 – *Nouria*, 7 – *Agardhella*, 8 – *Flabellammina*, 9 – *Triplasia,* ***10 a, b*** – *Ammomarginulina*(a – side view, b – apertural view). Subclass Nodosariana: 1 - *Lagena*, 2 - *Parafissurina*, 3 - *Syzrania*, 4, 5 – *Grigelis*, 6 - *Nodosaria*, 7 - *Multiseptida* (scheme of the longitudinal section), 8 a, b - *Lingulina* (a – microspheric, b – megalospheric forms), 9 – *Marginulina*, 10 a, b - *Tristix*, 11 – *Kyphopyxa*, 12 – *Dyofrondicularia*, 13 - *Flabellina*, 14 - *Polymorphina*, 15 - *Laryngosigma*, 16 - *Lenticulina,* 17 - *Planularia*, 18 – *Hemicristellaria*, 19 - *Saracenaria*.
V. Class Rotaliata. Subclass Textulariana: 1 – *Haplophragmoides*, 2 - *Minouxia*, 3 a, b - *Gaudryina*, 4 - *Pseudobolivina*, 5 a, b - *Clavulina* (a – outer view of the shell, b – section of the initial part), 6 a, b - *Asterotrochammina,* 7 – *Tiphotrocha* (view from the umbilical side).
Subclass Rotaliana: 1 - *Bermudezinella*, 2 - *Elphidium*, 3 – *Brizalina*, 4 – *Bolivinellina*, 5 - *Euuvigerina*, 6 – *Sporobulimina*, 7a,b – *Reussella*, 8 – *Pseudobuliminella*, 9a,b - *Discorbis*, 10 – *Neoconorbina* (view from the umbilical side), 11 – *Rotalia* (fragment of canal system), 12 a, b – *Ammonia*, 13 - *Baculogypsinoides* (horizontal section of the megalospheric individual with the canals),14 - *Eulinderina* (median section of the megalospheric form).
Subclass Globigerinana: 1 – *Orbulina*, 2 – *Globanomalina*, 3 – *Boliella*, 4 – *Hastigerina*, 5 – *Globigerinita*, 6 – *Globigerinoides*, 7 – *Globigerina*, 8 a, b – *Rotalipora*, 9 a, b – *Globorotalia*, 10a, b – *Rugoglobigerina*, 11a, b – *Globotruncana*, 12 a, b – *Plananomalina*, 13 – *Heterochelix*, 14 – *Gumbelitria*, 15 – *Laterostomella* (side view).

The Foraminiferal nuclear apparatus also have the features of the high level organization and peculiarity: their cells are often multinucleate and have nuclear dualism which is manifested in the differentiation of the structure and function of the nuclear apparatus in the generative and somatic nuclei. The foraminiferal nuclei undergo complicated changes during their life cycle which are characterized by heterophase (biphasic) alternating generations. At that the haploid- and diploid- phases usually constitute nearly even parts of the life cycle (See Parts VI. 4 and VI. 5). The classical diphasic foraminiferal life cycle usually takes one year and sometimes even two years or even more. Only in the primitive genus *Allogromia* it is completed within eight hours (Grell, 1973, Lee et al., 1979). Some cases are known where the parent continues to live following the spring release of gametes (Krueger et al., 1997). It is a widely-held opinion that the duration of the foraminiferal life cycle is adapted to the annual recurrence of nutrient abundance, particularly seaweed blooming. The long periods of the vegetative lifetime demands activity and life strategy comparable with such of the multicellular organisms. In this regard, the inherently low growth rate and unusually long lifetime of foraminifera are not typical among unicellular eukaryotes.

The multi-chamberedness of their advanced forms with their complex inner integrative systems, the differentiation of their nuclear apparatus along with the unusual among the unicellular Eukaryotes duration of their life span represents the evolutionary attempt to overcome a unicellular level of organization.

In the overwhelming majority of cases, foraminiferal gametes have two, or rarely three flagella. Occasionally, amoeboid gametes are encountered which obviously have lost their flagella. Mitochondria of both adult individuals and gametes possess tubular cristae.

The average size of the majority of the species is 0.5 mm and the weight 0.1 – 0.5 mg. Among the rotaliellids and glabratellids there are some groups of much smaller size (0.1 mm) (Pawlowski & Lee, 1991). The majority of the foraminifera are hardly visible with the naked eye. However, rarely fossil as well as recent forms may be found as large as a coin (2 – 3cm) or even a palm (12 – 16 cm up to as large as 20 cm) (e.g, the shells of some fossil and recent Nummulitids, recent *Cornuspiroides*, and rod-shaped *Bathysiphon*). These are real giants in the protozoan world. The mechanism by which such large cells function is not yet studied though it would be of great interest.

The overwhelming majority of foraminiferal species are sea bottom-dwelling forms moving along seaweed stolons, hydroids, and on the sediment surface. Others can bury themselves in the ground as deep as 2 – 10 cm, sometimes down to even 30 cm (Goldstein & Harben, 1993). Most of them are free living, although less commonly they may be attached or pseudocolonial. Attached forms of non-related remote groups have acquired convergent morphological similarities, they have adapted themselves to attachment and spreading on the substrate surface. Some commensal forms are also known but parasitic ones are extremely rare representing the cases of ectoparasitism.

A moderate number of Foraminifera have undergone profound morphologic changes to allow them to adapt a pelagic mode of life and float in the upper water mass (mostly not deeper than 200 m). In addition, forms which were previously considered as benthic were recently discovered to live a double life, partially as a benthic and also as a pelagic form (species of *Bolivina/Streptochilus*– Darling et al., 2009).

Foraminifera are practically omnivorous, using bacteria, yeast, algae, other protists, small metazoan animals sometimes exceeding them in size, their eggs and larvae, detritus, and organic matter dissolved in water as food sources. For a long time it was considered that foraminifera themselves only rarely served as food for other organisms and their role in the food chain was hard to estimate. In a few of the known cases (e.g., engulfment by fish or Echinodermata) they are more likely used as a source of lime. One study showed that the foraminifera were still living after being digested by fish (Debenay et al., 2011). However, predation on foraminifera by deep-sea scaphopods has been observed (Langer et al., 1995). The study of the role of foraminifera in the food webs is ongoing (Delaca et al., 1981, Langer et al., 1995, Gudmundsson et al., 2000, Nomaki et al., 2008, Rossignol et al., 2007, Nomaki et al., 2008, Dupuy et al., 2010, Debenay et al., 2011). Even the attempt to analyse the role of foraminiferal species *Ammonia tepida* in the complex primary and secondary consumer levels of marine communities was made (Dupuy et al., 2010). It is clear now that foraminifera represent an important intermediate links in the lower trophic levels of benthic and pelagic marine communities and it became possible to study their role in marine food webs (Lipps,1983, Lipps & Valentine, 1970, Gooday et al., 1992, M. Sokolova, 2000, Rivkin & Delaca, 1990).

Testate foraminifera are nearly exclusively sea organisms constituting one of the most important groups of the oceanic benthos in all the areas of the World Ocean – from the intertidal zone down to the abyssal planes. A rather small number of species inhabit the desalinated waters of estuaries and other paralic environments (Debenay, 2000, 2009, Debenay & Guillou, 2002, Debenay et al., 2001, 2003), surviving in fresh to hypersaline waters or in brackish water lakes (e.g. *Cribroelphidium gunteri* in Northern lake Winnipegosis, Manitoba, Canada – Robert et al., 2001; *Paratrochammina minutissima* in lakes of Vestfold Hills, East Antactic - Gibson et al., 2008), as well as in ground water of higher salinity (Brodsky, 1928, Gauthier-Lievre, 1935, Yankovskaya & Mikhalevich, 1972, Mikhalevich, 1976). The presence of two unilocular soft-walled foraminiferal groups was also discovered in sediment samples of fresh water environments with the use of molecular methods (Meisterfeld et al., 2001, Pawlowski & Holzmann, 2002, Holzmann & Pawlowski, 2002, Holzmann et al., 2003). According to the latter authors, the results of this study challenge the traditional view of Foraminifera as essentially marine organisms, suggesting they repeatedly colonized fresh-water environments. They were found in lakes in Europe, North America and Australia. Such forms need further investigations. Earlier, an advanced calcareous form (*Cribroelphidium gunteri*) was found living in such great abundance that it sometimes made up most of the sediment in a lake in Canada with salinities as low as 1 – 2 ‰ (Boundreau et al., 2001). The data of Lejzerowicz et al., 2010 showed that foraminifera of the early foraminiferal lineages may also be a component of soil communities,

The biomass of the sea benthic foraminifera may reach from 20 to 80%, sometimes even exceeding the biomass of the macrobenthos (Chusid, 1988). There may be so many that they resemble "living sand" (Lee, 1998, Hyams-Kaphzan & Lee, 2008). Attempts by Movellan & Howa (2012) to estimate the biomass of the planktonic foraminifera gave an approximate range of from 0.008 to 0.033 Gt yr^{-1}, even without considering the juveniles. Occupying rather vast spaces of the sea floor and upper horizons of the sea water masses, foraminifera represent an extremely significant reservoir of the reserve energy.

Many foraminiferal species can serve as indicators of water mass characteristics indicative of pollution or abundance of nutrients, which in their turn can be used to find perspective fishery regions. A special protective system was discovered in some benthic foraminiferal epibionts (Yanko et al., 1994). When present, this character may be used as a highly sensitive biomarker for the prediction of regions of anthropogenic pollution and its ecological consequences.

The shells of dead organisms contribute to the formation of sea sediment, composing one of the most important components of the thickness of the sedimentary cover of the World Ocean at all the stages of its formation. The presence of destroyed, eroded, and partially dissolved shells and calcitic foraminiferal shell elements (globular crystallites, rhombohedral elements, larger shell fragments) were shown associated with clay particles (Debenay et al., 1999, 2007). Taking into account that foraminifera can produce as much carbonate mass as 1 kg m^2 a year and that they contribute 50% of the total biomass of the environment (Snider et al., 1984), their contribution to sediment volume is considerable. Thus, foraminifera are a source of calcite, Mg/calcite and more rarely aragonite to the sediment. Not only are the benthic forms deposited in the sediment but the planktonic forms as well. Their dead shells rain down to the ocean floor, passing through the thick water masses with an average velocity of 2 cm/sec and forming so called blue or globigerine silts.

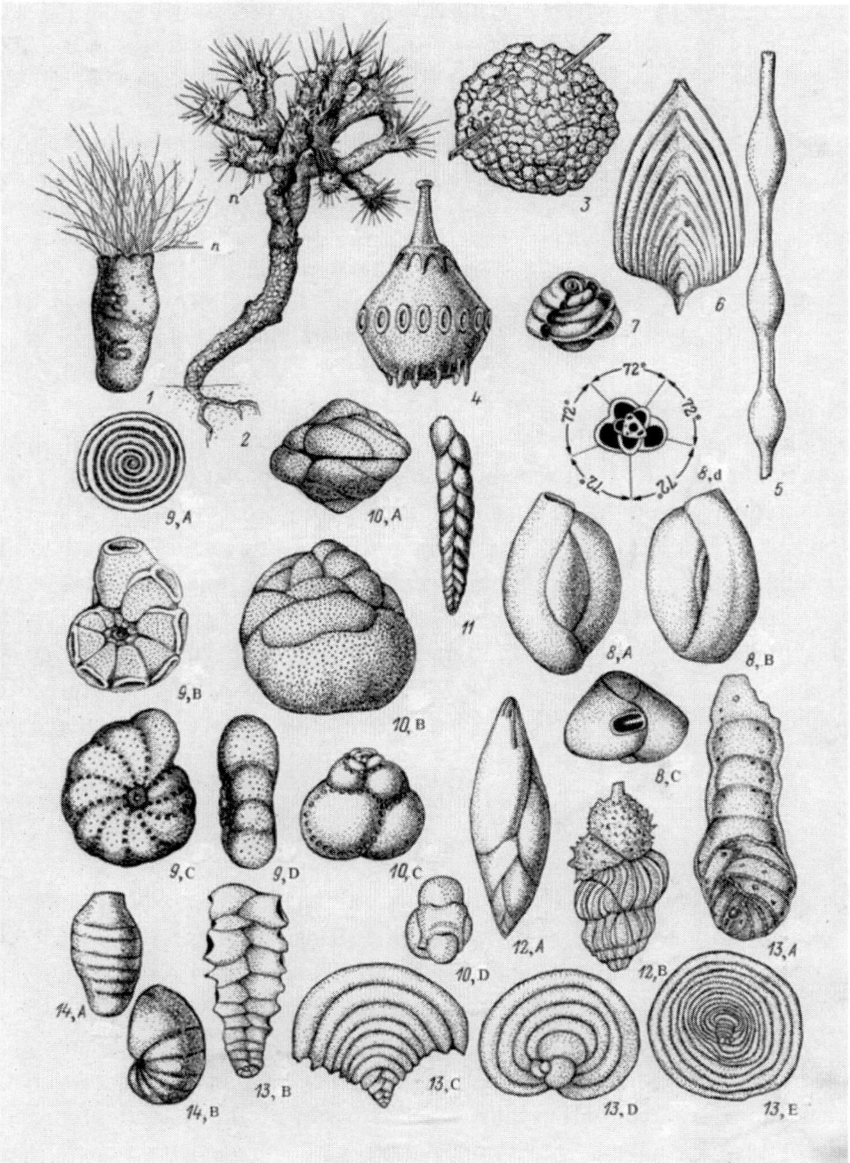

Plate 2 . Structure of the shell.

1 - Unilocular organic shell of *Plagiophrys* (Astrorhizata, Lagynana), P – podostyle; 2 - Branching unilocular agglutinated shell bored in the soil – *Halyphysema* (Astrorhizata, Astrorhizana), using Sponge spicules; 3 - *Psammosphaera* (Astrorhizata, Astrorhizana), unilocular agglutinated shell without definite aperture, with the Sponge spicule; 4 - *Lagena* (Nodosariata, Nodosariana), unilocular lime secreted shell with the sculpture; 5 - *Grigelis* (Nodosariata, Nodosariana), uniserial shell with the pseudochambers separated from each other by the shell wall constrictions; 6 - *Frondicularia* (Nodosariata, Nodosariana), uniserial shell with V- shaped chambers; 7 - *Glomospira* (Spirillinata, Ammodiscana), irregular-glomerate shell with the whorls formed by the long tubular chamber; 8 - *Quinqueloculina* (Miliolata, Miliolana), regular-glomerate shell with the two tubular chambers per whorl (a, b, c – outward appearance, d – transverse section); 9 - Planispiral shells: A- *Spirillina* (Spirillinata, Spirillinana), evolute shell with the long tubular second chamber; B – *Almaena* (Rotaliata, Rotaliana), partly evolute shell with additional apertures at the peripheral chamber ends; C, D –

Elphidium (Rotaliata, Rotaliana), involute shell with the canal processes opened in the sutures; 10 - Trochospiral shells: A – *Neoglabratella* (Rotaliata, Rotaliana), syzigy (Myers, 1940), B - *Tretomphalus* (Rotaliata, Rotaliana) with the floating chamber, C – *Candeina* (Globigerinana) with the sutural apertures, D – *Globigerinita* (Globigerinana) with bulla (additional skeletal plate); 11 - *Bolivina* (Rotaliata, Rotaliana), biserial shell; 12 - High trochospiral shells: A – *Polymorphina* (Nodosariata, Nodosariana), entosolenian tube is seen through the shell wall of the last chamber, B – *Euuvigerina* (Rotaliata, Rotaliana) with external structure; 13 - Heteromorphic shells: A – *Vulvulina* (Rotaliata, Textulariana), B – *Fissotextularia* (Rotaliata, Textulariana) with the additional apertures at the peripheral ends of the chambers , C – *Pavonina* (Rotaliata, Rotaliana), D, E - *Vandenbroeckia* (Miliolata, Miliolana), D – megalo- , E - microspheric forms; 14 - *Lingulina* (Nodosariata, Nodosariana), A – uniserial megalosperic form, B – coiled microspheric form.

They occupy approximately one third of the oceanic bottom surface (nearly 120 million square kilometres) forming thicknesses up to several meters in some locations. Belts of sediment enriched in foraminiferal remains stretch over the ocean bottom from pole to pole, changing in latitudinal direction and containing the predominant foraminiferal species in each of them (Beljaeva & Saidova, 1967). These foraminiferal silts eventually turn into limestones formed by chemical and tectonic processes over the course of time. Therefore, foraminifera serve as their constitutive rock components. They, along with the coccolitophorids, are also a part of chalk. In Cretaceous time, foraminifera became the basic part of limestone deposition in environments with a deficit of CO_2. For this reason, the World Oceans turned out to be saturated with dissolved calcium carbonate which is an active absorber of carbonic gas:

$$CaCO_3 + CO_2 \xrightarrow{H_2O} Ca(HCO_3)_2$$

Bicarbonate, Ca $(HCO_3)_2$ - is a well dissolved salt capable of being accumulated in the ocean for an arbitrarily long time, thus saving the planet from the greenhouse effect and overheating. Such is the biospheric role of foraminifera, contributing to the carbonate/bicarbonate balance of nature (Mikhalevich, 1999).

Foraminiferal shells are known since the Cambrian; there are also some records of their earlier occurrences (Plotnikova & Egorova, 1982, Gausher & Sprechmann, 1999, Bosak et al,, 2011). Foraminifera are the only group whose fossilized remains are represented and investigated so fully during the entirety of Earth's geological history. This circumstance, along with the features of their shell structure in which their ontogenesis often subsequently repeats the features of their ancestral form, provides a unique opportunity to investigate their phylogeny and the composition of their natural classification system. The soft-walled foraminifera with organic shell originated as early as Neo-Proterozoic time according to estimates received by molecular methods (Groussin et al., 2011) (see Part XI. 1).

Empty shell complexes at the oceanic floor (thanatocoenoses) reflect the environmental conditions of the living complexes and are used in paleo reconstructions of previous ecologic and climatic conditions, outlines of ancient continents, and in some cases, estimating coseismic events or tsunami layers (Williams, 1999). Foraminifera are indispensable in the determination of the age of sedimentary rocks, including oil- and gas-bearing ones. Because of their important role in the search for these energy sources, the last two centuries were only focused on studying their shells. The basic data on their living cell and biology were obtained only at the end of the 20th Century.

Chapter III

HISTORICAL SKETCH

Foraminifera were first mentioned in the 5th century B.C. when Herodot noticed some rounded matter resembling coins in Egyptian limestones, which later were called nummulites. In the 18th – 19th Centuries in Russia, such limestones were called “lentil stones”, and limestones with elongated fusulinid shells resembling grains were referred to as “bread stones”. Leonardo da Vinci described smaller foraminifera (*Elphidium*) which he examined by means of a tube with two magnifying glasses at the ends long before the microscope was invented. Several foraminiferal images exist from the 18th Century those of Beccarius(1731), Janus Plancus(1739; cited after Loeblich & Tappan, 1964) and some later pictures when foraminifera were already regarded as living organisms. At the end of 18th and beginning of the 19th Centuries, the first scientific works devoted to this group appeared, including Fisher, 1829, Batsch, 1791, Soldani, 1789-1798, Fichtel et Moll, 1798, Lamarck, 1801, and Montfort, 1808 (cited after Loeblich & Tappan, 1964). The first foraminiferal classifications based on their shell structure were devised at the beginning of the 19th Century - De Blainville, 1825, and d’Orbigny, 1826, 1839 (cited after Loeblich & Tappan, 1964). The name foraminifera, as bearing the openings (foramens), was introduced by d’ Orbigny (1839). He regarded them as a class containing 10 families, five of which are still valid. In all the publications mentioned above, foraminifera were regarded as mollusks up to Dujardin, 1835, who attributed them to the Protists and more specifically to the Rhizopode group based on the fact that they possess thread-like branching pseudopodia.

The concept of the composition of the group was widened owing to the number of works devoted to their morphology and classification (Schultze, 1854, Carpenter, Parker & Jones, 1862, Reuss, 1862, Jones, 1875 – cited after Loeblich & Tappan, 1987), Schwager, 1877, and many new taxa were described. Schwager (1877), Reiss (1862), and Schultze (1854) especially paid a significant attention to the shell wall structure. Schultze regarded this feature as having even higher taxonomic value than the shell structure and he also started the study of the living cell of these protists.

The works of the English scientist Brady (1884) initiated a new phase in the study of this group based on the vast materials of recent forms collected on the vessel “Challenger”. The atlas of Brady’s drawings which were accompanied by new taxonomic notes of Barker (1960) did not lose its significance until recently in spite of the fact that Brady’s understanding of the species volume was rather wide. The subsequent classifications of the end of the 19th and beginning of the 20th Centuries [Neymayr, 1887, 1889, Delage et Herouard, 1896, Eimer et

Fickert, 1899, Lister, 1903, Phumbler, 1911, Schubert, 1921 - (cited after Loeblich & Tappan, 1987)] mostly further developed the classification system offered by Brady. The most original among these are the works of Neymayr, 1887, Rhumbler (1911) and Lister (1903) (all cited after Loeblich & Tappan, 1987); the latter paid significant attention to the study of the shell's dimorphism which is connected with the alternating of generations discovered earlier by Munier-Chalmas (1880).

In the first half of the 20th century, Cushman [1925, 1927 - (cited after Loeblich & Tappan, 1987), 1948] published a three-part revision of Brady's classification system. He recognized fifty foraminiferal families, evoking a negative reaction from European specialists who called his system a "mad classification." For several decades, however, his approach has served as a basis for foraminiferal systematics and has been more popular than other schemes [Galloway, 1933, Chapman, et Parr, 1936, Glaessner, 1945, Hofker, 1951, Sigal, 1952, Pokorny, 1958 - (cited after Loeblich & Tappan, 1987)]. Two of these may be thought of as the most original: Hofker (1951) used apertural structures as basic taxonomic features, while Sigal (1952) emphasized chamber numbers; they recognized 23 and 62 families, respectively.

The first attempt to give the foraminiferal classification system a firm phylogenetic foundation was that of Neymayr (1887, 1889) (cited after Loeblich & Tappan, 1987). Later authors, starting with Cushman, paid much attention to this problem. However, the most consistent and profound phylogenetical evolutionary synthesis, based both on morphology and the geological record, was provided by Rauser-Chernousova and Fursenko (1959) in "Osnovy paleontologii". Unfortunately it did not cover all the diversity of the known forms and included only the groups found in the territories of the former USSR.

The second half of the 20th century was characterized by electron microscopic investigations of the shell wall ultrastructure. A number of such works became classical (Reiss, 1957, 1959, 1960, Reiss & Merling, 1958 (cited after Loeblich & Tappan, 1987), Towe & Cifelli, 1967, Hansen & Reiss, 1971 a, b, Bellemo, 1974 a, b, 1976, Hemleben et al., 1977, Spindler, 1978, Leutenegger & Hansen, 1979, Be et al., 1980, Debenay et al., 1996 b, 1998, 2000, Andersen, 1984, Gooday & Smart, 2000 and many others). Some studies were undertaken in Russia despite the technical limitation in the country at that time (Krasheninnikov, 1956, Grigyalis & Gorbatchik, 1980, Saidova, 1980, Maslakova, 1983, Mikhalevich et al., 1986, Gubenko, 1988, 1989, Plotnikova & Egorova, 1982, Afanasieva, 1982). The first steps in this direction were made by Reiss (1957, 1959 - cited after Loeblich & Tappan, 1987) who reorganized the Rotaliacea on the basis of observations of their shell wall (Reiss, 1963). The credit for the discovery of the bilamellar wall belongs to him. He placed rotaliaceans with a bilamellar wall in separate taxa, distinct from groups with a similar shell morphology. The same approach was adopted in the two volumes of the "Treatise" on foraminiferal classification of Loeblich & Tappan (1964). This monumental work became an indispensable tool for all foraminiferologists during a quarter of the century. Later, it was shown that all the Rotaliacea possess bilamellar walls, a fact that was not discovered initially because of methodical difficulties. As a result, the closely related groups within discorbids and cassidulinids were reunited. Although enthusiasm for discovering new ultrastructural details sometimes resulted in ill-conceived taxonomic changes, in general these technical advances have led to positive developments in foraminiferal taxonomy and a significant increase in the number of new taxa of different rank. While the 1964 'Treatise" included 127 foraminiferal families, Loeblich and Tappan's (1987) later publication incorporated an additional 104 new families and 207 subfamilies, including 40 families, 66 subfamilies and 173 genera that were described since 1980.

Along with electron-microscopic investigations, light microscope studies also continued. The works of Hottinger and his co-authors (1978 – 2006, see References) using the both microscopic methods are striking for their detailed observations and profound morpho-functional analysis. In the classification systems of the end of the 20th century the two approaches were also evident. The first places greatest emphasis on the shell-wall structure and the second on the structure of the organism and its shell. Loeblich & Tappan (1985, 1987, 1992) and their followers continued to develop the first approach. The position of some taxa in their classification may be considered questionable. Nevertheless, their two-volume 1987 publication, which includes diagnoses and illustrations of all known foraminiferal genera and revised synonymies of all generic and suprageneric taxa, is the most complete available treatise, the significance of which cannot be overemphasized. Without this book all the later progress in this field would hardly be possible. It also provides the basis for all other fields of foraminiferal studies. However, as Haynes (1990) justly believes, the use of this classification as a standard, as well as the success of applied paleontology, may impede new developments in taxonomy. This, in its turn, can lead to mistaken conclusions in stratigraphy. As Haynes points out, the separation of the former family Globigerinidae into globigerinids and globorotaliids helped to reveal sharp breaks in Cretaceous and Quaternary sequences. Similarly, the precise definition of the Cretaceous-Tertiary boundary was only possible when orbitoidal foraminifera, which initially had been placed in one family, were separated into distinct taxa. Haynes (1990) stresses that "changes of this kind are clearly a sign of scientific health' and complains about their being reminiscent of the lamentations that arose because Copernicus destroyed the stability of the Ptolemaic system".

The second approach to the classification of foraminifera is evident in the series of works of Russian authors (Rauser-Chernousova & Fursenko, 1959; Rauser-Chernousova et al., 1996; Fursenko, 1981; Subbotina et al., 1981; Mikhalevich, 1980 – 2010; Saidova, 1981; Grigyalis, 1978; Maslakova, 1990; V.A. Basov, 1990; Vdovenko et al., 1993 (with the participation of Rauser-Chernousova); Bugrova, 1990; Kulagina, 2012; Kamenskaya, 1987) and in the works of Hohenegger & Piller (1975), Brasier (1980 cited byr Lee, 1990), Hansen (1979), Hohenegger (1981, 1990), Jones (1984, 1994), Kaminski (2004), Pronina (1988, 1994), Pronina – Nestell & Nestell (2001), Nestell & Nestell (2006), Vachard et al. (2008, 2010) and. Unfortunately all of them are inferior to Loeblich & Tappan (1964, 1987) classification in wideness of their coverage.

A new era of taxonomic studies began with the application of molecular phylogenetic methods, which provide taxonomic criteria that are additional to, and independent of, morphological characters. The first steps in this field were made during the 1990s, since when knowledge has increased dramatically (Darling et al., 1996, 1997, Fahrni et al., 1997, Pawlowski, 2000, Pawlowski et al., 1995, 1997, 2002, 2003, Pawlowski & Gooday, 2009, Pawlowski & Burki, 2009, Gooday and Pawlowski 2004; Holzman et al., 2003, Habura et al., 2004, Flakowski et al., 2005, Bowser et al., 2006, Groussin et al., 2011, Holzmann et al., 2012, and others). These studies have generated a large body of data that are deposited in GenBank, and represent the molecular equivalent of museum collections of foraminiferal shells. Difficulties have arisen with the use of these new methods; for example, early studies suggested wrongly that foraminifera are related to the Alveolata (Wray et al., 1995). Nevertheless, significant progress has been achieved in understanding high-level phylogenetic relationships within the Protista (better to use term unicellular Eukaryota as the former Protista are shown to be heterogenous ecologic rather than taxonomic group – Krylov et al., 1980). In the revised classification of Eukaryotes Adl et al., (2012) placed Foraminifera in the super-group SAR (acronym introduced by Byrki et al., 2007). The new insights into the evolution of Rhizaria entering SAR group was provided by Burki et al., (2010) based on phylogenomic analyses. In these studies Foraminifera represent difficult subjects because of their hard shells, difficulties of cultivation as well as preservation of dead material in special solutions resulting in limitation of material, presence of symbionts and other contaminations, and also because of their high evolutionary rates. Nevertheless, within the Foraminifera, the monophyly of four of the five main phylogenetic lineages (the classes Astrorhizata, Miliolata, Spirillinata and Rotaliata), which emerged from the new approach to foraminiferal systematic and evolution based mainly on morphological features of the shell and on their development through time (Mikhalevich, 1980 – 2009; Saidova, 1981 and some other scientists of the Russian micropaleontological school), was confirmed by data from nuclear small subunit ribosomal RNA genes (SSU rDNA) (Fahrni et al., 1997, Pawlowski et al., 1995), and genes coding for actin (Flakowski et al., 2005) and tubuline (Bowser et al., 2006). Ertan et al. (2004), using insertions in SSU rDNA for taxonomic purposes, also concluded that "buliminids and rotaliids and the genera *Trochammina* and *Textularia* with agglutinated shells appear to be descendants of the stratigraphically older agglutinated foraminifers." The monophyly of the Foraminifera as a whole was also confirmed by molecular studies (Pawlowski et al., 1997, Pawlowski & Holzmann, 2002 and some others). Application of molecular-clocks suggests that Foraminifera originated in the Neo-Proterozoic (Pawlowski et al. 2003, Groussin et al., 2011) (see below – Part XI. 1).

The groups of different taxonomic rank within the foraminiferal classes and orders up to the generic and species level were also examined with the use of these new methods. The results of the phylogenetic analyses of the major taxa were often consistent with the morphological classification. For instance, the miliolid, soritid and nummulitid lines were shown to be homogenous and well supported and parallel evolution in nummulitid foraminifera was also documented (Holzmann et al., 2003).

The results of molecular studies at the generic level showed the genera *Brizalina, Cibicidoides, Fontbotia* and *Lobatula* to be invalid (Ertan et al., 2004, Scweizer et al., 2009) (it may be added that all these genera were doubtful and poorly defined morphologically as well). Similar investigations addressed other groups of benthic forms (Glabratellidae and monothalamous taxa). In the McMurdo Sound area of the Southern Ocean, where single-chambered (monothalamous) foraminifera are very diverse (Gooday et al., 1996), molecular studies have revealed a twofold greater diversity of phylotypes than morphotypes (Gooday et al. 1996, Bowser et al. 2002, Habura et al. 2004). Similar results were obtained at temperate sites (Habura et al., 2008). Recently, next generation metagenomic studies have revealed an extraordinary diversity of unknown foraminifera in deep-sea samples (Lecroq et al., 2011). Earlier previously unknown independent lineage within the evolutionary radiation of single-celled foraminifera (genus *Conqueria,* in Gooday & Pawlowski, 2004) was discovered. The study of intraspecific variations in planktonic forms, *Globigerina pachyderma, G. bulloides, Globigerinoides ruber, Neogloboquadrina pachyderma, Orbulina universa* (Darling et al., 1999, 2000) and *Globorotalia truncatulinoides* (de Vargas et al., 1999, 2001) also revealed their high genetic variability, the restriction of many of the recognised genotypes (cryptic species, formerly considered to represent one species) to limited geographical areas and environmental conditions (only a few of them were found to be cosmopolitan). At the same time, the impossibility to distinguish these cryptic species at the morphological level was also demonstrated in some cases. Thus Weiner et al., (2012) discovered multiple genetic types in *Globigerinella siphoniphera* and showed the need to reinvestigate its genotype.

Examination of the nine recognised genotypic groups of *Ammonia* showed that only one of the nine may be clearly distinguished by its morphology (Holzman & Pawlowski, 2000). The boundaries between more than 60 cosmopolitan morphospecies of *Elphidium* were found to be uncertain, and cryptic diversity was discovered among morphologically distinct variants (Evans et al., 2012). In *Cassidulina, Chilostomella, Cibicides* and some other benthic calcareous genera, high genetic diversity did not coincide with morphological differences. Conversely, in some species, e.g. *Bulimina marginata* and *Hyalinea balthica,* genetic diversity was low and corresponded to the known morphospecies (Schweizer et al., 2012). While these discoveries are of great importance, they nevertheless raise problems of species definitions in the lower eukaryotes as well as creating practical taxonomic difficulties for ecologists and especially for paleontologists, because these analyses are difficult for modern species and impossible for extinct forms. Among more advanced groups of organisms, especially mammals, such cases are rare but nevertheless exist. For example, two mouse species are very similar in appearance but differ genetically, while *vice versa,* genetically homologous species have some distinctions in their morphology (E. Mayer, Zoological institute RAS, personal communication, year 1990). A conference of mouse taxonomists agreed to base their classification on the morphological features. It is unclear how to classify such forms among Foraminifera (and perhaps among the other unicellular Eukaryota) where such situations are obviously more common. Thus, understanding of the species concept in foraminifera has become problematic.

Molecular methods are most useful in the case of genera and species in which few distinctive morphological features are available for their identification. The simple naked and soft-shell unilocular foraminifers are in the greatest need of such studies. The phylogenetic position of the advanced foraminiferal forms, especially those with complex shell structures providing wide opportunities of comparative morphological analysis, is more clearly defined. Collaboration between traditional foraminiferal taxonomists and molecular specialists has the potential to yield valuable results. It is best to use the two approaches in combination rather than in isolation as it was done, for example, to identify the new Astrorhizata genus *Conqueria* (Gooday & Pawlowski, 2004). As J. Pawlowski and F. Burki (2009) remark, “inferred SSU rDNA-based phylogenies should be considered with a lot of caution”. Altin et al. (2009) stress the need for information about cellular ultrastructure to verify “the fidelity of phylogenetic trees of allogromiids based solely on the sequences of a single gene”, the more so because various analyses “employ different statistic methods to interpret the SSU sequences”. The phylogenetic position of some of the taxa revealed by these new methods contributes much to our understanding of protistan phylogeny. Nevertheless, the relationships between major clades remain largely unresolved and need to be confirmed by more sequences derived from studies of a wider range of taxa. On the other hand, recent taxonomical works on the different groups of organisms clearly show that the possibilities of traditional morphological investigations are not exhausted yet. Indeed, they are undergoing something of a renaissance with electronic microscopy opening new opportunities for the study of even the smallest organisms.

Many foraminiferal taxa of subclass and lower rank require revision based on modern methods of investigation, including electron–microscopy. Every taxonomist understands, perhaps intuitively, that it is impossible to give a fully comprehensive diagnosis that encompasses all aspects of a particular taxon. In any grouping of objects there is always a core to which the characteristics that define the group apply precisely, and a periphery represented by relatively few objects that embody these characteristics to a much lesser extent (Chaikovskij, 2010) (hence the remarks often added to taxonomic diagnoses). The diagnoses given in this book for the high-ranked taxa are fully appropriate for the main components of the classes, subclasses, orders and superfamilies. They apply to a lesser extent to forms occupying the peripheries of each taxon, which may have some transitional features (as is always the case in any classification). The questionable position of such forms is stressed in the systematic part (Part XI) in order to focus the attention of specialists.

Foraminifera represent a unique group with a very diverse range of shell morphologies allowing them to occupy a wide range of ecological niches. The presence of relatively few chromosomes (usually less than 20 – see below, Part VI. 4) give the possibility to study the connection of particular genes with particular shell features. In this sense, foraminifera have the possibility to become, like *Drosophila,* a classic subject for genetic studies.

Until the 1970s, most publications devoted to foraminifera were devoted to the investigation of their shells, usually fossil ones. The bibliography of works concerned only with their taxonomy is so huge (over the 6000 of titles) that even a short overview of the authors addressing this subject constitutes a considerable list of references, 3456 of them were given in Loeblich & Tappan, 1987. The recent literature on the group is nearly limitless. From the early 1980s (see Haynes, 1981) until the present day, more than 500 publications have appeared per year. Thus catalogues and atlases acquire a special significance. The most widely known catalogue is that of Ellis & Messina, 1940, which began as early as 1940 and now contains over 80 bulky hard-copy volumes. In addition to this are the "Index to the genera and species of the foraminifera" of Sherborn (1893, 1896), reedited by Thalmann (1960), the catalog of planktonic foraminifera (Saito et al., 1976), the "Atlas of benthic shelf foraminifera of the Southern Atlantic" of Boltovskoy et al., 1980, the "Atlas of Eocene Planktonic Foraminiferas" of Pearson et al. (2006), the "Atlas of Holocene benthic foraminifers of Shimbara Bay, Kyushu, Southwest Japan" (Akimoto et al., 2002), the "Atlas of California Neogene foraminifera" (Finger, 1990), and the "Atlas of Palaeogene Cosmopolitan Deep-Water Agglutinated Foraminifera" (Kaminski & Gradstein, 2005), among others.

The story of micropaleontological studies in Russia is less well known. These began with Moeller (1878, 1880) (cited by Rauser-Chernousova & Fursenko, 1959). Between the 1930s and 1970s an important micropaleontological school was formed in the USSR under the leadership of A.V. Fursenko and D.M. Rauser-Chernousova. The latter celebrated her centennial jubilee with the publication of the reference book on Paleozoic small foraminifera: "Spravochnik po systematike melkich foraminifer Paleozoya" (edited by Vdovenko, 1993*)*. Significant contributions to our knowledge of the taxonomy and stratigraphy of fossil foraminifera from different regions of the former USSR was made by the works of O. S. Vialov, A.M. Voloshina, A.A. Krasheninnikov, M.S. Barash, E.A. Reytlinger, T.N. Gorbachik, V.N. Ben'jamovski, M.N. Solovieva and others. An important center of this school was formed in S. Petersburg (Leningrad at that time) in the Oil Institute (VNIGRI), issuing monographs in the form of the serial editions "The fossil foraminifera of the USSR",

"Trudy VNIGRI", "Microfauna of the USSR" (with the authors N.N. Subbotina, A.K. Bogdanovich, N. A. Voloshinova, V.P. Vasilenko, L.G. Dain, E.V. Bykova, E.V. Mjatljuk), as well as in the Geological Institute (VSEGEI). These investigations were later continued by the next generation of this school (S.P. Yakovleva, V.A. Basov, A.Ya. Azbel, L. A. Alekseychik-Mickevich, E. M. Bugrova, G. Pronina-Nestell, V. Vuks, T. Dmitrieva). Unfortunately at present, with the limitation of the financial support this school is somewhat in decline and has only a few scientists. Traditionally, the works of this school were based on the phylogenetic approach. It was not by chance that this school was the first to support the new approach to foraminiferal evolution and taxonomy and the new outline of foraminiferal classification, presented for the first time at a micropaleontological conference in Baku (Mikhalevich, 1980 a) and later in the series of micropaleotological seminars of the All-Russian (former All-Union) Paleontological Society (VPO) held in St. Petersburg (Leningrad) and Micropaleontological Commission held in Moscow during the 1980s. I'm deeply grateful for the support and development of these initial ideas in the works of Ch.M. Saidova, who was the first among others following and developing this new approach, and later after severe discussions also by N.I. Maslakova and T.N. Gorbatchik, who was an official opponent of my doctoral thesis defence in 1992. The latest work in this field in Russian is the monograph of Bugrova & Bugrova (2011).

The stratigraphic distribution, paleoecology and paleogeography of the different foraminiferal taxa and their historical development, as well as their taxonomy, were elucidated in the multiple works of Western and Russian schools, only of which can be mentioned here (Leven, 1963, 1970, 1997; Leven & Davydov, 2001; Loeblich & Tappan, 1964; Filon et al., 1974; Neagu, 1975; Hohenegger & Piller, 1977; Berger et al., 1979; Vuks, 1984; Hemleben et al., 1989; Ross & Haman, 1989; Kotlyar et al., 1989; Nagy et al., 1991; Stevens et al., 1991; Rettori et al., 1994; Davydov et al., 1996; Preece et al., 1999; Arnaud-Vanneau, 1988; Noe, 1988; Pronina, 1988, 1989, 1995; Pronina & Vuks, 1995; Pronina & Nestell, 1997; Klein & Mutterlose, 2001; Pronina – Nestell & Nestell, 2001; Nestell & Tolmacheva, 2004; Nestell & Nestell, 2006; Brenkle, 2005; Cozar et al., 2007; Groves, 2000; Groves et al., 2003, 2007; Richard et al., 2001; Vachard, 1991, 1994; Vachard & Krainer, 2001; Vachard & Fourcade, 1997; Holbourn & Kaminski, 1997; Vachard et al, 1994, 2002, 2010; Hayward et al., 2006, 2007, 2010, 2012; Kaminski et al., 1995, 1996, 2009; Kender et al., 2006; Hance et al., 2011; Fursenko & Rauser-Chernousova, 1959; Rauser-Chernousova, 1972; Kuznetzova & Basov, 2001; Dixon & Haig, 2004, and in many others). A significant contribution to the topic is also made in a set of publications in the serial editions of the Grzybowski Foundation (1995 – 2011), edited by M. Kaminski. In many of these publications, records of many of the taxa studied were extended compared to previous data. The rates of foraminiferal origination and extinction and the character of the acceleration of these processes were also investigated (Groves, J.R., Lee, A., 2008; Kaminski & Gradstein, 2005, Kaminski, Setoyama & Setean, 2008).

In the 20th century faunal, ecological and zoogeographical investigations were strongly developed (Natland, 1933-1951; Phleger, 1939-1973; Bandy, 1953-1971; Boltovskoy, 1953-1984; Be, 1957 – 1982; Berger et al., 1982; Matera & Lee, 1972; Todd & Low, 1957-1980; Murray, 1963 – 1996; Kennet, 1976; Tendal & Hessler, 1977; Colom, 1974; Delaca, 1980; Haynes, 1981; Lipps, 1983; Reiss & Hottinger, 1984; Hallock, 1987; Boersma, 1990; Gupta, 1994; Bernhard et al., 1997; Bronnimann et al.; 1992, Linke & Lutze; 1993; Loubere et al., 1995; Gooday et al., 1997; Sen Gupta et al., 1997; Williams et al., 1997; Bernhard et al.,

1997; Alve, 1999; Hallock, 1987, 2000; Langer & Hottinger, 2000; Alve & Murray, 2001; Schönfeld, 2001; Gustafsson, 2001; Alve & Goldstein, 2002, 2003; Murray, 2000; Rathburn et al., 2000; Goldstein et al., 1995; Ernst et al., 2005; Debenay et al., 1996, 2001, 2003, 2007, 2011; Debenay & Redois, 1997; McGann, 2002 – 2010; McGann & Sloan, 1996; Kathal, 2002; Holzmann et al., 2003; Scott, 2002; Leiter & Altenbach, 2010; Hayward et al., 1999; Zheng & Zhaoxian, 2001; Sen Gupta et al., 2009; Sergeeva et al., 2010; in Russia: Stschedrina, 1936-1982; Saidova, 1961-1988; Beljaeva, 1964-1989; Benjamovski, 1970 – 2010; Kopaevich, 1970 – 2010; Zakrevskaya, 1986 – 2010; Mikhalevich, 1966-1995; Barash 1964- 2010; Basov, 1974- 2000; Burmistrova, 1974-1978; Chusid, 1988 - 2009; and many others). It is impossible to identify the most important works in this field, even in this "short list". Very interesting observations were made on foraminifera as epiphytes on sea grass (Langer, 1993, Langer & Gehring, 1993, Fujita & Hallock, 1999, Ribes et al., 2000, Semeniuk, 2000, Debenay & Payri, 2010, Mateu-Vicens et al., 2010). Extremely important evidence has emerged for denitrification of sediments by benthic foraminifera, revealing their role in the removal of fixed nitrogen by respiration (Risgaard-Petersen et al., 2006, Pina-Ochoa et al., 2010). Lith et al. (2009) also showed experimentally the role of the benthic foraminifera in the process of denitrification. This role may be comparable with the bacteria role in the denitrification of the soil.

The new insight into foraminiferal zoogeography revealed by the use of molecular methods was mentioned above (this Part) (Darling et al., 1996 – 2000, Holzmann & Pawlowski, 2000; Gooday & Jorissen, 2012 and some others).

Investigations of the influence of pollution on foraminiferal faunas in coastal seas began in the 1970s (Seiglie, 1971; Lee et al., 1971) and have continued (Venec-Peyre, 1981; Berger et al., 1979; Ellison et al., 1986; Nagy & Alve, 1987; Debenay, 1990; Alve, 1991, 1995; Alve & Olsgard, 1999; Yanko, 1995; Yanko et al., 1994, 1998, 1999; Kaminski et al., 1995; Tyszka & Kaminski, 1995; Stouff et al., 1999 a, b; Angel et al., 2000; Ashraf M. Samir, 2000; Robert et al., 2001; Le Cadre & Debenay, 2005; Armynot du Chatelet & Debenay, 2010; Martins et al., 2010; Dabbous & Scott, 2012). The work of Bresler & Yanko (1995) initiated the study of the foraminiferal chemical ecology.

As mentioned above, even in modern forms, attention has focused mainly on the study of foraminiferal shells. The cytology, biology and life cycles of these protists are still insufficiently studied. Following early studies in this field (Schultze, 1854; Munier-Chalmas, 1880; Munier-Chalmas & Schlumberger, 1883, 1885; Lister, 1895; Shaudinn, 1894-1903; Rhumbler, 1911), the life cycles of several species were elucidated in more detail at the beginning of the 20th century (Winter, 1907; Myers, 1935, 1936, 1940; Foyn, 1936; Jepps, 1942; Le Calvez, 1938, 1946, 1950). An important contribution was made by Grell and his followers ((Grell - 1954-1988; Ross, 1972; Grell & Bardelle, 1977; Weber, 1965) as well as in a series of later investigations (Hedley et al., 1968; Sliter, 1965; Roettger et al., 1978; Roettger & Krueger, 1990; Arnold, 1982; Anderson, 1984; Angell, 1990; Lee et al., 1991; Faber & Lee, 1993; Goldstein & Moodley, 1993; Goldstein, 1997, 1999; Goldstein & Richardson, 2002; Goldstein et al., 2010; Kitazato & Matsushita, 1996; Krueger et al., 1997; Stouff et al., 1999a; Lehmann et al., 2006; Barras et al., 2009; Langer et al., 2009; and others – see below, Part VI).

Life cycles were investigated using light or electron microscopy by Voronova (1976, 1978a, 1978b, 1979) for *Cibicides lobatulus,* Goldstein (1988) for *Saccammina alba*, and Krueger et al., (1997) for *Cycloclipeus carpenteri*. The first cytophotometric investigations of the nuclei during reproduction were made by Zech (1964) and Voronova & Selivanova (1976). Electron-microscopic studies of the life cycles of other foraminifera are still fragmentary (Dahlgren, 1967 a; Schwab, 1969, 1973, 1976; Angell, 1971; Cezana, 1971-1978; Berthold, 1977; McEnery, Lee, 1981; Arnold, 1982; Matsushita & Kitazato, 1990; Kitazato & Matsushita, 1996; Pawlowski et al., 1995; Lehman et al., 2006). A series of works on the life cycles of the planktonic forms also appeared (Lee et al., 1965; Be, 1980; Be et al., 1976; Spindler et al., 1978; Anderson, 1984).

The end of the 20th century was marked by detailed electron-microscopic studies of foraminiferal cytoplasm (Allen, 1964; Febvre-Chevalier, 1971; Hedley & Adams, 1974; Lena & Freire, 1974; Nyholm & Nyholm, 1975 a, b; Hansen, 1975; Muller-Merz & Lee, 1976; McEnery & Lee, 1976; Schwab, 1977; Anderson & Be, 1976 b; Anderson & Tuntivate-Choi, 1984; Lee et al., 1965, 1991; Bowser et al., 1984, 1985, 1988, 1992; Bowser & Travis, 2002; and others). The information is reviewed in the works of Anderson & Be, 1978; Anderson, 1988; Lee, 1990, 1993; Goldstein, 1999; in monographs edited by Lee & Anderson, 1991 and Sen Gupta, 2002 and in some others). These previous investigations have yielded a general picture of foraminiferal cytoplasmic structure (the above mentioned works of Hedley & Adams, 1974; Schwab, 1977; Anderson & Be, 1976 b; Bowser et al., 1984, 1985, 1988, 1992; Bowser & Travis, 2002, and also Raikov et al., 1998) and the main features of their life cycles have been clarified (Anderson, 1984; Voronova & Mikhalevich, 1985; Lee et al., 1991; Goldstein, 1999).

The main data on foraminiferal biology were obtained from observations made on laboratory cultures (Arnold, 1954 – 1984, Spindler et al., 1984; Bowser et al., 1992; Hemleben & Kitazato, 1995; Goldstein, 1999; Goldstein & Corliss, 1994; Heinz et al., 2001; Barras et al., 2009; Allison et al., 2010; Dissard et al., 2010 and many others – see below, Part VI).

Special attention has been paid recently to groups of naked and soft-walled unilocular foraminifera that exhibit few morphological features and typically were neglected in previous studies. The use of new molecular methods, as well as detailed cytological studies, have revealed a rich fauna of such forms and some peculiarities of their cell and shell (Arnold, 1954; Delaca, 1986; Moodley, 1990; Bowser et al., 1992; Cedhagen et al., 2002; Bowser et al. 2002; Holzmann & Pawlowski, 2002; Pawlowski et al., 1999; Goldstein & Richardson, E., 2002; Goldstein et al., 2010; Souter et al., 2002; Habura et al. 2004; Gooday, 2002; Gooday et al., 1995, 1996, 2008, 2010 a, b, 2013; Cornelius & Gooday, 2004; Altin et al., 2009; Sergeeva et al., 2010; Sabbatini et al., 2010; and others). The study of this group is of considerable importance as all the other foraminiferal groups originated from this heterogenous assemblage of primitive forms.

A series of works have been devoted, at least to some extent, to foraminiferal biology (Hedley, 1964; Zmiri et al., 1972, 1974; Lipps & Ronan, 1974; Arnold et al., 1985; Hallock, 1981 a, 1986; Hickman & Lipps, 1983; Lee & Anderson, 1991; Lee et al., 1991; Bowser et al., 1995) and feeding (Muller & Lee, 1969; Murray, 1973; Lipps, 1975; Boltovskoy & Wright, 1976; Lee, 1980; Haynes, 1981; Brand & Lipps, 1982; Hemleben & Spindler, 1983; Spindler et al., 1984; Delaca 1982; Alexander & Delaca, 1987; Delaca et al., 1981; Lee et al, 1991 b; Faber, 1991; Faber & Lee, 1991 a, b; Bowser et al., 1992; Linke & Lutze, 1993; Goldstein & Corliss, 1994; Gooday et al., 2008; Dupuy et al., 2010; Suhr et al., 2008). A number of early authors studied foraminiferal symbionts (Butchli, 1886; Winter, 1907; Rhumbler, 1911; Doyli & Doyle, 1940) and these studies were later expanded considerably

(Haynes, 1965; Berthold, 1978; Roettger, 1972, 1974; Hallock, 1985, 2000; Lee et al., 1997; Chai & Lee, 2000; Bernhard, 2003; Bernhard & Bowser, 1999; Bernhard et al., 2000; Richardson, 2001; Pawlowski et al., 2001; Pochon et al., 2001, 2004; Hyams-Kaphzan & Lee, 2008; Siano et al., 2010; Hohenegger, 1994, 1995, 1999, 2011; and others). A special contribution to the field of foraminiferal symbionts was made in the works of Lee and Hallock and their colleagues (Lee and co-authors 1963 – 1991; Hallock and co-authors 1979 – 2008; Hohenegger, 1994, 1995, 1999, 2000, 2011; Hohenegger et al., 1999; Hohenegger and co-authors 1994 – 2011; Hottinger 1982, 2006; Leutenegger 1977, 1983, 1984). Nearly all aspects of foraminiferal biology (cytology, symbiosis, ecology, biology, physiology, feeding, taxonomy and even the DNA characteristic) were first addressed by J.J. Lee and his coworkers.

A significant number of publications are devoted to the relationship of foraminifera to organic carbon fluxes (Langer et al., 1997; Altenbach & Struck, 2001; Morigi et al., 2001; Allison et al., 2010; Gooday et al., 2012 and others).

Reviews of various aspects of foraminiferal research, and those focused on their taxonomy, are also of great importance (Lee, 1990; Lee & Anderson, 1991; Debenay et al., 1996 a; Gross, 2001; Sen Gupta, 2002; Bellier et al., 2010; Parker, J.H., 2009; Popescu et al., 1998; Hayward, 1990; Parker, 2009; Hayward et al, 1997, 2010, 2012; Kaminski, 2000, 2004 and some others). Unfortunately, most of them, especially those published during later years, were not available to the author of the present book.

Sources of information available on the Internet are becoming increasingly important for foraminiferal workers; some of these are listed below:

http://www.cushmanfoundation.org/jfr/jfrdata.php; www.micropress.org/;
http://www.es.ucl.ac.uk/Grzybowski/foramrefs.htm;
http://www.tmsoc.org/;
http://www.marinespecies.org/foraminifera/aphia.php?p=taxdetails&id=22528 on 2012-11-27; (WoRMS – World Register of Marine Species);
http://www.foraminifera.eu/morphoclass.html;
http://www.foraminifera.eu/astrorhizata.html;
http://www.foraminifera.eu/files/Mikhalevichdeutsch.pdf 2009 ; and some others.

This short overview of the literature on different aspect of foraminiferal studies only partly reflects the wide range of research in this field. Concerning the limited availability of literature accessible in our country, especially during recent years, it would not be out of place to express here my deep gratitude to my Western colleagues who have helped me with this problem: Mike Kaminski, Jean-Pierre Debenay, Pamela Hallock, Sev Kender, Jaroslav Tyszka, Elizabeth Alve, Theodor Neagu, Daniel Vachard, Bruce Hayward, Andrew Gooday and many, many others. The most severe difficulties were encountered with access to books. Thus the publication of Haynes (1981) was presented to me by M. Kaminski in 2005, while the book edited by Sen Gupta (2002) and some others are still unavailable. On my side I have tried to bring to the attention of readers some works published in Russian and less well known in the West. The more detailed literature on the shell and cell structure, biology, symbiosis, taxonomy can be found in the appropriate Parts (see below).

The investigation of Foraminifera is becoming ever more extensive and detailed; DNA sequences and proteins are studied (see above, this Part) including those associated with the mineral elements in the shell wall (King & Hare, 1972; Lee, 1993, 1995; Schroeder & Bada, 1977; Wray et al., 1993; Langer et al., 1993, Lee, 1993, 1995; Stathoplos & Tuross, 1994; Pawlowski and his school, 1995 – 2003; Bian & Martin, 2010; Dissard et al., 2010) and data on the amino-acid composition of the several species and genera are used in taxonomy (Haugen et al., 1989). The first biochemical works have appeared (Schwab & Schloboch, 1973; Anderson & Be, 1976 a; Schroeder & Bada, 1977; Laatsch & Schwab, 1984; Faber & Lee, 1991 b; Travis & Bowser, 1986, 1991; Bresler & Yanko, 1995). Physiologic investigations began using the *Allogromia laticollaris* as a favorite subject (Lee et al., 1961; Lee, 1980). Foraminiferal metabolism and processes of digestion were studied at the molecular level by Linke et al., (1995), microsensory studies of photosynthesis and respiration in larger symbiotic foraminifera were done by Kohler-Rink & Kuhl (2000). Interesting intracellular biochemical processes in the presence of chloroplasts and bacteria were analyzed by Bowser (2003). The study of cell structures (mitochondria in reticulopodia) and their functioning in the living foraminiferal cells (Cedhagen & Frimanson, 2002) reflects a new level of cytological and physiological studies in foraminifera.

Nevertheless, foraminiferal research still represents a vast field of investigation: the taxonomy within the main phyletic lines, the inner shell structure of many taxa, shell wall ultrastructure of some large taxa, life cycles, cytoplasmic and nuclear structure. The biology of the overwhelming majority of the species, especially in the Nodosariata and Textulariana, are unknown. Even the process describing of species and genera is far from being completed. The new methods of investigations (electron microscopy and genetic molecular methods, freezing and freeze substitution) provide new opportunities to widen our knowledge. But recent studies represent only the beginning of the new era, and the entire range of Foraminifera, particularly those described in the fundamental early taxonomic works, need to be restudied and revised at this new, more detailed level of investigations. This has already been started for the collections of D'Orbigny (Le Calvez, 1977, Papp & Scmid, 1985, Venec-Peyre, 2005), Brady (Barker, 1960, Jones, 1994), and Brady, Carpenter, T.R. Jones, Williamson, and W.K. Parker (Hodgkinsson, 1992), as well as for material of Karrer (Hornibrook, 1971). An important goal is to bring existing data obtained using traditional methods into closer agreement with data obtained using the newest approaches.

Chapter IV

CLASSIFICATION[1]

SUPERGROUP RHIZARIA CAVALIER-SMITH, 2002

PHYLUM FORAMINIFERA D'ORBIGNY, 1826,

nomen translat. Mikhalevich, 1980

CLASS ASTRORHIZATA SAIDOVA, 1981

(= subclass Astrorhizicae Saidova, 1981, part, = class Monothalamids Pawlowski et al., 2003, part)

Subclass *LAGYNANA Mikhalevich, 1980

Order LAGYNIDA Mikhalevich, 1980

Families: Lagynidae Schultze, 1854, Maylisoriidae E.V. Bykova, 1961, Blysmasphaerinae Bronnimann, 1988.

Order ALLOGROMIIDA Fursenko, 1958

Families: Allogromiidae Rhumbler, 1904, Shepheardellidae Loeblich et Tappan, 1948, Argillotubidae Avnimelech, 1952, Labyrinthochitinidae Mikhalevich, 1999.

Order AMMOSCALARIIDA Mikhalevich, 1980

Families: Phthanotrochidae Arnold, 1978, Ammoscalariidae Mikhalevich, 1982, Hospitellidae Loeblich, Tappan, 1984.

---

[1] for the subfamily composition of the calcareous taxa and some changes at the generic level see Systematic part (XI) of the present book, for the subfamily and generic composition of agglutinated taxa – Mikhalevich, 2004, 2005; Mikhalevich & Kaminski, 2008, of Miliolid taxa – Mikhalevich, 1998; the diagnoses of all the new taxa are also given in the Systematic part. Taxa with doubtful position are marked with the interrogative mark, those with strikingly reassigned position with mark *. The new genera published after Loeblich & Tappan (1987) were added in the systematic parts when the publications were available, especially those of the taxa characterized with the calcareous secreted wall as the new agglutinated genera are monitored in a set of works of M. Kaminski (1995 – 2012) and members of the International Working Group on Foraminiferal Systematics (IWGFS), those, published before 2004 year are also considered in Mikhalevich (2004 a, c). The mentioning of the new genera published after 1987 year is far from being full and this was not a goal of the present book. Unfortunately the majority of pdf files in Internet were not accessible for me. In the cases of the change of taxonomic rank the authorship of the authors is preserved within the same category (e.g. – subfamily, family, superfamily), and is changed with the transfer to the higher category (e.g. – from the family to the order rank).

Subclass ASTRORHIZANA Saidova, 1981
(= subclass Astrorhizicae Saidova, 1981)

Order ASTRORHIZIDA Haeckel, 1894

Superfamily Astrorhizoidea Brady, 1881
Families: Astrorhizidae Brady, 1881, Vanhoeffenellidae Saidova, 1981, Rhabdamminidae Brady, 1884, Hippocrepinellidae Loeblich et Tappan, 1984, Jugimuramminidae Zheng, 2001.

Order DENDROPHRYIDA Haeckel, 1894, emend.

Superfamily Dendrophryoidea Haeckel, 1894
Families: Dendrophryidae Haeckel, 1894, *Notodendrodidae Delaca, Lipps et Hessler, 1980, Dryorhizopsidae Loeblich et Tappan, 1984, ?Arboramminidae Shires, Gooday and Jones, 1994, *Komokiidae Tendal et Hessler, 1977, *Baculellidae Tendal et Hessler, 1977, *Normaninidae Mikhalevich, 1994.

Superfamily Schizamminoidea Norvang, 1961
Families Schizamminidae Norvang, 1961, Halyphysematidae Loeblich et Tappan, 1984, nom. corrected pro Halyphyseminae, Diffusilinidae Loeblich et Tappan, 1961.

Order PSAMMOSPHAERIDA Haeckel, 1894

Suborder STEGNAMMININA Mikhalevich, Pronina G.P., Nestell, M., 2000

Superfamily Stegnamminoidea Moreman, 1930
Families: Stegnamminidae Moreman, 1930, Colonamminidae Rauser-Chernousova et Reytlinger, 1993 in Vdovenko et al., 1993.

Superfamily Crithioninoidea Goes, 1894
Families: Crithionidae Goes, 1894, Oryctodermatidae Saidova, 1981
[= Subfam. Masonellinae Saidova, 1981 (as "Marsonellinae"), part].

Suborder PSAMMOSPHAERINA Haeckel, 1894
Families: Psammosphaeridae Haeckel, 1894, Polysaccamminidae Loeblich et Tappan, 1984; Lacustrinellidae Mikhalevich, 1995.

Order *PARATHURAMMINIDA Mikhalevich, 1980

Superfamily Parathuramminoidea E.V.Bykova, 1955.
Families: Archaesphaeridae Malakhova, 1956, Parathuramminidae E.V.Bykova, 1955.

Superfamily Marginarioidea Loeblich et Tappan, 1986
(= Ivanovelloidea Chuvashov et Yuferev, 1984).

Families Marginaridae Loeblich et Tappan, 1986, Tuberitinidae A.D.Miklukho-Maklay, 1958, Auroriidae Loeblich et Tappan, 1986, Eovolutinidae Loeblich et Tappan, 1986, Rauserinidae Sabirov, 1987, Cribrosphaeroididae Sabirov, 1987.

Order HIPPOCREPINIDA Saidova, 1981

Superfamily Hippocrepinoidea Rhumbler, 1895
Families: Hippocrepinidae Rhumbler, 1895, Hyperamminidae Eimer et Fickert, 1899, Ammovoluminidae Chernych, 1967.

Superfamily Botellinoidea Chapman et Parr, 1936
= Caligelloidea Reytlinger, 1959
Families: Botellinidae Chapman et Parr, 1936.

CLASS SPIRILLINATA MIKHALEVICH, 1992

(=Tubothalamea Pawlowski et al., 2012, part, in Adl et al., 2012)

Subclass AMMODISCANA Mikhalevich, 1980

Order AMMODISCIDA Mikhalevich, 1980
Families: Ammodiscidae Reuss, 1862, *Pseudoammodiscidae Conil et Lys, 1970, Turritellellidae Saidova, 1981 (fam. reinstated and given stat. nov), ?Tolypamminidae Cushman, 1928, Turriglomininae Zaninetti, in Limongi et al., 1987.

Order AMMOVERTELLINIDA Mikhalevich, 1999
Families: Usbekistaniidae Vyalov, 1968, Ammovertellinidae Saidova, 1981, *Pseudolituotubidae Conil et Longerstaey, 1980 from the Fusulinacea.

Order PLAGIORAPHIDA Mikhalevich, 2003.
Family: Plagioraphidae Mikhalevich, 1995.

Order *?ATAXOPHRAGMIIDA Fursenko, 1958

Superfamily Ataxophragmoidea Schwager, 1877
Families: *Duotaxidae Mikhalevich, 2003, Ataxophragmiidae Schwager, 1877 (with Ataxophragmiinae Schwager, 1877 - only genera having *Ataxophragmium*-like largely involute spiral coiling with not visible early whorls and with high chambers turned to the axis of coiling are retained here), Dobrogelinidae Mikhalevich, 1992.

Superfamily Textularielloidea Groenhagen & Luterbacher, 1966
Families: Pernerinidae Loeblich et Tappan, 1984, Textulariellidae Groenhagen et Luterbacher, 1966, ?Syrianidae Kaminski 2004, Cuneolinidae Saidova, 1981, Dicyclinidae Loeblich et Tappan, 1964, Pfenderinidae Smout et Sugden, 1962, Coskinolinidae Moullade,

1965, Hauraniidae Septfontaine, 1988, Parurgoninidae Septfontaine, 1988, Orbitolinidae Martin, 1980.

INCERTAE SEDIS:
A group of families which obviously is heterogenous: Pyreninidae Mikhalevich, 1999 (*Pyrenina*) and Nezzazatidae Hamaoui et Saint-Marc, 1970.

Subclass SPIRILLINANA Mikhalevich, 1992

Superorder ARCHAEDISCOIDA Pojarkov et Skvortsov, 1979

Order ARCHAEDISCIDA Pojarkov et Skvortsov, 1979

Superfamily Archediscidoidea Cushman, 1928
Families: Archaediscidae Cushman, 1928, Asteroarchaediscidae A.D. Miklukho-Maklay, 1957, Planoarchaediscidae Gubenko, 1989, Tubispirodiscidae Mikhalevich, 1996.

Superfamily Glomodiscidoidea Mikhalevich, 1998
Family: Glomodiscidae Mikhalevich, 1998.

Order *TETRATAXIDA Mikhalevich, 1981
Families: Tetrataxidae Galloway, 1933, Pseudotaxidae Mamet, 1974,?Valvulinellidae Loeblich et Tappan, 1984, Akcayidae Özdikmen 2009 pro Sabaudiidae Brönnimann, Decrouez et Zaninetti, 1983, Endotaxidae Bogush et Brazhnikova, 1996 in Rauser-Chernousova et al., 1996.

Order LASIODISCIDA Mikhalevich, 1993
Families: Lasiodiscidae Reytlinger, 1956, Monotaxinoididae Mikhalevich, 1999, Howchiniidae R. Martini et Zaninetti, 1988, Pseudovidalinidae Altiner, 1988.

Superorder INVOLUTINOIDA Hohenegger et Piller, 1977

Order INVOLUTINIDA Hohenegger et Piller, 1977

Superfamily Involutinoidea Butschli, 1880
Families: Involutinidae Butschli, 1880, Planispirillinidae Piller, 1978, Triadodiscidae Zaninetti, 1984, Aulotortidae Zaninetti, 1984, Hirsutospirellidae Zaninetti, Ciarapica, Cirilli et Cadet, 1985, ?Ventrolaminidae Weynschenck, 1950.

Superfamily Triasinoidea Loeblich et Tappan, 1986
Family: Triasinidae Loeblich et Tappan, 1986.

Order HOTTINGERELLIDA Mikhalevich, 1993
Family: Hottingerellidae Mikhalevich, 1993.

Superorder SPIRILLINOIDA Hohenegger et Piller, 1975

Order SPIRILLINIDA Hohenegger et Piller, 1975

Suborder SPIRILLININA Hohenegger et Piller, 1975
Families: Spirillinidae Reuss et Fritsch, 1861, Conicospirillinidae Mikhalevich, 1995.

Suborder SPIROTROCHOLININA Mikhalevich, 1993
Family: Spirotrocholinidae Mikhalevich, 1993.

Order PATELLINIDA Mikhalevich, 1992

Superfamily Hergottelloidea Loeblich et Tappan, 1984
Families: Hergottellidae Loeblich et Tappan, 1984, Ungulatellidae Seiglie, 1964, Placentulinidae G.K. Kasimova, Poroshina et Geodakchan, 1980, Pannellainidae Loeblich et Tappan, 1984.

Superfamily Patellinoidea Rhumbler, 1906
Families: Patellinidae Rhumbler, 1906, *Annulopatellinidae Loeblich et Tappan, 1964, Paleopatellinidae Mikhalevich, 1999, Patellinellidae Bugrova, 1990, *Chapmaninidae Thalmann, 1938, Dictyoconoidessidae Bugrova, 1990.

Order *?CYMBALOPORIDA ord. nov.

Superfamily Cymbaloporoidea Cushman, 1927, stat. elevated
Families: Cymbaloporidae Cushman, 1927, Cymbaloporettidae fam. nov., Halkyardiidae Kudo, 1931, stat. elevated.

Superfamily Fabianioidea Deloffre et Hamaoui, 1973, stat. elevated
Familiy: Fabianiidae Deloffre et Hmaoui, 1973 stat. elevated.

Order ?SEABROOKINIDA Mikhalevich, 1980
Family: Seabrookinidae Cushman,1927.

CLASS MILIOLATA SAIDOVA, 1981

(= class Miliolicea Saidova, 1981, = class Tubothalamea Pawlowski et al., 2012, part, in Adl et al., 2012)

Subclass MILIAMMINANA Mikhalevich, 1980

(= subclass Rzehakinicae Saidova, 1981, = Schlumbergerinana Mikhalevich, 1992)

Order LITUOTUBIDA Mikhalevich, 1992 (emend)
Families: Lituotubidae Loeblich et Tappan, 1984, Trochamminoididae Haynes et Nwabufo-Ene, 1998 (emend., stat. elevated).

Order SCHLUMBERGERINIDA Mikhalevich, 1980
(= ord. Rzehakinida Saidova, 1981)
Families: Rzehakinidae Cushman, 1933, Miliamminidae Saidova, 1981, stat. nov., Trilocularenidae Mikhalevich et Kaminski, 2008, Ammomassilinidae Mikhalevich et Kaminski, 2008, Sigmoilopsidae Vella, 1957, Schlumbergerinidae Mikhalevich, 1980, Ammoflintinidae Mikhalevich et Kaminski, 2008.

Order HAPLOPHRAGMIIDA Loeblich & Tappan, 1989

Superfamily Recurvoioidea Alekseychik-Mitskevich, 1973
Families: Cystamminidae Mikhalevich, 2003, Recurvoididae Alekseychik-Mitskevich, 1973, Ammobaculinidae Saidova, 1981, Acupeinidae Bronnimann et Zaninetti, 1984, Plectorecurvoididae Loeblich & Tappan, 1964.

Superfamily Haplophragmioidea Eimer & Fichert, 1899
Families: Haplophragmiidae Eimer & Fickert, 1899, *Mesoendothyridae Voloshinova, 1958.

Superfamily Labyrinthidomatidoidea Loeblich et Tappan, 1987
Family: Labyrinthidomatidae Loeblich et Tappan, 1987.

Order SPHAERAMMINIDA Mikhalevich et Kaminski, 2003 in Mikhalevich, 2003
Family: Sphaeramminidae Cushman, 1933.

Order LITUOLIDA Lankester, 1885

Superfamily Lituoloidea de Blainville, 1827
(Nom. corrected pro Lituolinoidea)
Families: Lituolidae de Blainville, 1827, Buzasinidae Mikhalevich, 2003, Mayncinidae Loeblich et Tappan, 1985, Placopsilinidae Rhumbler, 1913.

Superfamily Barkerinoidea Smout, 1956
Families: Barkerinidae Smout, 1956, Biplanatidae Mikhalevich, 1992, ?Haymanellidae Mikhalevich, 2003.

Superfamily Biokovinoidea Gusic, 1977
Families: Charentiidae Loeblich et Tappan, 1985, Lituoliporidae Gusic et Velic, 1978, Biokovinidae Gusic, 1977, Labyrinthinidae Septfontaine, 1988, Buccicrenatidae Loeblich et Tappan, 1985, Choffatellidae Maync, 1958, Hottingeritidae Loeblich et Tappan, 1985, Pseudochoffatellidae Loeblich et Tappan, 1946, Spirocyclinidae Munier-Chalmas, 1887.

? Superfamily Coscinophragmatoidea Thalmann, 1951
Families: Haddoniidae Saidova, 1981, Coscinophragmatidae Thalmann, 1951.

Order CYCLOLINIDA Mikhalevich, 1992

Superfamily Cyclolinoidea Loeblich et Tappan, 1964
Family: Cyclolinidae Loeblich et Tappan, 1964. .

Superfamily Cyclopsinelloidea Loeblich & Tappan, 1984
Families: Cyclopsinellidae Loeblich et Tappan, 1984, Illerdorbidae Hottinger et Caus, 1982, Orbitopsellidae Hottinger et Caus, 1982.

Superfamily Vaniinoidea Mikhalevich, 2003
Family: Vaniidae Mikhalevich, 2003.

Order LOFTUSIIDA Kaminski et Mikhalevich, 2003 in Mikhalevich, 2003
(= Loftusiida Kaminski et Mikhalevich, 2004 a)
Family: Loftusiidae Brady, 1884.

Subclass MILIOLANA Saidova, 1981

***Group of superorders FUSULINOIDS** Fursenko, 1958

Superorder ENDOTHYROIDA Fursenko, 1958

Order TOURNAYELLIDA Hohenegger et Piller, 1973
(= order Tournayellida Dain, 1953 in Vdovenko et al., 1996)

Superfamily Tournayelloidea Dain, 1953
Families: Tournayellidae Dain, 1953, Lituotubellidae A.D. Miklukho-Maklay, 1963.

Superfamily Chernyshinelloidea Reytlinger, 1958
Families: Chernyshinellidae Reytlinger, 1958, Mstiniidae Lipina, 1989, Palaeospiroplectamminidae Loeblich et Tappan, 1984.

Order ENDOTHYRIDA Fursenko, 1958

Superfamily Endothyroidea Brady, 1884
Families: Endothyridae Brady, 1884, Haplophragmellidae Reytlinger, 1959, Endothyranopsidae Reytlinger, 1958, Globoendothyridae Reytlinger in Voloshinova et Reytlinger, 1959, Dainellidae Cozar et Vachard, 2001, Loeblichiidae Cummings, 1955, Endostaffellidae Loeblich et Tappan, 1984, Endotebidae Vachard, Martini, Rettori et Zaninetti, 1994, Endotriadidae Vachard, Martini, Rettori, Zaninetti, 1994.

Superfamily Bradyinoidea Reytlinger, 1996
Families: Bradyinidae Reytlinger, 1950, Janischewskinidae Reytlinger, 1996.

Superder OZAWAINELLOIDEA Solovieva, 1980

Order OZAWAINELLIDA Solovieva, 1980
Families: Eostaffellidae Mamet, 1970, Ozawainellidae Thompson et Foster, 1937, Reichelinidae A.D. Miklukho-Maklay, 1959 stat. elevated, and Incertae family with the genera *Eostaffelloides, Kangvarella and Hubeiella*).

Superorder FUSULINOIDA Fursenko, 1958

Order STAFFELLIDA A.D. Miklukho-Maklay, 1949
Families: Pseudoendothyridae Mamet, 1970, Nankinellidae A.D. Miklukho-Maklay, 1963, Cheniidae F. Kahler et G. Kahler, 1966, Staffellidae A.D. Miklukho-Maklay, 1949, Pisolinidae Rauser, 1985, Kahlerinidae Leven, 1963, Pseudostaffellidae Putrja, 1956.

Order SCHUBERTELLIDA Skinner, 1931
Families: Shubertellidae Skinner, 1931, Boultoniidae Skinner et Wilde, 1954, Palaeofusulinidae A.D. Miklukho-Maklay, 1963, Yangchieniidae Leven, 1987.

Order FUSULINIDA Fursenko, 1958
Families: Profusulinellidae Solovieva, 1996, Aljutovellidae Solovieva, 1996, Fusulinidae Moeller, 1878, Hemifusulinidae Putrja, 1956, Fusulinellidae Staff et Wedekind, 1910, Wedekindellinidae F. Kahler et G. Kahler, 1966.

Order SCHWAGERINIDA Solovieva, 1985
Families:Triticitidae Davydov, 1986, Rugosofusulinidae Davydov, 1980, Schwagerinidae Dunbar et Henbest, 1930, Pseudofusulinidae Dutkevich, 1934, Polydiexodinidae A.D. Miklukho-Maklay, 1953.

Order NEOSCHWAGERINIDA Minato et Honjo, 1966
Families: Misellinidae A.D. Miklukho-Maklay, 1958, Verbeekinidae Staff et Wedekind, 1910, Neoschwagerinidae Dunbar et Condra, 1927, Sumatrinidae Silvestri, 1933, Pseudodoliolinidae Leven, 1963.

Group of superorders MILIOLOIDS Delage et Herouard, 1896

Superorder SQUAMULINOIDA Mikhalevich, 1988

Order SQUAMULINIDA Mikhalevich, 1988
Family: Squamulinidae Reuss et Fritsch, 1861.

Superorder CORNUSPIROIDA Jirovec, 1953
(= Order Cornuspirida Mikhalevich, 1980 (part), suborder Cyclogyrina Saidova, 1981)

Order CORNUSPIRIDA Jirovec, 1953
(=Order Cornuspirida Mikhalevich, 1980 (part), suborder Cyclogyrina Saidova, 1981)

Suborder CORNUSPIRINA Jirovec, 1953

Family: Cornuspiridae Cushman, 1919, Hoyenellidae Rettori, 1994.

Suborder HEMIGORDIOPSINA Mikhalevich, 1986

Superfamily Hemigordiopsidea Nikitina, 1969

Family: Hemigordiopsidae Nikitina, 1969.

Superfamily Shanitoidea Loeblich et Tappan, 1986 (nom corrected pro Shanitinidea Mikhalevich, 1988)

Families: Shanitidae Loeblich et Tappan, 1986, Baisalinidae Loeblich et Tappan, 1986, Neodiscidae Lin, 1984 – nomen translat. Gaillot & Vachard, 2007, pro subfamily, (after Vachard et al, 2008).

Superorder COSTIFEROIDA Mikhalevich, 1988

Order COSTIFERIDA Mikhalevich, 1988

Family: Costiferidae Senowbary-Daryan et Zaninetti, 1986.

Superorder MILIOLOIDA Delage et Herouard, 1896

Order NUBECULARIIDA Jones, 1875

(=order Cornuspirida Mikhalevich, 1980 (part), suborder Nubeculariina Saidova, 1981)

Superfamily NUBECULARIIDEA Jones, 1875

Families: Nubeculariidae Jones, 1875, Ophthalmidiidae Wiesner, 1920, Meandrolocu-linidae Bogdanovich, 1981, Wiesnerellidae Saidova, 1981, Nodobaculariellidae Bogdanovich, 1981, Spiriamphorellidae Senowbary - Daryan et Zaninetti, 1986.

Superfamily DISCOSPIRINOIDEA Wiesner, 1931

Family: Discospirinidae Wiesner, 1931

Order MILIOLIDA Delage et Herouard, 1896

Suborder MILIOLINA Delage et Herouard, 1896

Superfamily Quinqueloculinoidea Cushman, 1917

Families: Hauerinidae Schwager 1876, Tubinellidae Rhumbler, 1906, Pavoninoididae Saidova, 1981, Quinqueloculinidae Cushman, 1917, Spiroloculinidae Wiesner, 1920, Planispiroidinidae Saidova, 1981, Massilinidae Thalmann, 1941, Sigmoilinitidae Luczkowska, 1974, Glomulinidae Saidova, 1981, Triloculinidae Bogdanovich, 1981, Pyrgoidae Mikhalevich, 1983, Miliolechinidae Zaninetti, Ciarapica, Cirilli et Cadet, 1985.

Superfamily Milioloidea Ehrenberg, 1839

Families: Miliolidae Ehrenberg, 1839, Neaguitinidae Andersen, 1984.

Superfamily Riveroinoidea Saidova, 1981
[(= Riveroininea Saidova, 1981, p. 32 (nom.incorr.)]
Family: Riveroinidae Saidova, 1981.
Superfamily Austrotrillinoidea Loeblich et Tappan, October 9,1986
(=Mikhalevich, October 27, 1986)
Families: Austrotrillinidae Mikhalevich, 1986, Brebinidae Mikhalevich, 1986, Pseudohauerinidae Mikhalevich, 1988.

Suborder ALVEOLININA Mikhalevich, 1980
(= Alveolinidae Loeblich, Tappan, 1964; Alveolinellinina Saidova, 1981)
Families: Fabulariidae Ehrenberg, 1839, Alveolinidae Ehrenberg, 1839, *Keramosphaeridae Brady, 1884.

Отряд SORITIDA Schultze, 1854 (= Orbitolinida Ehrenberg, 1839, Orbitolitida Wedekind, 1937)

Superfamily Peneroplidea Schultze, 1854
Families: Peneroplidae Schultze, 1854, *Fischerinidae Millett, 1899, *Fischerinellidae Saidova, 1981, *Zoyaellidae Saidova, 1981.

Superfamily Meandropsinoidea Henson, 1848 (nom corr. pro Meandropsinidea Saidova, 1981 and Mikhalevich, 1988).
Families: Meandropsinidae Henson, 1848, *Rhapydionidae Keijer, 1945, Fusarchaiasidae Saidova, 1981, Hottingerinidae Mikhalevich, 1988, ?Orduellidae Sirel, 1998 (nom. corr. pro Orduellinidae Sirel, 1998).

Superfamily Soritidoidea Ehrenberg, 1839
Families: Soritidae Ehrenberg, 1839, Archaiasidae Cushman, 1927, Praerhapydionidae Hamaoui et Fourcade, 1973, Cycledomiidae Mikhalevich, 1988.

INCERTAE SEDIS:
Family: Milioliporidae Bronnimann et Zaninetti, 1971.

CLASS NODOSARIATA MIKHALEVICH, 1992

Subclass HORMOSINANA Mikhalevich, 1992

Order *SACCAMMINIDA Lankester, 1885
Families: Saccamminidae Brady, 1884, Thuramminidae A.D. Miklukho-Maklay, 1963.

Order HORMOSINIDA Mikhalevich, 1980

Superfamily Hormosinelloidea Rauser-Chernousova et Reytlinger, 1986, stat. elevated
Family: Hormosinellidae Rauser-Chernousova et Reytlinger, 1986.

Superfamily Hormosinoidea Haeckel, 1894
Families: Hormosinidae Haeckel, 1894, Ginesinidae Mikhalevich, 2003, (stat. elevated)., Aschemocellidae Vyalov, 1966, Reophacidae Cushman, 1910, *Oxinoxisidae Vyalov, 1968, *Glaucoamminidae Saidova, 1981, Dusenburyinidae Loeblich et Tappan, 1984, Cuneatidae Loeblich et Tappan, 1984, *Earlandinitidae Loeblich et Tappan, 1984, Kunklerinidae Rauser et Reytlinger, 1986, Ammolagenidae Kaminski, Henderson, Cetean et Waskowska, 2009.

Superfamily Cribratinoidea Loeblich et Tappan, 1984
Families: Cribratinidae Loeblich et Tappan, 1984, Thomasinellidae Loeblich et Tappan, 1984.

Order AMMOMARGINULINIDA Mikhalevich, 2002

Superfamily *Ammomarginulinoidea Podobina, 1978
Families: Ammomarginulinidae Podobina, 1978, Bykoviellidae Loeblich & Tappan, 1984 (reinstated from Vialoviinae in which it was placed in synonymy (Loeblich & Tappan, 1987) and given stat. nov.), Orectostominidae Mikhalevich, 2003, Lacroixinidae Saidova, 1981 (reinstated here from the sinonymy of Pseudobolivinidae in Loeblich &Tappan, 1987).

Superfamily *Flabellamminopsidoidea Mikhalevich, 2003
Families: Flabellamminopsidae Mikhalevich, 2002, *Spiropsammiidae Loeblich et Tappan, 1961.

Order *NOURIIDA Mikhalevich, 1980

Superfamily Nourioidea Chapman & Parr, 1936 (nom. corr. pro Nouriidoidea Mikhalevich, 2004)
Families: *Nouriidae Chapman et Parr, 1936, Reophacellidae Mikhalevich et Kaminski 2003 in Mikhalevich 2003, Barbourinellidae Saidova, 1981, Conotrochamminidae Saidova, 1981.

Superfamily *Liebuselloidea Saidova, 1981, stat. n.
Families: Liebusellidae Saidova, 1981, Marieitidae Loeblich et Tappan, 1986.

Order *PSEUDOPALMULIDA Mikhalevich, 1992
Families: Pseudopalmulidae E.V.Bykova, 1959,? Semitextulariidae Pokorny, 1956.

INCERTAE SEDIS:
family Agglutisolenidae Mikhalevich 2003.

Subclass NODOSARIANA Mikhalevich, 1992

Order LAGENIDA Delage et Herouard, 1896
Families: Lagenidae Reuss, 1862, Ellipsolagenidae A.Silvestri, 1923, Sipholagenidae Patterson et Richardson, 1987, Asipholagenidae Reddy et Rajshekhar, 2007.

Order NODOSARIIDA Calkins, 1926

Suborder SYZRANIINA Mikhalevich, 1992

Superfamily Caligelloidea Reytlinger, 1959
Family: Caligellidae Reytlinger, 1959.

Superfamily Syzranioidea Vachard, 1981
Family: Syzraniidae Vachard, 1981, Earlandiidae Cummings, 1955, Eogeinitzinidae Vachard, 1994, Paratikhinellidae Loeblich et Tappan, 1984.

Superfamily Eonodosarioidea Rauser, 1993 in Vdovenko et al., 1993
Family: Eonodosariidae Rauser, 1993 in Vdovenko et al., 1993.

Suborder NODOSARIINA Calkins, 1926

Superfamily Ptychocladioidea Elias, 1950
Family: Ptychocladiidae Elias, 1950.

Superfamily Nodosarioidea Ehrenberg, 1838
Families: Nodosariidae Ehrenberg, 1838, ?Dentalinidae Schwager, 1877, Rimulinidae R.W. Jones, 1984 (stat. elevated, reinstated from the synonymy of Lingulininae), Frondiculariidae Reuss, 1860, *Nodosinellidae Rhumber, 1895, Geinitzinidae Bozorgnia, 1973, Pachyphloiidae Loeblich et Tappan, 1984, Ichthyolariidae Loeblich et Tappan, 1986, Daucinoididae Mikhalevich, 1993, Protonodosariidae Mamet et Pinard, 1992.

Superfamily Stilostomelloidea Finlay, 1947
Family: Stilostomellidae Finlay, 1947.

Superfamily *Colanielloidea Fursenko, 1959
Family: Colaniellidae Fursenko, 1959.

Order *PALAEOTEXTULARIIDA Hohenegger et Piller, 1975

Superfamily Palaeotextularioidea Galloway, 1933
Families: Palaeotextulariidae Galloway, 1933.

Superfamily Louisettitioidea Loeblich et Tappan, 1984
Family: Louisettitidae Loeblich et Tappan, 1984.

Order *BISERIAMMINIDA Mikhalevich, 1981
Families Biseriamminidae N. Chernysheva, 1941, Globivalvulinidae Reitlinger, 1950.

Order POLYMORPHINIDA Mikhalevich, 1980

Suborder POLYMORPHININA Mikhalevich, 1980
Families: Polymorphinidae d'Orbigny, 1839, Glandulinidae Reuss, 1860, Plectofrondiculariidae Cushman, 1927, Partisaniidae Loeblich et Tappan, 1984, ?Ramulinidae Brady, 1884.

Suborder *PLEUROSTOMELLINA Mikhalevich, 1993
Family: Pleurostomellidae Reuss, 1860.

Order DELOSINIDA Revets, 1989, stat. elevated
Family: Delosinidae Parr, 1950

Order VAGINULINIDA Mikhalevich, 1993
Families: Vaginulinidae Reuss, 1860, Lenticulinidae Chapman, Parr, Collins, 1934, Marginulinidae Wedekind, 1937, Robuloididae Reiss, 1963, Palmulidae Saidova, 1981, Lingulinidae Loeblich et Tappan, 1961, stat. elevated, Nanicellidae Fursenko, 1959.

CLASS ROTALIATA MIKHALEVICH, 1980

(= class Globothalamea Pawlowski et al., 2012)

Subclass TEXTULARIANA Mikhalevich, 1980

Order NAUTILOCULINIDA Mikhalevich, 2003

Superfamily Nautiloculinoidea Loeblich et Tappan, 1985
Families: Nautiloculinidae Loeblich et Tappan, 1985, Haplophragmoididae Maync, 1952.

Superfamily Alveolophragmioidea Stschedrina, 1936
Families: Alveolophragmiidae Saidova, 1981, Discamminidae Mikhalevich, 1980, Cyclamminidae Marie, 1941.

Order SPIROPLECTAMMINIDA Mikhalevich, 1992

Superfamily Spiroplectamminoidea Cushman, 1927
Family: Spiroplectamminidae Cushman, 1927.

Superfamily Spirotextularioidea Saidova, 1975
Families: Spirotextulariidae Saidova, 1975, Novalesiidae Loeblich et Tappan, 1984, ?Ecougellidae Loeblich et Tappan, 1985, ?Pavonitinidae, Loeblich et Tappan, 1961.

Order TEXTULARIIDA Delage et Herouard, 1896

Superfamily Textularioidea Ehrenberg, 1938
Families: Pseudobolivinidae Wiesner, 1931, Textulariopsidae Loeblich et Tappan, 1982, Kaminskiidae Neagu, 1999, Textulariidae Ehrenberg, 1938, Bigenerinidae Saidova, 1981, Textularioididae Loeblich et Tappan, 1984.

Superfamily Septotextularioidea Loeblich et Tappan, 1985
Families: Septotextulariidae Saidova, 1975, Tawitawiidae Loeblich et Tappan, 1961.

Order VERNEUILINIDA Mikhalevich et Kaminski, 2003 in Mikhalevich, 2003

Superfamily Verneuilinoidea Cushman, 1911
Families: Verneuilinidae Cushman, 1911, Prolixoplectidae Loeblich et Tappan, 1985, Piallinidae Rettori et Zaninetti, 1993, Globotextulariidae Cushman, 1927, Tritaxidae Plotnikova, 1979, Plectoverneuilinellidae Cetean et Kaminski, 2011.

Superfamily Crenaverneuilinoidea Mikhalevich, 2004
Family: Crenaverneuilinidae Mikhalevich, 2003

Superfamily Valvulinoidea Berthelin, 1880
Families: Valvulinidae Berthelin, 1880, Eggerellidae Cushman, 1937, Pseudogaudryinidae Loeblich et Tappan, 1985, Clavulinidae Balakhmatova, 1973.

?Superfamily Chrysalidininoidea d"Orbigny, 1839
Family: Chrysalidinidae Neagu, 1968, Olgiidae Mikhalevich, 2011.

Order TROCHAMMINIDA Saidova, 1981

Superfamily Trochamminoidea Schwager, 1877
Families: Trochamminidae Schwager, 1877, Vialoviidae Suleymanov, 1983, Rotaliamminidae Saidova, 1981, Adercotrymidae Brönnimann et Whittaker, 1988, Ammosphaeroidinidae Cushman, 1927, Valvulamminidae Loeblich et Tappan, 1986.

Superfamily Remaneicioidea Loeblich et Tappan, 1964
Families: Remaneicidae Loeblich et Tappan, 1964, Carterinidae, Asterotrochamminidae Bronnimann, Zaninetti et Whittaker, 1983, Dictyopsellidae Bronnimann, Zaninetti et Whittaker, 1983.

Subclass ROTALIANA Mikhalevich, 1980

Superorder ?ROBERTINOIDA Mikhalevich, 1980

Order ROBERTINIDA Mikhalevich, 1980
(= Robertinida Haynes, 1981)

Suborder DUOSTOMININA subord. nov.

Families: Duostominidae Brotzen, 1963, Asymmetrinidae Brotzen, 1963, Oberhauserellidae Fuchs, 1970.

Suborder CERATOBULIMININA Mikhalevich, 1980

Families: Ceratobuliminidae Cushman, 1927, Epistominidae Wedekind, 1937, Reinholdellidae Seiglie et Bermudez, 1965, Epistominoididae Saidova, 1981.

Suborder ASTERIGERINIDA d'Orbigny, 1839

Superfamily Asterigerinoidea d'Orbigny, 1839

Families:Asterigerinidae d'Orbigny, 1839, Epistomariidae Hofker, 1954, Asterigerinatidae Reiss, 1963, Amphisteginidae Reiss, 1963, Alfredinidae S.N.Singh et Kalia, 1972, Actinosiphonidae Adams, 1987.

Superfamily Clypeorbioidea Sigal, 1952

Families:*Clypeorbidae Sigal, 1952, Boreloididae Reiss, 1963.

Suborder CONORBOIDINA Mikhalevich, 1992

Family: Conorboididae Thalmann, 1952.

Suborder ROBERTININA Mikhalevich, 1980

Families: Robertinidae Reuss, 1850, Alliatinidae McGowran, 1966.

Superorder DISCORBOIDA Ehrenberg, 1838

Order ROSALINIDA Delage et Herouard, 1896

Superfamily Rosalinoidea Reiss, 1963

Families: Bagginidae Cushman, 1927, Cancrisidae Chapman, Parr et Collins, 1934, Eponididae Hofker, 1951, Neoeponididae Hauser et Keller Grunig, 1993, ?Heleninidae Loeblich et Tappan, 1987, Mississippinidae Saidova, 1981, Rosalinidae Reiss, 1963, Bronnimanniidae Loeblich et Tappan, 1984.

Superfamily Discorbinelloidea Sigal, 1952

Families: Discorbinellidae Sigal, 1952, Pseudoparrellidae Voloshinova, 1952, Planulinoididae Saidova, 1981, Svratkinidae Bugrova, 1989.

Superfamily Gavelinelloidea Hofker, 1956, stat. elevated.

Families: Gavelinellidae Hofker, 1956, ?Anomalinidae Cushman, 1927 [(with Parrelloididae Hofker, 1956, Heterolepidae Gonzales-Donoso, 1969 and Oridorsalidae Loeblich et Tappan, 1984 (part) as synonyms of Anomalinidae (after Revets, 1996)], Lingulogavelinellidae Scheibnerova, 1972 (fam. reinstated by Bugrova, 1990), ?Karreriidae Saidova, 1981, Trichohyalidae Saidova, 1981.

Superfamily Sphaeroidinoidea Cushman, 1927
Families: Sphaeroidinidae Cushman, 1927, ?Pegidiidae Heron-Allen et Earland, 1928, Tremachoridae Lipps et Lipps, 1967, Bueningiidae Saidova, 1981.

Superfamily ?Siphoninidoidea Cushman, 1927
Family: Siphoninidae Cushman, 1927.

Order DISCORBIDA Ehrenberg, 1838

Superfamily Discorbinoidea Ehrenberg, 1838
Families: Discorbidae Ehrenberg, 1838, Torresinidae Loeblich et Tappan, 1984.

Superfamily Alabaminoidea Hofker, 1951
Families: Globorotalitidae Loeblich et Tappan, 1984, Osangulariidae Loeblich et Tappan, 1964, Alabaminidae Hofker, 1951 (with Oridorsalidae Loeblich et Tappan, 1984 placed by Revets, 1996 as synonym of this family (with the type genus *Oridorsalis*) and partially of the family Anomalinidae – see above), Coleitidae Loeblich et Tappan, 1984, ?Quadrimorphinidae Saidova, 1981.

Order GLABRATELLIDA Mikhalevich, 1994
Families: Glabratellidae Loeblich et Tappan, 1964, Rotaliellidae Loeblich et Tappan, 1964 (family transferred to glabratellids by Pawlowski et Zaninetti from Discorbacea in Loeblich & Tappan, 1987), ?Buliminoididae Seiglie, 1970, ?Heronallenidae Loeblich et Tappan, 1986.

Order PLANORBULINIDA Mikhalevich, 1992

Suborder PLANORBULININA Mikhalevich, 1992
Families:Planorbulinidae Schwager, 1877, Cibicididae Cushman, 1927, Victoriellidae Chapman et Crespin, 1930, Planulinidae Bermudez, 1952, Stichocibicididae Saidova, 1981, Annulocibicididae Saidova, 1981, Stensioeinidae Bugrova, 1990 (nom. correct. pro Stensioinidae Bugrova, 1990).

Suborder ACERVULININA Schultze, 1854
Families: Acervulinidae Schultze, 1854, Menoellidae Cherchi et Schroeder, 2005, Homotrematidae Cushman, 1927.

Order ROTALIIDA Lankester, 1885 (nom. corr. Calkins, 1909 (after Loeblich et Tappan, 1992)
Families: Rotaliidae Ehrenberg, 1839, Calcarinidae Schwager, 1876, Miogypsinidae Vaughan, 1928, Pseudorbitoididae M.G. Rutten, 1935 , Faujasinidae Bermudez, 1952, Notorotaliidae Hornibrook, 1961, Cuvillierinidae Loeblich et Tappan, 1964, Ammoniidae Saidova, 1981 (with *Gavelinopsis* Hofker, 1951 according to Revets, 2002 c).

Order ? CHILOSTOMELLIDA Haeckel, 1894
Families: Chilostomellidae Brady, 1881, Pallaimorphinidae Loeblich et Tappan, 1987, stat. nov., Chilostominidae Finger et Gaponoff, 1986, Lublinidae Gawor-Biedowa, 1987.

Superorder NONIONOIDA Saidova, 1981

Order NONIONIDA Saidova, 1981

Suborder NONIONINA Saidova, 1981

Superfamily Nonionoidea Schultze, 1854
Families: Nonionidae Schultze, 1854, Nonionellidae Voloshinova, 1958, ?Spirotectinidae Saidova, 1981.

Superfamily Almaenoidea Mjatliuk, 1959
Family: Almaenidae Mjatliuk, 1959.

Suborder ASTRONONIONINA Saidova, 1981
Families: Astrononionidae Saidova, 1981, *Bisacciidae Loeblich et Tappan, 1988.

Order ELPHIDIIDA Saidova, 1981
Families: Elphidiidae Galloway, 1933; Haynesinidae fam. nov., Cribroelphidiidae Voloshinova, 1958.

Order ORBITOIDIDA Copeland, 1956
Families: Orbitoididae Schwager, 1876, Lepidorbitoididae Vaughan, 1933, Lepidocyclinidae Scheffen, 1932, Linderinidae Loeblich et Tappan, 1984, Droogerinellidae Popescu et Brotea, 1995, Orbitosiphonidae Matsumaru & Jauhri 2003.

Order NUMMULITIDA Carpenter, Parker et Jones, 1862
Families: Nummulitidae de Blainville, 1827, Discocyclinidae Galloway, 1928, Pellatispiridae Hanzawa, 1937, Asterocyclinidae Bronnimann, 1951.

Superorder BULIMINOIDA Saidova, 1981

Order BOLIVINITIDA Saidova, 1981

Suborder SPIROBOLIVINA Mikhalevich, 1998
Families: Lacosteinidae Sigal, 1952, Spirobolivinidae Saidova, 1981.

Suborder BOLIVINITINA Saidova, 1981

Superfamily Loxostomatoidea Loeblich et Tappan, 1962
Families: Loxostomatidae Loeblich et Tappan, 1962, Bolivinellidae Hayward, 1980, Tortoplectellidae Loeblich et Tappan, 1985.

Superfamily Bolivinitoidea Cushman, 1927
Families: Bolivinitidae Cushman, 1927, Bolivinidae Glaessner, 1937, Bolivinoididae Loeblich et Tappan, 1984.

Superfamily Parabrizalininoidea Revets, 1996, stat. elevated.
Family: Parabrizalinidae Revets, 1996, stat. elevated.

Superfamily Eouvigerinoidea Cushman, 1927
Families: Eouvigerinidae Cushman, 1927, Siphogenerinoididae Saidova, 1981.

Superfamily Pavonininoidea Eimer et Fickert, 1899, nom. correct. pro Pavoninoidea Mikhalevich, 2000
Family: Pavoninidae Eimer et Fickert, 1899

Superfamily Millettioidea Saidova, 1981
Family: Millettiidae Saidova, 1981

Order CASSIDULINIDA d'Orbigny, 1839
Families: Cassidulinidae d'Orbigny, 1839, Ehrenberginidae Cushman, 1927, Cassidulinitidae Saidova, 1981, Orthoplectidae Loeblich et Tappan, 1984, Cassidinidae Boukhary et Scheibner, 2011.

Order BULIMINIDA Saidova, 1981

Suborder TURRILININA Saidova, 1981
Families: Turrilinidae Cushman, 1927, Tosaiidae Saidova, 1981, Virgulinellidae Loeblich et Tappan, 1984, ?Praeplanctoniidae Georgescu, 2009 (nom corr. pro Praeplanctonidae).

Suborder BULIMININA Saidova, 1981
Families: Buliminidae Jones, 1875, Uvigerinidae Haeckel, 1894, Reussellidae Cushman, 1933, Buliminellidae Hofker, 1951, ?Biedafranciszkinidae Gawor-Biedowa, 1992, Fursenkoinidae Loeblich et Tappan, 1961, Stainforthiidae Reiss, 1963, Trimosinidae Saidova, 1981, Pappinidae Haunold, 1990.

Suborder TUBULOGENERININA Mikhalevich, 1998
Family: Tubulogenerinidae Saidova, 1981.

Suborder CAUCASININA Mikhalevich, 1998
Family: Caucasinidae N.K. Bykova, 1959.

Subclass GLOBIGERINANA Mikhalevich, 1980

Order HETEROHELICIDA Mikhalevich, 1992

Superfamily Heterohelicidoidea Cushman, 1927

Families: Heterohelicidae Cushman, 1927, Guembelitriidae Montanaro Gallitelli, 1957, Chiloguembelinidae Reiss, 1963.

Order GLOBOROTALIIDA Mikhalevich, 1980

Superfamily Rotaliporoidea Sigal, 1958
Families: Rotaliporidae Sigal, 1958, Hedbergellidae Loeblich et Tappan, 1961, Praehedbergellidae Banner et Desai, 1988.

Superfamily Globotruncanoidea Brotzen, 1942
Families: Globotruncanidae Brotzen, 1942, Rugoglobigerinidae Subbotina, 1959.

Superfamily Globorotalioidea Cushman, 1927
Families: Globorotaliidae Cushman, 1927, Pulleniatinidae Bronnimann et Bermudez, 1953, Truncorotaloididae Loeblich et Tappan, 1961, Candeinidae Cushman, 1927, Catapsydracidae Bolli, Loeblich et Tappan, 1957.

Order HANTKENINIDA Mikhalevich, 1980

Superfamily Hantkenininoidea Cushman, 1927
Families: Hantkeninidae Cushman, 1927, Globanomalinidae Loeblich et Tappan, 1984.

Superfamily Planomalinoidea Bolli, Loeblich et Tappan, 1957
Families: Planomalinidae Bolli, Loeblich et Tappan, 1957, Globigerinelloididae Longoria, 1974, Schackoinidae Pokorny, 1958.

Order CASSIGERINELLIDA, ord. nov.

Superfamily Cassigerinellidoidea Bolli, Loeblich et Tappan, 1957
Family: Cassigerinellidae Bolli, Loeblich et Tappan, 1957.

Order GLOBIGERINIDA Carpenter, Parker et Jones, 1862

Superfamily Favuselloidea Longoria, 1974
Families: Favusellidae Longoria, 1974, Globuligerinidae Loeblich et Tappan, 1984

Superfamily Globigerinoidea Carpenter, Parker et Jones, 1862
Families: Globigerinidae Carpenter, Parker et Jones, 1862, Marginotruncanidae Pessagno, 1967, Orbulinidae Schultze, 1854, Hastigerinidae Bolli, Loeblich et Tappan, 1957, Globigerinitidae Bermudez, 1961.

Chapter V

SHELL MORPHOLOGY AND SHELL WALL ULTRASTRUCTURE

V.1. SHELL STRUCTURE

V.1.1. Shell Morphology

Foraminifera possess unilocular or multi-chambered shell, the cytoplasm is overflowing it from the outside (Plate **16**, figure – s 2, 3, Plate 18, figure 5). Thus, in a strict sense, their skeleton may be considered as being an inner rather than outer structure. Shell morphological types are an extremely variable and distinctive feature of the whole group. The main types are the following (Plate **2**).

1) branching, curved or of other irregular forms, 2) uniserial (with the chambers disposed in one or several (2 – 5) parallel rows; 3) irregularly coiled (whorls coiled at random) and regularly coiled (whorls coiled at the definite angles to each other, the angle depending on the number of chambers in each whorl and their position), 4) spirally coiled (the spire may be flat – planispiral, conical – trochoid, or with a more strongly elongated spire – polimorphine (loose spire of some authors; such forms are called in Russian screw-spired). Coiled shells may be evolute (all the whorls visible externally) or involute (last whorl or the last chamber embracing the previous whorls). The involute character may be expressed only partially (only part of the whorls visible externally, often one of the shell sides is involute, the other one evolute, as for example, in trochoid shells, which have an evolute spiral side and an involute umbilical side).

These morphological types do not exhaust the variability of foraminiferal morphotypes. There are many heteromorphic forms in which the shell begins as one morphological type and then changes to a second type during ontogenetic development. As a result, such shells encompass two or more different structural types. For example, the initial part of the shell may be planispiral, the later part uncoiled (*Ammobaculites*, *Ammotium*, microsphaeric form of *Lingulina* (see Mikhalevich, 1983, p. 168, figure – s 260, 261), *Amphicorina*, - Plate **2**, 13, A-E, 14, B, Plate **1 a**, III, Miliolana: figure – s 10, 15) or the initial part may be trochoid, later becoming triserial, biserial or uniserial (many of Ataxophragmiids, of representatives of Textulariana: Plate.**1a**, V, figure – s 2 – 5, Rotaliana: Plate.**1a**, V, figure – s 3 - 7). In miliolids the initial section may be glomerate, one of the miliolide types of coiling, or planispiral, the later part planispiral or free (Plate **1a**, III, Miliolana: figure – s 14, 15, 16, Plate **4**, figure – s A, B). In many unrelated forms the later chambers are often arranged cyclically (Plate **1a**, III, Miliolana: 7, 16, V, Rotaliana: 14, Plate **2**, 13, E). Such heteromorphic forms of one species have often been attributed to different genera; for example, the microsphaeric form of *Lingulina carinata,* which has an initial coil (Plate **2**, figure 14 a, b), was described as *Marginulina*, while *Hauerina fragilissima,* in which the planispiral part is strongly developed, was initially described as *Spiroloculina* in Brady (1884, Plate 9, figures 12 – 14) (Mikhalevich, 1983, p. 142, figure - s 244a, b).

The heteromorphism may be expressed not only in the chamber arrangement but also in the chamber form (Plate **1**, III, *Miliamminana*: figure – s 2 - 4, Miliolana: figure – s 7, 15), Plate **2**, figure 13, Plate **9**) and in the apertural structures as well. The type of shell structure (along with the other morphological features of the shell, such as apertural structure and chamber shape) may be considered as the most important taxonomic features at the class level, while the less significant details of their structure are characteristic of the lower rank taxa such as orders and superfamilies.

In the case of the two-chambered or multi-chambered shell the initial chamber is called the proloculus. In many miliolats there is a second tubular chamber following the proloculus and forming a full or partial whorl around it; this chamber is called the flexostyle (Plate **4**). The form of foraminiferal chambers is extremely variable; they may be spherical or subspherical, oval, tubular, fan-shaped, crescent-shaped, trapeziform, star-shaped, V-shaped or some other shape (Plates **1, 1 a, 2**). In this regard, foraminifera are a good group in which to study morphogenesis. As a rule, chamber form is characteristic at the generic and species level but in some cases it is specific for higher level taxa. Thus miliolids are characterized by tubular chambers of different lengths, spirillinats by long tubular chambers forming several revolutions, astrorhizats are usually characterized by branching or irregularly-shaped chambers, while in nodosariats V-shaped chambers are occasionally found.

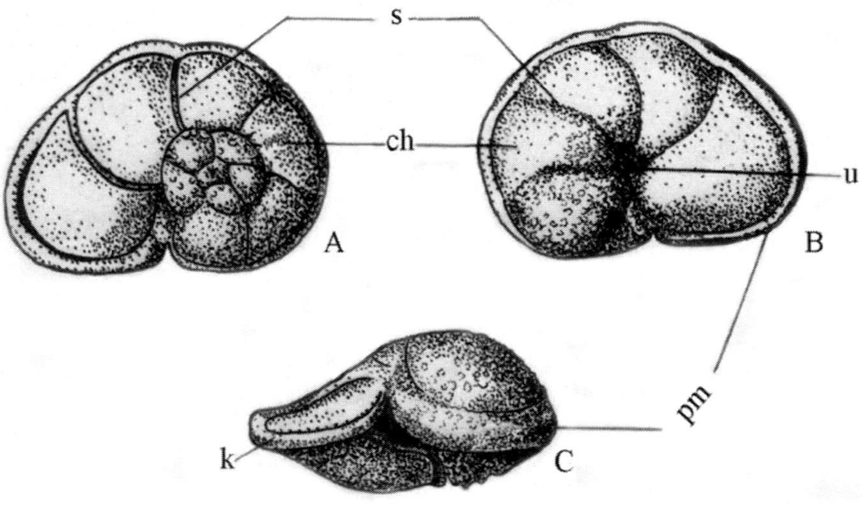

Plate 3. Designation of the shell parts (illustrated by the trochospiral multi-chambered shell of *Globorotalia*).
A – spiral side, B – umbilical side, C - lateral side, ch – chambers, s - sutures, u – umbilicus, pm – peripheral margin, k – keel.

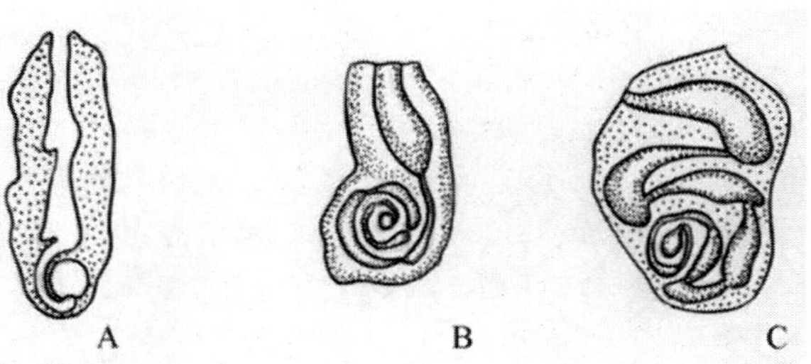

Plate 4. Shell sections of some Miliolata.
A – *Sarmatiella*, B, C – *Nubeculinella* (rounded proloculus followed by the second tubular chamber – flexostyle, then by the subsequent chambers) (drawings of A. Bogdanovich kindly granted by A. Azbel, VNIGRI, St. Petersburg).

The chamber lumen may be complicated with additional inner structures, cellular or net-like features, pillars, full or partial septa and valves (Plate .**8**, figure – s 5,6 A, B, Plate **10**, figure 4, Plate **11**, figure 3). The number of chambers in a shell usually varies from 4 -5 to 20 – 30, but may exceed several tens or hundreds (Plate **1 a**, III, Miliolana: figure - s 13, 16, V, Rotaliana: 14, Plate **8**, figure 1, Plate **9**). The chambers are separated from one another by the septae, which create sutures where they intersect with the chamber wall. The sutures may be flush with the shell surface, excavated or convex and they may be straight, oblique, curved to different degrees or sometimes zigzag in form (Plates **1 a, 2**).

V.1.2. Aperture

The communication of the cytoplasm with the outer environment is realized through apertures in the shell (Plate **5**). The apertures of the chambers previous to the last one, which are internal, are called foramens. In some foraminiferal groups (Ceratobuliminids) the primary foramens are overgrown by the shell material and replaced by new openings in a different location but performing the same function. These are called deuteroforamens. The apertures may be single or multiple, may occupy different position on the last final chamber (at its base or top) and in relation to the entire shell (at the end of the shell, on its spiral or umbilical side, more rarely in a peripheral position – Plate **2**, figure 9 B). In the evolutionary advanced genera additional (supplementary) apertures may present, more often being placed on the umbilical side of the shell or at the peripheral margin, along the sutures (sutural openings), rarely on the spiral side (Plate **6**, figure 7, Plate **48**, figure – s 12 a, b, Text. fig. 4). The apertural form varies greatly; it may be circular, oval, loop-shaped, fissured, sieve-like, dendritic or radial. The aperture may be complicated with additional skeletal formations – both external (neck, collar, lip) and internal (ribs, plates, teeth, groves, tubes, apertural chamberlets). Ribs, plates and teeth perform a basic supporting function being situated downstream of the strong cytoplasmic flow. That is why they are especially strongly developed in elongated forms, achieving greatest complexity in miliolats, buliminids and nodosariats. The so-called miliolid lamellar tooth (flap) represents a fully external formation and ought to be distinguished from miliolid inner teeth (Mikhalevich, 1981). Apertural structures may change during ontogeny (for instance, a single opening may change into multiple openings). The miliolid tooth may change from a simple to a complex structure

(Plate **5**, figure – s 3, c– m, Plate **7**, Plate **43**, figure – s 3 a, b, 4). Inner miliolid teeth represent unique structures and exhibit considerable complexity and diversity. In agglutinated forms, in which the shell wall, in contrast to that of calcareous forms, does not permit the formation of additional strong support elements, the supporting function is sometimes performed by sponge spicules incorporated into the shell wall near the aperture (Plate **2**, figure 2, Astrorhizana). The diversity of these additional apertural structures is linked to their important and varied functional role and associated with intensive cytoplasmic movement in the apertural area. In some primitive unilocular forms with an organic or agglutinated shell wall (subclasses Lagynana, Astrorhizana) an obvious aperture is absent (Plate **2**, figure 3). The various apertural types found in the Astrorhizana are shown in Plate **45**.

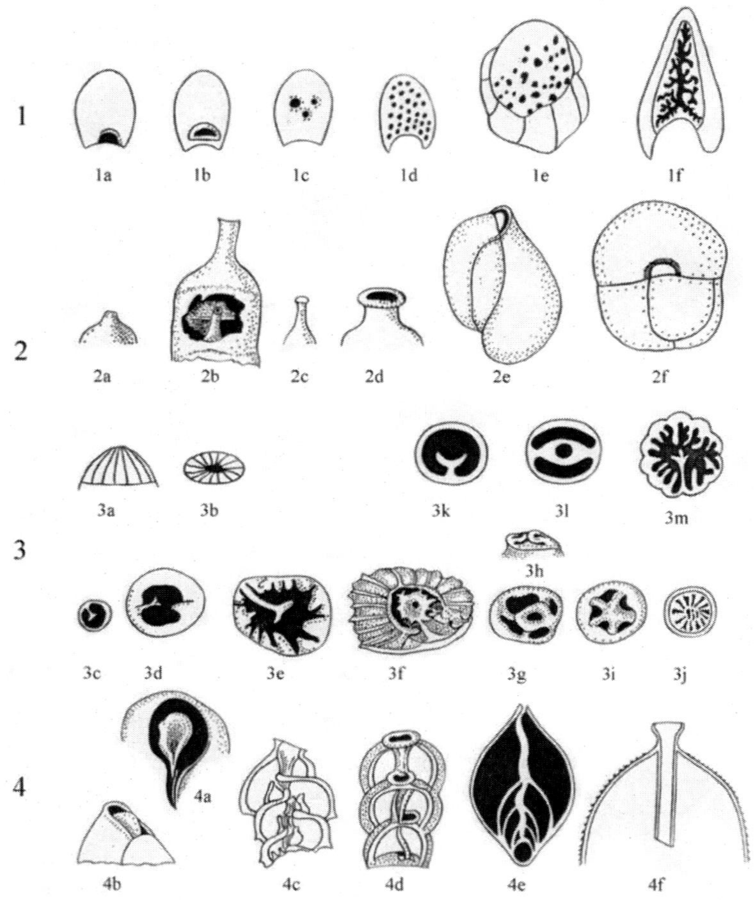

Plate 5. Structure of the aperture.
1 - Position of the aperture: 1a – aperture at the base of the septal surface, 1 b – f - elevated above the base of the septal surface (b - singular, c – e - multiple, f – dendritic); 2 - External elements of apertural structure (a – projection, b – neck, c – neck with a collar, d – neck with a lip, e, f - plate-like tooth; 3, 4 - Inner skeletal formations of the aperture (3 a, b - radial aperture of Nodosariata, 3 c – m - apertural teeth of Miliolata, simple (3 b, 3 k) and complex, 4 a – d - inner teeth, plates and grooves of Rotaliata, e, f - entosolenian tubes of Nodosariata.

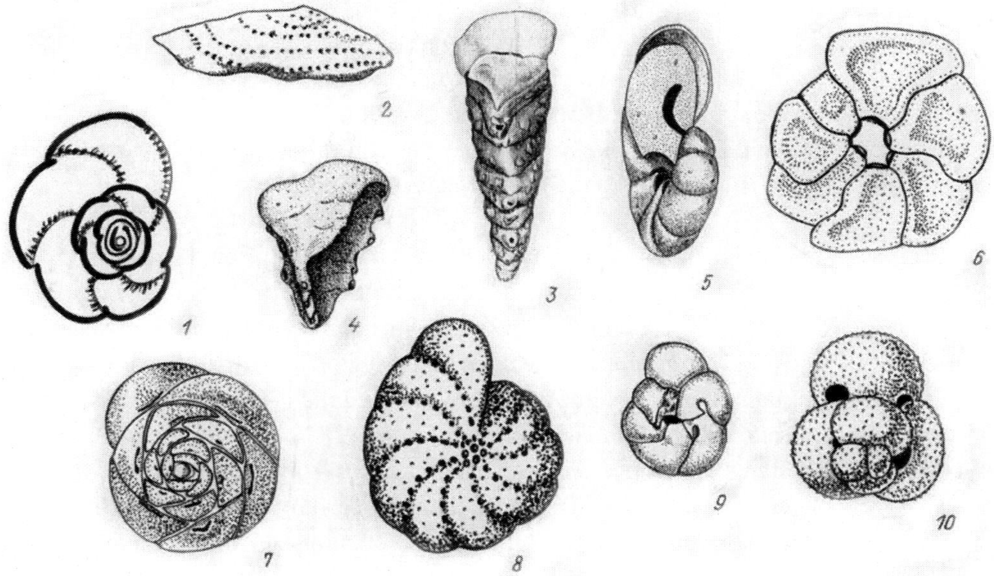

Plate 6. Additional apertures (from Mikhalevich, 2000).
1 - *Polysegmentina* (Miliolana), tubular chambers of the initial part glomerately (sigmoidally) coiled, lacking additional apertures, later chambers broad, planispirally coiled, having additional apertures; 2 - *Trocholinopsis* (Spirillinana) – additional sutural apertures in the spiral suture between the volutions of the long tubular chamber; 3 – 6 - Additional apertures of Textulariana: 3, 4 – additional apertures at the peripheral chamber margins (3 - *Norvanganina*, 4 – *Plotnikovina*), umbilical additional apertures (5 – in *Polystomammina*, 6 – in *Rotaliammina*); 7 – 9 - Additional apertures of Rotaliana: 7 – in *Pseudoeponides* (on the spiral side, in parallel to the spiral suture), 8 - in *Cribroelphidium* (sutural apertures), 9 - in *Strebloides* (sutural apertures on the umbilical side); 10 - Additional sutural apertures of Globigerinana (*Globigerinoides*).

Different features of the apertural structures have different taxonomic weight. Thus, the inner teeth of Miliolata are characteristic of this class only. Entosolenian tubes and the special radial apertures of Nodosariata (Plate **1 a**, IV, **5**, 3, a, b, 4 e, f, Plate **46**), present in their agglutinated (*Nodosinum*) and calcareous (*Dentalina*) forms, are also very important diagnostic features at the class rank. Ampula-like apertural formations are also characteristic of this class only. The strongly developed tooth-plates, grooves and valves are characteristic of the class Rotaliata and are not found in any other taxa. In planktonic forms (subclass Globigerinana) the apertures are strongly enlarged (Plate **16**, figure 1, Plate **49**, II).

The aperture of the last chamber is located terminally in the majority of the foraminiferal classes. Only in the class Rotaliata is it initially situated at the base of the apertural face of the last formed chamber (Plate **1 a**, V, Plate **16**, figure 1, Plate **48**). However, in some elongate rotaliats, it may come to occupy a secondarily terminal position during ontogenesis. The apertural position in relation to the whole shell is often used in the taxonomy of the different classes at the family level, while the more minute details of the apertural structures are used as generic features. Thus, many of the genera in the subclasses Textulariana, Rotaliana and Globigerinana are distinguished on the basis of the presence of additional shell apertures (Plate. **1 a**, V, Plate **2**, figure – s 9 B, C, 10 C, 13 B, Plate **6**, Plate **43**, figure – s 12 a, b). At the same time the ability to form additional apertures is characteristic of some classes only.

V.1.3. Integrative Systems of the Shell

In foraminifera with numerous chambers and whorls, communication between chambers as well as between the inner chambers and the exterior, is difficult. This problem is eliminated with the development of integrative systems with different degrees of complexity (Plate **8**, figure – s 1 – 4), in addition to the basic system of foramens.

1. Stolons – tubular openings in the chamber wall between adjacent chambers in multi-chambered specimens; these connect the chambers and facilitate more active communication between them. System of stolons forms interconnected inner ways providing and facilitating communication of the cytoplasm of the closely disposed and rather remote chambers (e.g. in alveolinids, lituolids – Miliolata, orbitolinids – Spirillinata, orbitoidids – Rotaliata) (Plate **8**, figure 4).
2. Tunnels – systems of empty spaces between the skeletal structures of the inner chambers. They may extend between deposits of skeletal material at the base of the chambers (chomata) and between the resorbed openings in the lower parts of the septae (Plate **8**, figure 1, Plate **10**, figure 5 – in fusulinids), or between the spiral partitions that do not reach the interchamber septa (alveolinids). In the both cases they are situated between the inner skeletal elements (endogenous origin) and have organic linings, thus not possessing their own wall. In some other representatives of the class Miliolana the passages between adjacent chambers situated in a vertical row one above another in the final part of the shell have a similar function and origin (e.g. *Rhapydionina*, see Loeblich & Tappan, 1987, plate 370, figure - s 9, 10; Haynes, 1981, fig. 8. 6). The cytoplasm streams through the spaces between the multiple apertural openings and pillars of these consecutive chambers. These passages, contrary to the canals, don't have their own walls. The term "tunnels" was originally introduced for fusulinids (in which they were formed by resorbtion) but has been extended to pfenderinids (Hottinger, 1978) and later to nummulitids by Hottinger (2006) in his "Illustrated glossary of terms used in foraminiferal research".
3. The inner apertural integrative systems - strongly developed inner apertural structures extending along the whole chamber from the aperture to the previous foramen and joining in a series the foramens of all the preceding chambers. These are strongly developed in the class Rotaliata, especially in its subclass Rotaliana (Plate **5**, figure – s 4, c, d). They are also encountered in the monoaxial representatives of the subclass Globigerinana and occasionally in the subclass Nodosariana (Plate **5**, figure 4 e) (Mikhalevich, 1993a, p. 7).

   In each of the classes, the inner apertural integrative systems differ in their structure. Rotaliana possess the most complex inner apertural formations (tooth plates, foraminal plates, cover plates) as their bilamellar wall provides more extensive constructive opportunities. According to Revets (1993b), the tooth plate of the Rotaliana is a wholly internal structure formed by the inner lamella (its upper part, considered previously to be the part of the toothplate visible free above the apertural opening, in fact represents the apertural lip capping the toothplate from above). As shown by Hofker (1951), and supported by some researches including Revets (1993 b), the rotaliid "toothplates", foraminal plates and coverplates are essentially similar to the typical buliminid toothplates in their pattern of laminations. Further ultrastructural investigations and comparative morphological analysis will help to clarify the terminology of these structures and facilitate their use for taxonomic purposes. System of tube pillars of some Spirillinata opening outside as apertures on the aperural face and extending inside to the previous septum obviously also may be considered as integrative apertural systems (e.g. – *Ferayina,* Plate **42,** figure 10; Loeblich & Tappan, 1987, pl. 776).

4. Interlocular (interlamellar) spaces (cavities) - represent (according Hottinger, 2001, Hottinger et. al., 2001 a, Hottinger et al., 2001 b) deep open slits between the chambers or whorls, which often may be covered by later lamellation. These spaces also form a system within which the cytoplasm circulates. The additional passages necessary for this circulation may develop as a result of partial resorbtion.
5. Canals - these differ from stolons and tunnels both in their morphology, origin (having an exogenous rather than endogenous origin and passing between the chambers instead of inside them – Hottinger, 1978) and in their function; according to Hottinger it is the ectoplasm circulating in canals which supplies the inner parts of the cell with the oxygen and facilitates a connection with the surrounding environment. The ectoplasm is separated from the inner cytoplasm by special structures (thin tubes) that lack, according the recent data an inner organic lining. Thus a high degree of cytoplasmic differentiation is achieved. Reticulopodia protrude to the exterior from the outer openings of the canals. Gamete release has also been observed through the peripheral openings of the canal system (Krueger et al., 1996). In the Rotaliata, canals may be formed from the preceding system of interlocular cavities and both systems may be connected by gaps or smaller tubular passages (Hottinger, 2001), and Hottinger et al., 2001 b have shown that the spiral canal may also be formed as a result of fusion of the posterior foliar chamberlets (term of Hottinger et al., 2001 b).

The first descriptions of these fine architectural details began in the 19th century (Carpenter, 1861), but it was only at the end of the 20th century that new and modified techniques made it possible to study them in more detail. The works of Hofker (1951, 1971), Hansen & Lykke-Andersen (1976), Spindler (1978), Roettger et al., 1984 and some others laid the foundations for further research, culminating in the studies of Hottinger, Hottinger and co-authors, Muller-Merz & Hansen (1976 – 2006). Their skill mastery and striking results revealed the staggeringly detailed complexity of these systems and made it possible to compare in the finest details the inner shell structures at the generic and specific level. The complexity revealed by these studies resembles to some extent the comparative anatomical analysis of multicellular organisms. Canal systems may be developed to different extents and may be regarded as an example of parallelism in closely related taxa among the Rotaliana. Usually these systems have the spiral and radial canals with a multitude of narrower branches (Plate **8**, figure – s 2, 3). The canals described in representatives of the subclass Spirillinana (Hottinger, 1976, Mantzurova & Gorbatchik, 1982, Azbel, 1986) reveal parallelism in these more phylogenetically remote taxa. They do not achieve the same complexity as the canal systems in Rotaliata and in the more simple cases represent system of fissures (e. g. Lasiodiscidae – Plate **1a**, figure – s II, 4 – 6). A spiral canal and fissured passages connecting it with the lumen of the tubular chamber are developed in the more complex spirillinats (Hottingerellidae) (Plate **1a**, figure – s II, 10, b). Since the presence of the canal system is regarded as a taxonomic feature at the order level, it was necessary to establish new orders of Spirillinata (Plate **1a**, figure – s II, 5, 10, b and Mikhalevich, 1993 b, figure 1 – 3). At the same time the presence of the canal systems is characteristic of only some classes [Rotaliata, Spirillinata, and as an exception in one genus of the class Nodosariata, *Delosina* (Revets, 1989). Because Spirillinata and Nodosariata have monolamellar walls (see below, Part V. 2. 2), the canals are formed by the other elements of the wall and in a somewhat different way than those found in the bilamellar Rotaliata.

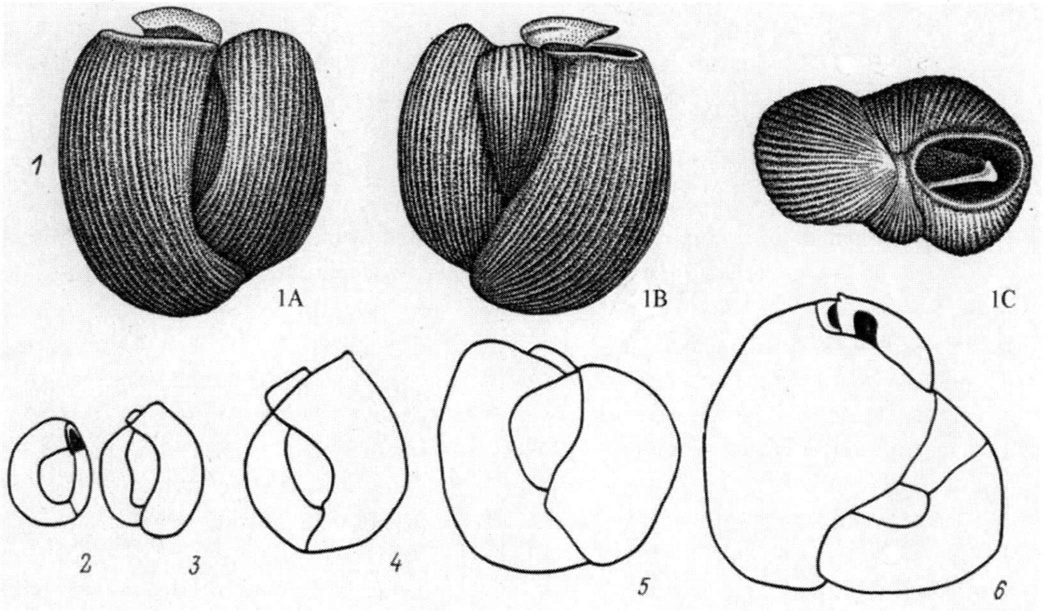

Plate 7. Growth stages of the apertural structures in *Flintina lingulata* (from Mikhalevich, 2000).
1 - Outward appearance of the shell (A, B - from the lateral sides, C - from the apertural side); 2 – 6 - Lateral view at the different growth stages.

The spiral Hottingerellid canal is positioned also in the sutural depression; it possibly may be covered from above by additional lamellae sometimes developed in Spirillinana, the radial connections between the whorls of radial canal are formed (according to Hottinger, 1976) by the fissures, which possibly may develop as a result of resorbtion. In the peculiar and unique canal system of the Nodosariana, the canals are formed by the subsutural tube; they underlie the sutures and are connected with the apertural trematophore of each chamber (Revets, 1989, Text figure 1). In all cases, the presence of the canal system facilitates communication of the cytoplasm of the inner chambers with the environment, shortening lines of communication. The further comparative anatomical investigation and analysis of the origin of different shell-wall elements used in the formation of canal systems in different foraminiferal groups would be of extreme interest.

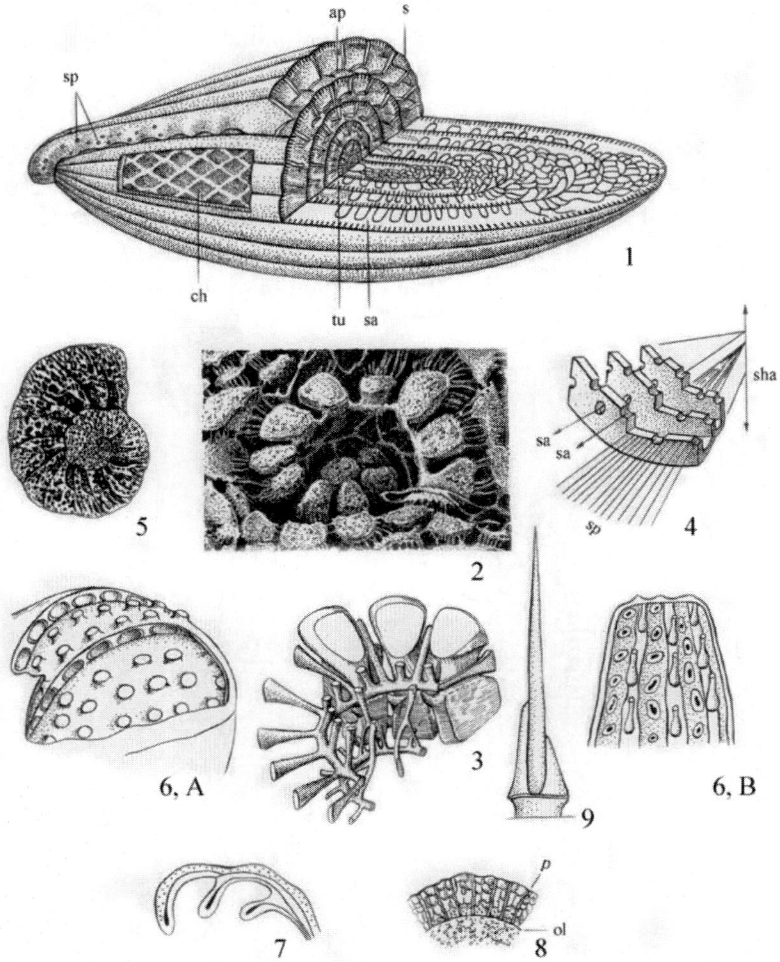

Plate 8. Details of the inner shell structure and of the wall ultrastructure (After Mikhalevich, 2000).
1 - *Praefusulina* (Miliolana, Fusulinoida) (from Fursenko, Rauser- Chernousova, 1959) – spirally coiled involute shell with complex inner structure: ap – aperture, s – septa, sp – septal pores, sa – septal archs, tu – tunnel, ch – chamberlets; 2, 3 - Fragments of Rotaliata's canal system (from Hottinger, 1978) (2 – *Baculogypsina* (x 240), 3 – *Rotalia* (x 240); 4 - The scheme of the Foraminiferal stolon system (from Hottinger, 1978) (sha – shell axis, sa– stolon axes, sp - plane of stolons); 5 - Cancellate wall of *Cyclammina* (section) (from Brady, 1884); 6 - Pillars of the inner shell wall of Miliolana: A – *Shanita*, B – *Archaias;* 7 - Bifontinal Rotaliana's wall with bilamellar septa (from Loeblich, Tappan, 1964); 8 - Transverse section of the wall of some Textulariana (ol– organic layer, p – pseudopore canaliculi (from Loeblich, Tappan, 1964); 9 - Spines of the planktonic species *Prosphaeroidinella*, loosely conjuncted (set into the hole) with the chamber wall (Hofker, 1971).

Several types of integrative systems may be present in the shell of the same species, for instance, canals and stolons in many of nummulitids. In *Heterostegina depressa,* different systems of cavities were described by Spindler (1978). In addition to the canals, there are tubes going inside the chambers as well as the presence of umbilical, marginal and septal connections, all of these systems being interconnected. Revets (1993 b) suggested that in low trochospiral rotaliids the toothplate "is linked to the presence of a canal". I suppose that the presence of additional sutural apertures may also be such a link, and that some more representatives possessing the canal system may be discovered among Spirillinana. Verhallen (1986) called the integrative apertural systems of the buliminid type the "internal canal". He also showed that it contains ectoplasm (as in rotaliid canals).

Canals may be regarded as a complex apertural apparatus. Unique and complex features such as the canal systems of Rotaliana do not exist in any other protistan group. The integrative systems described above represent a new level of organisation. They permit to eliminate the disconnection of different parts of the shell, which increase with the enlargement of the number of chambers, whorls and the size of the shell. At the same time they enhance the compartmentalization and differentiation of the cytoplasm and heighten the organizational level of the whole cell system.

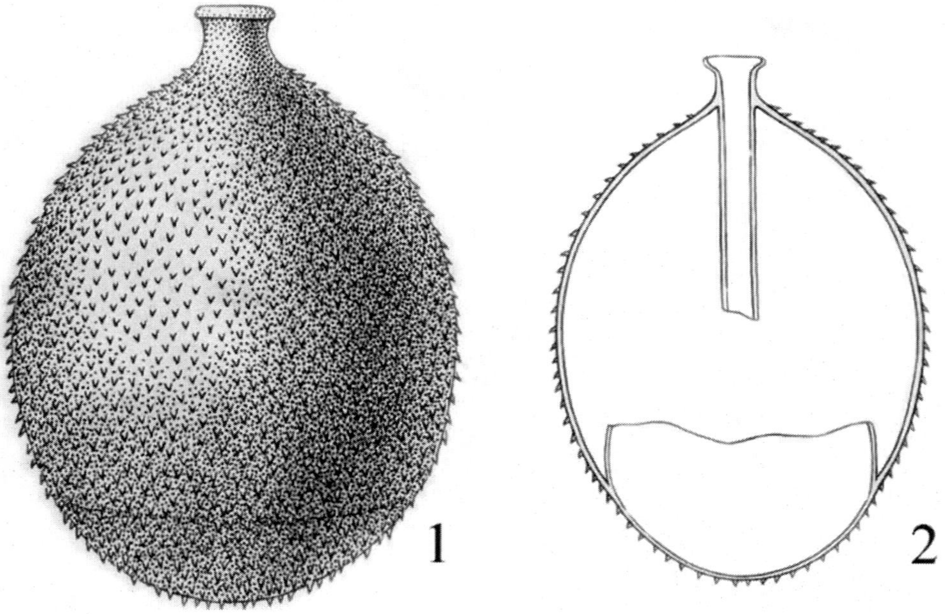

Text Figure 1. *Bombulina* (Nodosariana).
1. Outer view of the shell. 2. Longitudinal section, partly resorbed first chamber is seen (after Mikhalevich, 2000).

V.2. SHELL WALL

V.2.1. Chemical Composition

The organic (tectinous) shell wall of the more primitive foraminiferal groups (subclass Lagynana), and the organic base of the wall of all the other foraminiferal groups, is constructed mainly from mucopolysaccharides and scleroproteins. Its composition has been studied in greatest detail in *Heterostegina depressa* (Rotaliana) (Weiner & Erez, 1984). Data of Allen et al., 2000 have demonstrated that it consists of glycoproteinaceous material (sulphated glycoproteins), resembling collagen, with some contribution from specific amino acid and carbohydrate components.

Proteins account for about 0.5% of its mass. The macromolecules constituting the wall are of two main types, one dominated by proteins rich in asparagine acid and the other by proteins rich in serine and glycine, probably bound with polysaccharides. The biochemical properties of the macromolecules of the matrix in *Heterostegina* are similar to the properties of organic molecules in the matrices of other taxa. The mucopolysaccharide base of the shell wall (glycocalix) forms a barrier protecting the organism from the influence of the harsh external environment. This phenomenon has been observed at the biochemical level (Bresler & Yanko, 1995).

The agglutinated shell wall incorporates sand grains, sponge spicules and other foreign particles, which are bound together by organic material, and in more advanced forms by mineral cement secreted by the organism itself (Bender & Hemleben, 1988).

The secreted calcareous elements could originate from vacuoles in which the granules are formed as a result of the transport of HCO^{-}_3 in combination with the exchange of Ca^{++} (into the vacuole) for H^+ (into the cytoplasm). Ca++/H+ ATF-asa takes part in this process since the alkalescence of the vacuolar medium is the necessary condition of the process of calcification (Labas & Mikhalevich, 1987). The vacuoles originate from the Golgi apparatus.

The overwhelming majority of foraminifera have secreted calcareous skeletons, usually calcitic but rarely composed of aragonite, which is richer in strontium because of the peculiarities of its crystal lattice. An aragonitic shell is developed in the large group of robertinids and in some other groups – involutinids, some planktonic foraminifera (favusellids) (Grigyalis & Gorbachik, 1980). The taxonomic significance of an aragonitic shell is not clear. For example, in the genus *Trocholina* (Involutinida) one species has a calcitic and the other an aragonitic shell. In some foraminifera, calcitic and aragonitic regions have been observed within a single shell (Archambault-Guezou & Lefevre, 1981). Moreover, in fossilized forms aragonite easily turns into calcite.

Calcareous shells include also small amounts of Mg, Mn, Fe, Al, Sr and some other elements. Their composition is variable and insufficiently studied. However, a high Mg content in the calcareous wall of the Miliolana (up to 20% - Chave, 1954, Henbest, 1963) may serve as an additional character for this subclass (Bender & Hemleben, 1988), while the prevalence of ferruginous cement is characteristic of some agglutinated genera and species. The chemical composition and Ca isotope composition may be controlled by temperature (Gussone & Filippsson, 2010). Enrichment in metallic elements sometimes may be caused by abnormal concentrations of these metals in the surrounding water and their absorption by the protistan shell (Yalcin et al., 2008). In the agglutinated species *Textularia hauerii,* the enchanced concentration of the titanium-bearing mineral ilmenite ($FeTiO_3$) was observed (Makled & Langer, 2010). This species showed a preferential selection for this mineral, particles of which it arranged along sutures and on the apertural face. In other cases, the change in the mineralogical composition of the shell does not reflect changes in ambient water chemistry but depends on genotype and phenotype of the species; for instance, Mg/Ca variations in the shell of *Elphidium williamsoni* (Allison et al., 2010). Dissard et al. (2010) showed that the magnitude of changes in Mg/Ca and Sr/Ca ratio in *Ammonia tepida* were linked to water temperature and salinity.

Some unilocular soft-walled foraminifera include mineral particles in their cytoplasm (Gooday et al., 2005, Majewski et al., 2005). In two *Psammophaga* species such mineral particles were predominantly magnetite and titanoferous magnetite (Pawlowski & Majewski, 2011).

Among foraminifera there are also several agglutinated genera, and the genus *Silicoloculinina* with a secreted wall, having a siliceous composition. Since the mineral composition of the skeleton is determined by its organic matrix and its amino acid composition is different and strictly specific in taxa of high taxonomic rank (Foraminifera, Radiolaria, Acantharia, Chrysophita, Haptophyta, Diatomea, Spongia and others – King, 1977), these forms with siliceous shells deserve to be separated as independent taxa of at least class rank. But the secreted origin of silica in the foraminiferal shell has not yet been convincingly demonstrated (inspite of the work of Hansen & Hanzlikova, 1974) and hence the taxonomic position of all such foraminifera is still problematic. Most likely, silification of the shell wall is the result of secondarily processes that occur during fossilization. Golubev (1981) showed that in the process of destroying of the organic sheets encircling the crystals the latter became poorly soluble. This process takes place not only during fossilization but in the process of burning. If there is some silica nearby (if the baking goes on the glass, for instance, as it was in the Golubev's experiments when he put the calcareous foraminiferal shell on the glass and heated it on the spirit-lamp flame) – the crystals of the calcareous shell become silicificated.

V.2.2. The Ultrastructure of the Shell Wall

During recent decades electron-microscopic investigations of the foraminiferal shell wall have contributed substantially to our knowledge of its ultrastructure (see Part III). Below the basic wall types are described.

V.2.2.1. Tectinous Wall

The tectinous (organic) wall ultrastructure is studied mainly in the unilocular representatives from the subclasses Lagynana (e. g. in *Allogromia*, *Boderia*, *Myxotheca*, *Shepheardella, Xiphophaga*) and Astrorhizana (the agglutinated shell of *Iridia*) (Hedley, 1963, Hedley et al., 1973, Lengsfeld, 1969, Pierce et al., 1974, Cezana, 1981, Schwab & Plapp, 1983, Weiner & Erez, 1984, Cartwright, 1989, Bowser et al., 1995, 2002, Goldstein et al., 2010). It usually consists of an inner fibrillar and outer electronic-dense layers. Sometimes an outer granular layer or some other special formation may also be present. The thickness of the fibrillar layer differs not only in different species but also in the same individual, depending on its physiological state, varying mostly between 1 and 5 – 8 μm. In *Abyssotherma pacifica* it is only 0.5 μm thick (Lee at al., 1991). Previously, the structural features of the organic wall were not considered useful in taxonomy and perhaps for this reason did not attract much attention from scientists. In some cases, the tectinous wall may be covered with mineral particles and some parts of it may be fully agglutinated (e.g. *Amphifenestrella, Xiphophaga*). This type of the wall is the most primitive and undoubtedly the most ancient, although poorly preserved in the fossil record. Nevertheless, some genera with a wall of this type are reported from the Upper Cambrian (*Chitinodendron*), Ordovician (*Chitinolagena, Maylisoria, Archaeochitosa*) and Silurian (*Archaechitinia*). Bell & Winchester-Seeto (1999) reported organic linings in several species of Devonian agglutinated foraminifera.

The transition from the tectinous to the agglutinated shell observed in recent forms in which the tectinous shell has some agglutinated particles (as for instance in *Amphifenestrella* mentioned above and in some other genera) provide the basis to suppose that the process was going in the same direction during the process of historical development (Rauser-Chernousova & Fursenko, 1959, Arnold, 1978 a, b, Mikhalevich, 1992, 2000 and some others). The new classification proposed was composed leaning on the idea that the rise of agglutinated wall happened in each foraminiferal class independently and in parallel (Mikhalevich, 1980 – 2000, Saidova, 1981). Molecular data of Pawlowski et al., 2003 brilliantly confirmed this supposition. Thus two different methods of investigation (morphological and molecular-genetic) resulted in the similar conclusions.

The organic coating serves as the lining in all the other types of foraminiferal shell wall, and in this context it is called the Inner Organic Lining (IOL) (see below). Its ultrastructure may differ in different taxonomic groups (e.g. in *Xiphophaga* – Goldstein et al., 2010, in amphisteginids and miliolids – Leutenegger, 1977). If the calcitic shell is dissolved, this soft lining preserves the original shell form. Some examples of these organic linings have been considered to be the remains of multi-chambered calcareous shells. Their mass abundance in Callovian-Oxfordian limestones permitted the recognition of more than 30 multi-chambered foraminiferal taxa (Misik & Sotak, 1998).

Recent intensive molecular studies of soft-walled foraminifera reveal that some genetically distinct species are indistinguishable morphologically (Goldstein et al., 2010). The diversity of Antarctic allogromiids revealed by molecular analyses (Habura et al., 2004) often exceeds their morphological diversity (Gooday et al. 1996). This fact may stimulate detailed EM studies of the cell and its organic coating, as previously carried out for other protists in which the characteristics of the organic coating differ in several taxa at a high taxonomic level. It is also interesting to note that in the pellicule of some Mastigophora (Mylnikova & Mylnikov, 2010), some algae and in all the Sporozoa (Krylov, 1996), the rows of alveoli are adjacent to the upper membrane of the coating, similar to the arrangement seen in *Xiphophaga* (Goldstein et al., 2010), although the alveoli are flattened rather than ampula-like as in *Xiphophaga*. Variable thickenings and the structural peculiarities of the organic coating were observed in Amoebae and some other 'rhizopods' (Page, 1988). It cannot be excluded that in lagynins and allogromiins some variable structural features of their organic theca (shell) may provide additional taxonomic characters for these groups, which are so poor in morphological features. Such studies have already been initiated by Bowser et al. (1995, 2002), Goldstein et al. (2010) and Goldstein & Richardson (2002) in *Astrorhiza rara*, *Xiphophaga minuta*, and also in *Myxotheca* sp., in which the organic wall has a layer with a herringbone pattern, in *Hyperammina* sp., which has a wall consisting of band of crescent-shaped fibers, and in *Cribrothalammina alba.* The pioneering work in this field was done by Headley et al., 1972. With the accumulation of such data these morphologic peculiarities of the organic wall might serve in the future as reliable taxonomic features.

V.2.2.2. Agglutinated Wall

An organic layer of 0.5 microns thickness underlies agglutinated as well as secreted calcareous shell walls. It represents the IOL schematically described in the following plates: Plate **8**, figure 8, Plate **10**, figure 1, Plate **11**, figure 1, Plate **12**, figures 1, 3, Text- figure 3. The inner organic layer separates the cytoplasm from the shell wall. The outside surface of the agglutinated wall may also be covered by the thin organic layer (Text figure 12 (1)).

These organic layers can be seen in some EM photographs of the present study (Plate **10**, figure 7, Plate **60**, figure 3, Plate **61**, figure 5, Text figure 12 (1, 4)) and in some other works (Mikhalevich, 2011, Mikhalevich & Peryt, 2007, plate 1, figure 2, plate 2, figure 3, Plate 3, figure 5). The outer organic layer is absent in *Xiphophaga* (Goldstein et al., 2010). In the most simple case the agglutinated wall (Plate **1 a**, subclasses with agglutinated shell wall in I, II, III, IV, V, Plate **2**, figure –s 1 – 3, 13 A, B, Plate **8**, figure – s 5, 8, Plate **22**, figure –s 1 – 4, 8, Plate **23**, figure –s 2 A, B, F, L) consists of quartz grains cemented by the same organic matter as the inner organic lining (Lipps, 1973, Murray, 1973, Bronnimann et al., 1983, Loeblich & Tappan, 1989) (in representatives of the Plate **66**). The superfine coating of this material envelops each mineral particle and may form thin strands between them. These strands are occasionally irregular in character. Fine fibrils attached to the clay grains were also observed in *Xiphophaga minuta* (Goldstein et al., 2010). In addition to the quartz grains the empty shells of foraminifera and other organisms, mica particles, sponge spicules and other foreign material may be incorporated into the walls of agglutinated shells. Some genera (e.g. *Technitella*) build their shell exclusively from sponge spicules. *Carterina* builds its shell from calcite needles of elongate-oval form.

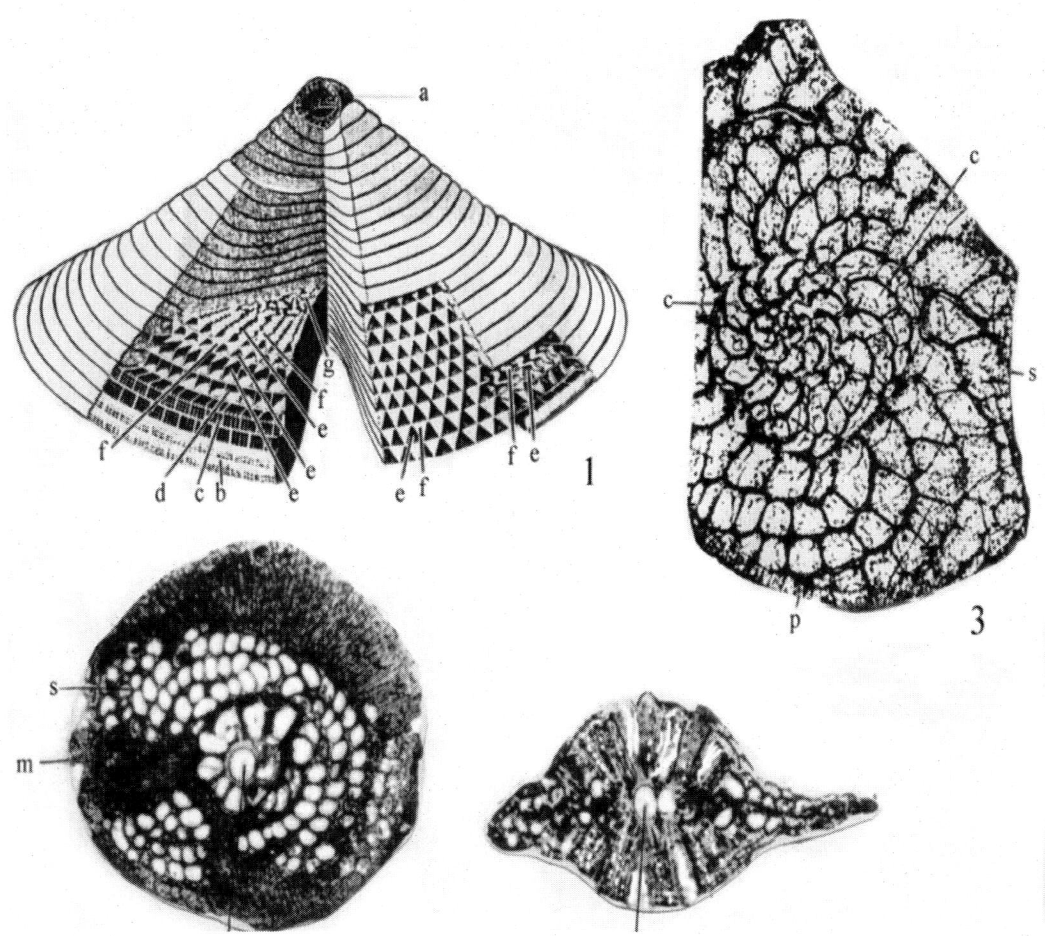

Plate 9. Differentiation of the chambers in the multi-chambered Foraminifera.

1 - *Orbitolina* (Ammodiscana) (according to Douglass, 1960, from Loeblich,Tappan,1964), (with large embryonal chamber – a, earlier part trochospiral, later chambers discoidal, than cyclical, with different internal structure: g – central cancellate zone formed by ramificating and coalescing partitions , f – radial partititions of the radial zone with the radial passages (e), c – marginal zone with the rectangular chambers, subdivided by the second order beams into smaller cellules (from 4 to 16), d – inframarginal zone with the chamberlets, b – alveolar wall layer of the marginal zone;
2 - *Biplanispira* (Pellatispiridae, Rotaliana) (according Cole, 1957),
A – transverse section, B – longitudinal section. Large embryonic chambers in the center (a), middle size chambers small in number in the next whorl (m), numerous small chambers in the successive whorls (s);
3 - *Heterostegina* (Nummulitidae, Rotaliana) (according to Eames et al., 1968). Chambers of the initial whorls crescentiform (c), not subdivided, in subsequent whorls chambers subdivided into chamberlets (s), in the chamberlets situated nearer to the peripheral margin, the interseptal pillars may be present (p).

In laboratory cultures in the absence of natural sediment, agglutinated foraminifera will use any available material (pieces of glass, coal, carborundum grains) (Bender & Hemleben, 1988, Kitazato & Matsushita, 1996).

In the natural environment different agglutinated species living in the same sediment often select particles according to size and mineralogy. Grain and mineralogical selectivity (quartz grains or platy alumosilicate minerals) has been demonstrated for several species (Scott et al., 1998). In *Abyssotrema pacifica* the agglutinated particles are mainly obsidian (volcanic glass, 65 to 75 % of SiO_2) (Lee et al., 1991). Some species are known to agglutinate heavy minerals.

In *Titanotheka* the wall is formed by the elongated rutile grains but we cannot judge wheather it was owing to the species selectivity or the character of the sediment in this Upper Vendian form (Gaucher & Sprechmann, 1999). Curiously enough, in *Psammosphaera fusca* and *Reophax cf. parvulus,* in addition to the predominant quartz and feldspar grains, microdiamonds as well as Co and Ni-rich grains were identified with the use of spectroscopic techniques (Kaminski et al., 2008).

Sand particles in the shell wall may be disposed in one or several layers and particles in these layers may be weakly or firmly cemented. Many species with agglutinated shell walls are capable of sorting the particles according to the size of grains; in these cases the wall may be coarse- or fine-grained.

The degree of cementation of the foreign wall material may also differ; for instance in *Rhabdammina parabyssorum* there is a hard inner layer and friable outer one (Gooday & Smart, 2000). Usually the inner layers and those around the aperture are finer and better sorted (Plate **60**, figure 2, Text figure 16).

In many agglutinated taxa (e.g. representatives of the subclasses Textulariana, Ammodiscana, Miliamminana) the foreign particles of the wall may be cemented also by the biogenic calcareous cement (Bender & Hemleben, 1988, Loeblich & Tappan, 1989 and others), and the ratio of organic matter to biogenetic carbonate is variable.

The presence of the agglutinated shell wall can be used as a taxonomic character of the subclass rank and its particular features as characters of the lower rank taxa, often only at the generic or specific level.

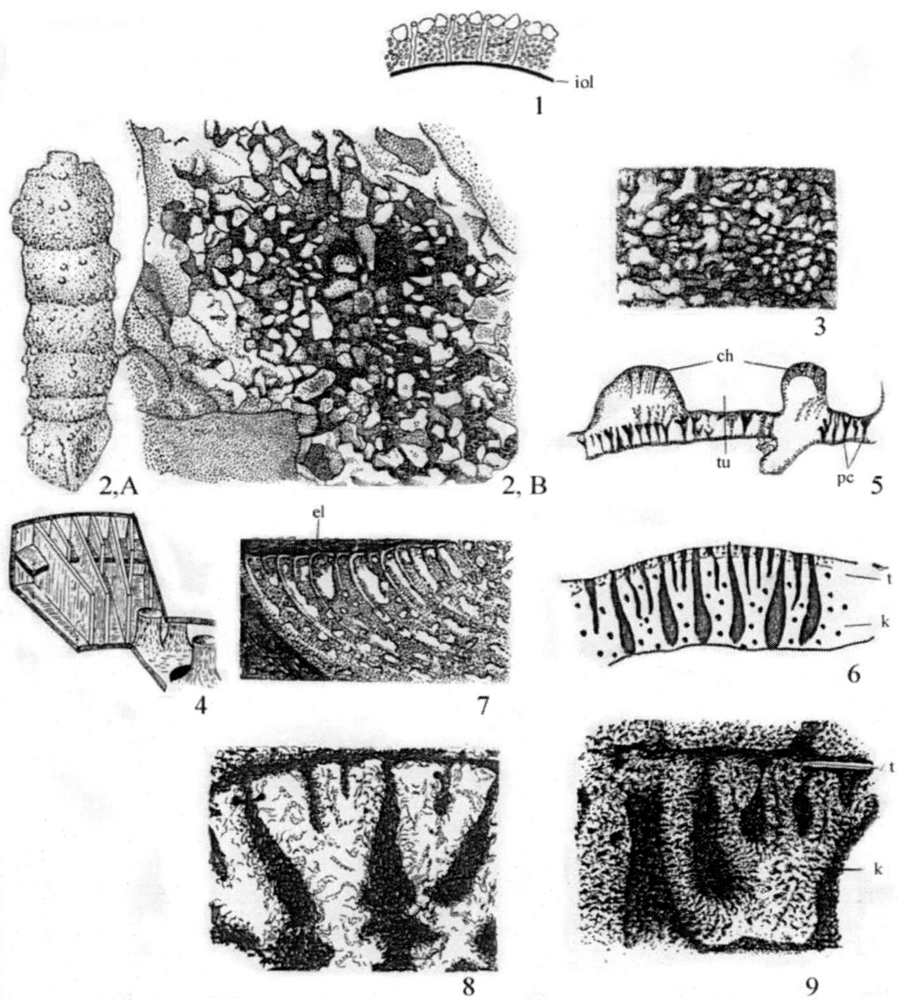

Plate 10. Micrograular wall of Foraminifera (after Mikhalevich, 2000).
1 - Scheme of the ultrastructure of the monofontinal micrograular wall (IOL – Inner Organic Lining); 2 - *Clavulina* (Textularioida, Textulariana), A – external view of the shell , B – ultrastructure of its micrograular wall (x 2000); 3 - Ultrastructure of the micrograular wall of (Fusulinoida, Miliamminana), (x 1800) (according to T.N. Isakova, personal communication); 4 - Scheme of the complication of the inner chamber lumen (transversal and radial partititions and pillars) in *Abrardia* (Orbitolinidae, Ammodiscana) (from Neumann, Daotte, 1960); 5 - Complication of the inner wall and chamber lumen in *Schwagerina* (Fusulinoida, Miliamminana) - additional deposition of the skeletal matter (chomata) (ch), tunnel (tu) passing between them, pc – pseudopore canaliculi (according to Thompson, 1961 – fragment, from Loeblich, Tappan, 1964); 6 - Scheme of the transverse wall section of Fusulinoida with the ampoule-like canaliculi of the pseudopores: t – tectum, k – keriotheca (from Fursenko, Rauser-Chernousova, 1959); 7 - Axial section of *Mesorbitolina* (Orbitolinidae, Ammodiscana) with the distinct outer epidermal layer (light one) (el) (x 50) (according to Douglass, 1960); 8 - Transverse wall section of *Simplorbitolina* (Orbitolinidae, Ammodiscana): the ampoule-like pseudopore canaliculi are seen (x 100) (according to Douglass, 1964); 9 - Axial section of *Sumatrina* (Neoschwagerinidae, Fusulinoida, Miliamminana), thick keriotheca (k), thin layer of tectum (t) and ampoule-like pseudopore canaliculi are seen (x 300) (according to Ozawa, 1927).

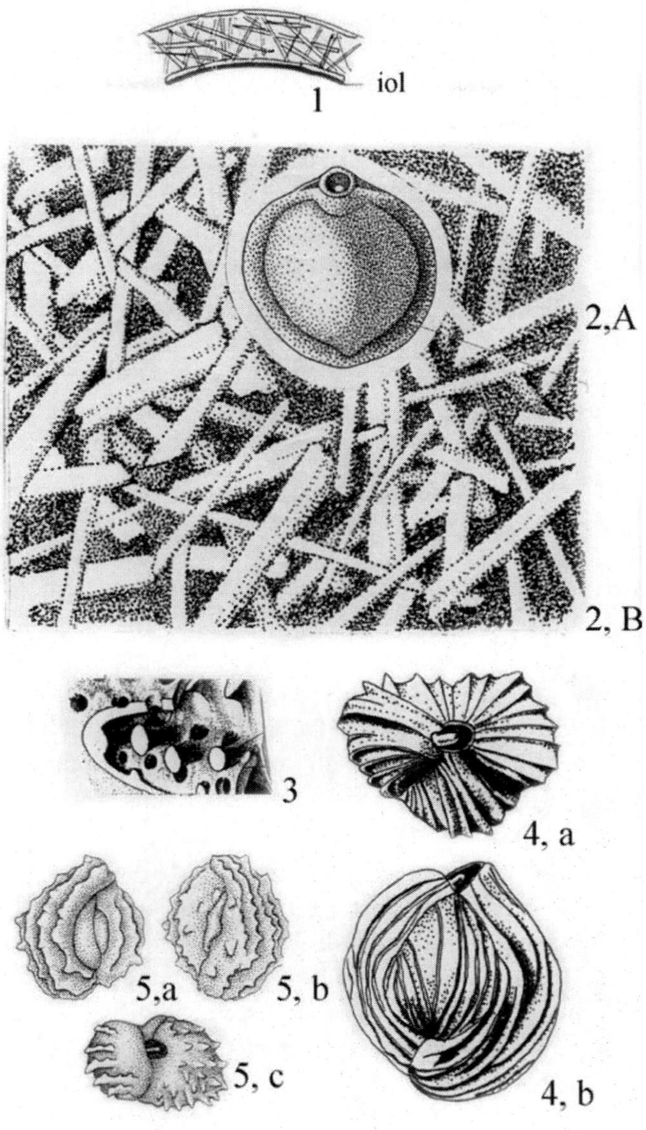

Plate 11. Porcelaneous (composed from needles) Foraminiferal wall (after Mikhalevich, 2000).
1 - Scheme of the ultrastructure of monofontinal porcelaneous wall (IOL – Inner Organic Layer); 2 - *Pyrgo* (Miliolana), A – outer view, B – its shell ultrastructure (x 39000) (according to Mikhalevich, 1988); 3 - Pillars in the inner chamber lumen of *Fusarchaias* (Miliolana) (according to Reichel, 1952); 4, 5 - Deposition of the additional shell material on the shell surface of some species of *Quinqueloculina* (Miliolana) (4 a, 5 a, b – side views, 4 b, 5 c – apertural views) (from Mikhalevich, 1983).

V.2.2.3. Calcareous secreted wall

The calcareous secreted shell wall is also underlain by the IOL. It consists of the same material as that present in taxa with tectinous and agglutinated walls (Angell, 1967, Banner et al., 1973). As was mentioned above, this organic lining is preserved if the calcareous shell is dissolved in acid. It is usually visible as a thin pellicle in the empty dissolved shell and it reproduces the shell morphology.

The diversity of secreted calcareous shell walls encompasses several basic types.

1. Microgranular wall (Plate **10**, figure–s 1– 9) (term of Rauser-Chernousova and Gerke, 1971). This type is characterized by rather large calcareous grains (1 – 5 microns) of oval, subcircular, angular or another shape, which are placed randomly and irregularly above the IOL (Plate **10**, figure- s1– 3). "Subangular", "equidimentional" etc. represent different types of microgranular wall. Each of the grains of this wall type is encircled by a fine organic envelope, and somewhat thicker strands of the same organic matter weave irregularly and discontinuously between the groups of grains, in a similar fashion to those of the agglutinated wall. Sometimes the grains are sorted according to their sizes. This wall appears rough and 'sugary' in the light microscope. It occurs, along with the agglutinated type, in the subclasses Astrorhizana, Ammodiscana, Hormosinana, Textulariana. There is no distinct boundary between the agglutinated and microgranular walls and the forms with a mixture of both are often encountered (Mikhalevich, 1981, 1992, 2000, 2003). Among the 142 agglutinated species studied by Bender (1995), forms with microgranular calcareous cement constituted 21%. Since some of the agglutinated species are able to secrete calcareous grains themselves (Mikhalevich, 1995, figure 10, 2) it is possible to envisage a gradual evolutionary transition from agglutinated to microgranular calcareous walls. In this case, the microgranular wall may not have the high taxonomic value assigned to it in the majority of the widely used classifications.
 Agglutinated and microgranular wall have much in common. They often are massive, thick and heavy. Additional skeletal structures and complications of the shell wall are developed almost exclusively inside the shell (reticular, alveolar, cellular, spongy layers, pillars and groups of merged pillars, mounds, chomata) (Plate **8**, figure – s5, 6 A, B, Plate **10**, figure – s 4, 5) and are developed independently and in parallel in the representatives of different phyletic lines (classes) (Mikhalevich, 1981, 1992, 2003) (see Part VIII.10.12). The clear-cut distinction between these two wall types (agglutinated and microgranular) seemed artificial (Piller, 1990). Our own studies convinced us that higher taxa based on this distinction would be dubious (Mikhalevich, 1980, 1992).
2. The porcellaneous wall of Miliolana (Plate **11**, Plate **56**, figure – s 4, 7, 8, Text figure 14). This consists of elongated calcareous crystals arranged above the IOL in the three different layers. A thick friable middle layer formed by randomly crossing long needles (or shorter cylindrical crystals) is bordered inside and outside by flat plate-like crystals orientated parallel to the surface as a parquet layer. The outer parquet layer is responsible for shiny white porcellaneous appearance of miliolids in the light microscope. This upper layer is sometimes destroyed, in which case the wall loses its shine and looks rough and lusterless (Text figure 14). Each of the crystals has a thin organic envelope as in the previously described types of shell (Bender et Hemleben, 1988). In the case of extremely thick miliolate shells, for instance in the giant Antarctic miliolids *Pyrgo sarsii, Pyrgoella sphaera* (in Mikhalevich, 2004 b), the needle crystals are arranged in lines radiating from the inner to the outer surface of the wall (Plate **56**, figure – s 7, 8, Plate **57**, figure – s 4 – 6). These lines are visible under lower magnifications (x 200) than the crystals themselves (x 7000). The

fusulinoid shell wall obviously belongs to this wall type ans was later changed as a result of recristallization (see Part XI.2.2.).

Electron-microscopic investigations of the shell of some representatives of the lower Miliolata (*Dentostomina* – subclass Miliamminana) (Mikhalevich et al., 1988) revealed calcareous secreted elements characteristic of the wall of the most advanced Miliolata of the subclass Miliolana (long needles along the inner shell wall and along the tooth; short cylindrical crystals in the middle wall layer – Plate **13**, figure I). The thin inner white layer underlying the thick agglutinated wall is clearly seen in broken sections of *Dentostomina* (Plate **13**, figure 1D, Plate **66**, figure 4). This fact, along with the close morphological similarity of the shells (quinqueloculine type of the chamber arrangement, tooth structure), supports the close relationship of the both groups and the validity of their elimination from the subclass Textulariana (Saidova, 1981). It also provides evidence for the gradual transition of the agglutinated into the secreted type of shell wall, as was also the case with the microgranular wall. Previously, the morphological similarity of the shell structure of the representatives of the subclass Miliamminana and Miliolana was regarded as convergent. This porcellaneous wall type can easily turn into a microgranular one in living (Macintyre, Reid, 1998) as well as in fossil forms (Gubenko, 1988, 1989) (Plate **56**, figure - s 5, 6 of the present study).

Unlike the shell wall of microgranular forms, the porcellaneous type may often develop different types of ornamentation on its outer side (cancellate network, ribs, bosses – Plate **11**, figure – s 4, 5, Plate **46**, figure – s 1, 2, 20, 21, 24, 27, 29). Internal pillars and deposits shell material are also found at the base of the chambers (e.g., *Shanita*, *Archaias* –Plate. **8**, figure 6 b, Plate **11**, figure 3).

3. The hyaline (or radial) type of wall characteristic of the Nodosariana comprises elongated columnar crystals (which most likely are formed of smaller crystallites) disposed above the IOL and oriented perpendicularly to the wall surface (radial disposition) (Plate **12**, figure – s 1, 2). Each of the radially disposed crystal is also enveloped in a very fine organic coating. Small vertical spaces between the radial crystals (1 – 1.5 microns in diameter) represent pseudopore openings (Plate. **12**, figure 1, Plate **13**, figure 3, Plate **64**, figure 3, Plate **65**, figure 5).

Distinct layers of growth are often well expressed in the initial chambers of the Nodosriana (Plate **12**, figure 2). This wall type appears vitreous in the light microscope. It has not been sufficiently studied.

The wall of recent Nodosariana is described as monolamellar. In some ancient forms (Permian, Triassic) it is monolamellar while other (*Eonodosaria*, *Eogeinitzina*) exhibit an inner microgranular and an outer radial layer (Gubenko, 1989). The former most likely is recrystallized. In any case it disappears in the evolution of this group. This combination of microgranular and radial hyaline layers has not been observed in recent species of Nodasariana. How this third shell wall type originated is unclear. However, it differs from the microgranular wall in its well-ordered disposition of radial crystals and also differs from the bilamellar wall of Rotaliana described below. This wall type is accompanied nearly exclusively by external structures (ribs, bosses, spines – Plate **2**, figure 4, Plate **15**, figure – s G – J, Plate **22**, figure 6); inner septa are encountered only in some fossil forms (e.g. - Colaniellidae (Plate. **35**, figure 2, K).

4. The monocrystalline calcareous secreted wall of Spirillinana represents a unique wall type. It also is insufficiently studied. Most probably the radial crystals that constitute the wall are disposed in a very well-ordered manner strictly perpendicular to the IOL. As a result they behave optically as one single crystal or two or three crystals. As Berthold (1976) showed in *Patellina corrugata,* the outer as well as the inner surface of the wall is covered by an organic layer. The inner granular calcareous layer observed in some fossil forms (e.g., *Archaediscus*) is probably a result of diagenetic processes; it is absent in recent representatives of the Spirillinana. This type of wall appears vitreous in the light microscope and often has a sculptured outer surface; inner septula may also be present. It possesses advanced structural features in some genera, a rather primitive canal system sometimes being developed present, for instance, in the Lasiodiscida, Hottingerellida (Hottinger, 1976, Azbel, 1986).

   In all four of the secreted calcareous wall types discussed above, the secreted material is formed above the IOL. None of them, including the poorly studied 3rd and 4th types (Nodosariana, Spirillinana) possess the middle organic lining (POM – see below). Both the 3rd and 4th wall types exhibit similar properties. Thus, the crystals forming the outer ribs of *Nodosaria affinis* behave like several large crystals (Hay et al., 1963). In both subclasses the capability to form growth lamellae repeatedly encircling the whole shell with each new growth stage is strongly developed (ortholamellar layers).

5. The bilamellar (= bifontinal[1] – see below) wall of the subclasses Rotaliana and Globigerinana (class Rotaliata) differs in principal from all the previous types (Plate **12**, figure– s 3, 4, Plate **13**, figure– s 4, 5, Plate **14**). Its ultrastructure is clear revealed in the classical works of Hemleben and co-authors (Be & Hemleben, 1970, Hemleben et al., 1977, Be et al., 1980). In addition to the IOL underlying the shell, the wall incorporates a middle organic layer (Primary Organic Membrane – POM) (Plate **12**, figure – s3, 4, Plate **14**). The inner and outer calcareous layers are simultaneously initiated and then formed from two sources of calcification on either side of the POM, which therefore represents the initial calcite crystal nucleation layer. The inner septa are also usually formed as double structures (Plate **8**, figure 7, Plate **14**). The inner calcareous layer below the POM remains of constant thickness during the individual's life. However, the number of the additionally formed layers on the outer side of the POM may be different, and they may be of variable thickness during the growth of the individual. These layers are also delimited by additional organic membranes (OM). The newly formed layers extend over the previously formed chambers of the shell, and as a result the walls of the preceding chambers are thicker than those of the subsequent ones. These additional layers reach their greatest development in planktonic forms (Plate **12**, figure 4, Plate **13**, figure 4, B) (see below – Part VI. 5, gametogenic calcification). This type of shell wall has a fine-grained, uniform and well-ordered ultrastructure. The finest crystallites encircled by the finest envelopes of the organic matter are united into small plates packed one above the other into vertical columns perpendicular to the wall surface (Hansen, 1970, Lipps,

[1] This term was introduced by Mikhalevich (1992) to avoid terminological confusion, since the term *bilamellar* is used in Russian language literature to designate a double-layered wall, while layers of different genesis occur in both agglutinated and microgranular walls. The term "bifontinal" reflects a double source of secretory activity and calcite nucleation on both sides of POM, bidirectional calcification around the POM.

1973, Nomura, 1983, Debenay et al., 1996). Thus these columns represent aggregates of the third order (Debenay et al., 1996). The wall appears transparent and hyaline in the light microscope. In contrast to the wall of members of the Nodosariana, which is completely colourless, it often is slightly yellowish or pink (especially in the early chambers), obviously owing to the higher amount of the organic matter. In cases where crystals in the bilamellar wall are orientated obliquely (at an angle 35°, 45° to one side or the other), rather than perpendicularly, with respect to the outer surface, the wall looses its shine and transparency and appears dull. This is similar to what is seen in the shells of mollusks – bright pearl layers are formed by the perpendicularly disposed crystals and dull, sugary ones by crystals that are orientated in different directions. Dull walls of this type in the subclass Rotaliana (former Rotaliacea) were said to be composed of granular calcite. The recognition of high rank taxa on the basis of this feature resulted in the separation of closely related groups among cassidulinids and discorbids (see the Part III, "Historical sketch"). Indeed, granular and clearly hyaline walls may occur within one genus, e. g. *Turrilina* (Hansen, 1972). Moreover, a single shell may encompass parts with perpendicular and obliquely orientated crystals, as for instance in *Globorotalia* (Maslakova, 1978).

Additional skeletal structures are highly developed and achieve most diversity and complexity in taxa with this wall type. These include inner valves, plates, grooves and complex canal systems (Plate **5**, figure – s 4, a - d, Plate **8**, figure – s 2, 3), external ribs, bosses and spines as well as the additional skeletal plates covering the umbilical area or the sutures in rotaliids and globigerinids (figure **1a**, V, Rotaliana, figure 9, Globigerinana, figure 5, Plate **2**, figure 10D, 12, B, Plate **8**, figure 9, Plate **15**, figure – s A – E, Plate **16**, figure – s 1, 2). The latter do not occur in other foraminiferal classes except of Spirillinata where, nevertheless, they are less developed. It is possible that the external shell structures make the shell stronger, as suggested previously, but usually the position and orientation of these features reflect cytoplasmic flows and suggest a connection with strong cytoplasmic activity (Hemleben, 1975, Hemleben et al., 1977), while pustules and keels may serve as places for reticulopodial attachment. Although they are similar externally, the long thick solid spines of some Nodosariana species (Plate **15**, figure F) are not homologous with the spines of *Calcarina*, *Baculogypsina* or *Baculogypsinoides* (figure **1 a**, V, Rotaliana: 13), as the latter are pierced by canals. Neither are they homologous with the needle-like spines of planktonic forms (Plate **16**, 1, 2), which serve as a support for their reticulopodia and the cytoplasm stretched between the spines (Plate **16**, figure 2). The spines of *Globigerina* are very specialized with a joint-like basal structure that gives them some mobility (Plate **8**, figure 9, Plate **16**, figure 1) (Hofker, 1971). These spines may change their orientation when necessary or be discarded when the shell sinks to the seafloor. Their outer ends are only slightly calcified, if at all, and as a result are flexible and do not break.

This fifth shell type is a product of centers of secretion on both sides of the POM, resulting in the simultaneous formation of two lamellae. That is why it is called bilamellar (= bifontinal in Russian, Mikhalevich 1992, 1995, 1999). This term stresses the contrast with the four other calcareous wall types, which are formed as a single layer and are therefore termed monofontinal (or monolamellar).

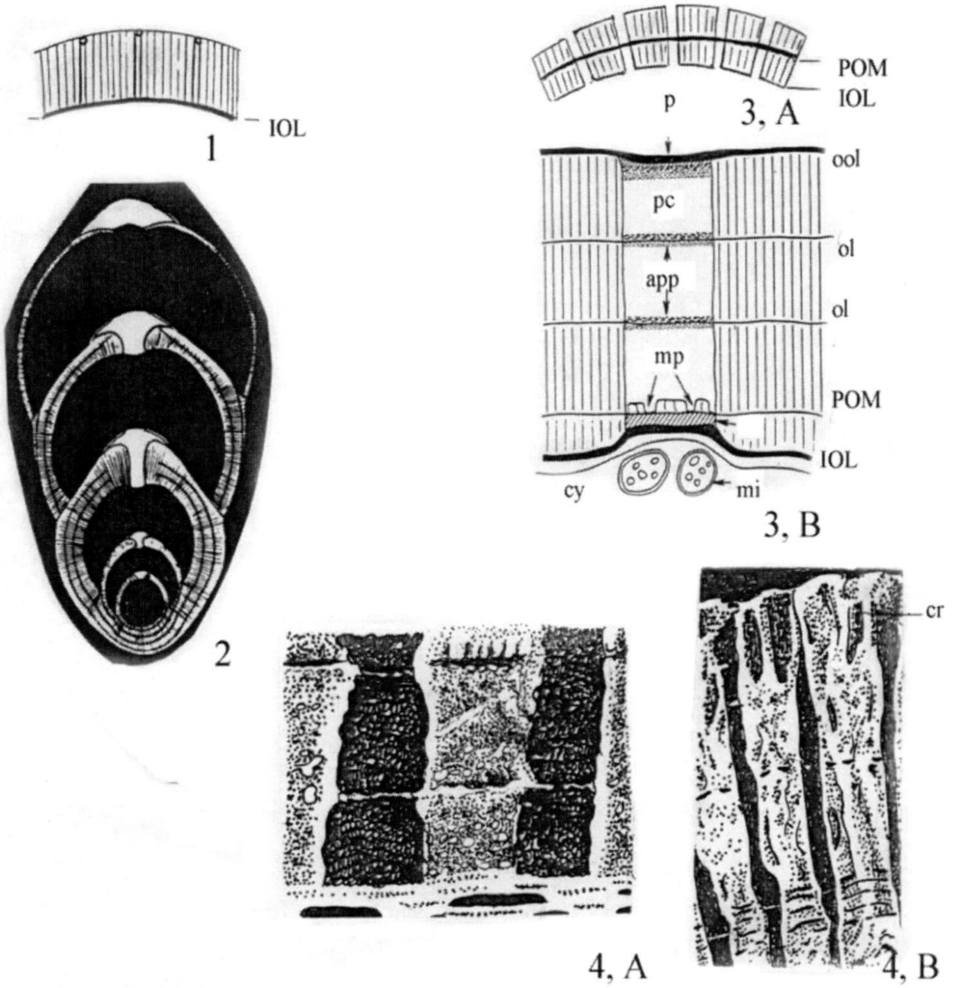

Plate 12. Monolamellar (monofontinal) hyaline-radial (radial) and bilamellar (bifontinal) wall of Foraminifera (after Mikhalevich, 2000).

1 - Schematic structure of monofontinal hyaline-radial (radial) wall (IOL – Inner Organic Lining); 2 - *Lingulina* (Nodosariata) (x 42), longitudinal section, layers of growth are visible (according to Ellis & Messina, 1940); 3 - Schematic structure of bifontinal wall in the newly formed chamber: A – two secreted lamella at the both sides of the POM [Middle (Primary) Organic Membrane], formed simultaneously, B – schematic structure of the wall and pore canaliculi illustrated by the earlier chamber of the benthic *Amphistegina* (Rotaliana): upper and lower lime layers on the both sides of POM, which are formed at the very beginning of chamber formation, lower lime layer is separated from the cytoplasm (cy) by the Inner Organic Lining (IOL), subsequent layers of growth of lime wall are separated from each other by the Organic Layers (OL), OOL - Outer Organic Layer, p – pore, pc – pore canal, app – additional pore plate, mp – micropores in the pore plate, mi – mitochondria); 4 - Shell and pore ultrastructure in *Globorotalia* (Globigerinana) (transverse section of the wall): A – in the young chamber (x 4000), note funnel-like pore widening in their upper part (left side), B – in the more younger chamber (x 1600), layers of growth and crust (cr) narrowing the pore openings from the outside are seen (3 and 4 - scheme according to Hemleben et al., 1977).

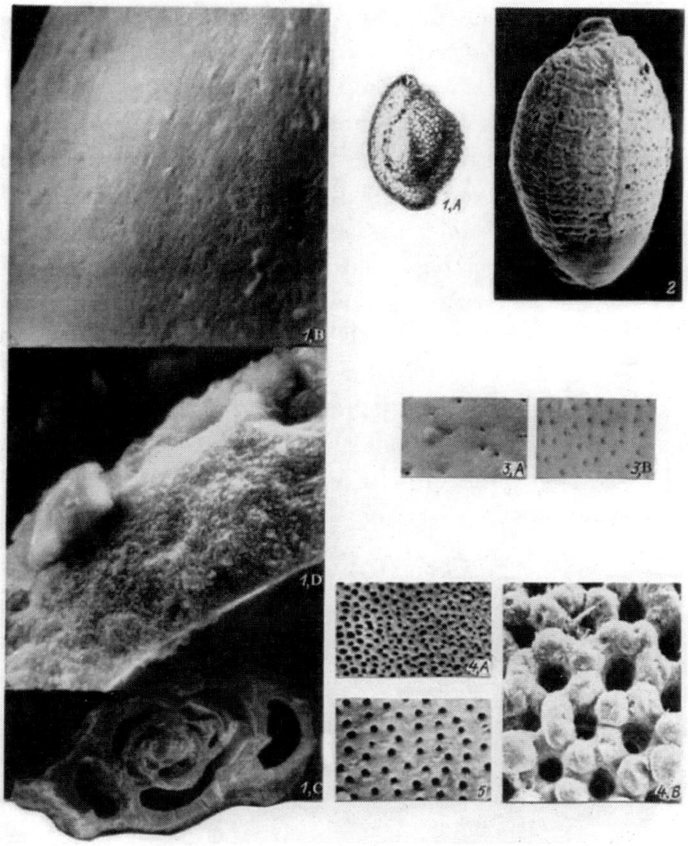

Plate 13. Structural details of the monolamellar (monofontinal) and bilammelar (bifontinal) wall (from Mikhalevich, 2000).
1 - *Dentostomina* (Schlumbergerinana) (monofontinal wall), A – outer view of the shell, B - needle ultrastructure of the inner shell surface and tooth (x 10000), C – transverse section of the shell (x 200), D – microgranular ultrastructure (transverse section) (x 3000); 2 - *Borelis* (Miliolana) (x 220) (monofontinal wall), pseudopore openings seen from the outside; 3 - Pseudopore openings at the shell surface of some Nodosariana, monolamellar (monofontinal wall) : A - *Laevidentalina* (x 7000), B - *Percultazonaria* (x 3000); 4 - Pore openings at the shell surface of some Globigerinana, bilamellar (bifontinal wall): A - *Globorotalia* (x 1500), B - *Globigerinoides* (x 4000) (crust); 5 - Pore openings at the shell surface of *Eponides* (Rotaliana) (x 1500) (bifontinal wall).

Between the main shell wall types there are also differences in the mode of secretion of the calcareous material. In the Miliolata the wall is formed as a result of the heterogenous nucleation of calcite, which occurs rarely in living organisms. The growth of the wall in *Patellina* (and all other Spirillinana) is also unusual in that there is no template but only a gradual accretion at the edge of the tubular chamber and the outer shell edge (although the process is studied insufficiently). Secondary lamellae are not deposited in the shell of *Patellina*. This mode is opposite to that observed in the other foraminiferal groups with a calcareous skeleton. Thus microgranular and agglutinated walls with calcareous cement, the walls of Nodosariana, supposidely Spirillinana and the bilamellar Rotaliana wall arise in an epitaxial way on the organic matrix as a result of polarized secretions. Their calcified extracellular matrices have been compared with the biomineralized matrices of higher metazoans (Weiner, 1985, Weiner & Erez, 1984), and the secreted material of the extracellular matrix of one of the *Halyphysema* species - whith collagen (Lipps, 1973).

In spite of all the differences between these foraminiferal wall types, nearly all of them have some common architectural features that reflect the following tendencies. 1. To acquire radial passages going through the wall from the cell to the surface, thereby connecting the cell to the environment (from irregular to more regular radial canaliculated pseudopore passages in agglutinated forms, irregularly-radial passages in the thick wall of most miliolans, to regular radial pseudopores in calcareous nodosarians and spirillinans and regular pores in rotalians). 2. To strengthen and protect the wall from the outside by covering its main thickness with a thinner outer pavement, orientated perpendicularly to the radial or randomly arranged elements; this includes the well-known pavement of the miliolana wall ("parquet layer")(Text. figure 14), the superficial flat aggregates with sutures resembling jig-saw-puzzle patterns of rotaliana (e.g. *Helenina anderseni* surface in Debenay et al., 1996, Plate 4, figure 2), similar superficial flat aggregates in the *Karreriella bradyi* (Text figure 15, 4) with the microgranular wall, and even attempts to construct something similar in some agglutinated taxa (Text figure 15, 1 – 3 - *Norvanganina pseudorugosa fistulosa*).

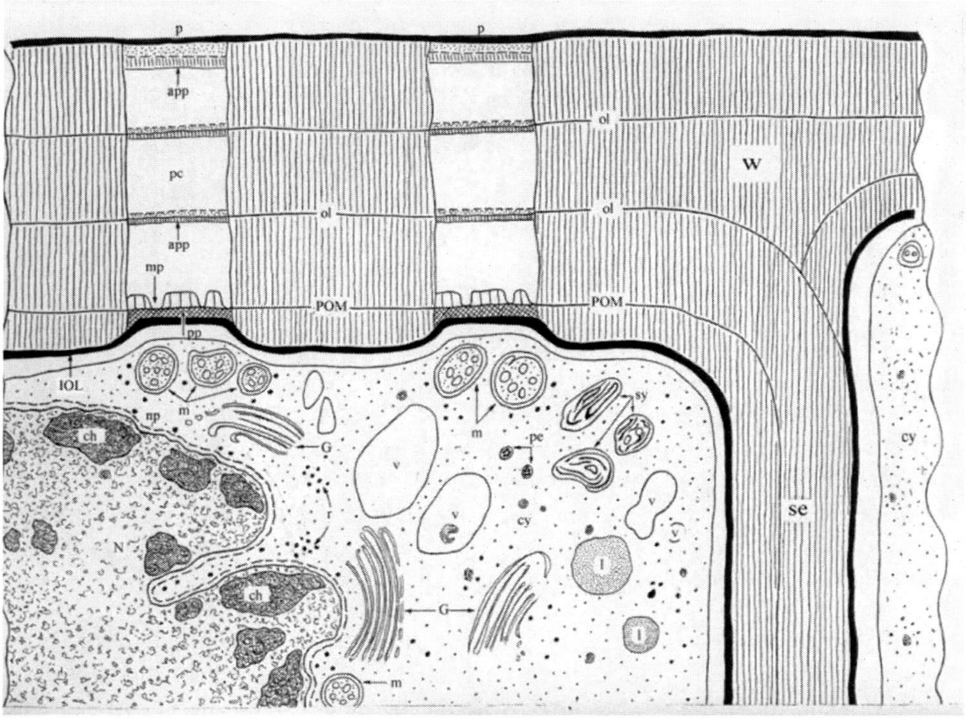

Plate 14. Schematic structure of bifontinal wall, pores and cytoplasm of the benthic Rotaliana (after Mikhalevich, 2000).

p – pore, mp– micropores, pp – pore plate, app- additional pore plate, pc - pore canal, POM - Middle (Primary) Organic Membrane, ol - Organic Layer, IOl - Inner Organic Lining, se - septa, w – wall, cy - cytoplasm, N - nucleus, np - nuclear pores, ch - chromatin, G - Golgi apparatus, m - mitochondria, pe - peroxisomes, r - ribosomes, v - vacuole, l – lipids, sy – symbiont algae (combined scheme according to Leutenegger, 1977 and Leutenegger, Hansen, 1979).

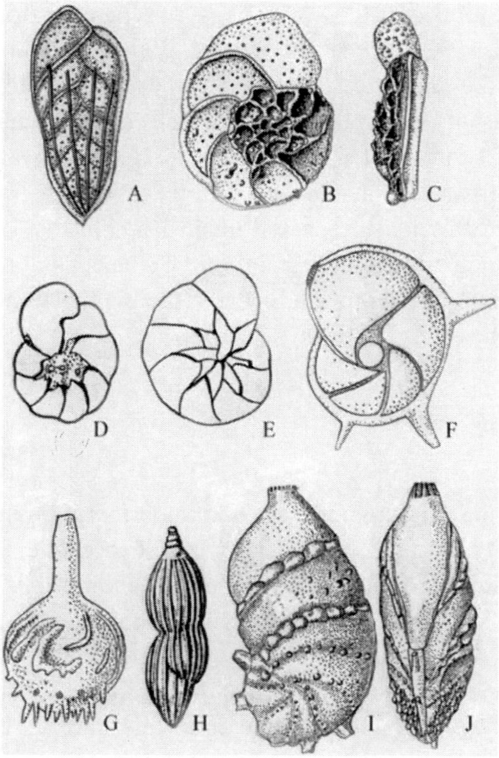

Plate 15. Outer additional skeletal formations in Rotaliata (Rotaliana) (A – E) and Nodosariata (Nodosariana) (F – J) (after Mikhalevich, 2000).
Rotaliana: [A – costae of *Brizalina*, B, C - granules and thickened sutures of *Planulina*, views from the spiral and lateral sides; D, E - additional skeletal plates (D – *Lamellodiscorbis*, E - *Astrononion*)], Nodosariata (Nodosariana) [F – peripheral spines in *Lenticulina*, G - spines and semicancellate structures of the upper and basal parts of *Lagena*, H - longitudinal costae of *Amphicorina*, I, J - spines and bosses of *Marginulina* (I – face view, J – side view)].

The transition from the tectinous shell wall to the agglutinated one and from the latter to the microgranular and porcellaneous or hyaline-radial wall may have proceeded gradually and obviously was going in parallel in defferent foraminiferal phyletic branches (classes) during their geologic history (Mikhalevich, 1981 – 2004) (see above). However, the transition from the agglutinated and microgranular to the bilamellar wall could have happened only as a result of an abrupt change in the mechanism of formation, when the second center of secretion suddenly arose, leading to the simultaneous formation of two lamellae (Labas & Mikhalevich, 1987).

The five types of calcareous foraminiferal walls described above are all represented and studied in detail among modern forms. The same types were present in ancient representatives of the subclasses Nodosariana, Spirillinana and Miliolana (Paleozoic fusulinoids – see below, Part XI. 2. 2), although strongly changed as a result of diagenetic processes. Microgranular calcareous walls may be of primary or secondary origin (as a result of metamorphosis). Among all the wall types examined above tectinous and agglutinated types are the more ancient and known in geologic strata as far back as the Pre-Cambrian. Calcareous microgranular types, along with the Nodosariana and Spirillinana wall types, are known from the Devonian [the ancient groups of these classes were separated from the former Fusulinina sensu Loeblich & Tappan, 1992 (Mikhalevich, 1992, 2000)]. The porcellaneous walls of the Miliolana first appear in the geological record in the Carboniferous and Permian, although this wall type may be present among the more ancient Devonian and perhaps even Silurian foraminifera (supergroup of Fusulinoids – see Part XI. 2. 2), but changed into a microgranular structure (this type of wall is easily changed even in recent forms – see below, Part XI. 2. 2). The bilamellar wall of the Rotaliana appeared much later, in the Triassic and Jurassic, but reached its fullest development beginning in the Cretaceous. Foraminifera as a whole represent a unique group in which to study the biomineralization and evolution of their skeletons over geological history.

V.2.3. Pore Structure

The partition of foraminifera into perforate and imperforate groups, suggested during the 19th Century (Carpenter et al., 1862, Reuss, 1862), is still accepted by many authors and used in foraminiferal diagnoses. Nevertheless, the wall of all foraminiferal species is penetrated to some extent and has different types of fine openings. These can be found even in tectinous shells. In many species considered to be imperforate there are spaces between the agglutinated particles, even in the solid walls of *Pseudonodosinella*, *Recurvoides* and *Miliammina* species (Mikhalevich, 2011 a, Plate **66** of this study). The nature of perforations differs and is closely linked to the wall type. In agglutinated and microgranular shells, they are represented either by occasional irregular spaces between the agglutinated or calcareous secreted grains or by very thin and often branching canaliculi with an inner organic lining (in some textularians, for example). In the fusulinid wall, these canaliculi may be represented by larger ampula-like or branching spaces and cavities (Plate **10**, figure – s 6, 9). Straight and thin canaliculi are present in the wall of some Astrorhizata (*Masonella* –Plate **1 a**, I, Astrorhizana: figure 7) and in representatives of the Textulariana (Text figure 12, 2 – 4 of this study, Mikhalevuch & Peryt, 2007 and in many other works of different authors). In the Textularielloidea (*Simplorbitolina*) these spaces may be rather wide and branching (Plate **10**, figure 8). The canaliculi pierce the outer shell wall and are absent in the septae (Mikhalevich, 2011 a, b, 2012, Mikhalevich & Peryt, 2007). In the Cretaceous *Gaudryina laevigata* Franke, 1914, slight traces of canaliculi are visible in some places within the recrystallized wall (Mikhalevuch & Peryt, 2007), while most parts of the shell appear noncanaliculate. This shows the need to be careful in interpreting the shell-wall ultrastracture of ancient species.

In fact, all the systems of fissures and alveoli in the thick spongy microgranular wall may be regarded as promoting interaction with the outer environment. Although these cavities are more often closed at one end the connection is provided by the spaces between the particles and by the penetrability of the organic membranes. All these structural details are used in foraminiferal taxonomy but in the new classification (Mikhalevich, 1980, 1992 – 2009), they are assigned less taxonomic significance than they were by Banner & Pereira (1981) and in the Loeblich & Tappan (1987, 1992) classification. In their latest version, isomorphic forms differing in the presence or absence of fine pore canaliculi in the agglutinated and microgranular wall were separated into unrelated taxa (for instance homeomorphic Prolixoplectidae and Eggerellidae). Since these peculiarities are closely associated with the microgranular wall generally, their character in different genera and their taxonomic value needs further investigations.

Straight and extremely narrow pore passages (pore canals) lacking additional pore plates (only a basal pore sieve plate being present) exist in the hyaline-radial shell wall of Nodosariana (Plate **12**, figure 1, Plate **63**, figure – s 6 – 8, Plate **65**, figure – s 5 – 7). An inner organic lining in the pore canals of this wall type have not been observed. Pore openings in species of Spirillinana are insufficiently studied. They were investigated in detail in *Patellina corrugata* (Berthold, 1976). The pores of this species are represented by rather wide funnel-like depressions looking like indentations. Their inner walls have an organic lining and they are closed from beneath with sieve plates with the extremely small micropores (0. 01 microns). The inner cavities of these pseudopores are filled with granular organic matter.

The wall of Miliolata species is usually described as imperforate. Nevertheless, it is possible to observe some unevenly scattered openings of irregular form on the shell surface of *Borelis pulchra*) (Plate **13**, figure 2), *Pyrgo sarsii* (Plate **56**, figure 8, Plate **57**, figure – s 4, 5) and of some other species of the subclass Miliolana (Mikhalevich, 2004 a, Pl. 1, figure - s a, g). Similar openings are also present on the surface of some agglutinated representatives of the subclass Miliamminana (Mikhalevich, 1983, Mikhalevich et al., 1986) (Plate **66**, figure 4). In transverse sections of the thick wall of the species mentioned the wall is pierced by radial elongated empty spaces (Mikhalevich, 2004 a, Plate 1, figure – s h, I, k, l, Mikhalevich, 2009, Text figure - s 9, 7, 8, Plate **56**, figure - s 7 – 8, Plate **57**, figure – s 4, 5). Irregular and branching pore canaliculi where observed earlier in the shell of *Peneroplis*. Similar pore openings are known also in some Triassic and Upper Permian miliolats (Milioliporinae, Kamuraninae). In one *Nubeculina* species Clark (1993) observed regular pore openings in the inner thinner and compact part of the wall, which were closed blindly in its thicker and more friable outer wall, which had a typical miliolid wall structure. Pits penetrating porcellaneous walls incompletely and to different depths have often been observed (Hottinger et al., 1993, Debenay et al., 2000, Langer et al., 2009). In all these cases we face the same phenomenon as in the many thick-walled astrorhizans and textularians, where pore canaliculi are formed only partially and in the rest of the wall penetrability is provided by the more friable nature of its structural elements. In the majority of Miliolana species, pore function is fulfilled by the empty spaces between long, loosely-packed crystals. When this becomes unsufficient they develop wider and more organized passages through their shell wall, although these never approach the perfect pore structure seen in Rotaliana.

In all the cases described above we are dealing with pseudopores. True pores are present only in the bilamellar shell wall (Plate **12**, figure – s 3, 4, Plate **13**, figure– s 4, 5, Plate **14**). These are usually evenly situated, straight, unbranching cylindrical canaliculi with their own organic lining. They have a rather thick spongy sieve plate pierced by pore openings (micropores) at the POM level and thinner plates associated with the organic wall layers (Arnold, 1954, Hemleben et al., 1977, Leutenegger, 1977, Spindler, 1978). The number of pore sieve plates and their micropores cannot be used yet in taxonomy because observations are incomplete. The largest pores are found in Rotaliana among the discorbids (2 – 2.5 microns in *Discorbis vesicularis*) as well as in *Bulimina marginata* (0.25 – 0.5 microns) and *Robulus midwayensis* (0.5 microns). In planktonic forms of the subclass Globigerinana pores are usually significantly enlarged; thus in *Globigerina bulloides* the funnel-like pores are 0. 5 microns in size in their lower parts, and 2 microns in their widened upper parts. For comparison, the pseudopores of Nodosariana species are much smaller (Plate **13**, figure 3);

the oval pores of *Nodosaria affinis* measure 0.16 microns in width and 0.33 microns in length. Pore sizes in the different Nodosariana and Rotaliana orders may serve only as a subsidiary taxonomic feature. In planktonic forms the outer part of the pore is usually resorbed, the secondary pore sieve plates disappears, the outer entrance of the pore widens into a funnel-like form (Plate **12**, figure 4, A) resembling the channels in the radiolarian capsular membrane, an example of convergent similarity between planktonic organisms. Before gametogenesis, the pores of many planktonic foraminifera are sealed with secondary shell material and stop functioning (Be at al., 1980). A special Globigerinana pore structure (Plate **12**, figure 4, Plate **13**, figure 4, B) may serve as a taxonomic feature at the class rank.

It is known that pore plates are permeable to O_2, CO_2, salt ions, low-molecular organic matters dissolved in the water. As a result, they play an important role in respiration, osmoregulation, maintenance of the ion equilibrium of the cell with the environment, and some other metabolic processes. The presence of pores therefore allows the cell to avoid excessive isolation. In many benthic species (*Nonionella stella*, *Buliminella tenuata*, *Bolivina argentea*, *Cassidulina cornuta* and others) accumulations of mitochondria are present in the cytoplasm near the inner pore openings (Plate **12**, figure 3, B, Plate **14**) as well as symbiotic algae (Leutenegger, Hansen, 1979, Hottinger, 1972), which confirms the respiration function of pores. Pores play also a protective role preventing the penetration of bacteria.

Before the work of Berthold (1976), understanding of pore function was based mainly on their morphology. Berthold carried out an original experiment. A conical shell of *Patellina corrugata* was placed in sea water with its top projecting through the opening in a thin aluminium plate. The wide base of the shell was raised above this plate, and the aperture situated at this base was sealed with silicone oil and paraffin. The pores that were in contact with the sea water were the only open shell openings. The stuff (neutral rot) dissolved in the water actively penetrated the pores, after some time the concentration of the stuff in the shell exceeded many times its concentration in the sea water. It was possible to see cytoplasmic flows through the translucent shell wall, confirming the living state of the cell.

Light microscopic observations by Myers (1935) led him to the conclusion that the pores, like the apertures, are the places where reticulopodia emerge from the shell. However, the early works of Le Calvez (1947) and Arnold (1954), and the later works of the other scientists (Towe & Wright, 1963, Sliter, 1974, Berthold, 1976, Leutenegger, 1977) demonstrated that the pores are closed from the inside either by the organic layer or by the sieve plate, and thus they can not serve as exits for the reticulopodia. These can emerge only through the apertures or through the openings of the canals. In some planktonic foraminifera (*Orbulina*, which at its final globular stage lack a main shell aperture), in addition to small pore openings there are the larger openings between them providing the reticulopodia with access to the exterior. The larger openings represent true multiple apertures (Berthold, 1976) (Plate **35**, figure 2, M). Well developed large tubulopores, for instance those in *Tubulogenerina ferox* (in Gibson, 1989) may possibly be in fact additional apertures.

Chapter VI

CYTOLOGY AND PHYSIOLOGY

The accumulated data on foraminiferal cell ultrastructure (Allen, 1964, Febvre-Chevalier, 1971, Hedley & Adams, 1974, Lena & Freire, 1974, Nyholm & Nyholm, 1975 a, b, Hansen, 1975, Muller-Merz & Lee, 1976, McEnery & Lee, 1976, Schwab, 1977, Anderson & Be, 1976 b, Anderson & Tuntivate-Choi, 1984, Lee et al., 1965, 1991, Bowser et al., 1984, 1985, 1988, 1992, Bowser & Travis, 2002, and others) have already found applications in taxonomy.

The cytoplasm in living individuals is usually colored rather brightly and may appear yellowish-green, brown, brownish-yellow (fulvous), orange, rose or wine-red depending on the presence of different inclusions and algae. During the reproductive period it is purified and becomes translucent or milky-white. Sudden mechanical, chemical, temperature or light disturbances cause the organism to draw its cytoplasm quickly into its shell. The final chambers become free of cytoplasm and occupied with fluid discharged from vacuoles. This fluid possibly plays a protective role, buffering the organism from its environment. In the majority of foraminifera a distinct boundary between the endo- and ectoplasm is not observed. Only in the most highly evolved groups possessing a canal system with constantly circulating ectoplasm is there a clear division between endo- and ectoplasm. The outer cytoplasmic layer flows over the shell from the outside forming a thin outer layer in benthic forms. In planktonic forms, this outer cytoplasmic layer is significantly wider and thicker; in *Hastigerina* it forms a specialized bubbled capsule resembling calimma in Radiolaria (Plate **16**, figure 2).

VI.1. RETICULOPODIA, THEIR STRUCTURE AND ROLE IN FORAMINIFERAL LIVE STRATEGIES

A reticulate net of thin cytoplasmic threads extending from the apertures or the outer cytoplasmic layer constitutes the reticulopodia (pseudopodia) (Plate **2**, figure 1; Plate **16**, figure- s 2, 3; Plate **23**, figure– s 3 – 7; Plate **25**, figure1; Mikhalevich, 1999, figure 28, 1). In benthic forms, this net (reticulum) may be elevated above the protist or be spread fan-shaped over the substrate. In sessile species it may be spread as a wide strip (lamellopodia), while in planktonic forms it extends radially from the shell in all directions, thus surrounding the whole organism like a “cloud”. The first observations of foraminiferal reticulopodia were made by Dujardin (1835), Carpenter et al. (1862), Lister (1895), Winter (1907) and some others. Recent studies employing new methods and technique include those of Allen (1964), Bowser et al. (1988, 1992, 1995, 2002), Bowser & Travis (2000, 2002), Travis & Bowser (1986), Weiner (1985), Weiner & Erez (1984), Welnhofer & Travis (1996, 2003), Orokos et al. (2000), Travis et al. (2002), Habura et al. (2005) and Hottinger (2006) in collaboration with other authors. The diameter of the rhizopodial threads varies from 30 nm up to 1- 2 µm (being thinner in planktonic forms), while their length exceeds the size of the shells by several times up to a factor of tens times, sometimes reaching 1 cm in length. In the elongated reticulopodia (Plate **16**, figure 4A) the outer part of the cytoplasm, termed reoplasma, moves quickly in two opposite directions (Allen, 1964) with a speed of up to 11 µm/sec (Bowser et al., 1984). The minute granules (0.5 x 5 µm) contained in reticulopodia give them a granular appearance, hence the term granuloreticulopodia. Mitochondria, xanthosomes and lysosomes may also occur in reticulopodia together with dense dark particles (possibly phospholipids – Travis & Bowser, 1991) and even more minute elements including particles that organize their movement (MOV – motility organizing vesicles). These vesicles are encircled with fibrillar material (Bowser & Rieder, 1985; Bowser et al., 1985, 1988). The inner, axial part of the reticulopodia – stereoplasma – is darker and denser and is formed by bundles of fine microtubules that extend along the reticulopodia (Plate **16**, figure 4A). These microtubules may be connected by 1 or 2 short transverse outgrowths (Plate **16**, figure 4B) (Febvre-Chevalier, 1971; Travis & Bowser, 1986) but they do not form such spatially complex structures as the multivalent tubules in radiolarian axopodia. The transverse outgrowths (bridges) join the microtubules with other organelles encircled by their membranes as well as with the membrane of the whole cell. Hottinger (2006) had observed such bridges connecting the microtubules with the membrane covering the reticulopodia in perpendicular direction of several *Amphistegina* species. The mechanism of movement (Travis & Bowser, 1991, in Lee & Anderson, 1991) involves the following processes. The protein dynein first transforms the free energy of adenosine triphosphate (ATP) hydrolyzate into mechanical work allowing the microtubules to rotate in sequence as in a ratchet. The movement is then transferred from the rotating microtubules to the cytoplasm. This complicated process still needs additional study. As shown by Bowser et al. (1988), the microtubules and filamentous actins are localized and more active in the regions of reticulopodial contact with the substratum. Their observations also confirm the supposed mechanism of interaction between microtubules and actin-containing filaments in various kinds of reticulopodial motility and cell body locomotion.

In planktonic foraminifera, the reticulopodia become rigid during the latest stages of ontogeny, losing their anastomoses and exhibiting some convergent similarity to the axopodia of other planktonic protists (Heliozoa, Radiolaria, Acantharia). The rigid reticulopodia incorporate the calcareous spines of the shell. Thus, here also Foraminifera have their own special developmental characteristics - when hard structures are necessary they form mineral (calcareous) structures, such as calcareous spines and shells, rather than organic ones.

A special mode of organization of pseudopodial microtubules was described by Bowser et al. (1992). In *Astrammina rara,* coiled bundles of microtubules form the cytoskeleton trunk, which acts like a spring and creates higher tension. These authors also demonstrated the role of the extracellular fibrous matrix vesting the pseudopodial surface of this species. These give the pseudopodia elastic and tensile properties, improving their ability to capture and retain prey. Reticulopodia isolated from the cell body preserved their motility for a period of up to several hours (Travis & Bowser, 1991). It is of extraordinary interest that, in some ways, the reticulopodial extracellular matrices of *Astrammina rara* have the “hallmarks of metazoan organization” (Bowser et al., 1992), while the calcified biomineralized matrices of other foraminifera are comparable to biomineralized matrices secreted by higher metazoans (Bowser et al., 1992, Weiner, 1985, Weiner et Erez, 1984).

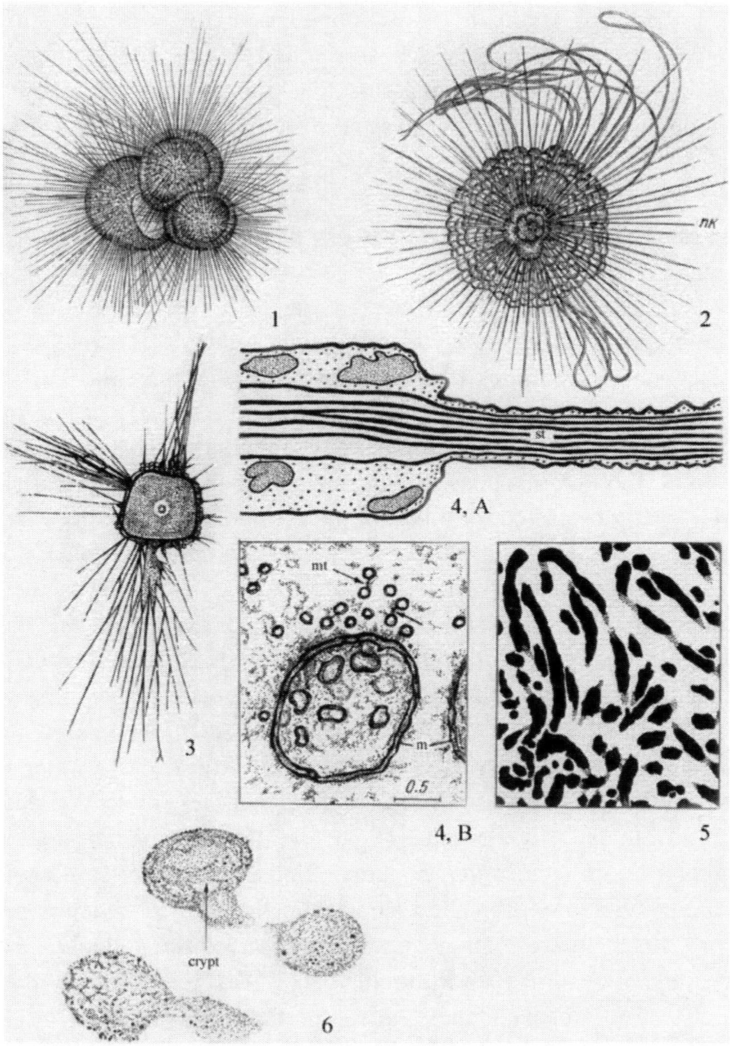

Plate 16. Peculiar structure of some planktonic and benthic forms (after Mikhalevich, 2000).
1. Shell of planktonic *Globigerinoides* with the spines and brood chamber (the last one); 2 - Shell of planktonic *Hastigerina* [with spines, filose pseudopodia and cytoplasmic bubble capsule (bc)]; 3 - Organic shell of the benthic species *Myxotheca* (Lagynana) with reticulopodia; 4 - Microtubules in pseudopodia of planktonic *Globigerina* (Globigerinana): A - longitudinal section (x 35000), B - transverse section (x 38000), [ST - stereoplasma (bundle of longitudinal microtubules), mt - microtubules, mi - mitochondria, fibrillar bridges between microtubules are showed by arrows] (from Febvre-Chevalier, 1971); 5. Microvillies in the cytoplasm of *Globigerina* (Globigerinana) (x 24000) (fragment of the inner content of vacuole) (after Mikhalevich, 2000, scheme according Leutenegger, 1977); 6 - *Globigerina* (Globigerinana) (x 600): crypt - cryptosomes, situated in the cytoplasm of the chambers, preserving their form after special dissolution of the shell (scheme after Lee et al., 1965).

The presence of reticulopodia is a very important taxonomic feature of the phylum Foraminifera clearly distinguishing it from other unicellular eukaryotes. Reticulopodia have multiple functions: movement, food capture, sense of touch, formation of chambers and different kinds of cysts (feeding cysts, reproductive cysts) and respiration. Some observations suggest that they have a role in the partial solution of the parent shell, allowing juveniles to escape (Winter, 1907). The complex of pseudopodia in the shell aperture is called the podostyle (Plate **2**, figure 1). As discussed above and below, foraminiferal reticulopodia constitute a multifunctional dynamic apparatus that plays a role in many different vital activities, fulfilling physiological, chemical, respiratory as well as mechanical functions.

VI.1.1. Role of Reticulopodia in the Process of Cell Movement

The reticulopodial threads stick to the substrate and their contraction causes the shell to move. A considerable effort is necessary to drag along rather large shells in this way or to hold them in place on a substrate (e.g. algae) against the force of a water current. If foraminifera are dislodged from algae, they are able to find their way back to the same place (Zmiri et al., 1974). Positive thermotaxis and negative phototaxis have been observed in some shallow-water species. In the forms possessing canals, the main aperture is usually overgrown with shell material (being reopened only before the formation of the new chamber), and the reticulopodia emerge through the outer canal openings – through the "back door" according to the Hottinger's picturesque expression (Hottinger, 1982). This gives the organism the possibility to remain mobile even when its cytoplasm is drawn inside the shell in order to escape harmful chemical effects. Thus it is able to position its shell on the keel, turning either of its side surfaces to the light, which is important for the species containing symbionts. We have observed in Petri dishes how the benthic species *Massilina secans*, and some other miliolids, attach the ends of their reticulopodia onto the surface of the water film with their aperture turned upwards, and "float" in this position, sometimes for several days in search of food items scattered on the water surface (Plate **26**, figure 9). Their reticulopodia were extended onto the surface film in order to gather food particles. In some species the shell moves ahead, being pushed along by the cytoplasmic flow of the reticulopodia. But in *M. secans* the shell moves in the direction opposite to that of the reticulopodial elongation, giving the impression that the organism is moving backwards (jet principle). Foraminifera can move along the flat substrate using their reticulopodia at a rate of 11 – 188 mm a day (Wetmore, 1988). *Laticarinina pauperata* covered 12 -16 mm a day on the sediment surface and 5 mm a day when moving within the sediment (Weinberg, 1991). Kitazato and Matsushita (1996) observed *Trochammina hadai* moving with a velocity of 14 μm/minute, which is equivalent to only 2 cm per day. According to our observations, *Massilina secans* advanced 100 μm per minute and then covered a distance of 215 μm in the next 15 minutes. During its progress it leaves behind obvious traces of light-coloured organic matter at the bottom of the Petri dish (Plate **24**, figure– s 1 – 6). As they move across algal surfaces, the reticulopodia of *Textularia bocki* and *Quinqueloculina ungeriana* leave behind traces of secreted organic matter (supposedly sulfate-containing glucoseaminoglucans). This provides a nutrient medium for different bacteria and fungi. The latter, in their turn, serve as food for the foraminiferal species. The purposeful repeated return of these epiphytic foraminifera to their "farms" is possible on the basis of chemotaxis. Thus, the interaction between the foraminifera and algae provide mutual benefit; the algae gain from the cleaning their surface, and the foraminifera use this surface as the source of food. This process is described in detail in Langer & Gehring (1993). The substances excreted by decaying algae diminish the harmful influence of pollution (including the heavy metals) on foraminifera (Bresler, Yanko, 1995).

Reticulopodia may be active in the process of reproduction. Thus the reticulopodia of two adjacent individuals may anastomose, and then, becoming thicker, draw both shells together to form syzygy. The residual cytoplasm of the mother shell in *Cycloclypeus carpenteri* removes the small shells of young gamonts from its surface (Krueger et al., 1996).

VI.1.2. Role of Reticulopodia in the Process of Feeding

In addition to facilitating movement and attachment, reticulopodia are involved in the capture of minute particles or small organisms. Faber and Lee (1991a) had even called them "an apparatus for the food capture". Reticulopodia extend over the sediment penetrating the interstitial spaces between the particles up to 2 – 3 mm depth, or form an elevated net in the overlying water. In the latter case they need some support. In sessile benthic species this is provided by elevated, branching parts of the shell, or sometimes sponge spicules placed near the aperture (Plate **2**, figure 2); in many of the planktonic species long skeletal spines perform this function (see part V. 2. 2, (Plate **16**, figure - s 1, 2).

The choice, capture and displacement of food particles may involve rather complex manipulations (Anderson, 1988), including rotation, passing from one reticulopodial thread to another one, elevation to the top of the extended net. Touching the outer surface of the prey, the reticulopodia increase the pressure, and the surface of the victim (small crustaceans, diatoms) cracks at its weakest points like a nut. Then some of the reticulopodian threads penetrate the interior and detach the soft inner parts in a very precise manner resembling intelligent action. After this the reticulopodia carry the pieces of food into the intracellular digestive vacuoles located inside the shell or in the reticulum. Undigested particles resembling grayish-brown globules are transported outwards by reverse cytoplasmic flow to the extremities of the reticulopodia and discharged into the surrounding medium. Another possible way to extract the soft parts of the prey may be the penetration of its carapace by a dissolving medium, as suggested by Bowser et al. (1985). The modes of prey capture by benthic foraminifera described by Anderson (1988) were later reported for several other species (Bowser et al., 1992; Austin et al., 2005; Rossignol et al., 2007). Recently, Duplay et al. (2010) described observations on *Ammonia tepida* in culture showing that this species, known previously as a herbivore, preyed on mobile metazoans such as nematodes, copepods, and a larval gastropod, which are bigger and more mobile than the foraminifera itself. *Astrammina rara* may capture a dozen brine shrimp (*Artemia*) within an hour, as well as using its reticulopodia to feed on bacteria, algae and assimilating dissolved organic matter (Bowser et al., 1992).

The selection of food particles possibly reflects chemosensory processes (Anderson, 1988). In cultures of *Massilina secans,* we observed how dead *Chlorella* cells were gathered by the reticulopodia. In the thin reticulopodian threads these inclusions looked like small inflated nodules (Plate **25**, figure 1) and were transferred in the direction of the aperture at a rather high speed – approximately 20 – 25 μm/sec. With the microscope used it was impossible to discern whether these nodules move inside or over the reticulopodia. Near the aperture the globules formed a lump exceeding the size of the apertural opening (Mikhalevich, 1999, Plate **28**, figure 2, Plate **25**, figure – s 2 – 5 of this study). In some time this lump becomes dark green and than gradually disappeared (see below, part VII.1.2.).

Some observations (Schultze, 1854; Christiansen, 1971) suggest that the reticulopodia of certain foraminiferal species (*Elphidium*, for instance) are able to secrete a substance that paralyzes the prey. When the caught animal (quickly moving Copepods, Caprellids, Cumacea or slowly moving Echiuroidea, Amoebae) sticks to the reticulopodial threads it first struggles for rather a long time, making convulsive attempts to tear itself away, but becoming gradually weaker and weaker, entangling itself in the adhesive threads and at last perishing. The photographs look dramatic. Some animals manage to free themselves but they fall to the bottom of the container and blankly move their limbs, incapable of normal movement.

An interesting observations were made by Cedhagen and Pawlowski (2002) on the behavior of their new genus *Toxisarcon* (Cedhagen & Pawlowski, 2002) which is able to abandon its shell like some allogromiids. When the nake body of protist was touched it became blue and died in a minute. A toxic compound released into the water caused the death of the individuals of the same species and small invertebrates. The authors suggest that the toxin was derived from the algal dinoflagellate diet and when the cell membrane was injured the toxin was released into the water. They also suggest that usually *Toxisarcon* is able to break down the toxin thus detoxificating the natural surroundings.

The agglutinated *Pilulina argentea* has an original way of obtaining food (Christiansen, 1971). It builds a kind of house in the sediment, which represents a cavity paved inside by large sand grains. This cavity is closed from above by a cover ("roof") built from silt and mud and flush with the level of the sediment surface. These 'houses' are 3 - 4 times bigger than the protist itself. *Pilulina* extends its reticulopodia through the "roof" and lies in wait for its prey. Capturing some little crustacean incautiously swimming close to the swaying threads it pulls it into its house. When such houses were unsealed the protists are discovered holding its prey tightly. *Astrammina rara* is also considered to capture prey (Delaca, 1986; Bowser et al., 1992; Suhr et al., 2008). Its diet includes not only crustaceans and mollusks but also polychaetes and echinoderms, being predominantly carnivorous. However, this species also consumes bacteria, diatoms and dissolved organic matter (Bowser et al., 1992, Bernhard & Bowser, 1992). It is supposed that foraminiferal reticulopodia secrete some substance, possibly attracting or paralyzing prey animals. Thus it has been observed that *Dunaliella* stops its movement on reaching the reticulopodial net of *Ammonia* (Lee at al., 1961). It has also been suggested that the matrix covering the reticulopodia of *Astrammina rara* have adhesive properties (Bowser et al., 1992). Pseudopodia possibly may assume adhesive or non-adhesive states depending on the necessity (for instance, proximity of prey) (Buchanan & Hedley, 1960).

Foraminifera may be predators of other foraminifera. *Floresina* (Rotaliana), with its high conical shell, has been observed to crawl up the lenticular shells of three larger *Amphistegina* species (Rotaliana). It made up to 10 holes in the amphisteginid shell wall and extracted the cytoplasm from several chambers. In 3 to 7 days the prey usually died but some restored themselves after *Floresina* became detatched from their shell. Sometimes *Amphistegina* managed to repulse an attack of the predator (data of Hallock & Talge, 1994). *Floresina* attacked only adult shells, on the surface of which they later reproduced. Based on the presence of similar traces on the shell of *Globorotalia menardii,* Nielsen et al. (2002)

suggested that *Floresina* attacks this species also, and that predatory drilling and foraminifer-foraminifer interactions may be more widespread than was previously believed.

There is an accumulation of data on foraminiferal feeding in culture as well as some field observations (see also above – Part **III**). Some of foraminiferal species feed selectively but the majority of them are obviously omnivorous. The trophic diversity of some deep-sea benthic foraminifera was discussed by Gooday et al., (2008). Foraminifera utilize a variety of nutritional sources. They may use amino acids and the other dissolved organic substances (e.g. *Notodendrodes antarcticus* – Delaca et al., 1981; *Pararotalia spinigera* and *Rosalina macropora* – Bresler, & Yanko, 1995), some consume detritus or bacteria (usually below the photic zone) (Bernhard & Bowser, 1992), others eat faecal material, different algae (mostly *Diatomeae*, but *Dunaliella* , *Chlorella* and some others as well), yeast, different protists, eggs, larvae, minute Crustacea (for example the calanoid copepods – the main food component for Globigerinids) [carnivory in planktonic foraminifera (Boltovskoy and Wright, 1976, Bé et al., 1977) is known better than in benthic forms]. Koho et al. (2012) showed that uptake of particulate and dissolved organic matter by some agglutinated foraminifera was as high as that of bacteria : 6.1±0.1 nmol C m^{-2} and 5.0±4.6 nmol C m^{-2} [1] This indicates their major role in ecosystem functioning and stresses the need to study them. Interesting enough, the uptake of particulate and dissolved organic matter by some calcareous foraminiferal species was shown to be significantly lower. Some cases of cannibalism are also known; *Glabratella ornatissima* (according to Myers, 1940), *Patellina corrugata*, *Heterotheca lobata* (according to Grell, 1988) consume their own zygotes.

Modes of foraminiferal feeding may be either active (movement, the activation of the water flows through the activity of reticulopodia) or passive. The passive mode of feeding is especially characteristic of sessile forms or those rooted in the sediment (Plate **2**, figure 2; Plate **1a**, Astrorhizana, figure 9). Forms elevated above the sediment catch detritus deposited from the water. Foraminiferal shells may be epibenthic, lying on the sediment surface (*Astrorhiza*), or they may move within superficial layers, creating burrows at a depth of 1 -2 cm (infaunal forms). In the latter case the shells usually have an elongated form that facilitates their movement through the sediment with the least resistance (*Reophax*, many miliolids). But foraminiferal species with another shell form also may be found in the sediment as deep as 30 cm (*Arenoparrella mexicana* – Goldstein et al., 1995), or even living deeper than 30 cm in the case of *Elphidiella* sp. in wet near-shore coastal sediment of Issyk-Kul lake (Mikhalevich, 1976, Yankovskaya & Mikhalevich, 1972).

The presence and distribution of food influences many aspects of foraminiferal life, including the onset of sexual reproduction, the size of the shells and their distribution. Planktonic species possessing spines are mostly zooplankton feeders while those lacking spines are phytoplankton feeders (Hemleben & Spindler, 1983). The latter track precisely the displacement of the chlorophyll maximum up and down between 0 and 95 meters, as well as the decrease of chlorophyll in the high latitudes. These facts can be considered while interpreting the data of the fossil sediments. Seasonality in the alternation of generations in benthic foraminifera is also often linked to phytoplankton blooms and therefore to the abundance of food.

[1] According to Gooday (personal communication), the uptake of POM in bacteria is not direct.

VI.1.3. Foraminifera as Epiphytes, Commensals and Parasites

Foraminifera may be epiphytes on algae, usually *Enteromorpha, Zostera, Zannichellia, Ulva, Polysiphonia, Laminaria* (rarely on the *Fucus* and *Codium*) (Lee et al., 1969), on *Sargassum natans* (Spindler, 1980), on *Corallina* (Delaca & Lipps, 1972). Experiments with artificial seagrass leaves showed that the type of substrate did not significantly affect colonization by foraminifera, seasonal factors being more influential (Ribes et al., 2000). Debenay & Payri (2010) reported more than 150 foraminiferal species as epiphytes on > 80 algal species, most of which were macroalgae. The authors demonstrated the complexity of the distribution of foraminiferal assemblages on phytal substrata, prompting the need to be prudent in interpreting paleoenvironmental data.

Foraminifera also may live as epibionts on the shells of mollusks and crustaceans, on hydroids (Dobson, Haynes, 1973), bryozoans (Poag, 1969, Pozaryska & Voigt, 1985), corals, and polychaetes. I observed *Cibicides grossepunctatus* on the setae of the Antarctic polychaetes, as well as on the shells of other Foraminifera. On the grooved shell of the scallop *Chlamys opercularis* in the Irish Sea 51 epibiontic foraminiferal species were reported, some of them not characteristic of this environment (Harward & Haynes, 1976). Sometimes foraminifera themselves serve as a mobile substrate for other organisms; for example, Zapata et al., 1981 mentioned *Buccella frigida* as a substrate for some Turbellaria, or for their egg cases as supposed by Gooday (personal communication) (unfortunately the publication of Zapata et al., 1981 was not available). Commensal relationships are also known. Thus *Cibicides refulgens* scrapes bacteria and algae off the shells of scallops *(Pecten)*, at the same time building elevated agglutinated tubes through which its reticulopodia emerge, permitting it to filter feed (Voronova & Mikhalevich, 1985b). The same species behaves as a parasite, dissolving holes in the shells of Antarctic scallops and adsorbing concentrated amino acids in the extrapallial cavity of the bivalve (Alexander & Delaca, 1987) (the data were obtained using radioactively labelled amino-acids). There are few cases of true parasitism among the foraminifera. *Planorbulinopsis parasitica* was described as a parasite of *Alveolinella* (Banner, 1971), *Talpinella cunicularia* found in Cretaceous *Orbitoides* was considered by Baumfalk et al., 1982 to be a parasite. Some rosalinids parasitize bivalve mollusks (*Rosalina* - Todd, 1965, *Hyrrokkin sarcophaga* (syn. *Pulvinulina punctulata* – Cedhagen, 1994). All the parasitic forms mentioned above belong to the subclass Rotaliana. In the subclass Nodosariana, some cases of parasitism are known in the genera *Fissurina* and *Lagena,* (the former picks granules from the reticulopodia of *Discorbis mediterranensis;* Le Calvez, 1947) and in polymorphinids (e.g. *Vasiglobulina*) (Poag, 1971). Collen & Newell (1999) consider *Fissurina marginata* to be apparently an obligate parasite. Some species in the agglutinated genera (in *Globodendrina* – Plewes et al., 1993, *Rotaliammina, Siphotrochammina* – Venec-Peyre, 1987) are also able to bore into hard surfaces although this capacity is more pronounced in calcareous forms (Venec-Peyre, 1987). In the majority of known cases it is not fully evident that the boring foraminiferal species are true parasites that use the amino acids of their hosts. The problem needs further studies (Plewes et al., 1993). Some representatives of Lagynana (*Rhynchosaccus* and *Thalamophaga*) are also supposed to be parasitic (if they are not inhabitants of empty shells). All the above mentioned cases represent examples of ectoparasiticm. Data of Collen (1998) give some tint on endoparasitism of *Metarotaliella tuvaluensis* in some species of *Triloculina* and *Quinqueloculina*, but the phenomenon needs further study.

Some benthic foraminiferal epibionts possess a highly sensitive protective system (Yanko et al., 1994). It can be used as biomarker for predicting the ecological consequences of anthropologic pollution.

The capability of foraminifera to bore into the calcareous walls of various organisms is rather widely known. The surfaces of many recent and fossil metazoans, calcareous algal and large foraminifera (*Orbitoides*, *Orbitolina*, some alveolinids, nodosariids) may be pitted by foraminifera. The bored passages may be labyrinthic. Such traces were made by *Vasiglobulina, Rotaliammina* sp,. *Spirotrochammina* sp. and *Talpinella cunicularia* (cited after Plewes et al., 1993). It is not clear whether all these forms are really parasitic or use the bored organisms as substratum, shelter or a source of calcium. Plewes et al., (1993) suggested that the use of the calcareous hard parts of different organisms may be an initial stage leading to a full or partial parasitic mode of life.

VI.1.4. Role of Foraminifera in Food Webs

Many animals (fish, echinoderms) ingest foraminifera incidentally (Daniels & Lipps, 1978). Foraminifera may serve as food for the small species of Gastropoda and Polychaeta and for their young stages (Saidova & Beklemishev, 1953, Lipps & Ronan, 1974). Some species of scaphopods are able to extract foraminiferal cytoplasm from their shells. Scaphopoda as a group are obligatory consumers of foraminifera (Newell, 1970, Bilyard, 1974). Some gastropods, being the selective predators, may influence significantly the densities of the species consumed (e.g. *Islandiella islandica*). Another gastropod, *Littorina littorea,* bores into foraminiferal shells (Walker, 1971) as do nematodes, populations of which grow quickly in foraminiferal cultures. Fossil shells with the traces of these predators (circular openings in the shell wall) are not uncommon (Sliter, 1975, and some others). Arnold A.J. et al. (1985) mentioned that one such predator (probably nauticid gastropod) is highly species selective, feeding almost exclusively *Siphouvigerina auberiana* (d'Orbigny), which rarely survives attacks. The snail borings could be distinguished from those made by nematodes (Sliter, 1971, 1975). Some specific prey and predator behaviours were described (Arnold A.J.et al., 1985); these and similar biological observations hint at interesting and complex predator/prey relations. Foraminiferal shells have been found in the stomachs of Polychaeta, Sipunculoidea, Echinoidea, Holothuroidea, Echiuroidea, Ophiuroidea (Saidova, 1976, Chusid, 1988). Most likely in the majority of these cases they are used not as a source of energy but as a source of $CaCO_3$. At least it was shown that foraminifera were still living after being ingested by fish (Debenay et al., 2011). But asellote isopods (Crustacea) were surely shown as specialist predators of different benthic foraminifera crashing their hard agglutinated or calcareous shells with robust mandibles (Svavarsson et al. 1993). Multiple examples of predation on (and sometimes by) foraminifera were also described by Gudmundsson et al., 2000, M. Sokolova, 2000, Culver & Buzas 2003, Brokeland et al., 2010. The latest authors analysed gut contents of the four species of deep-see isopods.

At the beginning of my scientific career (the end of 1960s) my chief, Z.G. Stschedrina, used to say:"Foraminifera don't eat anybody, and nobody eats them". But it is clear now that they represent an important intermediate links in the lower trophic levels of benthic and pelagic marine communities, feeding on various organic substrates, microscopic bacteria and algae and even preying on metazoans that exceed them in size. In their own turn they are used as food by other organisms, and predators upon Foraminifera may be highly specialized or less selective (e.g. see Lipps & Valentine, 1970, Gooday et al., 1992). With the presently available data it is now possible to study their role in marine food webs. As it was mentioned above, biotic interactions in benthic foraminifera were reviewed by Lipps (1983), Lipps & Valentine, 1970, Gooday et al., 1992, M. Sokolova, 2000), the trophic dynamics of Antarctic benthic communities including foraminifera were studied by Rivkin & Delaca (1990). Lipps & Valentine wrote as early as 1970 that "Foraminifera form part of a key link in marine food chains, assimilating energy…" The participation of Komokiacea and Xenophyophoria in deep-sea benthic trophic chains was examined by Kamenskaya (1987) and also by some aforementioned authors (M. Sokolova, 2000; Brokeland et al., 2010). Dupuy et al. (2010) made the first attempt to analyse the complex primary and secondary consumer levels of marine communities and suggested that *Ammonia tepida* is a secondary consumer, changing its feeding mechanism according to the availability of different food types. These authors also demonstrated how to study carbon fluxes between small metazoans and foraminifera. The study of Koho et al., (2012), measuring the uptake of particulate and dissolved organic matter by some agglutinated foraminifera, further demonstrate their important role in the functioning of ecosystems.

Although they represent major advances in knowledge, these studies represent only the first steps towards a full understanding of the role of foraminifera in benthic and pelagic marine communities

VI.1.5. Role of Reticulopodia in Chamber Formation

Reticulopodia also take an active part in the formation of the new shell chambers. This process was studied in *Discorbis* (Le Calvez, 1938), *Calcituba polymorpha* (Arnold, 1967, Berthold, 1976, Hemleben et al., 1986), *Rosalina floridana* (Angell, 1967 a, b, 1979), *Clavulina* (Coleman, 1980), *Orbulina universa* (Spero, 1988), *Spiroloculina hyalina* (Angell, 1980), *Trochammina inflata* (Angell, 1990) and *Textularia candeiana* (Bender, 1992). In the latter species chamber formation proceeds in the following way: before the formation of the new chamber the organism feeds strenuously. The subsequent process can be divided into the several stages.

1. The reticulopodial net emerges from the aperture of the last chamber forming veneer, becomes somewhat shorter and forms a surrounding cyst of tightly packed particles of detritus. As the particles are not cemented, the cyst may be easily destroyed.
2. Formation inside the cyst, in front of the apertuture of the last chamber, of an empty space (cavity) having approximately the dimensions of the future chamber and providing a space for its assembly.
3. Selection of the sediment particles for building the new chamber and their arrangement within the cyst from the penultimate chamber to the aperture of the new chamber with simultaneous cementation of the particles by organic cement and the formation of the outer organic layer.
4. Deposition of the biomineralized calcareous material from the distal to the proximal part of the shell wall and arrangement of the crystals to form the cylindrical cavities of the pore system.

5. Formation of the layer underlying the pore openings.
6. Formation of the thin organic layer covering the inner chamber surface takes place after the protist leaves its cyst; during the subsequent growth this layer is augmented by addition of several additional layers.

This means that in the multi-chambered agglutinated shells of Textulariana, as well as Hormosinana, layers of growth exist only in the organic part of the wall (Bender, 1995). The entire process lasts for a period of 24 hours. *Textularia* belongs to the subclass Textulariana of the class Rotaliata.

In the higher, more advanced representatives of the class Rotaliata, subclass Rotaliana, the process of new chamber formation was studied in detail in *Rosalina floridana* (Angell, 1979). The duration of the process was also 24 hours. Its first stages – the formation of a reticulopodial veneer of the protruding cytoplasmic projection, coinciding in its outline with the form of the future chamber, and the formation of the cyst - proceed in a fashion similar to that observed in *Textularia*. Then the "membrane complex" of the cytoplasmic projection is formed, which obviously corresponds to the outer organic envelope in *Textularia.* The formation of this membrane accomplishes the formation of the cytoplasmic projection. Unlike *Textularia* the calcareous crystals are layered "on and in the substance of the membranes" (Angell, 1979). Although the calcareous wall of *Rosalina* is considered to be monolamellar (Goldstein, 1999), in the photographs (Angell, 1979, plate 4, figures 18, 20) the units of crystallites appear layered on both sides of the membrane resembling the POM (Primary Organic Membrane) (see above – Part V.2. 2.). In the bilamellar benthic rotalians and planktonic globigerinans the crystallites are definitely layered on the both sides of the membrane which represents the POM. The beginning and completion of the process of calcification were observed by Angell (1979) in great detail. The minute crystallites can be seen initially as myriads of dark dots, and then as larger and light-coloured centers of crystallizations. At this stage in the light microscope the scattered white speckles could be seen at the surface of the developing chamber. Approximately an hour after their formation the aperture of the future chamber also becomes visible. The centers of crystallizations gradually increase in size and their edges merge, but between these spaces filled with calcareous material the pore spaces stay unfilled. Apparently, at the surface of the cytoplasmic projection, there is some 'chemical marking' of the plasmatic membrane that differentiates between zones that are becoming calcified and "islands" (the future pores) which remain uncalcified (Labas & Mikhalevich, 1987). This pattern of calcification is obviously determined by the Golgi apparatus. The "patchy" character of the calcareous secretion, and its absence in areas occupied by future pores, could presumably be explained by the zonation of the discharge of the vesicles, where in membranes the carboanhydrase and Ca/H ATF-ase are functioning. Although it was not observed in Angell's work on *Rosalina floridana* the next step must have been the formation of the organic lining - of the inner organic lining (IOL) of the chambers, as observed in *Textularia.* The shell-forming secretion of the calcareous material alternates with the muccopolysacharid secretion of the organic elements (Labas & Mikhalevich, 1987), and finally, the reticulopodia destroy the cyst. The process of formation of the Inner Organic Lining (IOL) occurs later, when the next new chamber is already developing (Spindler, 1978). Thus, the disposition of the new chambers is defined by the disposition of foramens reflecting directions of cytoplasmic flow, and the form of the new chamber is controlled by the length and disposition of reticulopodia, which are defining the first outer envelope – the matrix of the future calcification. The extension of the cytoplasm beyond the bounds of the chamber, and formation of the new chamber, proceed cyclically and with a definite periodicity. It also occurs more frequently in the younger individuals than in the adults. For example, the young gamonts of *Heterostegina depressa* formed a new chamber nearly every day during the first 9 – 12 days (Roettger & Spindler, 1976). The formation of the new chamber and the formation of the new organic and calcareous layer in the preceding chambers (layers of growth – see above) occurs simultaneously as well as formation at the same time (in both benthic and planktonic forms) of the additional pore plates in the pore canaliculi at the level of the new organic shell layer (Plate **12**, figure 3; Plate **14**).

The formation of new chambers in planktonic forms of the subclass Globigerinana has some peculiar features (Hemleben, 1969, Towe, 1971, Spindler & Roettger, 1973). Species possessing spines do not construct a cyst; instead, their Primary Organic Membrane (POM) is stretched like a tent between the needle-like spines. After this, calcareous material is deposited on both sides of this membrane, as in the benthic bilamellar Rotaliata. In the non-spinose species of *Globorotalia,* an organic envelope resembling the cyst in *Rosalina* has been observed (Hemleben et al., 1977, 1989; Be et al., 1979). In *Orbulina universa,* the cytoplasm forms a ring around the shell rather than a veneer, prior to the formation of the last spherical chamber (Spero, 1988). However, this is not a fundamental difference as it is predetermined by the form of the future spherical chamber. In spinose species the spines are usually added after the chamber is formed (Hemleben et al., 1976). However, in *Orbulina universa* the bases of the spines are laid down simultaneously with the calcification of the chamber surface; they are initially smooth but after 12 – 36 hours acquire their triradiate character (Spero, 1988). Mitochondria were observed in the vicinity of the POM during the process of calcification in this species (Spero, 1988, figure 4: 4e). Secondary gametogamic calcification in the planktonic forms is considered below (see Part VI. 5).

The formation of the new chambers in the class Miliolata, which was observed in detail in *Spiroloculina hyalina* (Angell, 1980), differs in principal from the process described above. At the very beginning of the formation of the new chamber the reticulopodia also build the cyst encircling the whole shell. After the cyst is formed the reticulopodia remain inside it and a small amount of cytoplasm emerges through the apertural opening of the preceding chamber. This marks the beginning of the building of the new chamber. In contrast to *Rosalina* and *Textularia*, there is no evidence neither for the cytoplasmic projection, nor the membrane enveloping it, in *Spiroloculina.* In this genus, the cytoplasm transports two types of vacuoles to the place of shell formation; one type delivers the organic matter (sometimes flattened fibrillae are observed inside them) and the other type delivers elongate calcareous crystals. Both types of vacuoles discharge their contents into the space which is formed between the thin layer of cytoplasm and the reticulopodia. In this way a rather thick organic matrix is formed in which the calcareous crystals are embedded (in Rotaliata calcareous crystals are deposited mostly on the membrane rather then inside it). The thin cytoplasmic layer forms the outer part of the shell wall, and then the reticulopodia complete the wall from the inside, also delivering to it organic and calcareous components. The chamber is not formed in one process, as in Rotaliata, but segment by segment. Firstly the thin cytoplasmic layer moving across the previous chamber forms the basal part of the chamber, embedding the needles randomly; then, building on this base, it lengthens the chamber, continuing the process up to the formation of the apertural tooth. After the thin part of the new chamber wall is formed by the cytoplasm from the outside, the reticulopodia add new wall elements from the inside, in the same step by step manner by which the exterior part was formed. They pack the crystals in a regular way, parallel to the wall surface, and the shell wall becomes gradually thicker. The reticulopodia complete the process of constructing each successively built segment of the wall with the formation from the inside of the thick organic membrane underlying the wall. This membrane differs in its composition and density from the organic matrix (layer) in which the calcareous crystals are embedded. Each crystal has its own thin organic envelope which preserves its form and is clearly visible after the crystal has dissolved. The calcareous needles are 0.2 μm in length and 0.01 μm in breadth. During the formation of the last chamber, new organic and calcareous material continues to be added to the wall of the previously-formed chambers, as a result of which the earlier chambers are much thicker than the later ones (see Plate **30**, figure 6 – thin translucent newly formed chamber of *Massilina secans*), although layers of growth are not formed. The reticulopodia that complete the building of the shell wall contain only vesicles with organic and calcareous material, in contrast to reticulopodia building the shell in the cyst, which have multiple longitudinal tubules and filaments. Using these reticulopodia, the shell rotates itself in the cyst and emerges after the process of new chamber formation is completed. Building of the cyst takes several hours and the entire process, from the moment when the cytoplasm emerges from the chamber aperture up to the fully completed building of the new chamber, occupies a further six hours.

This process obviously may vary somewhat in different miliolid genera and species According to our observations of miliolids in culture (see below, Part VII.), during the process of new chamber formation the cyst of small-sized shells encircled the whole shell while in larger forms it covered the shell only partially. Sometimes the protist left its cyst before chamber formation was finished. In this case the final stages of the process can be clearly seen. Where new shell material was being created, a shapeless, swirling mass of cytoplasm was observed (Plate **30**, figure- s 1, 2, 5b, Mikhalevich, 1999, Plate 28, figure 3); reticulopodial activity was high and the shell often changed its position, turning itself from side to side. As already mentioned, the newly built chamber was thin and translucent (Plate **30**, figure 6; Mikhalevich, 1999, Plate 28, figure 4). While the new chamber remained unfinished, the borders of the successively formatted segments were clearly visible, the most recently formed elements being distinguishable owing to their thinness and transparency. The process of chamber building within the cyst lasted 8 hours; without the cyst it could be as long as two days.

In *Archais angulatus* (Soritida, Miliolata), which has rather wide rather than tubular chambers, the wide cytoplasmic anlage in front of the last chamber was formed first (Wetmore, 1999). But the formation of the organic matrix and the process of calcification were not observed in detail in this species. There is no reason to suppose that the typical monofontinal (monolamellar) wall of *A. angulatus* shell could be formed in a way similar to that of the bifontinal (bilamellar) wall of Rotaliata species, with its complex regular pore system. The similarity is clearly limited to the formation of the cytoplasmic anlage. The membranous envelop encircling this anlage was not observed. In all other respects the process of calcification in this species resembles that of other miliolids. (Uniquely among foraminifera, the formation of the porcellaneous shell wall involves the heterogenous nucleation of calcite, a process that is rare among living organisms generally; see above, part V. 2. 2).

In the representatives of the class Spirillinata (e.g., *Patellina corrugata*) the process of chamber growth also involves the participation of the reticulopodia but the formation of the cyst was not observed. Instead, growth was limited by the edges of the wall of the tubular chamber and the beginning of the process of calcification was not preceded by the formation of a cytoplasmic projection (anlage). The morphogenetic processes in this foraminiferal group differ fundamentally from those in the Rotaliata. The first structures to be formed are those nearest to the aperture, not distant from it. New material is accreted to existing structures, similar to the process of building a wall upwards. But the alternate secretion of calcareous and organic material, creating the inner and outer layers of the shell and underlying the pores, should also be feasible in this case as well.

Thus in the representatives of the three different evolutionary lineages (classes Rotaliata, Miliolata, Spirillinata) reviewed above, the mechanism of growth and chamber formation differ in principle and in detail. In my opinion (Mikhalevich, 1980, 1981. 2000), the similarity in the processes of new chamber formation in the Textulariana and Rotaliana confirms the close relationship of these groups and disproves the generally accepted view that they are convergent. The differences between them (formation of POM and deposition of calcareous material on both sides of it in the Rotaliata) are fewer than the similarities, namely the building of a cyst only around the newly formed chamber, the formation of the cytoplasmic projection and the organic layer encircling it, the alternating (rather than simultaneous as in the Miliolata) deposition of calcareous and organic matter, and the formation of entire new chambers in one process and in a direction towards the aperture rather than away from it.

The growth and formation of chambers may be strongly modified under extreme conditions, such as strong mechanical impacts, near lethal values of temperature or salinity, or in polluted settings (Yanko et al., 1998, Stouff et al, 1999 a, b, Geslin et al., 2002, Meric et al, 2008 a, b, Polovodova & Schoenfeld, 2008). For example, under such conditions the soritids, which usually have discoidal shells, may become cylindrical or form a semi-circular growth perpendicular to the primary basal disk. The chambers of regular-glomerate miliolids begin to grow irregularly, assuming an ugly form, often strongly curved and without the type of the symmetry peculiar to these taxa. The disorganization of the cytoplasmic flow is the reason of such changes. The polymorphism of the shells of *Peneroplis p;anatus* observed by Hottinger (2007) may also be caused by the environmental conditions. At least I didn't met such variability of this specied in my material. The influence of high salinity on the abnormal morphology of juveniles of *Ammonia tepida* was shown by Stouff et al. (1999 b). But sometimes a shell that is broken or destroyed on one side is able to recover and to complete its growth normally. This suggests the presence of a genetically fixed mechanism of growth control, regulating chamber formation according to an axis of symmetry. This regulation is obviously defined by the chemical gradient of the outer primary membranes of the cytoplasm and reticulopodia.

VI.2. CYTOPLASMIC ORGANELLES

The main foraminiferal cytoplasmic organelles are described on Plate **14**. Foraminiferal mitochondria (Plate **14**; Plate **16**, figure 4b; Plate **18**, figure - s 2, 3) have tubular cristae (according to the binary nomenclature) and are up to 0.5 μm in diameter and 1- 5 μm in length. Mitochondria take part in active cell processes. Their role in calcification was confirmed by ultrastructural studies of new chamber formation in *Rosalina floridana* and *Orbulina universa.* These observations showed them to be abundant (together with fibrillar material and dense bodies) in the cytoplasmic anlage (Angell, 1967, 1979, Caron et al, 1987) and in close proximity to the organic membranes, adjacent to the discontinuous section of POM (Spero, 1988). A similar situation was apparent in a light microscope study of *Globorotalia truncatulinoides* and *G. hirsuta* (Be et al., 1979). Cedhagen & Frimanson (2002) observed an increase in the velocity of mitochondria in the reticulopodia of seven foraminiferal species of the classes Astrorhizata and Miliolata with increased temperature They explained these facts with the increased rate of metabolism under the higher temperatures.

Peroxisomes (microbodies) perform the same function in the cell as mitochondria but are smaller in size and more primitive in their structure; their central part is filled with a dark granular mass (Plate **14**). The presence of peroxisomes in planktonic foraminifera was confirmed by the cytochemical studies of Anderson & Tuntivate- Choy (1984).

The dictyosomes of the Golgi apparatus secrete carboanhydrase and also mucus, which is transported in vesicles to the shell and to the reticulopodia and is involved in shell attachment and sometimes holding prey. The vacuoles with calcareous crystals for shell construction, enzymes (including the digestive ones) and lysosomes (digestive vacuoles) are also secreted in the Golgi apparatus. Digestive vacuoles are mostly found in the final chambers. Digestion is obviously occurring not only in these vacuoles but also in the cytoplasm (in lacunas) (Leutenegger, 1977). There is also evidence for extrathalamous reticulopodian digestion in some foraminifera (Faber & Lee, 1991a, Lee et al., 1991). The main reserve nutritive substances of the foraminiferal cell are in the form of fat globules. The starch grains present in some foraminiferal species (*Archaias angulatus*, *Cyclorbiculina compressa*) appear to result from the activity of their symbiotic algae.

The endoplasmic reticulum in foraminifera is well developed and xantosomes (pigmented granules) and annular membranes (which most likely serve as a reserve material for the building of the nuclear membranes during the reproductive process) are also present (Spindler & Hemleben, 1982).

Some organelles, namely cryptosomes (Lee et al., 1965) and microvillies, are found only in planktonic forms. The former are subcircular or irregular in form (Plate **16**, figure 6). Their role in the cell is unclear. The microvilly system (called also as fibrillar system, fibrillar units or vesicular reticulum) is a unique foraminiferal system formed by branching tubules up to 10-μm diameter, tightly packed in the vacuolar space (Plate **16**, figure 5) (Anderson & Be, 1976, Leutenegger, 1977). They are composed mainly of strands of protein (Lee et al., 1965). Their function is not yet fully understood. Some authors suggest that they play a role in buoyancy (Zucker, 1973, Anderson & Be, 1976), which might permit the shell to undergo slow vertical movements as a result of the replacement of heavy ions by lighter ones, changing its specific gravity (Hansen, 1975). Later, the detailed EM observations of Spero (1988) of the process of formation of the spherical *Orbulina* chamber showed that this fibrillar system may play an important role in biomineralizing. Its bundle aggregates were found in the area where the new chamber was being formed (within the final spherical and penultimate trochospiral chambers of *Orbulina*) during all the stages of its development.

The tubules of the microvilli system underwent three parallel stages of development, from loose to fibrillar and then dense material. Spero (1988) hypothesized that they serve as a storage site for the intracellular calcium pool. Further studies concerning the function of this system are necessary.

VI.3. Symbiosis

Many large benthic foraminifera with secreted calcareous shells living at depths of 10 – 70 meters (Alveolinidae, Soritidae – Miliolana; Amphisteginidae, Calcarinidae, Nummulitidae - Rotaliana), and in the planktonic foraminifera of the euphotic zone (Globigerinidae, Hastigerinidae, Globorotaliidae) host algae cells in their cytoplasm (Plate 14) (mostly Dinoflagellata and Diatomea but also Chrysophyta, Rhodophyta, Chlorophyta and obviously Haptophyta) (Haynes, 1965, Hansen & Dalberg, 1979, Hallock, 1985, Hallock & Peebles, 1993, Lee, 1990, Lee & Zucker, 1969, Lee et al., 1979, 1980a, b, c, 1985, 1991 a, b, 1997, Muller-Merz & Lee, 1976, Be et al., 1977, Leutenegger, 1977, 1983, 1984, 1990, Jorgensen et al., 1985, Lee, 1980, 1990, 2006, Kremer et al., 1980, McEnery & Lee, 1981, Hallock, 1984, Williams et al., 1981, Hemleben & Spindler, 1983, Anderson, 1984, Krueger et al., 1996, Bernhard, 2003, Bernhard & Bowser, 1999, Bernhard et al., 2000, Hohenegger, 2004, 2011 and others). These algae give the cytoplasm of the foraminiferal shell its color – rose, green. yellow, orange and others. In some representatives of Nonionida and Elphidiida, as well as in *Haynesina, Bulimina, Stainforthia, Virgulinella* and even the agglutinated genus *Reophax,* isolated chloroplasts were also found (Bernhard et al., 1999, Bernhard, 2003). Such foraminifera containing only plastids but not intact algae cells are termed kleptoplastic (the phenomenon of kleptoplasty). Bernhard (2003) investigated the role of intact plastids in the bathyal species *Virgulinella fragilis* living in sulfidic habitats at 375 m (below the photic zone). Hottinger et al. (2001 a) demonstrated the presence of symbiontic diatom chloroplasts in *Haynesina germanica*, in *Cribroelphidium salseyense* and in several species of *Elphidium*, although these authors considered that light-dependent chloroplasts are not obligatory to all the elphidiids. Some of these algae are commensals or parasites but, as a rule, particular algal species are symbionts of particular foraminiferal species. Bacterial symbionts were reported by Bernhard (1996, 2003). The digestion of algae by foraminiferal reticulopodia is suggested as a possible preadaptation to symbiosis (Lee et al., 1991 c).

Symbionts are presently known in more than 20 families from different foraminiferal subclasses. Many groups of closely related foraminifera are associated with the same, or closely related, symbionts. The symbionts of Rotaliata are diatoms, and those of Miliolata are either rhodophytes or chlorophytes (Peneroplidae, Archaiasinidae) or dinoflagellates (Soritidae). The similarity between the symbionts of nummulitids (Rotaliata) and alveolinids (Miliolata) is regarded as exceptional (Leutenegger, 1984). The soft-walled monothalamous species *Xiphophaga minuta* and *X. alominuta* ingest diatoms belonging to the genus *Pleurosigma* and their cells are packed with intact diatoms or sometimes with their chloroplasts, the photosynthetic pigments giving this foraminifer its golden-brownish color (Goldstein et al., 2010). Their role in the *Xiphophaga* cell is not entirely clear. Although they are partly digested, their unusually large quantity suggests possible symbiotic activity.

Interestingly, some carnivorous foraminifera have symbionts (Goldstein, 1999, Goldstein et al., 2006). Bernhard et al., 2006 also discovered symbiont-bearing allogromiids.

The light requirements of symbionts influence the depth distribution of their hosts. During asexual reproduction the daughter individuals may inherit symbionts from their parent cell, while after the sexual reproduction they acquire them from the immediate environment (Lee et al., 1997, Krueger et al., 1996). For *Cycloclypeus carpenteri* the highly organized process of transferring the diatoms from the mother cell to the two-chambered daughter (firstly to deuterococh, then to the proloculus) was shown in detail by Krueger et al. (1996). In *Hastigerina depressa* the composition of its symbionts changed seasonally. Seasonal variations in different *Symbiodinium* (dinoflagellate) phylotypes was also described in several soritid genera (Pochon et al., 2001), each specimen of which contained several lineages of symbionts. During this latter study one *Symbiodinium* phylotype was found to be common to soritid and coelenterates.

The host cell provides its symbionts with habitat and food, for example, nitrogenous compounds. The intracellular Diatomacea contained higher quantities of glucose and glycerin comparing their free living relatives (Kremer et al., 1980). The host, in its turn, receives the products of photosynthesis and nutritive material (carbohydrates and some amino acids, such as glutamine and alanine). This is beneficial in environments poor in organic matter (Lee, 1980). Overall, the symbiont-host system receives additional energy. It is likely that the symbiont and host utilize this energy in a different ways. In the foraminifera investigated the rate of calcification, and probably the production of organic membranes, increases substantially. According to the data of Hallock (1981 b) the rate of calcification in symbiont-bearing foraminifera is double that of species without symbionts. The rate of ^{14}C uptake in the two *Amphistegina* species in laboratory culture was 30 -80% in full sunlight and sharply decreased at lower light levels (Hallock, 1981 a). The calcium uptake and calcification rate in the planktonic species *Orbulina universa* was enchanced by its symbionts, resulting in a growth rate that was up to the three times higher under high light levels (Lea et al., 1995). The utilization of HCO_3- produced by the symbiont in the process of shell formation was demonstrated by the use of 14C tracer (Kuile, 1991, Kuile & Erez, 1987). These processes can be summarized by the following formula:

$$Ca^{++} + 2HCO_3^- = CaCO_3 + CO_2 + H_2O$$

calcification

phtosysnthesis

In sunlit areas, *Amphistegina lobifera* may accumulate a reserve of carbon in a special repository, which it then uses gradually. Miliolats do not have such a stock and, unlike the symbiont-dependent *Amphistegina*, they use carbon derived from the symbionts immediately. The symbionts in the host cell are usually encircled by membrane preventing the invasion of lysosomes (Plate **14**). But when starved, especially in darkness, foraminifera can digest these organelles. Krueger et al. (1996) showed that permanent digestion of symbionts in *Cycloclypeus carpenteri* constituted a consistant part of its normal nutrition (Krueger et al., 1996). In the rhodophicean symbiont of *Peneroplis,* this membrane is absent and symbionts lie freely in the host cytoplasm. Such a case may be regarded as more advanced form of symbiosis (Lee, 1990).

During the day, the symbionts are distributed in the ectoplasm while at night they are located inside the shells. The removal of the symbionts causes the same effect as the darkness: reduction in life span and a decrease in shell size.

These symbiotic relationships developed in foraminifera during the long process of their evolution (Hallock et al., 1991 and some others). During this time they influenced the size and the form of their shells, their inner morphology (shell complexity, increase in the number of chambers and chamberlets), the structure of the shell wall (the development of thinner translucent areas ("windows") or pits (in miliolids – Hohenegger, 2011) in the shell wall, partial resorbtion of the organic material). Hottinger et al. (2001b) suggest that the inflated interlamellar cavities of pellatispirines (Rotaliana) may serve as greenhouses for algal symbionts. The work of Hohenegger (1994, 1995, 1999, 2000, 2004, Hohenegger et al., 1999) revealed a variety of host strategies of this kind among the larger nummulitids; the monograph of Hohenegger (2011) describes greenhouses in large foraminifera, which he calls as "gardeners". The ecomorphology of another large nummulitid, *Operculina ammonoides,* was described by Pecheux (1995). In *Discocyclina* and some other related groups (asterocyclinids, lepidocyclinids, lepitorbitoids, orbitoids), the "pillars" growing on lateral chamber wall concentrate the light for their algal symbionts ((Fernandez-Canadell & Serra-Kiel., 1992). These authors propose the term 'crystalline cones' for these "pillars", in order to distinguish them from non-translucent pillars and to stress their function. Tightly bound vertical calcareous crystals in the foraminiferal shell wall may have the effect called "Stoletove's foot" which is conducting and comcentrating light (A.Bondar, Moscow, personal communication). All such changes in shell architecture promote a more convenient distribution of symbionts and increase shell wall transparency, enabling the host cell to regulate light intensity (Hallock & Hansen, 1979, Hohenegger, 2004). Similar morphological changes in the shells of the fossil forms (e.g. fusulinids) suggest that they also possessed symbionts.

The distribution of symbionts inside the shell depends of the morphology of the host shell. In the forms with bilateral symmetry that stand upright on the edge of their shells and orientate their lateral sides towards the light (*Amphistegina, Elphidium, Heterostegina*), their symbionts are located on both sides of the shell, whereas in flat forms that lie with one side on the substrate (*Sorites*), the symbionts are located on the upper side of the shell. In the both cases the symbionts take up a position in peripheral parts of the chambers, where they are often observed near the pores. In *Baculogypsina sphaerulata,* symbionts are distributed evenly throughout the chamberlets rather than being restricted to the particular parts of the shell (Hyams-Kaphzan & Lee, 2008).

The distribution of symbionts in some spinose planktonic species is rather interesting. At night, the symbionts are inside the shell but with the daybreak they leave the shell and take their places along the spines in the reticulopodial network of the host (Be et al., 1977).

The presence of the symbionts influences the shell isotopic composition (the fractionating of O^{18} and C^{13}). In addition to symbionts, parasitic Sporozoa have also been reported to occur in foraminiferal cytoplasm (Nyholm, 1962).

During the last decade the composition and distribution of symbionts within the benthic foraminifera has been studied with the use of molecular methods (Pochon et al., 2001, 2004, Pawlowski et al., 2001). These studies revealed the presence of different kinds of dinoflagellates in soritids from geographically distant regions (East Pacific and Carribean).

However, there was almost no evidence for the transmission of symbionts between foraminifera and other symbiont-bearing invertabrates in the same environment.

The new evidence of denitrification by the benthic foraminifera (Risgaard-Petersen et al., 2006, Pina-Ochoa et al., 2010) may be possibly connected with the symbiont activity. In later publications, Bernhard and co-authors (Bernhard et al., 2012 a, 2012 b) continue to explore the precise mechanisms of foraminiferal denitrification. They suggest that foraminiferal endobionts (denitrifying bacteria, chloroplasts) may participate in these processes. They report "nitrate respiration and/or denitrification"' in several benthic species of the subclass Rotaliana (*Bolivina argentea, Buliminella tenuata, Fursenkoina cornuta, Nonionella stella*) that either possess or lack endobionts. Their experiments demonstrate an important role for foraminifera in marine nitrogen budget and also may advance understanding of eukaryotic respiration and their surviving in anoxic conditions. The presence of prokaryotic endobionts in foraminifera, together with the fact that an allogromiid can survive in anoxic conditions, led Bernhard et al. (2006) to speculate on the role of symbiogenesis as a driving force in early foraminiferal evolution.

VI.4. NUCLEUS

The sizes of foraminiferal nuclei varies from very small (5 – 12 μm) to larger (40 – 100 μm) and very large (0.6 mm in giant *Bathysiphon*). For instance, the size of *Conqueria laevis* nucleus (Lagynana) is 35 – 45 μm (Gooday & Pawlowski, 2004).

The structure of the nuclei is similar to that typical of eukaryotes (Plate **14**; Plate **17**; Plate **18**, figure- s 1, 2, 3). The number and the size of the nuclear pores varies in different species and it also depends on the physiological state of the cell. In many of the unilocular agglutinated species the nuclear membrane is significantly thickened from the outside or from inside with fibrous or spongy (honeycomb, reticular) layers, for instance, in *Boderia turneri* (Plate **17**, figure 1), which makes it more rigid. The outer surface of the nuclear membrane of *Astrammina rara* is coated with a thin layer of material, presumably cytoplasmic in origin, and mitochondria are adhering to the nuclear surface (Habura et al., 2011). The thinner and more elastic nuclear membrane of many of the more advanced calcareous foraminifera permits them to change the shape of the nucleus and allows it to pass through the narrow passages between chambers (Plate **17**, figure 3). Their nuclei often have a lobate, rosary-like, amoeboid (term of McEnery, Lee, 1976) (Plate **17**, figure 2) or irregular form rather than a globular one (as in *Cibicides lobatulus, Planorbulina mediterranensis, Orbulina universa, Globigerina bulloides, Bolivina argentea*). The nuclei of the young *Elphidium ponticum* gamont are subcircular in outline while mature nuclei have a lobate form (Raykov et al., 1998; Mikhalevich, 1999). In the less advanced unilocular foraminifera, gamont nuclei of amoeboid form are sometimes observed (e.g. in *Allogromia laticollaris, Myxotheca arenilega* - Schwab, 1971). This shape obviously develops from the preceding subcircular form during gametogenic changes to the nuclei, as observed in *Allogromia laticollaris* by McEnery and Lee (1976) and in *Saccammina alba* by Goldstein (1988). An elongated nucleus was observed within the long tubular chamber of *Quinqueloculina* sp. (Plate **19**, figure 2) (Mikhalevich, 1999). In some species the nuclear envelope develops hair-like projections, folds and curves on its outer surface (*Ovammina opaca, Globigerinoides sacculifer*). Such structures result in the enlargement of the nuclear surface and the intensification of nuclear functions (Dahlgren, 1968). There are few histochemical studies of foraminiferal nuclei (Lee et al., 1965, Dahlgren, 1967, Schwab, 1971, Spindler et al., 1978, Raikov et al., 1996, Voronova, 1976 and some others). They revealed the accumulation inside the nuclei of chromatin material (DNA) and nucleoli, which usually represent larger and darker-colored granules containing protein and RNA. The accumulation of RNA molecules could be often seen outside the nuclear pores (data of Febvre-Chevalier, 1971 on *Globigerina bulloides*, Be et al., 1977 on *Globigerinoides sacculifer*, Cesana, 1978 on *Iridia lucida*, Voronova, Mikhalevich, 1985 on *Cibicides lobatulus*). The nucleoli may be scattered all over the nucleus (Plate **17**, figure 2; Plate **18**, figure 1) or may accumulate at its periphery (Plate **14**). Their disposition and their sizes obviously change according to the physiological state of the nucleus. The chromatin in the non-dividing nuclei is usually dispersed.

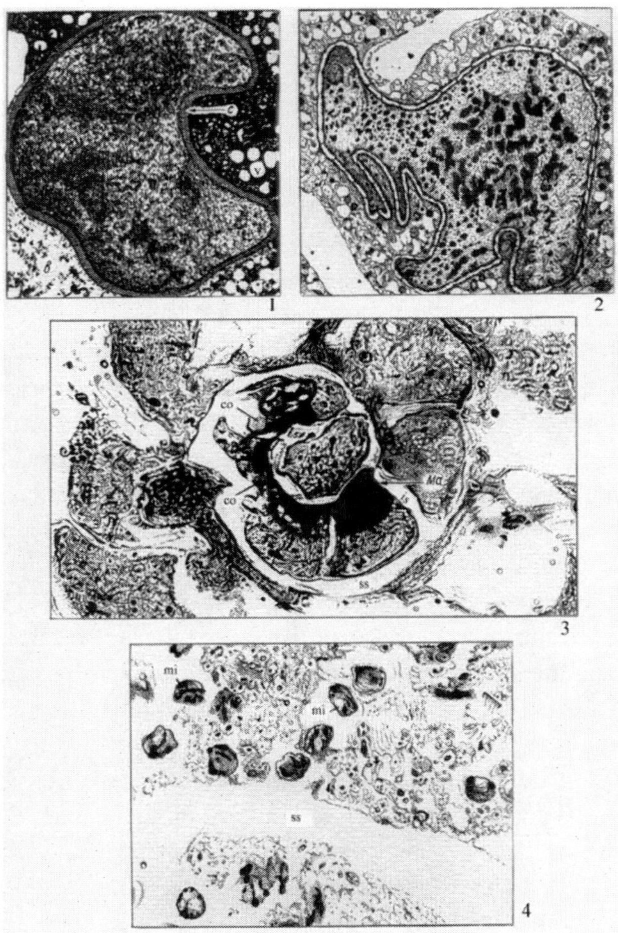

Plate 17. Foraminiferal nuclei (after Mikhalevich, 2000).
1 - Nucleus of *Boderia* (Astrorhizata) (x 1500) with honeycomb layer (C), v - vacuole (scheme after Schwab, 1977); 2 - Somatic nucleus of *Orbulina* (Globigerinana) «amoeboid « in form (x 2900) (scheme after Be et al., 1977); 3 - Rosary-like migrating somatic nucleus (macronucleus - Ma) of *Cibicides lobatulus* (Rotaliana), continuing along the four chambers, in the places of its passing through foramens from chamber to chamber - the constrictions (co) are seen, part of the macronucleus

does not have a clear outline as it is degenerating; cytoplasm preserves the form of the dissolved chambers, its previous position is marked by the spiral and interchamber sutures (ss, is) (x 1200) (scheme after Voronova, Mikhalevich, 1985); 4 - Generative nuclei (Mi) of the *C. lobatulus* agamont in the process of meiosis, (white space – the dissolved spiral suture (ss) of the shell) (x 1200) (scheme after Voronova, Mikhalevich, 1985).

Grell (1956) was the first to describe the two types of nuclei in foraminifera - vegetative (= somatic, = macronucleus – Ma) and generative (= micronucleus – Mi). Since this first description the number of the known species with nuclear dimorphism has increased significantly (Lee et al., 1965; Muller-Merz, Lee, 1976; Leutenegger, 1977; Voronova, 1976, 1978a; Grell, 1979 and many others). It is now clear that differentiation of the nuclei into the somatic types, which control the metabolic functions, and into the generative types, which give rise to the new generations (Plate **14**; Plate **17**, figure - s 1 – 3; Plate **19**, figure - s 2 – 3) (big vesicular and small vesicular nuclei according to Raykov 1978), took place during foraminiferal evolution. The somatic nuclei contain three times more RNA and protein than the generative ones, while the DNA quantity is the same in both. The differentiation of the somatic nuclei is irreversible. The generative nuclei may turn into somatic nuclei if the latter are lost.

All presently known species with nuclear dimorphism belong (according to the classification adapted here) to the two subclasses of the class Rotaliata (Rotaliana and Globigerinana) and to the class Miliolata (subclass Miliolana). Data on the Miliolata are rather poor but the discovery of somatic and generative nuclei in the evolutionary advanced genus *Sorites* (*S. marginalis* - Muller-Merz, Lee, 1976 and *S. orbiculus* - Leutenegger, 1977), and also in the more primitive *Quinqueloculina* sp. (Mikhalevich, 1992), suggest that this group also exhibits nuclear dimorphism.

In the less advanced foraminifera, such as unilocular Laginana and Astrorhizana (class Astrorhizata) as well as in such peculiar group as Spirillinata, nuclear dimorphism is not reported. Initial investigations of the life cycles and nuclei in the multi-chambered agglutinated trochammins (belonging to the lower subdivision of the class Rotaliata – subclass Textulariana) (Salami, 1976; Angell, 1990; Pawlowski et al., 1995) showed that the nuclei of their agamonts were also not differentiated. Thus it is possible to use nuclear homogeneity as an additional taxonomical feature of these taxa. The nuclear apparatus of Nodosariata is almost unknown.

Nuclear dimorphism is one of the most characteristic features of the phylum Foraminifera. This group also exhibits polymerization of the nuclei; the number of the generative nuclei may range from several tens to several thousands or even the tens of the thousands. In some species polymerization of the somatic nuclei also took place (4-6 in *Rosalina, Glabratella, Globigerinoides*, 40-60 in *Sorites*). The second way of increasing the amount of nuclear material – polyploidy, - is also known in foraminifera. Cytophotometric studies on *Patellina corrugata, Rotaliella heterocariotica* (Zech 1964) and *Cibicides lobatulus* (Voronova & Selivanova, 1976) revealed a 20-30-fold increase in nuclear ploidy. The polygenomic nucleus was also mentioned in *Psammophaga simplora* by Arnold (1984) and in *Saccammina alba* by Goldstein (1988).

The incomplete division of the nucleus obviously serves as a mechanism for the creation of polyploidy in the nucleus. The differentiation of the nuclei, followed by polymerization and polyploidization, could be regarded as an important step in the evolution of the nuclear apparatus leading to the intensification of its functions and an increase in its metabolic activity. In the primitive unilocular species *Allogromia laticollaris,* differentiation of generative (germline) and somatic genetic material occurs in the interior of the amplified nucleus during the uninucleate stage (Parfrey et al., 2012). Apart from foraminifera, nuclear heterogeneity is known among protists only in the phylum Infusoria, where it has its own peculiar characteristics (see below, Part XI. 1).

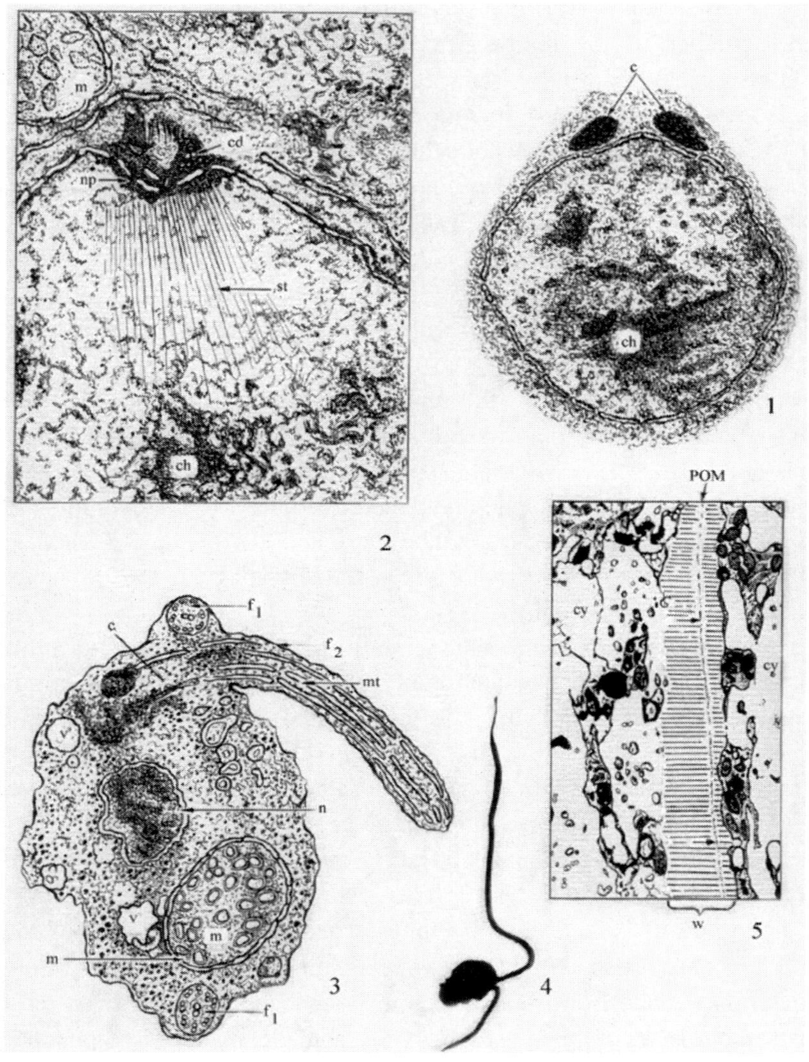

Plate 18. Structure of regenerating Foraminiferal nuclei and gametes.
1 - generative nucleus of *Iridia lucida* (Astrorhizata) (x 30000) at the beginning of division: c – centrosomes, ch – chromatin (scheme according to Cesana, 1978); 2 - Further process of development of generative nucleus of the same species (x 64000): cd - center of division, ch– chromatin, np – nuclear pores, st – spindle threads, m - mitochondria (scheme according Cesana, 1978); 3 - Gamete of *Iridia lucida* (Astrorhizata) (x 41000), longitudinal section: c – centrioles, f_1 – transverse section of the flagellum incorporated, f_2 – longitudinal section of flagellum, m - mitochondria, n - nucleus, mt – microtubules , v - vacuoles (scheme according to Cesana, 1972); 4 - Gamete of planktonic *Hastigerina pelagica* (Globigerinana): a - (x 8882), б – (x 4500), (scheme after Spindler et al., 1978); 5 - part of the

shell wall of the young chamber of *Heterostegina* (Rotaliana) (x 154000): w – wall, inner cytoplasm (cy) from the left, outer cytoplasm – from the right , POM – Middle (Primary) Organic Membrane (scheme according to Spindler et al., 1978); 6 – plastogamic *Planoglabratella opercularis* (d'Orbigny, 1839) containing off-spring shells of young embryonal agamonts in the parent gamont shell with dissolved umbilical area (x 256) (near Furugelma Island, Peter the Great Bay, Sea of Japan) (after Preobrazhenskaya & Tarasova, 2004).

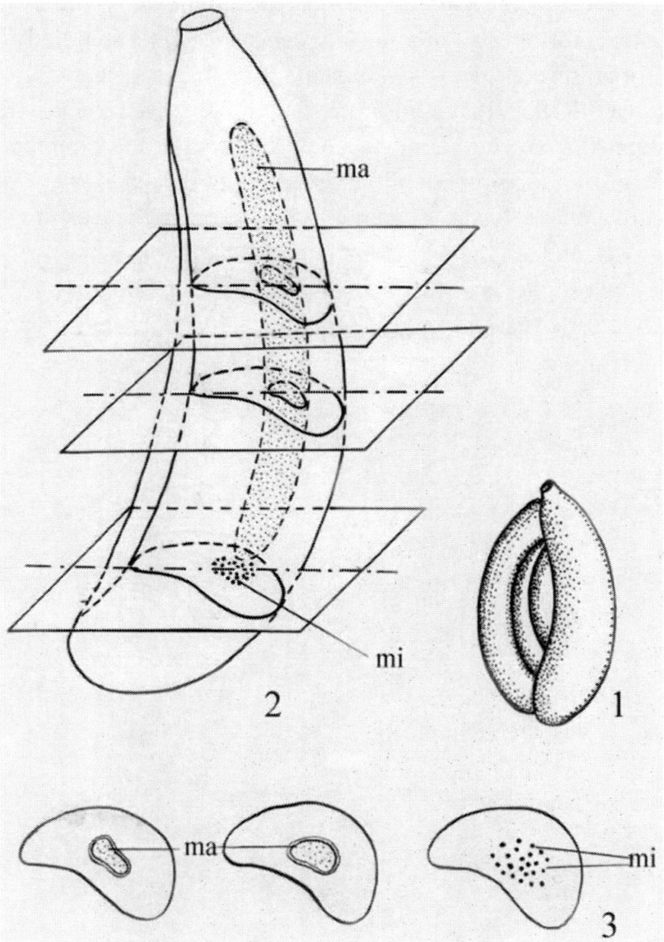

Plate 19. Nuclei of *Quinqueloculina sp.*
(1 – outer view of the shell, 2 – scheme of the successive transverse sections of the last chamber, 3 - view of the transverse sectioned nuclei, (Ma – somatic nucleus, Mi – generative nuclei) (x 120).

Foraminiferal mitosis follows the pattern of closed intranuclear pleuromitosis (Myers, 1935; Zech, 1964;Voronova, 1978; Raykov, 1978) but may also have some features of orthomitosis (Cesana, 1978). During the haploid mitosis the presence of centrosomes (Plate **18**, figure 1) and of centrioles was observed (Plate **20**, figure 1) in *Cibicides lobatulus* (Voronova, 1976; Voronova, Mikhalevich, 1985) and also in some other species (Le Calvez, 1938; Schwab, 1969, 1973, 1976). It is interesting that in the flagellate daughter cells of *Myxotheca arenilega* gamonts the centrosomes have centrioles while in the nonflagellate agametes the centrosomes lack these features. The cellular centers of division were studied in minute detail in *Iridia lucida* (Cesana, 1978) (Plate **18**, figure – s 1-2). The centrosomes with the centrioles are formed on the outer surface of the nuclear envelope and appear to be enclosed from beneath with a dense acroplasmatic cap. Rather large windows are formed in the nuclear envelope in exactly the same place. Then, opposite the centrosomes and inside the nuclear envelope, a layer of dense plasma develops, which serves as a center of organization for the microtubules of the spindle (COMT). These microtubules are linked to the kinetochores of the chromosomes (half spindles). At the later stages of mitosis the half spindles diverge to the opposite poles of the nucleus. Some of the microtubules join both COMTs with each other, forming an intranuclear central (lateral) spindle lying parallel to the nuclear envelope. Foraminiferal meiosis was studied in detail in the agamonts of *Cibicides lobatulus* by Voronova (1978b) (Plate **20**, figure 2). She observed the premeiotic duplication of DNA, the two stepped character of meiosis, and chiasma at the diplotene stage (Plate **20**, figure 2 E). Chiasma during the process of meiosis were observed also in *Patellina corrugata* (Berthold, 1977). In its general features foraminiferal meiosis proceeds in the same way as in the majority of protists and animals, but in *Cibicides lobatulus*, as well as in the other foraminiferal species whose meiosis was investigated (*Patellina corrugata, Iridia lucida*), the equatorial plate is not clearly marked and the pronounced metaphase is practically absent.

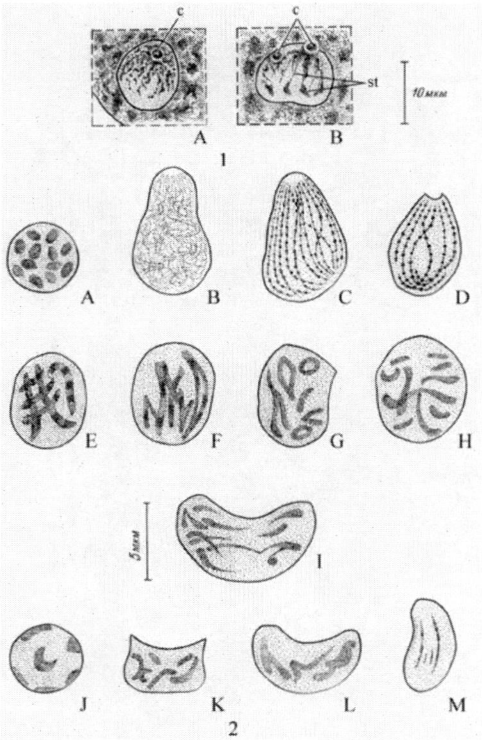

Plate 20. Processes of division in the nuclei of the *Cibicides lobatulus* (Rotaliana).
1 - Mitose of the agametes nuclei of *C. lobatulus* (drawings from the preparation): A – appearing of the centrosome (c) and the beginning of chromosome condensation , B - beginning of the division (prophase) (c - centrioles, st – spindle threads) (from Voronova, Mikhalevich, 1985); 2 - Meiosis of the agamont nucleus of the *C. lobatulus*: A - proleptotena, B - leptotena, C - zigotena, D, E - pachitena, F - diplotena, three chromosomes with chiasmata, in the center of the nucleus the mere crossing of the chromosomes, G – diakinesis, H – metaphase I, I – anaphase I, J - telophase I, K – prophase II, L - anaphase II, M - telophase II (according to Mikhalevich & Voronova, 1985).

The chromosome number is known only for a few species, where it is rather small (2, 3, 6, 9, 18, 24). Habura et al., 2011 studies investigating genomic composition of the giant nuclei of *Astrammina rara* provide the first window into organization of foraminiferal genome. They represent the first sizable body of sequence information about the non-coding regions of this f genome. The authors calculated an approximate genome size for *Astrammina rara as* ~79 Mb. It is comparable with haploid genome size of the foraminiferan *Allogromia laticollaris* (83 ± 29 Mb). Relatively little overlap between the three genomic-scale foraminiferal datasets was discovered (3% overlaps the *Reticulomyxa* dataset). The two EST (Expressed Sequence Tag - tiny portion of an entire gene) database share less than 10% of their siquences. According to Habura et al., (2011) this may be explained by "…the small sample sizes and the relatively long evolutionary distances between different foraminiferal species. The LCA of *Astrammina, Quinqueloculina* and *Reticulomyxa* probably dates to approximately 800 million years ago." The presence of introns in foraminiferal genes was also documented as well as the translocation of mitochondrial sequences to the nucleus. The latter phenomenon is known in many eukaryotic lineages.

The nuclear apparatus of foraminifera undergoes strong changes during their complex life cycle. Thus, even in the primitive species *Allogromia laticollaris,* the ribosomal DNA during its life cycle is amplified differentially. More complex changes during the life cycle are undergone by the nuclei of more advanced multi-chambered taxa that possess differentiated nuclei (see Part VI. 5).

VI.5. LIFE CYCLES

Foraminifera occupy a rather primitive position among eukaryotes in the way in which the nuclear phases alternate. Their life cycles can be considered to be heterophasic, that is the haploid and diploid phases constitute approximately equal parts of their life cycles (the antithetic alternation of generations) with the haploid phase somewhat prevailing. Such alternation of haploid and diploid phases occurs in some unicellular algae. In Foraminifera additional agamic generations (A2, A3) may be intercalated between the first agamic generation, which generates the gamont (A_1), and gamogony which generates the zygote and agamont (B). Thus the alternation of generations may be described as A-B-A-B, or as A1-A2-A3-B (usually intercalated agamic generations are called schizonts, although in the Russian-language literature all the three (including agamont) are often called "schizonts" – Rauser-Chernousova & Fursenko, 1959). The additional agamic generations are known in many of the primitive unilocular forms (e.g. *Allogromia*) and in *Spirillina vivipara*. Grell (1979) did not observe such repetitive agamic generations in the higher Rotaliata. However, they are reported in *Discorbis orbicularis* and in *Discorinopsis aguagui*, while in *Ammonia tepida* they are repeated many times, especially in culture (Goldstein & Moodley, 1993). They have also been observed in *Heterostegina depressa* (Roettger & Krueger, 1990). In cultures of some species, for instance, *Elphidium crispum* (data of Hofker, 1930) the gamont stage is entirely absent and prolonged agamic reproduction occurs. The cyclic production of schizonts directly from agamonts was observed in *Amphistegina gibbosa* in culture over several generations by Dettmering et al. (1998). These authors suggest that the result of this trimorphic cycle is a predominance of megalosphaeric shells in recent and fossil populations. A trimorphic life cycle in populations of the same species was also reported by Harney et al.

(1998), as well as in *Heterostegina depressa* by Roettger et al. (1990) and in *Trochammina inflata* by Lehmann et al. (2006). In *Cycloclypeus carpenteri* schizonts were not observed (Krueger et al., 1996). Repeating agamic generations were observed in some miliolats (*Cornuspira, Calcituba, Spiroloculina*) as well. Lee (1976) termed life cycles characterized by repeated agamic generations as apogamic. The opposite phenomenon, namely an exclusively sexual form of reproduction, was observed in planktonic species (Be, 1982). Hemleben et al. (1989) proposed the term "gamic" for such life cycles. It is possible that the advanced planktic forms have lost the asexual stage of the reproductive cycle. The exact position of meioses in this type of reproduction is currently unknown, but most likely occurs in gamont cells (rather than in the zygote).

Among the protistan taxa descended from the flagellates, the Foraminifera represent a very rare example of a group that preserves in their evolution this ratio between the diploid and haploid phases. Among the lower plants this form of life cycle is widespread. It is also described in the Mixosporidia (Uspenskaya, 1982). The conservation of the heterophase alternation of generations during the long process of foraminiferal evolution could be regarded as an additional evidence of their origin from flagellate forms.

At present time the life cycles of about a hundred of foraminiferal species are known to a greater or lesser extent, although only 26 of them at the electron-microscopic level. Despite interest in this topic and a number of recent investigations (Grell, 1988, Roettger et al., 1989, Angell, 1990, Krueger et al, 1996, Kitazato & Matsushita, 1996, Dettmering et al., 1998, Preobrazhenskaya & Tarasova, 2004, Lehman et al., 2006, Lee et al., 2009, Langer et al., 2009, Altin et al., 2009, Barras et al., 2009, Goldstein, 2010 and some others) the majority of taxonomic groups remain unstudied. The full life cycles are known in a little more than 20 species. In only a few cases have their nuclei been examined by EM or light microscopic studies : these are *Myxotheca arenilega* (Foyn, 1936); *Ovammina opaca* (Dahlgren, 1964, 1976, 1978); *Allogromia laticollaris* (Schwab, 1971; McEnery & Lee, 1976; Schwab, 1969); *Cibicides lobatulus* (Voronova, 1976, 1978); *Saccammina alba* (Goldstein, 1988); *Trochammina* sp. (Pawlowski et al., 1995); *Xiphophaga minuta* (Goldstein et al., 2010).

The foraminiferal life cycle typically takes a year and its stages are usually seasonal, although seasonality is strongly variable in different species. In winter, spring and summer reproduction is generally asexual (agamogony); as a result the haploid gamonts are formed from the agametes (Plate **21**, figure – s 9 – 13). The process of gamogony in the majority of studied foraminifera begins in autumn, with the lowering of temperatures, but it may occur during the summer also (in Preobrazhenskaya & Tarasova, 2004, in Triantaphyllou et al., 2012 and in some other works). In *Trochammina hadai,* long-term observations indicate that asexual reproduction took place in autumn while sexual reproduction occurred in spring (Kitazato & Matsushita, 1996). The widely-held opinion on the dependence of the beginning of these processes on changes in different environmental parameters and food availability was not confirmed by the observations of Kitazato & Matsushita (1996) on *Trochammina hadai,* where the nutrient levels, salinity and temperature seemed to have no influence on the beginning of the reproductive processes. The duration of both of these processes also varies. They usually last about 24 – 30 hours, but in a laboratory study of *Cycloclypeus carpenteri* the duration was from 0.7 to 15 hours (Krueger et al., 1996), while in *Trochammina hadai* it was 1.5 hours (Kitazato & Matsushita, 1996).

Gamogony is a sexual process in which the gamont forms gametes and then the fusion of two gametes gives rise to a zygote and the agamont. In some species (*Patellina corrugata* and many of the higher Rotaliata) the parent gamonts join in pairs on their ventral sides, thus forming syzygy (Plate **2**, figure 10A); the ventral shell wall between the joined shells becomes resorbed and gametes fuse in the enclosed space. This form of reproduction is called gamontogamy or plastogamy (e.g. in *Glabratella sulcata, Discorbis patelliformis, Rosalina leei,* in the different *Bolivinella* species) and contrasts with the form of gametogamy characteristic of the majority of foraminiferal species, in which gametes fuse in open water. The membrane uniting the parent shells in gamontogamy is composed of mucopoly-saccharides that lack sulfates, contrary to normal shell membranes (Lipps & Erskian, 1969). The latter participate in the process of calcification while the function of non-sulfate mucopolysaccharides is different; they constitute gel-like or semi-fluid intracellular cohesive material, which is analogous to that produced by animal taxa. The syzygy is enveloped externally by a tight organic membrane and often also by a cyst. In *Patellina corrugata,* associations of the several shells (mostly 3 - 4) may also be formed. Preobrazhenskaya & Tarasova (2004) discovered in surface sediment samples several gamont parent shells with dissolved umbilical areas filled with off-spring agamonts (Plate **18**, figure 6). Some plastogamic species undergo autogamy when only the gametes from one individual fuse with each other (*Rotaliella heterocariotica, R. roscoffensis*). The process of gamete formation has been studied more fully in gamontogamic (plastogamic) species (Myers, 1935; Le Calvez, 1950; Grell, 1954, 1979; Zech, 1964; Weber, 1965, Grell & Bardelle, 1977, Berthold, 1977, Hayward & Brasier, 1980) than in gametogamic species, where the gametes fuse in the surrounding water (Cezana, 1978; Schwab, 1969, 1973; Voronova, 1976, 1978a, b; Voronova & Mikhalevich, 1982; Arnold, 1982, 1984; Grell, 1988, Krueger et al., 1996, Altin et al., 2009, Goldstein et al., 2010). In this case the gamete numbers can reach several tens of thousands, ten or more times their quantity in plastogamic species. They are especially numerous in pelagic species. The autogamy has been demonstrated also in *Allogromia laticollaris* (Arnold, 1954). In *Hastigerina pelagica* (Roettger, 1971), after the multiple divisions and gamete formation the cytoplasmic remains of the parental individuals sometimes regenerate, reconstructing the resorbed chambers and continuing their life and even their growth. Similar observations have been made in some other species [*Spiroloculina hyalina* (Arnold, 1955) and *Bolivina doniezi* (Sliter, 1970)]. This demonstrates the high regenerative capabilities of foraminifera. But in most cases, the life of the mother is terminated after the gametes are released (e.g. *Orbulina universa* – Spero, 1988).

In addition to the usual life cycle, many primitive forms with tectinous shells (Lagynana) undergo budding and cytotomy (= plasmotomy), i.e. multiple budding (in *Allogromia* for instance). Simple fission into two parts also occurs. Agamogony may be considered as the next stage of asexual reproduction (multiple fission). During evolution, progressive changes in the sexual part of the foraminiferal life cycle also took place, from isogamy (*Discorbis patelliformis*) to anisogamy (*Rubratella*), sometimes with the selective sexual activity of the gametes. Thus in *Rubratella intermedia*, gamonts are of different rather than of equal sizes. The gametes are also anisogamic: smaller gamonts have smaller gametes with the smaller nuclei. This sexual differentiation was studied in experiments with *Metarotaliella parva* (Weber, 1965). The young gamonts of this species could not conjugate with each other but only with more mature and adult individuals. The adult gamonts are able to conjugate with each other but their reaction to each other is stimulated by the younger gamonts (Lee, 1990). In some rare cases, namely *Allogromia laticollaris* (Arnold, 1955) and *Rotaliella heterocaryotica* (Grell, 1954), autogamy was observed.

Plate 21. Life cycle of the *Cibicides lobatulus* (Rotaliana) (after Mikhalevich, 2000).
1 - gametes (n), 2 - zygote (2n), 3, 4 - unilocular agamonts (2n) (3 - two nucleate, 4 – three nucleate), 5 – 8 - growing multinucleate agamonts with the macronucleus (MA) and multiple small micronuclei migrating from the central chambers to the periphery, (R1, R2 - 1-st and 2-nd reductional (meiotic) divisions, MA is degenerating), c – cyst around the agamont (n), 9 – agametes (n), having leaving the cyst, 10 – 13 - growing gamonts (n).

Let us view foraminiferal life cycle in detail, taking as an example *Cibicides lobatulus* in which the lifecycle has been studied most fully (observations in culture with material taken in the Barentz Sea were performed by M. Voronova between 1972 and 1978). Let us begin from the agamogony (Plate **21**, figure - s 1, 2). After the fusion of the two biflagellate gametes the diploid zygote is formed, which grows into the young agamont. Its nucleus immediately initiates diploid mitosis. As a result of the very first division, two nuclei are formed (Plate **21**, figure 3), one of which becomes the somatic (or vegetative) macronucleus (Ma). It increases in size, enriches itself in karyoilymph, and nucleoli, rich in RNA, also develop within it. During subsequent divisions, the second nucleus generates firstly two and later up to several tens of generative nuclei (micronuclei -Mi). During the agamogonic process the macronucleus enlarges in size and moves from chamber to chamber towards the center of the shell, while the generative nuclei migrate from the center to the last-formed chambers at the shell periphery and stay there until the end of the process. Finally, they undergo two-stage meiosis, the 1st and the 2nd reductions (meiotic divisions) resulting in the formation of the haploid nuclei.

It is interesting to note that, as a result of the asynchrony of the meiotic divisions in one cell, both diploid and haploid nuclei are present simultaneously. It is possible that this phenomenon may be widely distributed in the other multinuclear protists. Asynchronous calcification of the juvenile gamont megalospheres observed by Langer et al., 2009 in *Peneroplis* sp. may possibly also be explained by preceding asynchronous meiotic divisions. Although Kitazato & Matsushita (1996) observed the synchronous division of the mother cytoplasm into the cells of young gamonts, these observations were only visual and did not include the process of division occurring inside the nuclei, which possibly might also be asynchronous. Only the generative nuclei participate in meiosis. Although it is sometimes possible to observe changes in the somatic nuclei that suggest attempts to perform the meiosis, the process is never completed. The somatic nucleus either only enlarges in size or, as in *Rotaliella roscoffensis* and in some other species (Grell, 1979), chromosome condensation and formation of the intranuclear spindle may also take place. The chromosomes, however, stay univalent. After the nuclear envelope is destroyed they emerge into the cytoplasm, become picnotic and then disappear (zerfall). [Interestingly, the process of nuclear zerfall, the elimination of the previously amplified DNA and other nuclear material, was observed in the unilocular *Allogromia laticollaris* prior to cell division at the uninucleate stage of its life cycle (Parfrey et al., 2012)].After the meiotic process, the generative nuclei may undergo repetitive haploid mitoses, providing a huge number of young in the gametogamic species. Towards the end of the mass micronuclei divisions, usually after the repetitive mitoses, the macronucleus degenerates: it loses its clear outline and then disintegrates into separate fragments, which are later resorbed in the cytoplasm. The cytoplasm is arranged around the newly formed haploid micronuclei and then the shell is formed. These individuals represent agametes they then give rise to gamonts, the sexual generation, which in turn undergoes gamogony. The nuclei of the adult gamont divide (haploid mitoses) and give rise to gametes.

In its general features, the life cycle of *C. lobatulus* reflects the life cycles of the other evolutionary advanced foraminifera. According to the data of Voronova (1978 a, b), it differs from the latter only in the following features. At the gamogony stage the macronucleus is formed in the same way that it was at the agamogony stage. Sometimes its formation coincides with that of the first chamber of the shell but more often it appears at the 4 – 5 chamber stage. During the sequence of haploid mitoses of gamogony of the generative nuclei the macronucleus becomes band-like, rosary-like, beaded (Plate **17**, figure 3) and is carried by cytoplasmic currents to the 5th and6th chambers where parts may break away. At end of the gamogony mitoses it fully disintegrates. The agamonts of all the studied Rotaliana, Globigerinana and Miliolana species are multinuclear and heterokaryotic. Nuclear dualism of the gamonts was observed only in *Cibicides lobatulus*, as demonstrated by cytophotometry. It differs from the nuclear dualism of the agamonts in its origin at the haploid base. Despite the extreme complexity of investigations of nuclear processes in the foraminifera, and the difficulties involved in interpreting the various forms of data, studies on foraminiferal cytology and life cycles (especially those made by light microscopy and in some cases electron-microscopy) provide important information but nevertheless do not yield a full understanding of the main nuclear processes. For this reason, it is not possible to say whether some species are homo- or heterokaryotic, and at which stage of the life cycle they were at when studied. Formation of the nuclei of future gametes was not observed in most studies, which could be explained by significant methodological difficulties. First, before gamogony the cytoplasm is full of various inclusions, and as macronucleus occupies less than one millionth of the cell volume it is hardly distinguishable from the nuclei of the other cell organisms (food remains, commensals, symbionts). Second, individuals with decayed micronuclei or amicronuclear individuals may also occur in populations. Third, in general, there are often a number of degenerated nuclei of different origins. Finally, staining with methyl-green pyronine and ferrous haemotoxylin may mask the nuclei.

Voronova (1978 a, b) also speculated that the multinuclear gamonts may easily be mistaken for agamonts because of their morphological similarity in many species. The micronuclei of the agamonts are brighter and are identified more easily. Only Foelgen staining of full series of sections that are thin enough permits evaluation of all nuclei, while cytophotometry confirms which generation they belong to. But hard inclusions in the foraminiferal cytoplasm make it difficult to prepare a full series of sections, and cytophotometric investigations are rare. For this reason it is impossible to confirm some previous cytological observations, in the same way that the "chromidial" theory could not be confirmed (Le Calvez, 1938).

Initially it was considered that the generative nuclei are formed from the so called "chromidia", which exist in the cell along with the main nucleus. Using some methods for preparing nuclear sections, the nuclear apparatus of *Cibicides lobatulus* resembles a combination of the macronucleus and "chromidia", which in fact contain at least three components: micronuclei, the remains of the degenerated nuclei, and the remains of the nuclear material of foreign organisms (Voronova, 1978 a, b). In thick sections, all these components are indistinguishable. At present it is possible to conclude that in some cases the chromidial net represents the cytoplasmic RNA (the ribosomal net). The dense "chromidia" consisting of DNA are, obviously, just the generative nuclei, as well as the "spherical bodies" described in *Polystomella* (=*Elphidium*) (Lister, 1895). The observation of Voronova outlined above suggests that the nuclear dimorphism of the foraminiferal gamonts discovered in the gamonts of *Cibicides lobatulus* may be more widely distributed among foraminifera.

Investigating the nuclei and life cycles of the Nodosariata, which are practically unknown despite the peculiar shell structure and shell wall ultrastructure in this group, is a top-priority task. Such a study will significantly contribute to our understanding of foraminiferal life-cycles in general, and possibly may help to clarify the phylogenetic relations of this group. This illustrates how a new approach to foraminiferal classification may help to highlight the groups requiring urgent investigation.

The representatives of the naked or unilocular forms with an organic shell (Lagynana), of the unilocular Astrorhizana with an agglutinated shell and of the peculiar, structurally rather primitive Spirillinana represent homokaryotic foraminifera. In their life cycle usually the mononuclear gamont forms the nuclei of the multiple gametes, and multinuclear agamont has homogenous nuclei not differentiated into the vegetative and generative ones. In *Iridia lucida* (Le Calvez, 1936, 1953; Cezana, 1972, 1975, 1978) the agamont is mononuclear up to the beginning of the process of formation of young gamonts. In the other agamonts of Lagynana and Astrorhizana species, the period during which the agamont is mononucleate is very short but nevertheless exists, unlike species of Rotaliana in which the cycle is known, where practically only the zygote is mononuclear (Voronova & Mikhalevich, 1985). Goldstein (1988) notes that from the 40 – 70 nuclei of the *Saccammina alba* agamont, only 3 – 30 young gamonts are formed; the rest of the nuclei perish. She cautiously calls them non-generative. She did not discover any other distinctions between these nuclei and the generative nuclei. Here we obviously meet the same phenomenon noted earlier for some other foraminifera (for *Allogromia laticollaris* by Schwab, 1976, for *Patellina corrugata* by Berthold, 1971, for *Cibicides lobatulus* by Voronova & Mikhalevich, 1985) when part of the nuclear and cytoplasmic material perishes and is not directly used in the formation of offspring. In the planktonic species *Hastigerina pelagica,* these remains contained vacuoles, several nuclei and flagella, and they formed sphaerical bodies which were able to move (Spindler et al., 1978). In *Trochammina* cf. *T. quadriloba,* the mother individual was destroyed after the process of asexual reproduction, while in *T. hadai* it survived but was weakened and fragile (Kitazato & Matsushita, 1996). In *Cycloclypeus carpenteri* only a small fragment of the agamont shell that contained some cytoplasm at the end of multiple fission was able to regenerate a regular shell and even to reproduce again (Krueger et al., 1996). Arnold (1984) investigated in detail features of the formation of the generative nuclei in the gamont of *Psammophaga simplora.* The comparison of this stage of the life cycle in *C. carpenteri* with the analogous phase in the heterokaryotic *Cibicides lobatulus,* and also with the heterokaryotic agamonts of some other species, shows that gamogony of *P. simplora* nevertheless could not be attributed to the heterokaryotic type since (as the author indicates) the generative nuclei formed in the primary nucleus replace this primary nucleus successively but do not coexist simultaneously with it. For most of its life, the *P. simplora* gamont has only one nucleus. After the generative nuclei escape the nuclear envelope, the latter becomes practically devoid of chromatin material and for some time continues its existence as a remnant of the full nucleus. Here again we encounter the incomplete utilization of the nuclear material during formation of offspring. In true heterokaryocity during the main part of the agamont phase (and in *Cibicides lobatulus* obviously during its gamont life also) the vegetative and the generative nuclei coexist simultaneously. Arnold's investigations show once more the complexity and variability of the nuclear changes in the course of the foraminiferal life cycle and the necessity of new studies. Nevertheless, analysis of the presently available data on the life cycles of the primitive unilocular foraminifera does not reveal heterokariocity but only the multiple nuclei of their agamont stage. *Phthanotrochus arcanus* (Arnold, 1978) represents an exception among the Lagynana. The author identifies, though with some doubt, its bigger degenerative nuclei (8 – 14 μm) as vegetative nuclei. They coexist simultaneously with a multitude of small (1.6 – 2.4 μm) generative nuclei.

Unlike other organic-walled Lagynana, the shell of *Phthanotrochus arcanus* has a multi-chambered trochoid structure of the Rotaliata type. If this species really exhibits nuclear dimorphism, then this would confirm its close relationship to species of Rotaliata with morphologically similar secreted calcareous shells, and cast light on the origin of the latter group. In this case, a taxonomic placement of *P. arcanus* closer to the lower representatives of the agglutinated Rotaliata (i.e. members of the subclass Textulariana) appear to be justified. Arnold's (1979) suggestion regarding the origin of the higher secreted foraminifera from forms with multi-chambered organic shells, which he elaborates with the examples of *Periptigma* and Miliolata, will receive further support.

This review of current knowledge concerning foraminiferal life cycles suggests that they can be divided into three groups:

1. with homokaryotic multinuclear agamonts and uninuclear gamonts (Lagynana, Astrorhizana, Spirillinana), (with *Iridia lucida* as an exclusion – uninuclear agamont);
2. with heterokaryotic agamonts and uninuclear gamonts (Miliolana, majority of the known Rotaliana, Globigerinana);
3. with nuclear dualism of both agamonts and gamonts (some Rotaliana: *Cibicides lobatulus*, possibly *Elphidium*).

It is clear that the first group represents the initial type, as confirmed by the works of Grell and his school (Weber, 1965; Grell, Bardell, 1977; Grell, 1979) in which the different phases of the differentiation of the somatic nuclei were observed. In the representatives with a lesser degree of nuclear differentiation the macronucleus not only enlarges in size in the course of meiosis but also undergoes chromosome condensation (while remaining univalent) and even the formation of a spindle (*Rotaliella roscoffensis*), as if repeating but not finishing the processes occurring in the micronuclei. At the more advanced stages of nuclear differentiation the macronucleus merely increases in size and in the DNA quantity that it contains (*R. heterocariotica*). Probably, this increase of the macronucleus DNA in parallel with the increase in the number of micronuclei recapitulates the evolutionary origin of the polyploid nucleus. In micronuclei this enhanced quantity of DNA is divided during the process of fission. In macronuclei on the contrary, the process of fission is not accomplished and the somatic nuclei remains polyploidizated.

Cibicides lobatulus provides an example of a species in which the differentiation of the nuclei is more advanced (Voronova & Mikhalevich, 1985).

The process of gametogenesis has its own special features in different foraminiferal groups. At the start of the reproductive process, some species build a reproductive cyst from the weakly cemented foreign material (*Cibicides lobatulus, Rosalina globularis, Trochammina inflata*). These fertilization (reproductive) cysts are similar to those build by *R. floridana* during the formation of new chambers. In the unilocular *Cribrothalamina alba,* 10-15 rather big openings are formed in its shell wall, these "pores" give the gametes the opportunity to leave the shell (Goldstein & Barker, 1988). Similar enlarged pore openings are formed during gamogony process in *Heterotheca lobata* (Grell, 1988). In another monothalamous foraminifera, the soft-walled finely agglutinated species *Xiphophaga minuta,* gamete were released through the aperture. In the benthic Rotaliata gametes may exit through pores widened by resorbtion (in *Cibicides lobatulus* pores widen with the beginning of the mitoses of gamogony) and also through the aperture (*Ammonia tepida, Trochammina hadai*) or through the openings of the canal system (in *Cycloclypeus carpenteri* – Krueger et al., 1996). In planktonic forms the gametes emerge through the wide apertures. The spherical chamber of *Orbulina universa,* which serves as a protective envelope enclosing the gametes, provides for their exit through the "large pores", which functionally are multiple circular apertures (Spero, 1988). The pelagic forms have some other peculiarities. Their release of gametes may be connected with the Moon cycles, for example, in *Hastigerina pelagica* (Spindler et al., 1978) and in *G. sacculifer* and *G. ruber* (Bijma et al., 1990) (lunar periodicity has also been reported for benthic *Trochammina hadai* by Kitazato & Matsushita, 1996). A link to the day/night cycle was also observed by Kitazato & Matsushita (1996). The shells of planktonic foraminifera also undergo significant changes. As a result of secondary gametogenetic calcification, the alveolar ultrastructure of the surface becomes smoothed

(Plate **12**, figure 4b; Plate **13**, figure 4b), the wall becomes thicker, a veneer of calcite is deposited over the terraced edges around pores, which become sealed nearly completely with calcareous material, the shell discards or resorbs spines (Plate **8**, figure 9; Plate **16**, figure 1) and settles to the seafloor. This process was described in detail by Be et al. (1976) and Be (1980). There are also some species in which the last reproductive chamber has a special form (*Globigerinoides sacculifer* (Plate **16**, figure 1). [Brood chambers may also occur in some benthic forms, e.g. *Sorites orbicularis* (Kloss, 1980) and *Cycloclypeus carpenteri* (Krueger et al., 1996), but their mode of formation is different (see below, this Part)].

Among benthic forms, there are some species with a short pelagic stage in their life cycle. For example, the floating gamonts of *Tretomphalus* (Plate **2**, figure 10 b), which have subsphaerical chamber beneath the shell, are good examples of such cycles. Their agamonts, which lack a floating chamber, were for a long time attributed to the genus *Rosalina.* The gamont of *Tretomphalus bulloides* encysts during the period of gamete formation. Before the release of the gametes it forms a big floating chamber filled with the gas beneath the trochoid shell. It then free itself from the cyst and floats to the water surface where the biflagellate gametes leave the mother shell through the multiple circular openings beneath this sphere and fuse in the water. Finally, the resulting zygotes sink to the bottom, completing the pelagic stage of the cycle. Recently, the presence of a pelagic stage of the benthic species *Bolivina variabilis* was documented by both morphological and molecular methods (Darling et al., 2009, Darling & Leckie, 2010). The pelagic stages of benthic foraminifera promote the wider distribution of species (Mikhalevich, 1978, 1983; Voronova & Mikhalevich, 1985). In agglutinated forms wide dispersal may be facilitated by the absence of an agglutinated covering of the chamber at the proloculus stage (Alve & Goldstein. 2010). These authors also showed the important role of propagules in dispersal of the benthic foraminifera.

The presence of two nuclear phases in the foraminiferal life cycle manifests itself not only in their cytological characteristics but also in their shell structure. As in other organisms with a similar alternation of generations, it may be isomorphic or heteromorphic, since both forms evolve more or less autonomous. The agamonts of the majority of foraminifera have a microsphaeric shell, while in the gamonts it is megalosphaeric, except in Spirillinana where the relationship is reversed. This inverted correlation of the size of the asexual and sexual generations is known also in *Glabratella sulcata, Heterotheca lobata*, *Saccammina alba* (although not in all the species of the latter genus) (Grell, 1988, Goldstein, 1988, 1999). In the simplest case, the dimorphic generations differ only in the size of their initial chamber (proloculus), as reflected in the names of the shell types, as well as in the size of the entire shell; i.e. the microsphaeric agamonts are usually bigger than the macrosphaeric gamonts. *Cibicides lobatulus, Rotaliella heterocariotica, Ammonia beccarii, Entzia (Jadammina zernovi)* and many other may serve as demonstrative examples, some of them described on the Plate **22**. If the agamic generation is repeated twice, the microsphaeric agamonts and the macrosphaeric schizonts are formed as a result (for instance in *Elphidium crispum* and *Heterostegina depressa*). In some highly organized foraminiferal groups with very high numbers of chambers (e.g. Nummulitacea, subclass Rotaliana – *Biplanispira, Cycloclypeus*) the two initial chambers of the megalosphaeric gamont (the proloculus and the second chamber – the deuteroconch) are rather big, sometimes significantly bigger than the other chambers (Plate **9**, figure - s 1, 2, A, B). Both chambers are encircled by a common nonperforate wall, which differs from the wall of the rest of the shell. These first large chambers, termed embryonic chambers, have one big central subcircular pore and the row of pores at the base of the septa. In *Cycloclypeus carpenteri* the difference in the dimensions of the shells of the two generations is striking: 4.5 cm in agamont and only 6.3 to 9.2 mm in gamont shell (Krueger et al., 1996). Embryonic chambers are also known in the large shells of Miliolata and Spirillinata.

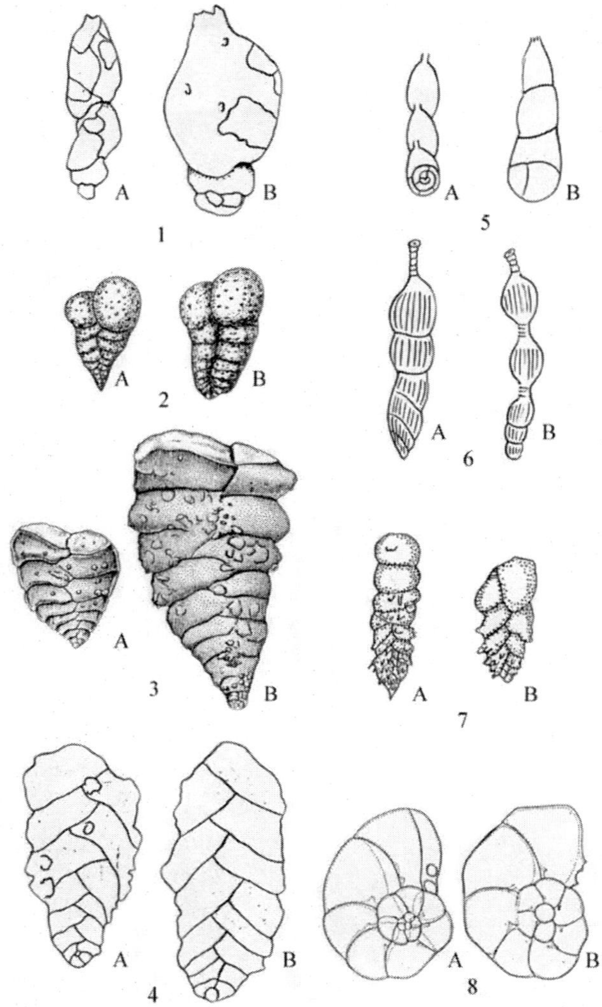

Plate 22. Differences in the structure of the micro – (A) and megalospheric (B) shells of the same species.
1 A, B - in Hormosinana (*Reophax bermudezi*), 2 – 4 - in Textulariana (2A, B - *Textilina*, 3A, B – *Norvanganina pseudorugosa fistulosa*, 4 A, B – *Textella foliacea occidentalis*), 5 - 6 - in Nodosariana (5 A, B – *Dimorphina tuberosa*, 6 A, B – *Amphicoryna scalaris*), 7 – 8 - in Rotaliata (7 A, B – *Rectobolivina advena*, 8 A, B – Entzia (*Jadammina) zernovi*).

There are also good examples of heteromorphism among agglutinated foraminifera, as well as the secreted calcareous species (Plate **2**, figure - s 13 a, b, 14 a , b; Plate **22**, figure- s 1- 8) (Mikhalevich, 1983, p. 142, figure – s 244 a, b; Voronova & Mikhalevich, 1985, figure– s 26 a – m, 27 a – t, 28 a – k, 29 a – d, 30 a - p). The shells of the different generations may differ in their outlines, their general proportions, the ratio of their parts, or in the presence or absence of some features. For example, in linear, elongated forms an initial planispiral, trochospiral or glomerate part may be present in the agamonts but absent in the gamonts. The microsphaeric asexual generation very often repeats the more ancestral features of shell morphology. Thus, the ontogenetic features of the shell cast light on the phylogenetic development of the group.

The processes of agamogony and gamogony, and their duration and timing (season, day-time), vary in different taxonomic groups. Krueger et al. (1996) notes some common features in *Heterostegina depressa* and *Cycloclypeus carpenteri* and suggests that they apply to nummulitids in general. In many distantly related taxa, such as *Peneroplis* among the Miliolana (Winter, 1907) and nummulitids among the Rotaliana (e.g. Krueger et al., 1996), the septae between the initial chambers of the agamont became resorbed, forming the brood chamber. The tiny young gamonts may be naked at the stage of the very first chambers and begin building their shell at the three-chamber stage (Krueger at al., 1996). In *Trochammina hadai,* young gamonts consisting of only one chamber when they leave mother cyst begin to build their agglutinated shell at this one-chamber stage (Kitazato & Matsushita, 1996). However, in *Cycloclypeus carpenteri* the young gamonts with three-chambered shells are formed from the cytoplasm on the surface of the mother shell. In some cases the absence of a hard shell may facilitate the exit of the offspring from the parent shell.

VI.6. Gamete Structure

In the overwhelming majority of the foraminiferal species the gametes are biflagellate, possessing two flagella of unequal length (Plate **18**, figure– s 3 – 4; Plate **21**, figure 1, Plate **31 a**, figure– s 13 - 17). They contain the nucleus, fatty drops, mitochondria, osmiophil granules, microbodies, dictiosomes of the Golgi apparatus, and in species with selective sexual activity – the acrosome (Cesana, 1972) (thus the presence of acrosome in *Iridia lucida* proves the existence of sexuality in this species. Differentiated sexual gametes are known in *Patellina corrugata* (Grell, 1958).

The gamontogamic species *Glabratella sulcata, Discorbis patelliformis, D. mediterranensis, D. opercularis* (Myers, 1940, Le Calvez, 1950, Grell 1958), and the gametogamic species *Nummulites venosus* (Roettger et al., 1998), have triflagellate gametes. Gametes with three flagella do not occur in sarcodins and are extremely rare in the protozoa in general, being known only in some Sporozoa. In forms with the amoeboid gametes (*Patellina corrugata, Spirillina vivipara, Metarotaliella simplex*, some *Rotaliella*) the absence of the flagella is obviously of secondary origin. They may be lost because they are not necessary for the gamontogamic mode of reproduction. The presence of both amoeboid and biflagellate gametes in the representatives of the primitive soft-walled unilocular Lagynana (both with gametogamic way of reproduction) needs further study along with more detailed investigations of their cytoplasmic peculiarities.

The gametes of some miliolids have been described as containing a haptonema-like structure resembling a straight prominent axostyle; however, its ultrastructure has not yet been studied. Recent investigations of the gametes of *Triloculina oblonga* (Goldstein, 1997) revealed no trace of such a structure, and they were not observed in the gametes of *Miliolina* sp. either (Plate **31** a, figure - s 13 – 18). In *Boderia turneri* (Hedley et al., 1968),

Amphistegina gibbosa (Dettmering et al., 1998) and *Operculina* cf. *complanata* (Roettger et al., 1998), one flagellum has superficial hairs, termed mastigonemes. These hairs are also developed in many classes of unicellular algae, fungi and mastigophora; in each case they differ in their fine structure and their position on the flagella. The ultrastructure of the mastigonemes in foraminifera has not been studied yet.

The biflagellate state in protists is usually considered as being primitive (Lynn & Small, 1981). Previously it seemed that the presence of three flagella on the gametes of glabratellids could be used as a taxonomic feature. In Sporozoa the presence of one, two or three flagella is used (along with the other morphologic details) as a taxonomic feature of classes and sometimes families, for instance of the family Eneriidae (Krylov, 1996, p. 134). However, the discovery of triflagellate gametes in *Nummulites venosus,* together with the existence of biflagellate gametes in most other species of this genus, undermines their taxonomic utility. Thus, although different types of flagella are present in different foraminiferal taxa, they cannot be used as diagnostic characters.

Foraminiferal flagella have the usual structure (2 central and 9 pairs of the peripheral fibrils), but their ultrastructure (e.g. root systems, kinetosomes (basal bodies), paraxial structures and some others) have not yet been studied. Although some of these elements may be reduced or simplified in specialized cells such as gametes, nevertheless some useful information might be obtained. The general scheme of the organization of the root system is also generally preserved in gametes, and similar information in other unicellular eukaryotic groups is widely used in their taxonomy and phylogeny.

Chapter VII

OBSERVATIONS ON SOME RECENT MILIOLIDS IN CULTURE

Till now, only a few of the approximately five thousands of the recent foraminiferal species were studied in culture or simply observed in living state. As far as it is known, less than twenty of them belong to the class Miliolata Saidova, 1981: *Nubecularia lucifuga* Defrance 1820 (in Le Calvez, 1953; Arnold, 1967), *Triloculina* (in Dujardin, 1835; Schultze, 1854), *Triloculina rotunda* d'Orbigny, 1826 (in Le Calvez, 1935, 1938), *Triloculina oblonga* (Montagu, 1803) (in Goldstein, 1997), *Triloculina barnardi* Rasheed 1977 (in Collen, 1998), *Calcituba polymorpha* Roboz, 1884 (in Schaudinn,1895, Berthold, 1976; Arnold, 1967), *Spiroloculina hyalina* Schultze 1875 (in Arnold, 1964; Angell, 1980), *Quinqueloculina fusca* Brady, 1870 (in Schneider, 1878), *Quinqueloculina annectens* (Schlumberger, 1893) (in Hofker, 1930a), *Quinqueloculina suborbicularis* d'Orbigny, 1826 (in Le Calvez, 1935, 1938), *Quinqueloculina seminula* (Linné, 1758) (in Le Calvez, 1953), *Miliolina circularis* (Bornemann, 1855) (in Hofker, 1930b), *Peneroplis* (in Dujardin, 1835; Schaudinn,1894, Dreyer, 1898, Langer et al., 2009), *Peneroplis pertusus* (Forskal, 1775) (in Le Calvez, 1935, 1938, 1953, Winter, 1907, Faber & Lee, 1993) *Sorites marginalis* (Lamarck, 1816) and *Archaias* sp. (in Muller-Mertz & Lee, 1976), *Marginopora vertebralis* Quoy et Gaimard, 1830 (in Ross, 1972, Lee et al., 1997) and *Marginopora* sp. (in Lee et al., 2009).

First detailed observations were carried out by Le Calvez (1935-1953) and Arnold (1964, 1967). Among more recent studies, Muller-Merz & Lee (1976) and Goldstein (1997) have stressed their studies on the life cycle and the reproductive process, while Angell (1980) devoted her study to the process of chamber building.

The miliolid *Massilina secans* (d'Orbigny, 1826) had never been a subject of such study. It belongs to the subfamily Massilininae Thalmann, 1941, family Massilinidae Thalmann, 1941, superfamily Quinqueloculinoidea Cushman, 1917, order Miliolida Delage & Herouard, 1896. It is a shallow-water tropical, subtropical and even temperate species widely distributed in the Atlantic Ocean.

The material was collected from the littoral off Yeu Island, Biscay Bay, on the 27 of March1998. Living specimens of *M. secans* were transported in plastic shell tubes with natural sea water to the laboratory of geology of Angers University (France) directed by the Prof. J.-P. Debenay. Twenty specimens were selected with a brush and put into watch glasses. In these glasses each of the specimens were washed several times in distilled water and later in microfiltered seawater coming from Yeu Island. The water was dropped into the glasses and removed by means of a pipette. After washing, the specimens were put into two Petri dishes, 19 cm in diameter, containing microfiltered seawater. During the observation processes, some specimens were transferred for some hours or days into small glass crystallizing jars, 70 mm in diameter, also containing microfiltered sea water. The rather large size of the specimens (from 0.55 to 1mm in length) permitted them to be easily selected with a brush. Salinity was checked every week with a hand refractometer (Atago S10), and regulated by the addition of distilled water, or by renewing the whole volume of water when necessary. During very hot days and after very long observations under strong lamps which accelerated the evaporation, it was necessary to check the salinity more often (every two or three days). The cultures were incubated at 20 °C. The incubator was subject to a vertical lighting of 2 x 20W, twelve hours every day.

Each two or three days the *Massilina* were given a drop or two of dead diatom (*Dunaliella*), by means of a pipette. The methods of this culture only slightly differ from those described by Stouff et al. (1999). The most interesting observations were photographed using a Nikon Coolpix camera.

VII.1. Observations on *Massilina Secans* (d'Orbigny), 1826

The living specimens of *Massilina secans* (d'Orbigny), 1826 were observed in culture during two months, from March 23 to May 31, 1998. Twenty specimens were shared into two Petri dishes. The color of living cells was orangy-pink, with different degrees of brightness. The last chamber was more often devoid of cytoplasm and white. It was sometimes yellowish or rosy, but paler than the central part. Dead specimens had a white shell although there were cases when, after being for some time white, the shell became rosy again.

Nearly half of the specimens had partly or fully deformed chambers that result from mechanical damages during their transport in the shell tubes.

VII.1.1. Reticulopodia, Position of the Shell and Movement

A small amount of cytoplasm, which is sometimes visible in the apertural area (Plate. **23**, figure - s 1, 2, 3; Plate **29**, figure – s 2, 3), as well as reticulopodia (Plate **23**, figure - s 4 - 10; Plate **25**, figure 1), are colorless. It was rather difficult to fix and photograph the reticulopodia due to their translucency in the water and against the glass. The reticulopodia of the observed specimens strongly varied in length, but were usually rather short, not exceeding 2.5 times the size of the shell. They were only slightly branching. They generally formed one or two bunches (Plate **23**, figure - s 4-6), but were sometimes distributed all around the aperture (Plate **23**, figure 10) or all around the shell (Plate **23**, figure 9). In some cases the cell formed a very thin single reticulopodial thread exceeding the size of the shell up to ten times (Plate **23**, figure - s 7, 8), a flat band, or a rather thick single reticulopodial trunk. Later the same cell may form a bunch of branching reticulopodia. Reticulopodia may not only spread on the bottom but may also go upwards in the water. When the Petri dish was shaken the reticulopodian threads curved and waved in the water resembling flagella movement. The individuals with the protruded reticulopodia were either lying on their lateral sides (Plate **23**, figure- s 7-10) or standing vertically-obliquely with their aperture turned to the bottom (Plate **23**, figure11; Plate **25**, figure 1), in the later position they could move with a velocity of 4 cm an hour or even more. The direction of their movement is opposite to the emission of reticulopodia and it looks as if the protist moves back. Very rarely the shell stands on its base with its aperture turned up. Some individuals climbed up the vertical boarding wall of the Petri dishes and then got over the water, moving along its surface by extruding the reticulopodia. Dujardin (1835) had also watched living milioline foraminifera climbing the wall of his vessels. In our case the *Massilina* specimens could stay motionless on the water surface for some time and even for some days. They built a small cyst close to their apertural area (Plate **23**, figure 12; Plate 26, figure 9) and stayed in the slightly deepen water surface like in a hammock, swaying with the shaking water. Individuals with a broken shell were also capable to float on the water surface. Some of the floating specimens could sink deeper into the water and stayed hanging, attached to the water surface from beneath by their aperture. Their reticulopodia gathered food from the water surface. Their cytoplasm and the tooth were visible from above (Plate **23**, figure - s 13, 14). It was possible to see how the cell moved its short branching reticulopodia and just at that same moment the shell twitched and sharply went down. It may sink to the bottom at once or swim slowly some time in the water mass. The massilins began to climb up the dish wall only when nearly all the food at the bottom was eaten. They could feed on some residual food particles remaining on the water surface. Only some individuals used to rise up to the water surface; the majority of them have never been observed to try to do so. Such individual behavior at the unicellular level was surprising and inclined me to accept the theory of panpsychism.

When they moved on the bottom, the cells left tracks on the glass, made up of a thin layer of light organic matter (Plate **24**, figure - s 1–6). Langer & Gehring (1993) had described such organic tracks left by *Textularia bocki* and *Quinqueloculina ungeriana*. They supposed that this organic matter is used as a food for bacteria, which in turn are eaten by the foraminifera. They called farming this mode of relationship. After a massilin had stopped for some time at the same position, the reticulopodia may have left linear tracks looking like rays around the shell (Plate **24**, figure - s 3, 4).

VII.1.2. Feeding

Every two or three days the *Massilina* were given one or two drops of heat-killed *Dunaliella*. Initially *Amphypora*, or more rarely *Chlorella* had been given, but it turned out that *Dunaliella* were preferred. Just after the food was dropped, the massilins lay almost motionless on the bottom before becoming more active, extruding reticulopodia and gathering food particles, or moving in the direction of the new food. In the latter case, they stood up almost vertically on the bottom with their reticulopodia more often dividing into two butches. Sometimes they stood on the heap of diatoms, which were consumed and disappeared within a day. Food particles, looking like small dots (Plate **25**, figure 1) or rarely bigger lumps, could be distinctly seen, streaming along the reticulopodia in the direction of the aperture. After three-four seconds of movement from the end of the reticulopodia, these particles reached the aperture. In a short time a whitish mass of dead *Dunaliella* accumulated outside, at the entrance of the aperture (Plate **25**, figure 2). It gradually became dark or bright green (Plate

25, figure - s 3, 4). Such green accumulations disappeared after one to three days. In those specimens with strongly damaged chambers (Plate **25**, figure 5) such green mass was deposited along all the broken area of the shell (as if it was eating with its side). It is difficult to say definitely whether this green mass was a process of digestion or its result - containing the undigested residues - but the food particles streaming along the reticulopodia were always whitish, never green. With our microscope it was not possible to determine whether these particles moving along the reticulopodia were on the reticulopodial surface or inside them, but most likely the picture was similar to that one photographed and showed by Bowser et al., (1992) where prey organisms were carried along the reticulopodia adhered to a fine array of extracellular filaments (Bowser et al., 1992, figure –s 8, 9, 12).

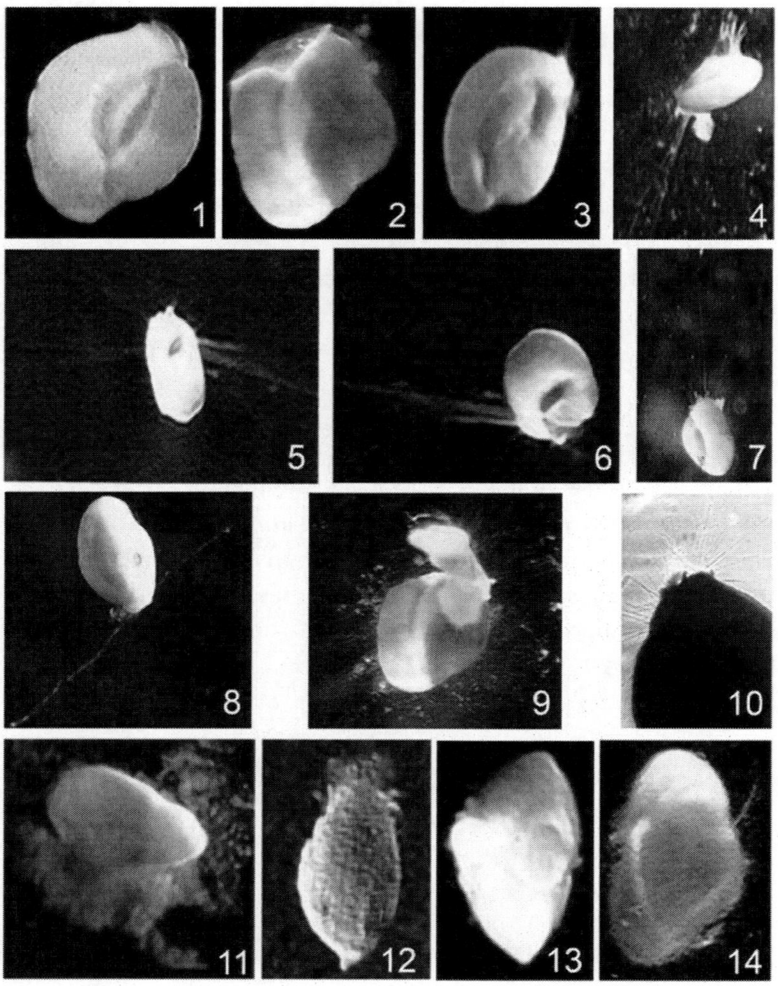

Plate 23. Intact and broken living specimens of *Massilina secans* showing cytoplasm and/or pseudopodia. 1 - intact shell; 2 - shell with a broken last chamber showing the tooth and the partly extruded cytoplasm; 3-6: reticulopodia of different length in bunches; 7 - rather thick straight pseudopodial trunk; 8 -long singular thread-like reticulopodia; 9 - reticulopodia all around the shell; 10 - reticulopodia all around the aperture; 11- specimen standing vertical-obliquely to the bottom; 12 - broken shell on the surface of the water film; 13, 14 - specimens attached to the water film from beneath.

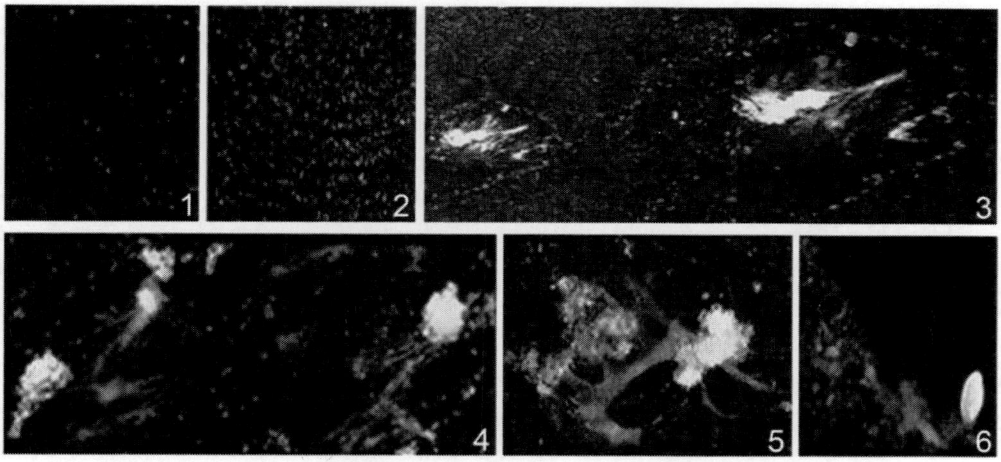

Plate 24. Protist traces (*Massilina secans).*
1- first traces on the nearly clean bottom of the dish; 2 - multiple traces of the individuals; 3, 4 -thick and wide traces of organic matter and matter contouring the previous position of the shell; 5 - thick traces of organic matter; 6: traces of a shell moving in the vertical-oblique position (shell to the right).

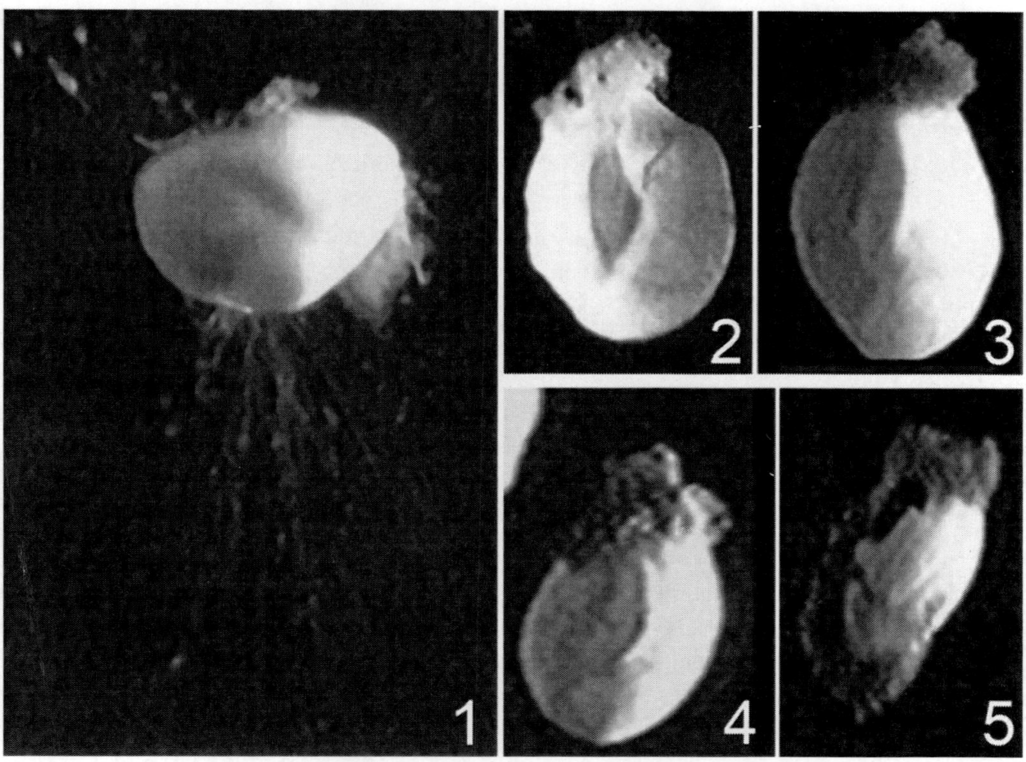

Plate 25. Feeding of *Massilina secans*.
1 - specimen in the vertical-oblique position to the bottom, food particles are visible streaming along reticulopodia in the direction of the aperture; 2 - whitish plug of dead Dunaliella gathered by the reticulopodia in the apertural area; 3, 4 - plug of food in the aperture getting dark green; 5 - dark green border along the open cytoplasmic area of the broken shell.

VII.1.3. Cyst Building

Cyst building is widely known among foraminifera. They construct their cysts prior to reproduction or during the processes of creating new chambers. Feeding cysts are also known. In our culture we could observe the process of cyst building from the very beginning (Plate **26**, figure - s 1–8). Firstly, a single translucent thread appeared in the aperture (Plate **26**, figure 1). We don't know whether this thread resulted from a secretion of the cell or if such very thin and delicate threads may fall into the clean culture dish from the air, then being drawn towards the aperture by the reticulopodia. However, most likely they are excreted by the cell, as their concentration in the apertural area with their lower ends placed in the apertural neck gives more reasons to consider them to be produced by the organism. Location of the cyst around the aperture, at least at the initial stages of its building, also shows the active role of the cytoplasm in the process. At the next stage, some more threads of different length were then added by the foraminifera that interweaved the previous ones to make a frame (Plate **26**, figure 2). The process resembles to some extent the building of the nest by a bird. Afterwards, the cytoplasm extruded some glue matter, which sticks together the thread frame and other material, mostly detritus particles gathered by the cell from the bottom (although there are a few of the foreign particles in the clean culture). It could be easily seen how the adhesive matter came out from the aperture, solidified and stretched. The cyst progressively covered all the apertural part of the shell (Plate **26**, figure - s 3, 4) and gradually extended over the whole shell (Plate **26**, figure - s 5–7). It was thick, soft and friable, of indefinite outline, in some parts it was half-translucent which permitted the contour of the shell to be distinguished, but mostly the shell was not visible. Sometimes it was possible to see that the individual changed its position inside the cyst several times from side to side. The cyst was usually slightly greenish-yellowish in color, sometimes singular threads piercing it were visible. Dark claret-colored spots could be seen somewhere in the cyst, which were presumably bacteria (Plate **26**, figure– s 3–7). The massilins floating on the water surface could build a cyst also (Plate **26**, figure 9). After their sinking to the bottom, the remains of their cysts stay on the water surface for a long time. The empty cysts were also scattered here and there at the bottom of the dish. The process of construction of a cyst could take several hours but sometimes the cyst was practically growing before our eyes. The cell may live in the cyst during several days, afterwards it turns inside it and comes out. Deep holes could be seen in the empty cyst, marking the space of the previous position of the shell in it (Plate **26**, figure 8). All the cysts observed in the culture could be considered as feeding cysts and cysts for chamber building. Once, the protist built its cyst and subsequently went out of it with a new chamber five hours later. Feeding cysts were observed in many foraminifera earlier (Hofker, 1927; Nyholm, 1957; Goldstein & Corliss, 1994).

We did not notice reproductive cysts in the culture of *Massilina*. Once, a cloudy material came out from the aperture of one of the *Massilina*'s shells without a cyst, and small dots fell to the bottom around the apertural area. It looked like a reproduction process, but no young *Massilina* developed afterwards. As there were claret-coloured dots inside this cloudy material, it is possible to assume that it was the excretion of bacteria or may be juveniles infected with bacteria. However, the nature of this event remains unclear.

The *Massilina*'s cysts observed under this study were built in a clean culture in the absence of the sediment. The whitish color of the cyst obviously was due to the use of the food particles, algae frustules, as constructive material in the absence of sediment. In a natural environment these cysts would be the sooner completed with the inclusion of the sand grains on their surface. The chemical or electron microscopic study of the cyst was not carried out. Such studies in future and comparison of its results might throw light on the composition of the cyst material and the nature of the theca (cyst? shell?) of *Astrammina rara* studied in detail by Bowser et al., (1995). The multiple empty cysts of *Massilina* were abundant in the culture resembling on this point the empty *Astrammina*'s "shells". The main difference is that in *Massilina* the activity of pseudopodia is located inside the cyst, not forming a special extrusion ("arm" in *Astrammina* – Bowser et al., 1995). Moreover, in cysts of *Miliolina* sp. (see below – Part VII. 2.1) formed around the young first chamber (proloculus) the long outgrowth was formed giving the cyst the pear-like form (Plate **31 a**, figure – s 12 a, b). Formation of such outgrowth obviously was connected with the necessity of the growth of the next tubular chamber – flexostyle. Such cysts may have a protective role similarly to the cysts formed during the new chamber formation process.

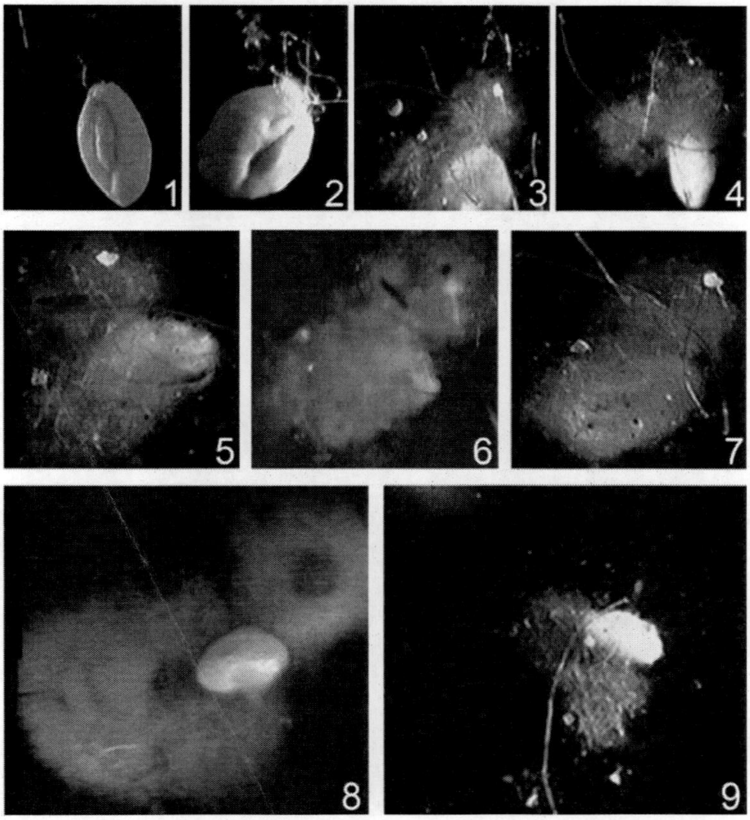

Plate 26. Cyst building of *Massilina secans* specimens.
1-7 - stages of the cyst building of one individual: 1 - the beginning of the process with only one thread attached to the aperture; 2 - several threads attached forming the frame of the future cyst; 3 - organic matter secreted by the cytoplasm and detrital material added to the frame of the cyst; 4: cyst covering the apertural area; 5: growing cyst covering 2/3 of the shell; 6, 7 - the final cyst covering the whole protist (notice the dark claret-coloured matter); 8 - two halves of the cyst after this protist left it, the space of the previous position of the shell can be seen; 9 - initial part of the cyst of another specimen laying on the water film (the reticulopodia spread along the water film and gathering food particles are not visible on the photograph but clearly seen in microscope).

VII.1.4. Reaction to Some Environmental Factors

We have already mentioned the immediate reaction of the protists to the given food. As *Dunaliella* was provided as dead and motionless material, it must have produced a solute or solutes that attracted the foraminifera.

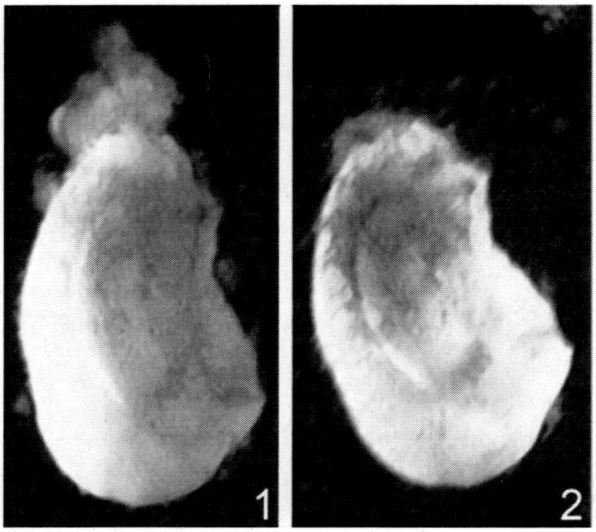

Plate 27. Surviving of *Massilina secans* in hypersalinity. 1 - individual with a broken shell in water of 45 ‰ salinity with its cytoplasm partly extruded, 2 – the same individual in normal salinity with its cytoplasm intruded back into the shell.

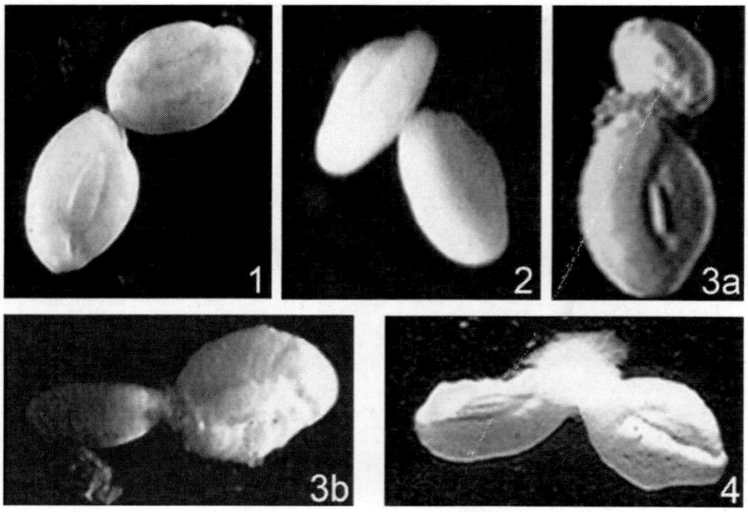

Plate 28. Contact between the shells of *Massilina secans*.
1, 2 - two different pairs of individuals, one of them touching the surface of the other by its aperture; 3a, b - two stages of the approaching process of one pair of individuals: 3a - contact between one aperture and the side area of the other individual; 3b - contact between the apertures (notice the dark green plug of food in the apertures); 4 - contact by the apertures between an undamaged specimen and a specimen with a broken side (the later was living till the end of experiment, the first one died earlier).

The massilins had a negative reaction to the strong light of the microscope lamp and they moved rather quickly from the center of the dish to its shadowed margins, where they preferred to stay. If they were brought back into the center of the dish exposed to the strong light again, they immediately and rather quickly moved to the shadowed margins of the dish. When as a result of evaporation the salinity increased up to 45 and even 65 per mille, the cytoplasm was extruded outside of the shell through the aperture (Plate **27**, figure 1), with sometimes nearly all the cytoplasm being extruded. When salinity was changed back to normal conditions (Plate **27**, figure 2) the cytoplasm was intruded back and the protist continued as normal.

VII.1.5. Contacts between Individuals

Some of the observed shells moved to meet each other and stayed in contact for up to several hours (Plate **28**, figure 1-4). In some of such couples, one of the shells touched by its aperture one of the lateral sides of the other (Plate **28**, figure - s 1, 2). Many of the shells were covered externally with very faint slightly greenish, slightly fluffy coating (may be algae or bacteria), and it looked as if one of them was eating this coating from the surface of the other, thereby cleaning it. This position should also be a transitional stage before a contact between the apertures (Plate **28**, figure 3a), which lastly happened (Plate **28**, figure3b). In one case, the two shells touched each other by their sides and 15 minutes later one of them was touching the other's side by its aperture. In the case photographed in Plate **28**, figure 3b, each shell of the pair had a dark green plug near their aperture formed after their previous separate feeding. In the other case (Plate **28**, figure 4), there was a cytoplasmic matter between them (Plate **28**, figure 4). Sometimes we saw three shells joined by their apertures. We don't know what happened between these "kissing" individuals – was it some kind of sexual exchange or cannibalism? Sometimes one of the paired specimens was left empty, sometimes both continued to live. In the case which was photographed in figure 4 (Plate 28), the broken shell stayed living till the end of the experiment (the end of May), while the intact shell died soon. The reason of its death could be the bacteria growing on its surface (claret-coloured dots). In some cases, both shells remained rose after their separation. More often one of them became white and soon died, but one of such white individuals which we considered dead during 10 days suddenly extruded its reticulopodia and began to eat actively.

VII.1.6. Infection

Shells of many of the specimens studied in our culture had on their surface dark claret-coloured marks that look as small dots or somewhat bigger spots made by some parasites, supposedly bacteria (Plate **23**, figure 8; Plate **29**, figure- s 1–6, 8, 9; Plate **28**, figure 4; Plate **30**, figure 6). In one specimen with such a spot (Plate **30**, figure 6) it was possible to observe the growth of a soft claret-coloured thread. Progressively, other threads grew, two or more of them sticking out of the shell (Plate **29**, figure - s 1–4). Later, deep winding borrows growing from such dark claret-coloured spots destroyed the surface of the shell (Plate **29**, figure 6) and finally the cell died. Nevertheless, the individual photographed in figures 1–4 (Plate **29**) and figure 6 (Plate **30**) stayed living till the end of the observations, and, at the initial stage of this disease, it was capable of building a new chamber (Plate **30**, figure6) while the individuals with multiple small dots died earlier (Plate **29**, figure 9). At the end of the experiment, only nine of the twenty initial specimens remained living. On some other shells there was a thin greenish coating with circular deepened areas (Plate **29**, figure 7). Such specimens were less active and their behavior seemed disrupted. All these features of parasitical damage appeared during the observations in culture, as initially the shells put into the dishes were clean.

VII.1.7. Broken Shells, Chamber Restoring and Building of a New Chamber

As it was written above, several of the twenty observed specimens had chambers damaged mechanically during the transport from the Bay to the Laboratory. Some of them can be seen in the photographs (Plate **23**, figure 2; Plate **23**, figure 11; Plate **29**, figure - s 8, 9; Plate **30**, figure - s 1-6).

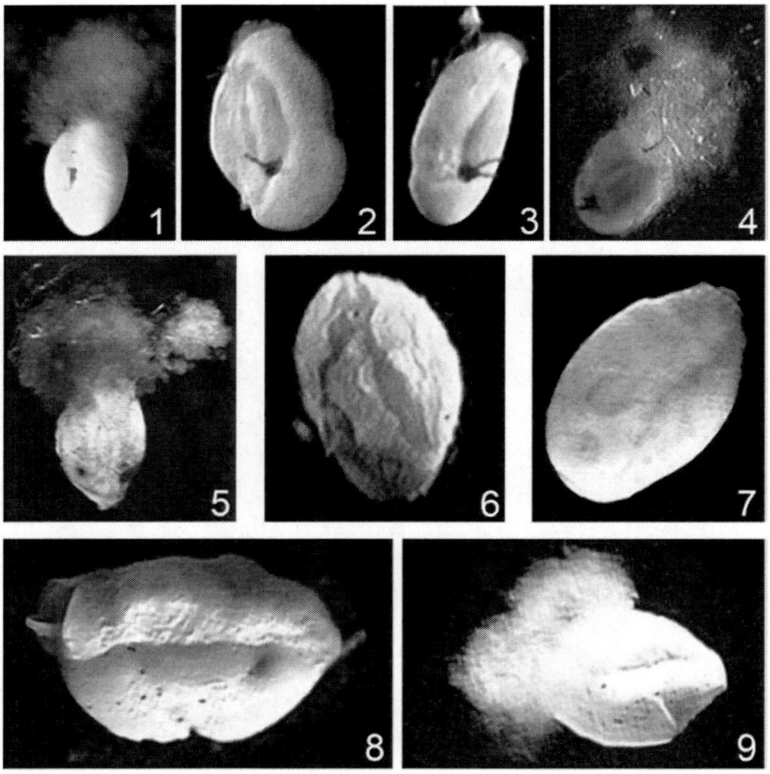

Plate 29. Damage made in *Massilina secans* by parasitic organisms.
1 - a dark claret-coloured spot on the shell surface; 2-4 - a dark claret-coloured spot on the surface of the same individual with claret-coloured threads gradually growing out from this spot (fig. 4 – building the cyst); 5 - claret-coloured spots at the surface of another protist building a cyst (two bigger and many small dots); 6 - shell fully destroyed by the claret-coloured organisms; 7 - a specimen with a thin greenish coating (supposedly bacteria) with circular depressed areas on its shell; 8 - broken shell with a protruding tooth - the surface of the shell is covered with multiple dark claret-coloured dots; 9 - dead protist whose shell bears dark claret-coloured dots, with the residual cyst near its aperture.

The broken last chamber usually revealed the tooth of the previous chamber (Plate **29**, figure 8; Plate **30**, figure - s 2, 3), which is stronger than the rest of the wall and withstands the destruction. In figures 4a and 5 (Plate **30**) in the shell with the broken peripheral side of the chamber the inner cytoplasm becomes visible, and it gaped like a bright rosy wound. All the broken specimens from this plate survived till the end of the observations. They (as well as the other damaged individuals) extruded their reticulopodia (Plate **23**, figure 9); thus, they were capable of moving and even of eating (Plate **25**, figure 5). In some of them, the broken areas of the damaged chamber were repaired (Plate **30**, figure 4a, b). This process was somewhat different in its features from the process of building new chambers, leading to a different aspect to the surface. The newly built parts of the wall of the repaired and the new chamber are extremely thin and translucent (Plate **30**, figure – s 4b, 6).

VII.2. OBSERVATIONS ON *MILIOLINA SP.*

The material of *Miliolina* sp. was collected at the same place as *Massilina secans* and were cultured in the same regime of salinity and temperature, with the same feeding (see above) and the same equipment with the addition of observations in the microscope "Naschet", made in France, NS 400 700 863. The number of the individuals initially selected for the culture was 30. In the rare hot days, for a short time salinity went up to 65 and even 70 promiles but the protists could survive without any visible damage.

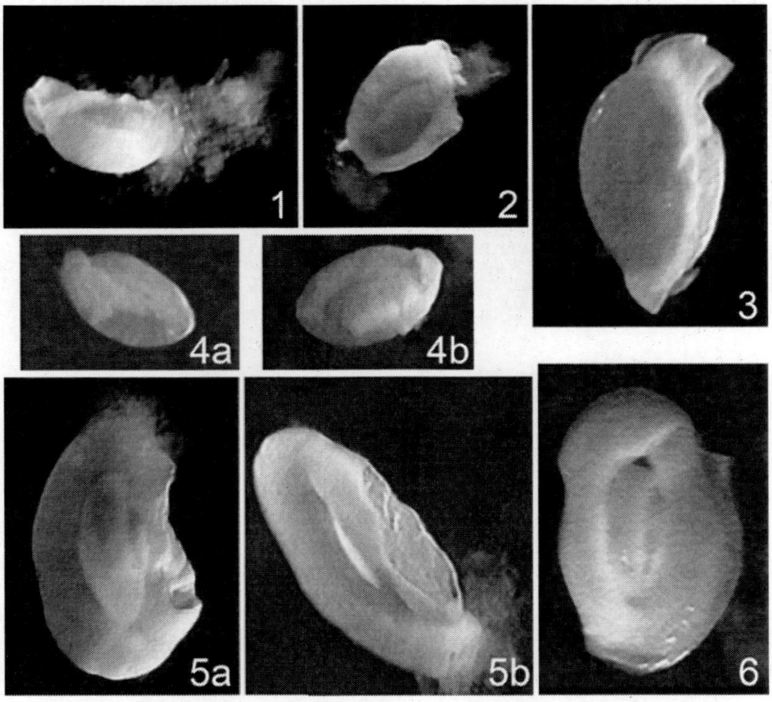

Plate 30. Specimens of *Massilina secans* with the broken shells, restoring and building a new chamber. 1, 2, 3, 4a, 5a,b - broken shells (in 2 and 3 the tooth of the last broken chamber is exposed, in 4a, 5a, 5b, the rosy cytoplasm is visible as a pulp); fig. 4b - shell with a repaired side; 6 - shell with a translucent newly build chamber (right side).

The gathering was made at the littoral during the low-tide. It is interesting to note that the distribution of this species as well as of some other littoral species of *Ammonia*, *Elphidiella, Nonionina* has patchy character. The shells of the protists could be met not only on the ground but hanging on the algae stolons like the toys on the fur tree (Plate **31a**, figure - s 2 a, 2b) (Similarly the specimens of *Nonionina* sp. were also observed on the algae (Plate **31a**, figure 1). Falling from the algae stolons, all these species could also live on and in the surrounding sediment. Frequently, they could be encountered at the southern side of the algae community, being absent at its North side. One needs to have a nose in the search for them, like habitual mushroom-gatherer knowing the best place for mushrooms to fruit. The relations between foraminifera and algae seem to be specific, resembling the relations of insects and plants. However, it is still an unknown field waiting for its researchers.

VII.2.1. Reticulopodia, Feeding, Cyst Building

During our observations the small specimens of *Miliolina* sp.(Plate **31a**, figure 3) move and feed extracting their reticulopodia similarly to the *Massilina secans* representatives (Plate **31a**, figure - s 4a- c). The length of the reticulopodian threads exceeds the size of their shells up to ten times, reaching 2 mm in length comparing the 180 - 200 μ of the shell's in breadth and approximately 300 μ in length. Many of their reticulopodia were branching. Frequently, the shells were observed floating on the water film like it was in *Massilina* culture. Sometimes two or more (3 – 4) individuals at the bottom came nearer to each other for a short time and then parted (Plate **31 a**, figure- s 7a –d). Some of them built cysts in culture (Plate **31 a**, figure 6 a, b) and on the algae also (Plate **31 a**, figure 2b). The latter have narrowing ends attached to the surface of the algae stolon and look as sitting on a small leg. At the Plate **31 a**, figure 6 a, the protist in cyst is building a new chamber, the light translucent end of which could be seen protruding out of the opened end of the cyst (Plate **31 a**, figure 6a). After the several hours when the process is finished the protist came out of the cyst (Plate **31 a**, figure 6 b). Sometimes in a short period the same individual could built a new cyst. The new chamber always looks thinner than the previous ones and is translucent. Sometimes it is possible to see in it the small white dots which later became bigger and coalesce. The surface of such chambers looks spotty.

VII.2.2. Process of Reproduction

During the end of the March and till the end of the April the behavior of the observed protists was rather monotonous, as described above. However, on 19th April suddenly all the dish bottom turned out to be sown with the myriads of small cysts (Plate **31 a**, figure - s 8a, 8 b, figure 9). Some of them were translucent fully or at their upper part only and it was possible to see inside small oval translucent cells (Plate **31 a**, 8 b) which soon began to add the new chambers inside or, more often, outside the cysts (Plate **31 a**, figure 10). The same small oval translucent cells were often lying openly (Plate **31 a**, figure 20). Some small cysts had a pear form (Plate **31 a**, figure - s 12 a, 12 b). Some of the small cysts were not translucent, yellowish and covered with small particles (possibly minute sand grains). The most of these small cysts were hemispheric, all of them were elastic, resilient and adhesive.

They stuck to the needle with which they were detached from the bottom. The lower side of the cyst was flat, translucent and had a central opening. It was possible to compose the multiple successive rows of the shells beginning from the unilocular ones (=oval) cells inside and outside the small cysts – proloculi; then proloculi with the tubular bent outgrowths – flexostyles; and up to 2-3-4 chambered shells (Plate **31 a**, figure 19). The pear-form cysts with the elongated "tails" (Plate **31 a**, figure - s 12 a, b) possibly reflect the beginning of the formation of the flexostyle (Plate **31 a**, figure - s 21 a, b). All these events were a reason to check the water mass under the microscope "Nachet" under the bigger magnifications (up to x 400 and 1000 times). The myriads of the biflagellate gametes were discovered each having the two flagella of uneven length (Plate **31 a**, figure - s 13 – 18). Gamete form varied from the subcircular (more often) to oval or rarely pear-form. The photographs were made under the blue oculars and at the different levels of the gamete volume. That's why their flagella were not visible on some photographs. The movement of gametes was both in the direction of their flagellate end and also opposite to it, possibly depending of the character of flagella work. Some gametes were later discovered at the bottom of the dish, where they soon had lost their flagella. Some short pieces of flagella could be seen in the water mass and at the bottom of the dish where the oval cells with the small opening (Plate **31 a**, figure- s 22 a – d) were later covered with the small hemispheric cysts or lying openly. These are the same as on the plates described above (Plate **31 a**, figure- s 9 – 12, 20), and could be regarded as zygotes giving rise to proloculus, and to the future multi-chambered shell, presumably agamonts. During the later days of 23rd and 27th of April and till the 25th of May, the waves of the dense gamete presence in the water and their rare presence interacted with each other. In the case of the more dense gametes presence the water became muddy for a short time and then translucent again. After pipetting of some water with gametes and dropping it into the other sterile dish, a similar successive picture was observed with the deposition of zygotes soon covered by the cysts, and the successive stages of growth of the proloculi with the flexostyles, and of the 2-3-4 chambered shells.

After the release of gametes, small groups of the shapeless cytoplasm with irregular pieces of the shell wall prominent from them were observed – obviously the residuals of the mother shells.

After the zygotes were formed, the successive stages of the growth of the agamont shells from the first stage of the proloculus with the flexostyle (Plate **31 b**, figure- s 2 – 5) and till the formation of the second and third chambers (Plate **31 b**, figure- s 1, 6 – 11) could be observed. At the stage of the formation of the second and third chambers (Plate **31 b**, figure - s 7 – 10) the newly formed tooth could be clearly seen. The very earlier chambers lack any teeth. The duration of the formation of each new chamber took several hours. The young specimens looked colorless and translucent till the stage of the 2 to 4 chambers when, firstly, the proloculus became rose-coloured. Some of the newly formed shells with the 3 and 4 chambers built the cysts (Plate **31 b**, figure 11). As in *Massilina secans* the long thin colorless and rather hard threads protruding from the aperture formed the carcass of the future cyst.

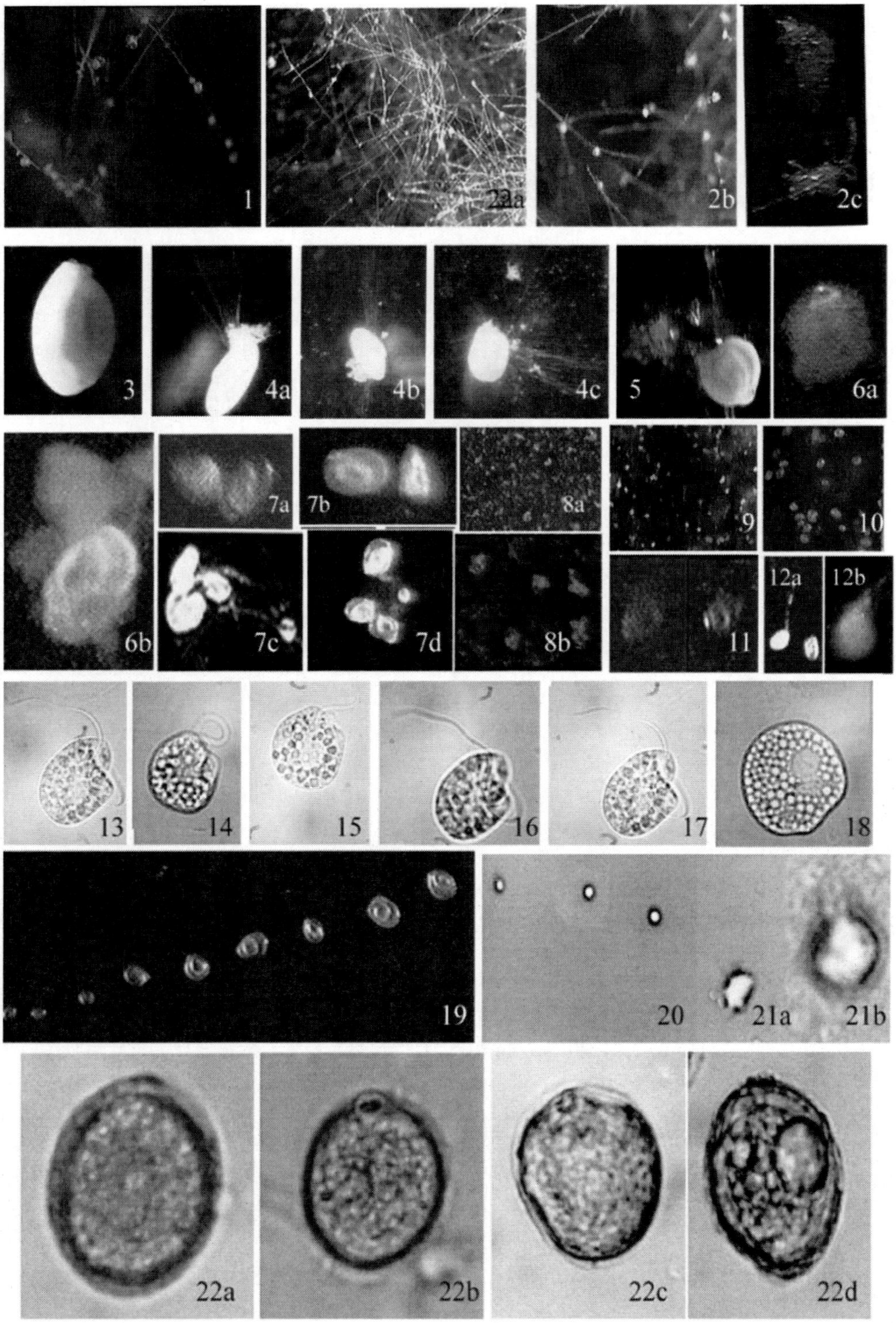

Plates 31 a, b - Observations on *Miliolina* sp. in culture (figures 1 – 12 and 19 - 21 of the Plate 31 a and 1 – 16 of the Plate 31 b photographed under stereo binocular, figures 13 – 18, 22 and 17 – 20 of the Plate 31 b – in the microscope "Nachet") .

Plate 31 a.
1 – 3 - Foraminifera on littoral algae *Rhodothamniella floridula* (1 – *Nonionina* sp., 2a -c *Miliolina* sp. on littoral algae *Rhodothamniella floridula (*2b – one of the branches enlarged, 2c – cysts sitting on the elongated outgrowths like on the leg); 3 - *Miliolina* sp. (x 110); 4 a – c - *Miliolina* sp. with pseudopodia; 5 – the beginning of the cyst building; 6 a, b –formation of the new chamber (6a – inside the cyst, note the translucent end of the new chamber not fully covered by the cyst, 6b – the same specimen going out of the cyst, note the new translucent chamber (right side); 7 a- d – closing in of the two or several specimens; 8 a, b – zygotes under the cysts at the bottom of the dish (8b – enlarged); 9 – proloculi under the cysts (x 110); 10 – 2 and 3 chambered shells under the cysts or already free (x 110); 11 – part of the figure 9, enlarged; 12 a, b - pear-form cysts, may be covering the proloculus beginning the formation of the flexostyle; 13 – 18 – biflagellate gametes photographed at the different levels of their volume, in some optical cuttings (as in figure 18) flagella are not seen (x 1000); 19 – the successive stages of growing of agamonts from zygotes (x 180); 20 – zygotes not covered by the cysts (x 110); 21 a, b – possibly formation of the proloculus from zygotes; 22 a – d – zygotes with microopening (x 400).

On the 18th of May the protists showed their new unexpected surprise: the groups of young cells were extruded from the apertures of agamonts, and not only of the 2- 4 chambered individuals, but also of the youngest ones having only proloculus and flexostyle (Plate **31 b**, figure- s 12, 13). The process looked viviparous. Some of the more grown 2 to 4 chambered agamonts extruded the branched bands of cytoplasm (Plate **31 b**, figure 14) and in some cases the young cells were deposited in lines in such cytoplasmic bands (Plate **31 b**, figure- s 15 – 18). The cytoplasmic bands around the rows of youth could be clearly seen in the microscope “Nachet” (Plate **31 b**, figure- s 17-18). In the groups of the youth extruded from the apertures of the shells having only proloculus with the flexostyle some pairs of extruded cells were also encircled with the common cytoplasm ((Plate **31 b**, figure 12 (upper pair), figure 13 (pair at the right part of the picture)). The growth of these separate young cells photographed in “Nachet” in the yellow and green oculars are shown at the Plate **31 b**, figure- s 19, 20. It is not clear whether these represent proloculi of the gamonts or of the second generation of agamonts (agamont-2 or schizont) as there was no opportunity to investigate their nuclei and to continue the observations (the author needed to go back to St. Petersburg). Nevertheless, some differences in the proloculus form and size [more large in dimensions, more circular and having micropyle in the first agamic generations (Plate **31 a**, figure– s 22 a – d) and 2 -3 times smaller, more elongated, with a short and very narrow tubular projection in the second generation born through the agamont aperture(Plate **31 b**, figure 19)] give reason to consider the latter to be gamonts rather than to suppose the trimorphic cycle. The specimen 6 of the plate **31 b** having the smaller proloculus and the thinner flexostyle possibly represents the young gamont. The author’s three months work in the laboratory of the University of Anger (France) (headed by the professor J.-P. Debenay) where the modern equipment as well as the kind aid of the colleagues Veronique Stauff, Florence Sylvestre and Fabrice Redois permitted the fulfilment of these observations on miliolids in culture.

Long-term laboratory observations on *Massilina secans* (d’Orbigny, 1826) and *Miliolina* sp. were carried out for the first time. These observations of living foraminifera in laboratory conditions helps to better understanding of this group and its ecology, its communication with the environment, and to provide new information on the biology of these foraminifera: the use of pseudopodia for moving and nutrition; the construction of new chambers and reparation of broken shell; the possibility of their active floating on the water film, the relationship between different individuals, and even the features of some individual behavior of some of them, which encourages consideration of the theory of panpsychism.

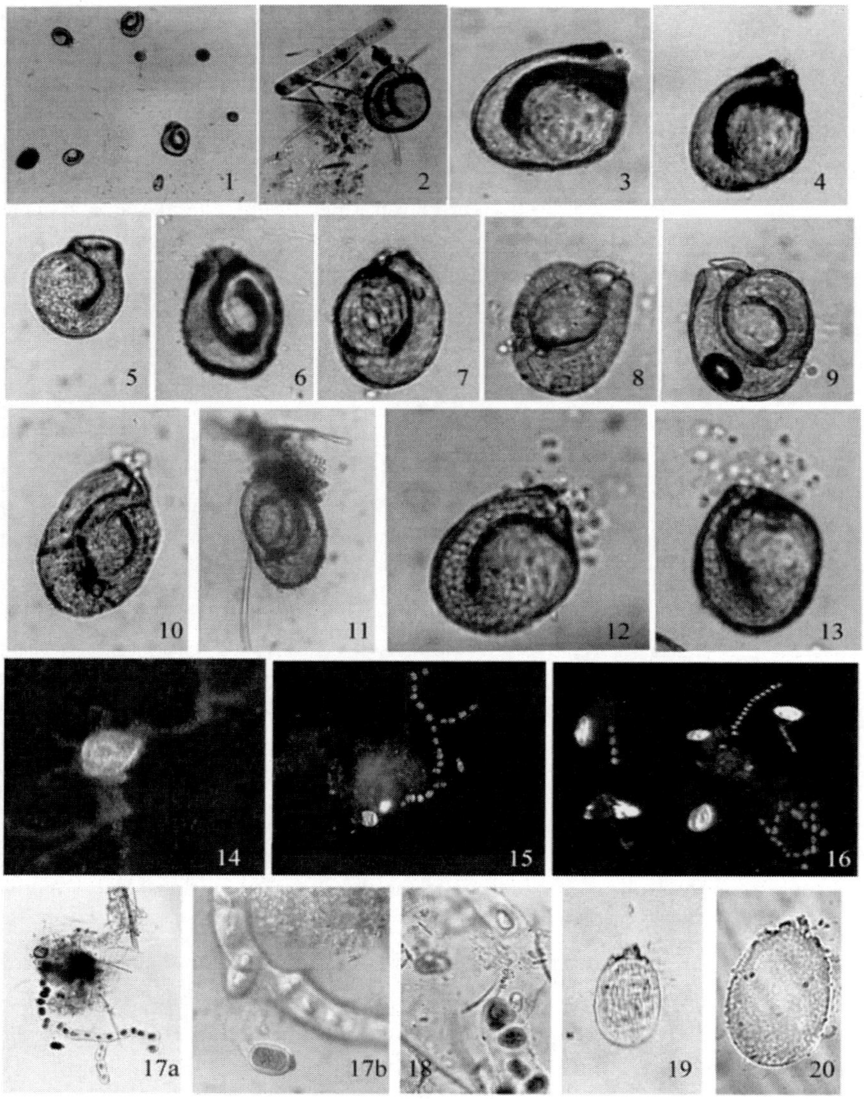

Plate 31 b.
1 – one to 3 chambered agamonts without cysts (x 60); 2 – 5 - agamonts with the flexostyle (x 220); 6 – proloculus with the flexostyle and the second chamber (x 110); 7 – 10 – young agamonts with the 2 and three chambers, note the tooth (in the figure 9 – the second chamber of the other side is not visible) (x 110); 11 – 3-chambered specimen building the cyst (note long translucent thread used later as carcass of the future cyst (x 110); 12, 13 – specimens at the stage of proloculus and flexostyle releasing the youth (it is possible to see that the two upper young cells of the figure 12 and two lower pairs of the figure 13 are encircled by the common cytoplasmic envelope (x 110); 14 – specimen with the branching bands of the cytoplasm going from the aperture (x 60); 15, 16 – bands of cytoplasm of different pattern with the youth inside (x 60); 17a, b – figure 15 enlarged in the microscope, cytoplasmic band is clearly seen (15 a - x 120, 15 b – x 250); 18 – part of the band of another specimen – x 250); 19, 20 – growth of the youth from the bands (gamonts ?), x 400.

This study also provides new information that may have implications for ecological and paleoecological interpretations. "Infection" of the foraminifera by parasites, supposedly bacteria that destroyed the surface of the shell, suggests that foraminiferal shells, even of thick benthic forms, may be destroyed very rapidly, which may bias paleoecological, or even ecological interpretations. The extrusion of cytoplasm with the increase of salinity suggests a first step towards cryptic forms of resistance and/or cryptic propagules that could provide an effective mechanism of resistance and dispersion of foraminifera. This study also shows that these solid benthic foraminifera are capable of floating on the water surface, suggesting that they can use this mechanism for dispersal, at least over short distances. Further studies would be necessary to better understand all these aspects.

In foraminiferology, mostly the dead shells are studied. To promote that we study living organisms, that they have their own choice where to move, what to eat, their own individual preferences and individual behavior I dare to place in the scientific text some lines of poetry as a result of my scientific observations.

ДВАДЦАТЬ ЧЕТЫРЕ ЧАСА ИЗ ЖИЗНИ *MASSILINA*

«Милейшие простейшие, сложнейшие из всех»

А. Кушнер

Я подсмотрела как они живут,
я столько дней за ними наблюдала,
и прежде, чем их поместить в сосуд,
я кисточкой отмыла их сначала.

Они по дну стеклянному ползли,
развертывая веер псевдоподий
как будто по поверхности земли,
по дну залива в их родной природе.

Я каждую запомнила в лицо,
отдельными назвала именами.
Одна - с разбитым устьевым концом,
другая – с поврежденными боками,

в которых цитоплазма – словно кровь
сквозь раковину с выбоиной рваной
и сквозь мембраны тоненький покров
зияла ярко-розовою раной.

А та - по стенке поднялась со дна,
где пищи оставалось слишком мало,
на пленке натяжения одна
остатки этой пищи собирала.

Две пары, опираясь на стекло,
в отличие от остальных – сближались.
Не знаю, что друг к другу их влекло –
соприкасаясь устьем, целовались.

Я с них ночами не сводила глаз,
узнала их характер и желанья –
немного не похожие на нас,
но все-таки милейшие созданья.

Alla Valeria Mikhalevich

Twenty Four Hours from the Life of *Massilina*

"Dearest, simplest creatures, the most complex of all".

Alexander Kushner

I spied on them to learn how they live,
for so many days observing them,
and, before placing them in a Petri dish,
at first I washed them with my little brush.

They crawled along the glass bottom of the dish,
extending the wide fan of pseudopodia
as if they still were on the surface of the earth,
in their native bay and natural environment.

I memorized the appearance of each one
and gave to each one individual name.
This one - with its smashed mouth ending,
That one - with its damaged sides,

in which the cytoplasm - as if it were blood -
gaped like a bright-red pulsating wound
through the shell with its uneven dent
and through the membrane's delicate cover.

One of them crawled from the bottom up the dish-wall,
as at the bottom too little food remained,
and, moving alone on the tension film,
was gathering the residuary food droplets.

Two pairs, slightly leaning on the glass,
unlike the others, were coming together.
I do not know what attracted them to one another -
but, their mouths touching, they were kissing.

For nights on end I could not take my eyes off them,
I learned their characters and their desires.
They are rather different from us,
but nevertheless are the dearest creatures.

Translated from Russian by Marcus Wheeler

Chapter VIII

THE MAIN TRENDS IN THE FORAMINIFERAL EVOLUTION

VIII.1. INTRODUCTION

Phylogenetic changes of organisms are chiefly seen in their morphological organization. As far back as 1859 Darwin had stressed the necessity of study of morphological regularities of the evolutionary process. What are those general principles according to which phylogenetic changes of organs, organelles and whole organisms occur, which give these organisms the high degree of complexity and plasticity of their organization?

The initial studies were founded by Cuvier (1801) and later developed by a number of evolutionists. In the 20-th century especially wide investigation of morphological regularities in the evolutionary process of living organisms was done by Severtzov (1925, 1939) and by his followers (Schmalchgausen (1939, 1946), Dogiel (1929, 1954), Beklemishev, 1964, Golubowski, 1994 and others. All of these studies were mainly based on biological objects and predominantly on multicellular organisms. Severtzov (1925, 1939) had elaborated the basic principles of comparative morphology, of morphological and biological progress, of phylembryogenesis. The progressive biological development which led to the general increase of animal and any living organism organization and activation of the vital functions, opening the way to widening of ecological niches was called by him a morpho-physiological process or aromorphosis (for which some recent authors introduced the term "phase transfer"). In the English-language literature this term is close to a " key evolutionary change" "(see below, Part VIII.9.). Dogiel (1929, 1954) showed the wide occurrence of aromorphoses among Protista. Dogiel was also the creator of the concepts of polymerization and oligomerization. These both represent important morphological regularities of the evolutionary process. He understood polymerization as the formation and the presence in the organism of a great number of homologous organs in Metazoa or organelles in Protozoa. Oligomerization he understood as the diminution of the number of homologous structures. This last is possible only on the basis of the preceding polymerization. Both processes are connected with differentiation in the organism in the process of evolution and they are usually interconnected.

Later the polymerization and oligomerization theory was developed in studies by Podlipaev, Naumov, Borkin, 1974 and Naumov, Borkin, Podlipaev, 1977. They warned of the confusion of these processes and their results seen in polymerouse or oligomerouse structures and so made more precise the notion itself. The process of polymerization they defined as the process of increase of the number of homologous formations connected with the intensification of the functions of the system of these formations, the process of oligomerization – as the process of decrease of the number of the homologous and homodinamic formations up to some definite number connected with the intensification of the functions of the system of these formations. The increased number of elements gives the summarized effect of their work.

Revising Dogiel's ideas they established three modes of polymerization contrary to the previous four offered by Dogiel. The forth one - reduction – they excluded from this phenomenon. These three modes according to Podlipaev, Naumov, Borkin, 1974 are: 1. Fission – the enlarging of the number of organs or organoids as a result of the division of structures existing previously; 2 – Additional polymerization – enlargement of the number of new organs or organoids by the gradual addition of new units similar to ones already existing; 3. Similation – contrary to the two previous modes, in this case, in the process there are involved organs and organoids having a common origin, but differing in their functions when heterogenous structures take on similar functions.

Multiple examples of polymerization of organs in Metazoa are widely known (firstly metameric organization in the higher worms Annelids and in the Arthropoda, their multiple limbs and multiple repetitive branchial outgrowths, nervous, excretory and blood systems, and even more often multiple gonads, disintegration of the cutaneous-muscular envelope of the lower worms into separate muscles in the higher ones, the faceted organization of the compound Insecta eyes and many others). There are also other examples of oligomerization (the fusion of the metameric parts of the body and the metameric systems of organs – for example of the head and breast segments in the Insecta, of all the body segments in the Acarina).

But among Protista these phenomena were not studied in detail, only among the most advanced protistan groups such as Infusoria were significant observations made (Poljanskij & Raikov, 1977). Meanwhile polymerization at the initial and lower levels of organization plays the main role in the evolutionary process providing the basis for all subsequent evolutionary stages. It represents one of the forms of primary self-organization of matter, thus molecular polycondensation became the basis for the origin of life.

Polymerization and oligomerization - they both represent two sides of a more general regularity – the intensification of the functions and of the reliability of all live organism systems. In the western scientific literature the term polymerization is mostly used to describe only the chemical processes of formation of polymeric compounds. V. A. Dogiel (1954) had indicated the wider character of the processes of oligomerization and polymerization. Actually these concepts have an extremely general character which oversteps the limits of biology and are essentially systemic notions. The basic regularities of the general system theory permit to apply them to all structural levels of the organisms as well as to the different levels of organization of matter in the Universe (see below, Part X).

VIII.2. Polymerization and Oligomerization in the Various Protistan Groups

As was earlier shown by Dogiel (1929, 1954), the main evolutionary regularities can also be observed among different Protista. Thus the flagella could be polymerized in Mastigophora from one or two to 4 and 8 and can become multiple in Hypermastigida (Plate **32**, figure - s 1 - 11). Enlarging the number of flagella gives the protist the possibility of more quick and complex movements. In *Trichomonas* differentiation of the flagella takes place and one of them is transformed into the undulate membrane thus permitting more variable actions (Plate **32**, figure 8). The multiple infusorian cilia giving the entire ciliate covering are also considered as polymerized flagella. (Plate **32**, figure 12; Plate **33**, figure 15). It is also possible to note the polymerization not only of the kinetid structures but also of the chromatophores and pirenoids in Mastigophora, of the mitochondria and centrioli in many protistan groups, enlarging of the number of skeletal crossbars and of the number of the chambers in Paleozoic Spumellaria and Nassellaria (Radiolaria), of the number of the inserted rims in Diatomea (Plate **32**, figure 17) and of the number of nuclei in many protistan groups (Plate **33**, figure- s 1 – 15): in some Amoebae (*Nuclearia simplex, N. delicatula, Arachnula impatiens* - from 2 to 50 – Plate **33**, figure- s 1 - 4), in Shellacea (1 or 2) (Plate **33**, figure- s 5, 6), in Acantharia, in Opalinata (Plate **33**, figure- s 7 – 11), in Diatomea (from 1 to 8), in lower Infusoria (*Stephanopogon, Trachelonema*) (Plate **33**, figure- s 12 - 14). Very interesting phenomenon can be observed in some Radiolarians (e.g. *Sphaerellaria* – Plate **33**, figure 17) when the axoplast (the system of the tough microtubular axes of the radial pseudopodia) penetrates the very nucleus thus forming the complex specific structure of the higher level of organization.

The processes of polymerization of different cell and skeletal structures took place in different protistan groups in parallel although having peculiar features in each of them. The role of these processes is especially high in unicellular organisms.

VIII.3. Polymerization of the Nuclei and Some Cytoplasmic Structures in the Foraminifera

Foraminiferal nuclei are also affected by the process of polymerization (Voronova & Mikhalevich, 1985; Mikhalevich, 2005). Although the nuclear apparatus was observed in approximately only two hundred species and the detailed study of the gradual changes of this apparatus during the generative cycles (at least partial) was made in less than half of them nevertheless it is possible to make some conclusions. The majority of the representatives of the subclasses Lagynana and Astrorhizana (*Saccammina alba, Iridia lucida* – Plate **34**, figure - s 4, 5) (class Astrorhizata) as far as is known are mononucleate, but in some genera polymerization of the nuclei took place, for example in subclass Lagynana in *Boderia* (from 1 to 10), in *Ophiotuba* (many small nuclei), in *Dactylosaccus* (1-2 nuclei), in *Penardogromia* (1 to 20 sphaerical nuclei) (Plate **34**, figure 1) (Loeblich & Tappan, 1987). In recent Spirillinata in the agamonts of *Patellina* and *Spirillina* polymerized nuclei were also observed (Berthold, 1977) (Plate **34**, figure - s 6, 7) as well as in the lower agglutinated representative of Rotaliata, subclass Textulariana – in *Trochammina* sp. with up to four similar nuclei in its proloculus (Pawlowski & al., 1995) (Plate **34**, figure 10). All these multinucleate species are homokaryotic. In the higher calcareous representatives of the classes Rotaliata (subclasses Rotaliana, Globigerinana) and Miliolata (subclass Miliolana), as was mentioned above, there is another more advanced type of polymerized nuclei when the nuclear apparatus is differentiated into somatic and generative nuclei. The somatic nucleus (Macronucleus) is usually singular and large, while the generative nuclei are small and multiple, from tens to hundreds and thousands in different species. The volume of the somatic nucleus can exceed that of the generative by two, ten and even a thousand times. The macronucleus also contains three times more RNA than the micronucleus, more proteins, and there are also some other differences in their structure. The DNA quantity does not usually differ in the both types of the foraminiferal nuclei. Separation of functions results in morphological differences of these differentiated nuclei. The well known examples of nuclear dualism in the foraminiferal agamonts are *Cibicides lobatulus, Rotaliella heterocariotica, Metarotaliella parva* (Rotaliana*), Globigerina bulloides* (Globigerinana), *Peneroplis pertusus, Sorites marginalis* and *Quinqueloculina* sp. (Miliolana) (Plate **34**, figure- s 8, 9, 11 - 14). In some representatives of both classes polymerization of the somatic nuclei also took place. The agamonts of *Sorites marginalis* (Miliolana), of *Rosalina columbiensis, Glabratella sulcata* (Rotaliana), possibly of *Globigerinoides ruber* (see Lee et al., 1965) (Globigerinana) have not one, but several somatic nuclei (Plate **34**, figure – s 9, 14).

Comparison of the structure of the nuclear apparatus at the different stages of the life cycles in the lower agglutinated unilocular foraminifera (*Myxotheca arenilega, Saccammina sphaerica, S. alba, Ovammina opaca, Allogromia laticollaris*) (Grell, 1973; Hedley, 1962; Dahlgren, 1967 a) with that of the advanced calcareous Rotaliata (*Cibicides lobatulus* for instance) shows significant differences between them. The mononucleate stage in the life of the agamont of the lower agglutinated unilocular forms is more prolonged. Their young agamonts are mononucleate and sometimes continues to be so in the adult stage, and only before the formation of the gamonts after the serials of the mitotic divisions does it become multinuclear (*Myxotheca arenilega* (Plate **34**, figure 2). While in the forms of the higher organization only the zygote is mononucleate, the agamont has several nuclei beginning from its earliest stages (*Cibicides lobatulus, Metarotaliella parva, Rotaliella heterocariotica*) (see also Part VI, 4 and 5). It is interesting to note that in the unilocular tectinous *Allogromia laticollaris* the agamont I is mononucleate, the agamont II – multinucleate (McEnery & Lee, 1976). Obviously the stage with regenerating nuclei in the process of preparing for cell division became fixed during evolutionary development in the process of natural selection as the more durable stage of the life cycle.

Polymerization of the nuclear material is taking place in Foraminifera as well as in some other groups of unicellular eukaryotes not only at the organelles level (polymerization of the nuclei) but also at the molecular level –polymerization of the quantity of DNA in the nucleus. In the *Cibicides lobatulus* agamonts the DNA content could increase 20 – 30 times (Voronova, 1979, Voronova & Mikhalevich, 1985). The similar phenomenon was reported for the *Patellina corrugata* and *Rotaliella heterocariotica* (Zech, 1964), for the primitive unicellular agglutinated species *Psammophaga simplora* (Arnold, 1982) (Plate **34**, figure 3). In all of these cases it is possible to speak of polyploidization – the polymerization of the genomes within the nucleus. The endoreplication of parts of the genome of *Astrammina rara* during certain phases of the life cycle was documented by Habura et al., (2011).

Some advanced Radiolaria also have a highly polyploid nucleus (Plate **33**, figure - s 16, 17). In Infuzoria the polyploidy of the macronucleus also takes place.

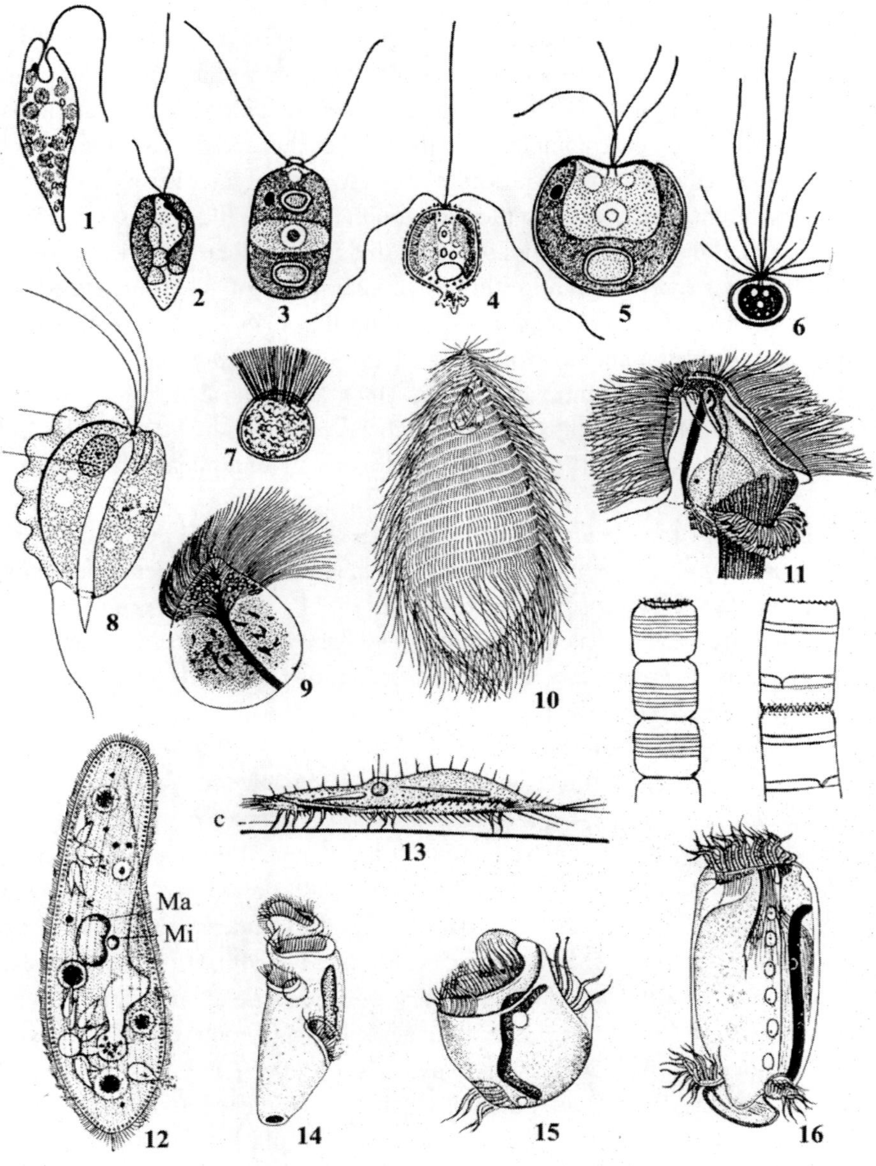

Plate 32. The polymerized and oligomerazed structures in the different protistan groups (after Mikhalevich, 2005).

1 – 11 – the polymerization of flagella (1 – *Euglena proxima*, 2 – *Ochromonas ludibunda*, 3 – *Chlamydomonas pertusa*, 4 – *Chrysochromulina kappa*, 5 – *Platymonas cordiformis*, 6 – *Schizochlamys galatinosa*, 7 - *Derbesia neglecta* (zoospore), 8 – *Trichomonas*, 9 – *Calonympha grassii*, 10 – *Holomastigotoides tusitala*, 11 – *Joenia annectens* (anterior part); 12 – polymerized cilia in Infusoria *Paramecium caudatum (*Ma – macronucleus, Mi – micronucleus), 13 -16 – oligomerization of cilia into cirri (13 – *Stylonichia mytilus*, c – cirri, 14 - *Spirodinium equi*, 15 – *Triadinium galea,* 16 – *Cycloposthium bipalmatum*); 17 – the polymerized inserted rims in Diatomeae (a – annular rims in *Lauderia*, b – collar-like rims in *Stephanodiscus*).

The usual way of nuclear apparatus evolution is by polymerization of the nuclei, then polymerization of the genomes (polyploidy) and subsequent oligomerization of the nuclei. In comparing the Foraminifera with the Infusoria the oligomerization of their nuclei is not known. Primitive Infusoria have several polymerized nuclei (Plate **33**, figure - s 12, 13) or polymerized macronuclei and micronuclei (*Stephanopogon*, *Trachelonema*) (Plate **33**, figure 14). In the higher Infusoria only one macronucleus and one micronucleus exist as a result of oligomerization (Plate **33**, figure 15). Polyploidy as well as nuclear polymerization also arose as the result of division not being completed to the end. In this second case the processes are taking place inside the nucleus. Both processes of nuclear polymerization in Foraminifera occurring at two different levels result in the activation and immobilization of functions of the nuclei and the whole organism. Obviously just the multinuclear and polyploid nuclear apparatus permits some representatives of this unicellular group of eukaryotes to achieve such large dimensions as several centimeters and such long life spans as one year or even more which are so unusual among unicellular organisms (e.g. – *Cycloclypeus carpenteri* in Krueger et al., 1997). The necessity of centralization and integration in such large protists having above all this rather big and complex multi-chambered shell with the chambers relatively autonomous also demands intensification of functions and complication of their nuclear apparatus.

Some peculiarities of the nuclei and their changes during the life cycle in foraminiferal taxa of the different degree of their evolutionary advancement were mentioned above (See above, Part VI.4 and VI.5, see also Mikhalevich, 1999). In the lower primitive Lagynana, along with the usual life cycle simple division into two parts also occurs. Agamogony may be considered as the next step of the asexual reproduction (multiple fission). During the evolutionary process some progressive changes happens in the sexual part of their life cycle as well: from the copulation of the gametes even coming from one parent individual (*Myxotheca*) to the gamete copulation coming from different parental specimens with firstly isogamy (*Glabratella patelliformis*) and then anisogamy (*Rubratella*), sometimes with the gamete sexual selectivity. As shown by analysis of the available present-day data on the foraminiferal cytology and their life cycles, the trends of the main evolutionary changes of these structures goes in foraminifera in parallel to the tendencies of the progressive development of their skeletons (Mikhalevich, 1980, 1981, 1990, 1999, 2000). Thus, the heterokaryocity evolved in the more evolutionary advanced taxa: in the higher Miliolata (Miliolana) and Rotaliata (Rotaliana, Globigerinana). The agamonts of the lower Rotaliata (Textulariana) are homokaryotic. At least the Pawlowski group (Pawlowski et al., 1995 a, b) and Kitasato & Matsushita, 1996 did not manage to discover nuclear dualism in *Trochammina*. It is not impossible that the same picture would be discovered in the higher and lower Nodosariata for which such data are absent as yet, maybe because earlier they didn't attract any special interest because of their wrong taxonomic position (they were included in the rotaliata group). In Spirillinata which are more primitive and in which even the agamonts of their higher representatives (Spirillinana) are homokaryotic it also would be difficult to expect to discover nuclear dualism in their lower representatives (Ammodiscana). Thus a definite taxonomic approach helps to plan the strategy of future investigations of the foraminiferal nuclear apparatus.

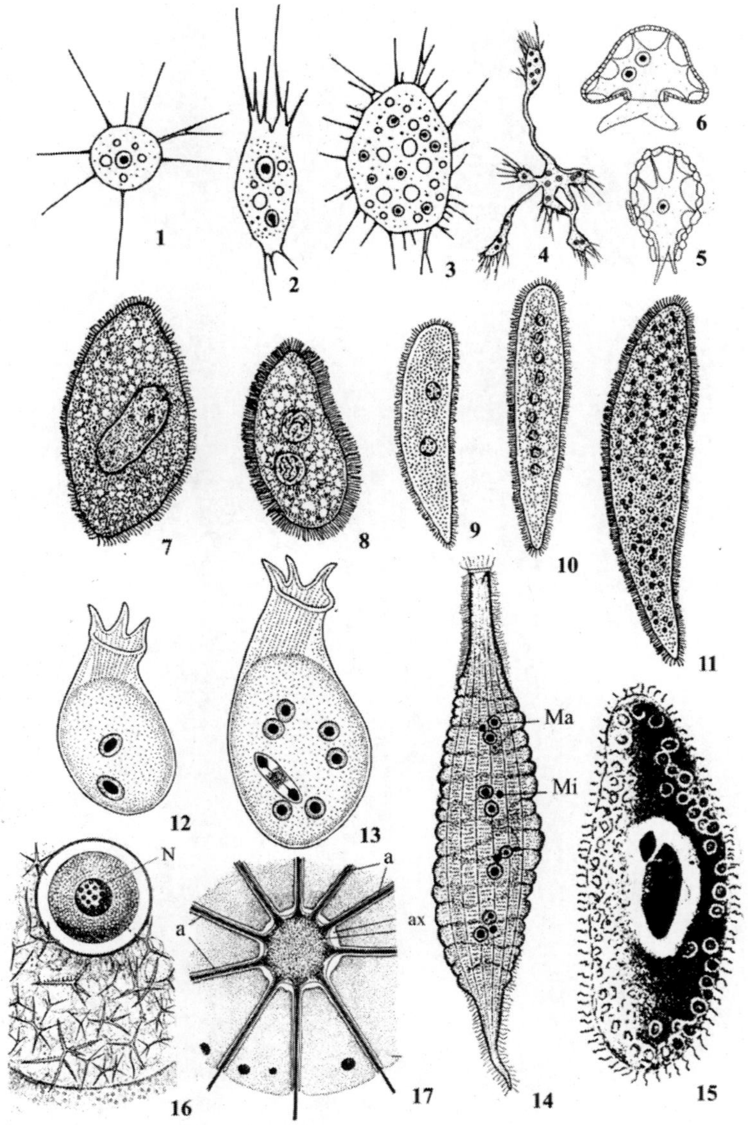

Plate 33. Polymerization and differentiation of the nuclei in the different protistan groups (after Mikhalevich, 2005).

1 – 4 – polymerization of the nuclei in Amoebae (homokaryotic), cl. Filosea (1, 2 – *Nuclearia simplex*, 3 – *N. delicatula*, 4 - *Arachnula impatiens*); 5, 6 - polymerization of the nuclei in Testacealobosea (homokaryotic), cl. Lobosea (5 – *Difflugia* sp., 6 – *Arcella* sp.); 7 – 11 - polymerization of the nuclei in Opalinata (homokaryotic) (7 – *Hegneriella dobelli*, 8 – *Zelleriella macronucleata*, 9 – *Protoopalina intestinalis*, 10 – *Protoopalina axonucleata*, 11 – *Cepedea dimidiata vulgaris*); 12 – 15 - polymerization of the nuclei in Infusoria (12, 13 – *Stephanopogon colpoda*, homokaryotic, 14 – *Trachelonema oligostriata*, heterokaryotic, with multiple nuclei combined in groups of two Ma and one Mi, 15 – *Paramecium bursaria*, heterokaryotic, with oligomerized macronucleus and micronucleus); 16, 17 – polyploidized nuclei of Radiolaria (16 – *Lamoroxanthium pandora*, N – nucleus, 17 – axoplast inside the nucleus of *Sphaerellaria,* a – axopodia).

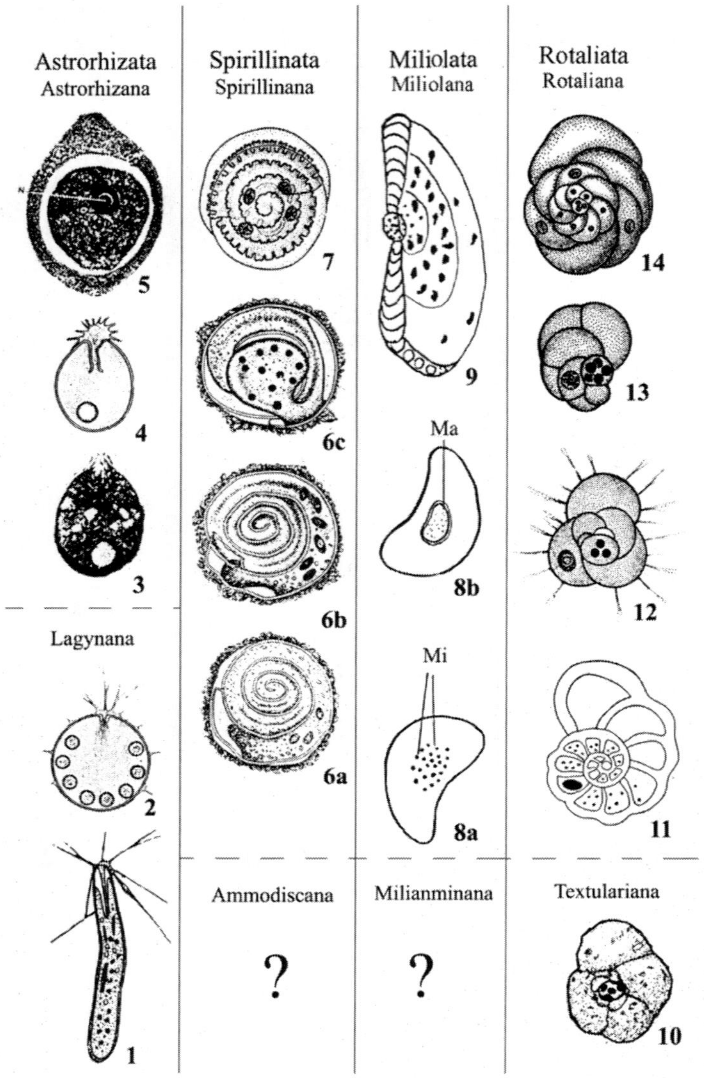

Plate 34. Polymerization of the nuclei in the different classes of Foraminifera (here and further the boarders between the classes are marked with the unbroken line, between the subclasses – with the dotted line) (after Mikhalevich, 2005).

1 – 5 – class Astrorhizata (1, 2 – subclass Lagynana, 3 – 5 – subclass Astrorhizana): 1 – *Penardogromia linearis*, multinucleate; 2 – *Myxotheca arenilega,* multinucleate agamont); 3 - *Psammophaga simplora*, polygenomic (polyploidized) nucleus, 4 – *Iridia lucida,* agamont nucleus, 5 – *Saccammina alba*, supposedly agamont; 6 – 9 – class Spirillinata, subclass Spirillinana, homokaryotic, 6 – *Spirillina vivipara*, a, b, c – the enlarging of the number of nuclei during the agamont (schizont) stages); 7 -*Patellina corrugata*, agamont; 8 - class Miliolata, subclass Miliolana, 8 – *Quinqueloculina* sp., transverse slice of the chamber cytoplasm, a – micronuclei, b – slice of the elongated macronucleus, 9 - *Sorites marginalis*, schematic section of the shell with the hundreds of micronuclei in the embryonic apparatus and the inner chambers (small) and several tens of the macronuclei (large), mostly in the intermediate chambers, a few in the inner or outer chambers; 10 – 14 – class Rotaliata (10 – subclass Textulariana, 11 – 14 – subclass Rotaliana), 10 – *Trochammina* sp., homokaryotic, 11 – *Cibicides lobatulus*, 12 – *Rotaliella heterocaryotica*, 13 – *Rubratella intermedia*, 14 – *Glabratella sulcata*, with three macronuclei and several micronuclei (11 – 14 – heterokaryotic).

Another cytological structure in the foraminiferal cell, involved to some extent in the process of polymerization, is the cytoplasm itself. In the lower unilocular foraminifera it is not differentiated into endo- and ectoplasm (see above – Part VI) and only very weakly so in more advanced species, usually without a definite border between its inner and outer parts. But in the highest specialized forms of the class Rotaliata (subclass Rotaliana - Rotaliidae, Elphidiidae, Nummulitidae), in some forms the ectoplasm is strongly separated and specialized in function: in the forms having a canal system the ectoplasm circulates in canals thus even having its own walls (Plate **18**, figure – s 2, 3) (the similar ectoplasm is obviously circulating in the canals of *Delosina* (subclass Nodosariana) (Revets, 1989) – the unique representative of this group owning a canal system).

The primitive canal system exists also in some pseudo-two-chambered higher Spirillinata (Plate **1a**, II – 12c; Plate **40**, figure 15). As it was mentioned above (Part V.1.3.) the ectoplasm circulating in canals plays a physiological role comparable with that of the blood system of higher animals. In the other Rotaliata subclass – in Globigerinana the wide outer ectoplasmic layer playing the floating role and forming in *Hastigerina* the bubbled capsule, resembles the radiolarian kalymma (Plate **16**, figure 2).

Thus the two cytoplasmic layers are also differentiated in their functions. The separation of the cytoplasm into two layers represents at the same time the polymerization resulting in the activation of special functions. The multiplication of the number of microtubules along the length of granuloreticulopodia, giving them higher rigidity, also represents an example of multiplication of cytoplasmic structures.

It is also necessary to note a particularity of foraminiferal evolution as their co-evolution with symbionts (see Part IV. 3). In many of the calcareous secreted foraminifera (in the advanced groups of rotaliats – *Amphistegina, Calcarina, Nummulites,* and miliolats – in *Alveolina, Sorites*) special relations with the symbionts were elaborated during the process of their long time development (*Amphistegina* even became the symbiont dependent species). These relations influence the form and the size of the shell and its inner morphology (the complication of its structure, formation of a huge number of the chambers and chamberlets), the structure of the shell wall (partial resorbtion of the organic matter, appearance of the more thin and translucent parts of the wall permitting the light to reach the symbionts). The whole symbiont-host system receives additional energy.

Data on the species investigated show that such relations significantly increase the rate of calcification and probably the production of the organic membranes as well as the growth of the individual shell size (Hallock, 1981b). In the different phylogenetic foraminiferal lineages the forms that evolved in cooperation with the symbionts achieved a higher degree of complexity of their organization. When symbiosis is facultative – the symbionts do not constitute a single organism with their host and may not be regarded as a single biologic system.

The processes of polymerization of nuclei and other cytoplasmic structures occurred in many foraminiferal groups in parallel, in some of them having peculiar features. The higher degree of polymerization of the structures indicates the higher evolutionary advance of the group.

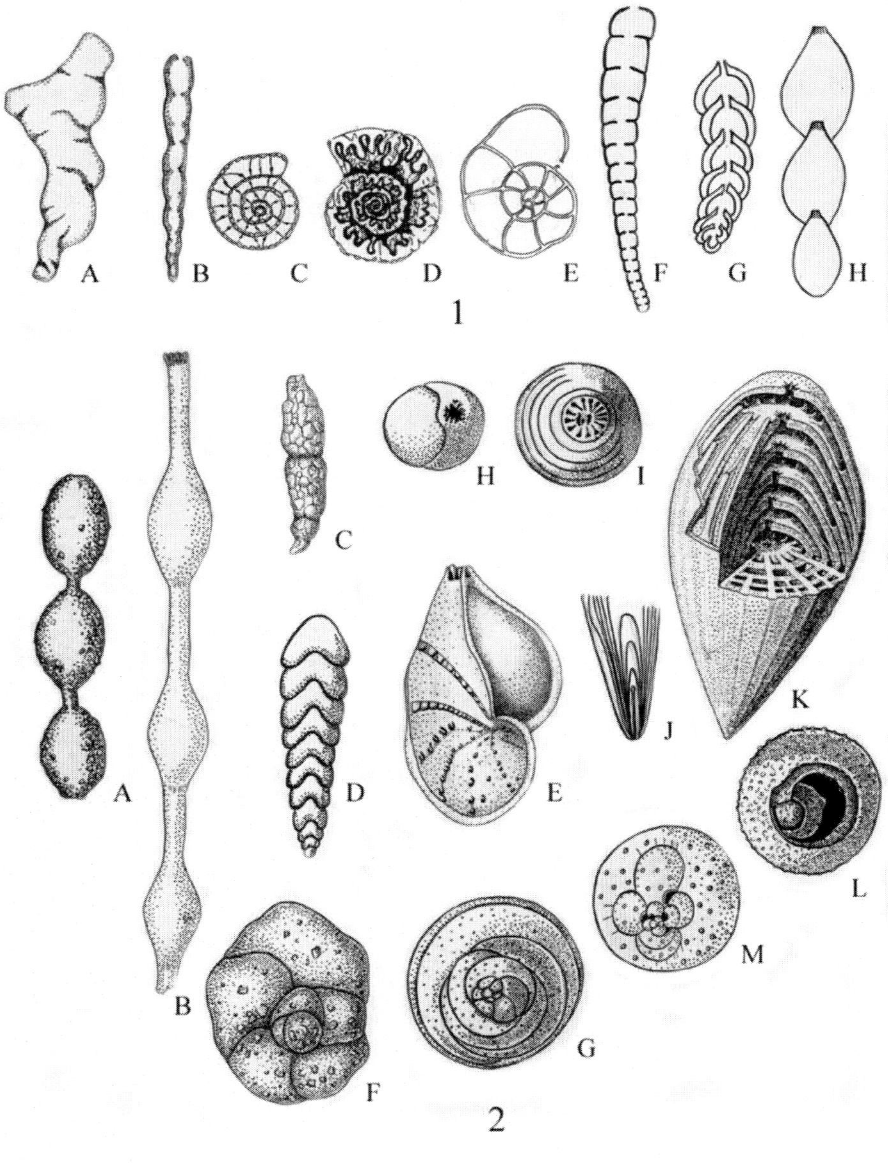

Plate 35. Processes of polymerization, differentiation and oligomerization in the Foraminiferal evolution (after Mikhalevich, 2000).

1. Modes of polymerization: fission [by the way of wall intrusions into inner chamber lumen or by the wall constrictions - subdivision of previously formed unit (A – *Caligella*, B - *Paratikhinella*, C - *Ammoscalaria*)], additional polymerization (growth by the addition of the new units) (D - *Discammina*, E - *Haplophragmoides*, F - *Earlandinita*, G - *Dusenburyina*, H - *Nodosaria*).

2. Differentiation of the chambers and their subsequent oligomerization in different Foraminiferal phylogenetic lineages (classes): A, B - chambers not differentiated in dimensions [A – *Saccamminopsis* (Hormosinana), B - *Grigelis* (Nodosariana)] , C – M - chambers differentiated in dimensions [C – *Reophax* (Hormosinana), D - *Lunucammina* (Nodosariana), E - *Saracenaria* (Nodosariana), F - *Trochammina* (Textulariana), G - *Neoconorbina* (Rotaliana), H, I и - *Idalina* (Miliolana), J - *Daucinoides* (Nodosariana), K - *Colaniella* (Nodosariana), L - *Ammosphaerulina* (Textulariana), M - *Orbulina* (Globigerinana) ; G – M - oligomerization of the chambers at the final stage (up to the fully embracing last chamber)].

VIII.4. POLYMERIZATION OF CHAMBERS IN THE FORAMINIFERA

The foraminiferal skeleton reflects the main building plan structure of their unicellular organism in its connection with the functions; during the evolutionary process it undergoes the most significant changes.

Investigation of the evolutionary regularities of unicellular organisms was difficult because of the minute sizes of these objects and the relative lack of structural morphological characters. Foraminifera—the vast group of unicellular eukaryotes —represents in this regard one of lucky exceptions owing to the variability and complexity of their skeletons, which is unusual at the unicellular level. The abundant occurrence of their agglutinated and calcareous shells in the geological strata since the Cambrian also provides an opportunity to study their evolutionary development, to search for changes of the foraminiferal skeletons through time.

For the right and appropriate understanding of evolution of the group and for composing its taxonomic system, the question – which of the features we need to consider as primary ones, which ones evolve in the different groups independently and in parallel, and which ones are convergent features – is of primary and principal significance. Changes of the foraminiferal skeleton through time occur in two directions: changes of the shell structure and of the composition and ultrastructure of their shell wall. Let us begin to view each of them separately beginning with the shell structure.

What were the main ways and the main regularities of foraminiferal development that resulted in such variable and complex forms?

The foraminiferal shells are unilocular or multi-chambered. The first stage of foraminiferal evolution manifested in the transition from the unilocular to the multi-chambered and even super-multi-chambered shell was the process of polymerization. Among the three main modes of polymerization defined by Podlipaev et al. (1974) (see above, Part VIII. 1.), foraminifera use two of them: fission and additional polymerization. As an example of the first it is possible to indicate the partial subdivision or full division of the chamber lumen by constrictions of the wall (Plate **35**, figure- s 1, A,B, 2 A, B; Plate **36**, figure- s 1,5, 14, 15, 25, 26, 29), by the invasion of the outer wall inside the chamber lumen (Plate **36**, figure- s 7,10, 37), by the formation of the septa or hemisepta dividing or subdividing the chamber lumen (Plate **35**, figure 1,C); Plate **36**, figure- s 2, 3, 6, 17, 21, 23, 24, 35, 44 – 46), by the inner deposition of skeletal material (Plate **36**, figure– s 13, 22). These ways of separating the chamber lumen arose independently and in parallel in the different foraminiferal classes (Astrorhizata, Spirillinata, Miliolata, Nodosariata) and subclasses with more primitive tectinous or agglutinated walls as well as in subclasses with different types of calcareous walls. It may result in the formation of new chambers [(which would be better to call pseudochambers (Plate **35**, figure- s 1 A, B, C, 2 A, B; Plate **36**, figure- s 5, 8, 14, 15, 25, 26, 29) in contrary to the true chambers formed under additional polymerization (Plate **35**, 1 D – H, 2 C - M)] or only the new parts of the chambers formed as chamberlets or simply representing part of the more or less isolated chamber space (Plate 35, 1D; Plate **36**, figure- s 10, 21, 23, 24, 44–46). The fission mode of polymerization dominates as the basic mode in the more primitive groups (Plate **36**): in the class Astrorhizata, in the lower representatives of the subclasses Spirillinana, Miliolana, Nodosariana (not in Rotaliana). In the higher representatives of the subclasses Miliolana and Rotaliana (Plate **36**, figure- s 16–24, 36–45), more rarely in Spirillinana (Plate **36**, figure10), only in one or two cases in Nodosariana

(Plate **36**, figure 35) it occurs along with additional polymerization as the second supplemental mode of polymerization when true chambers are divided into chamberlets. In this case it is possible to speak of a mixed or multiphase ("multistored" in terms of Russian scientists) polymerization. The comparative morphological differences of the formation of the inner chamber compartments in some definite groups having a different character of shell wall were analyzed by Hottinger (1978, 2000). Here, the parallel evolutionary development mostly of the fission mode of polymerization in the different foraminiferal groups is generalized.

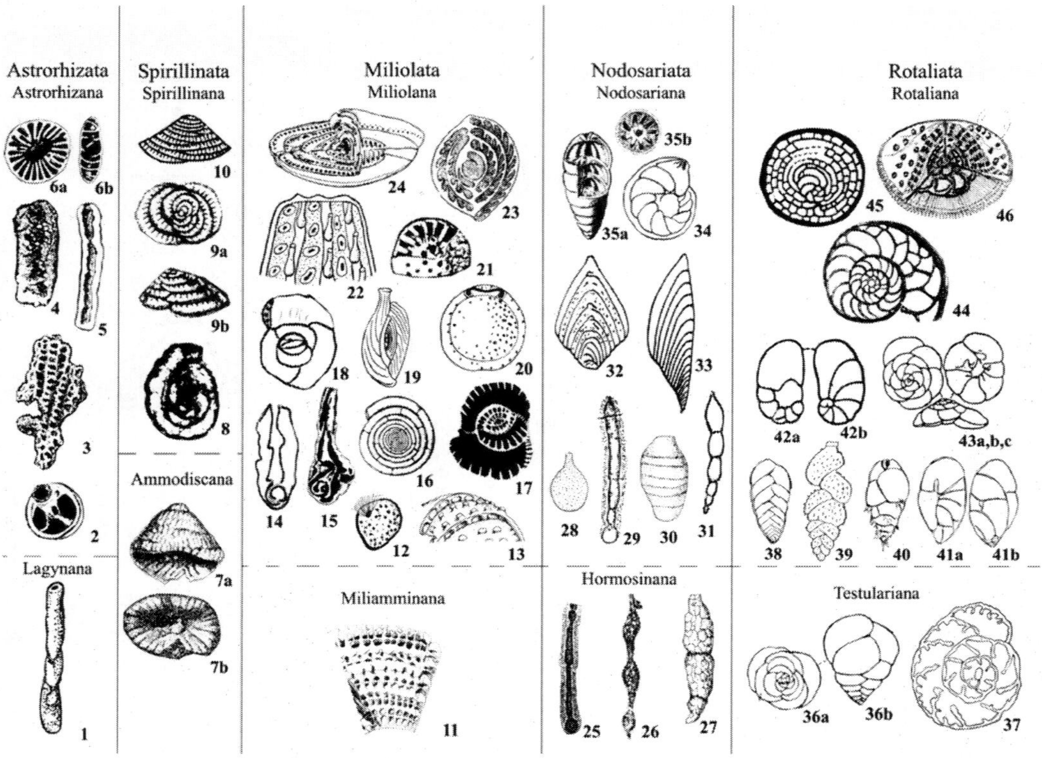

Plate 36. Polymerization of the chambers in the different classes of Foraminifera (formation of pseudochambers, true chambers and chamber subdivisions).

1 – 6 – class Astrorhizata (1 – subclass Lagynana, 2 – 6 – subclass Astrorhizana), inner divisions and subdivisions of the unilocular shells (fission): 1 – *Xenothekella* with growth constrictions, 2 – *Pseudowebbinella* subdivided by short radial partitions, 3 – *Kerionammina*, irregular partitions forms subquadrate chamberlets, 4 – *Botellina* subdivided by the protrusion of the sponge spicules of the wall, 5 – *Protobotellina* subdivided by the protrusion of the sponge spicules of the wall and wall constrictions, 6a,b – *Daitrona* nearly completely subdivided by radial semisepta; 7 – 10 – class Spirillinata (7 – subclass Ammodiscana, 8 – 10 – subclass Spirillinana): 7 – *Valvulinella*, chambers subdivided into chamberlets, 8 - *Tournarchaediscus* with intercameral partial subdivisions by the constrictions of the wall in the last whorl of tubular chamber, 9a,b – *Pseudopatellina*, multi-chambered, 10 –*Patellina*, proloculus followed by the tubular chamber, then multi-chambered with short radial septula of the first and second order; 11 - 24 - class Miliolata, (11 – subclass Miliamminana, 12 – 24 - subclass Miliolana), new chambers formed under additional mode of polymerization and by the subdivision of the chambers as a result of fission, beginning from unilocular shell: 11 – *Broeckinella* (Biokovinoidea), scheme showing the complex system of inner vertical and transverse partitions, in the

marginal zone of the shell with short secondary partitions, resulting in a polygonal meshwork (after Hottinger, 1967), *12 - Squamulina*, unilocular, 13 – *Shanita*, with the inner pillars, 14 – *Sarmatiella*, with pseudochambers, 15 – *Orthella*, with pseudochambers, 16 – *Dolosella*, 17 – *Pseudohauerina*, chambers subdivided by short radial septula, chambers in one whorl nearly equal, 18 – *Hauerina*, 19 – *Sigmoilina*, 20 – *Pyrgo*, 21 – *Rhapydionina*, with vertical radial septula subdividing the chambers, 22 – inner pillars in the *Archaias* chamber, 23 – *Riveroina*, chambers subdivided by the oblique radial septula, 24 – *Alveolina*, super-multi-chambered shell, with tunnels, numerous septula perpendicular to the septum form chamberlets; 25 – 35 - class Nodosariata, (25 – 27 - subclass Hormosinana, 28 – 35 - subclass Nodosariana), new chambers formed mostly by the additional mode of polymerization and by the fission, beginning from unilocular shell : 25 – *Hyperammina*, with pseudochambers, 26 – *Hormosinella*, with pseudochambers, 27 – *Reophax*, with true chambers, 28 – *Lagena*, unilocular, 29 – *Syzrania*, with pseudochambers, 30 – *Lingulina*, megalosphaeric form, 31 – *Nodosaria*, 32 – *Flabellina*, 33 – *Citharina*, 34 – *Robulus*, (30 – 34 – with true chambers), 35 – *Multiseptida*, multi-chambered, with vertical radial partitions of the chambers; figures 36 – 46 - class Rotaliata (36, 37 – subclass Textulariana, 38 – 46 – subclass Rotaliana), exclusively multi-chambered, the mode of fission present as an additional way of chamber subdivision): 36a,b – *Eggerella*, 37 – *Remaneica*, with the secondary septula formed by the invagination of the outer wall, 38 – *Bolivina*, 39 – *Eouvigerina*, 40 – *Bulimina*, 41a,b – *Robertina*, 42a,b – *Cancris*, 43a,b,c – *Discorbis*, 44 - *Heterostegina*, super-multi-chambered, with chamberlets formed by complete secondary septa in the elder whorls , 45 – *Crespinina*, microsphaeric form with finally annular chambers, divided by the radial partitions formed by infolding of the outer wall, 46 – *Sulcorbitoides*, super-multi-chambered, with pillars and radial plates (scheme after Frost, 1974) (44 – 46 – super-multi-chambered).

Under the second mode of polymerization (additional polymerization) the chambers can be of equal or nearly equal size (which occurs more rarely, e.g., *Ammoscalaria* - Plate **35**, figure 1 C, - *Saccamminopsis* - Plate **35**, figure 2 A; Plate **37**, figure 4; *Dolosella* – Plate **36**, figure 16; Plate **37**, figure 13; *Loeblichia* - Plate **37**, figure 14) or differ significantly in dimensions, usually rapidly increasing in size in the process of growth (in the majority of cases, Plate **35**, figure- s 1 A, B, D – H, 2 C - M; Plate **36**, figure- s 18–20, 23, 24,31–35, 36–45). The first case may be considered to be more primitive, as in the second case the size differentiation takes place on the basis of the preceding polymerization. The majority of species of the classes Miliolata and Nodosariata, as well as all representatives of the class Rotaliata used this second mode in their evolution. In some higher Spirillinata, this mode of formation of new chambers also exists [in *Patellina* – Plate **36**, figure 10, in *Patellinella, Paleopatellina* and *Coleiconus* (supposedly belonging to the lower Spirillinata, see Mikhalevich, 2004, further investigations are necessary)] although their two-chambered forms with the long tubular second chamber are referred to as pseudo-two-chambered (= pseudobilocular) shells because their proloculus is separated from the second chamber only by a slight constriction of the wall.

In two groups – in the higher Miliolata (order Alveolinina) and the higher Rotaliata (orders Rotaliida, Nummulitida) – polymerization of the chamber number leads to higher progress and may result in an extremely large number of chambers, achieving hundreds and thousands of them as compared to the usual numbers of several tens. As a result super-multi-chambered large shells are formed (e.g., Plate **36**, figure- s 24, 44, 46).

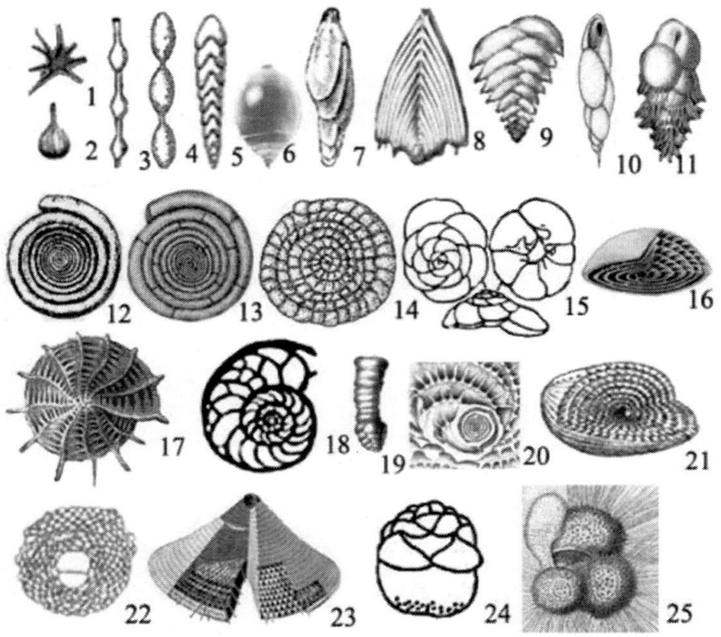

Plate 37. Polymerization and differentiation of the chambers of the shell in the different foraminiferal classes.
1, 2 - unilocular shells: 1 - *Astrorhiza*, 2 - *Lagena*; 3, 4 - polymerized undifferentiated chambers of equal size: 3 - *Grigelis*, 4 - *Saccamminopsis*; 5 - 25 - chambers differentiated in size: 5 - *Lunucammina*, 6 - *Pseudonodosaria*, 7 - *Gorisella*, 8 - *Frondicularia*, 9 - *Bolivinella*, 10 - *Stainforthia*, 11 - *Bulimina*, 12 – *Cornuspira (* circular proloculus followed by tubular chamber), 13 - *Dolosella*, 14 - *Loeblichia*, 15 - *Discorbis*, 16 - *Nummulites*, 17 - *Elphidium*, 18 - *Heterostegina*, 19 - *Clavulina*, 20 - *Discospirina*, 21 - *Neoschwagerina*, 22 - *Lepidocyclina*, 23 - *Orbitolina*, 24 - *Tretomphalus*, 25 - *Globigerinoides* (18-25 the chambers are differentiated also in form; 21-25 - the chambers are differentiated also in function; 21-23 - initial large embryonal chamber; 24 - a big floating chamber at the lower part of the shell; 25 - the last elongated broad chamber.

Formation of the different types of the integrative systems in such complex super-multi-chambered forms (see below) as well as formation in some of them of layers of chambers which differ in their form and function (see below) provides examples of the multiplication (polymerization) of structural levels at the organism level.

Polymerization can be simultaneous, when homologous structures are formed at the same time (moment of growth) [(the primary multiple apertures of *Astrorhiza, Guanduella* and others (Plate **39**, figure- s 1– 4); the secondary multiple aperture on the apertural face (Plate **39**, figure- s 25 – 28 (see below); some septula inside the chamber]. It also can be successive, such as the repetitive addition of new chambers in multi-chambered foraminiferal shells during their growth, extended in time.

The polymerization of chambers in the foraminiferal shell also results in the intensification of functions and in an increase of the energy of the whole organism. The subdivision of a large chamber lumen into smaller compartments plays the same metabolic role (Hottinger, 2000).

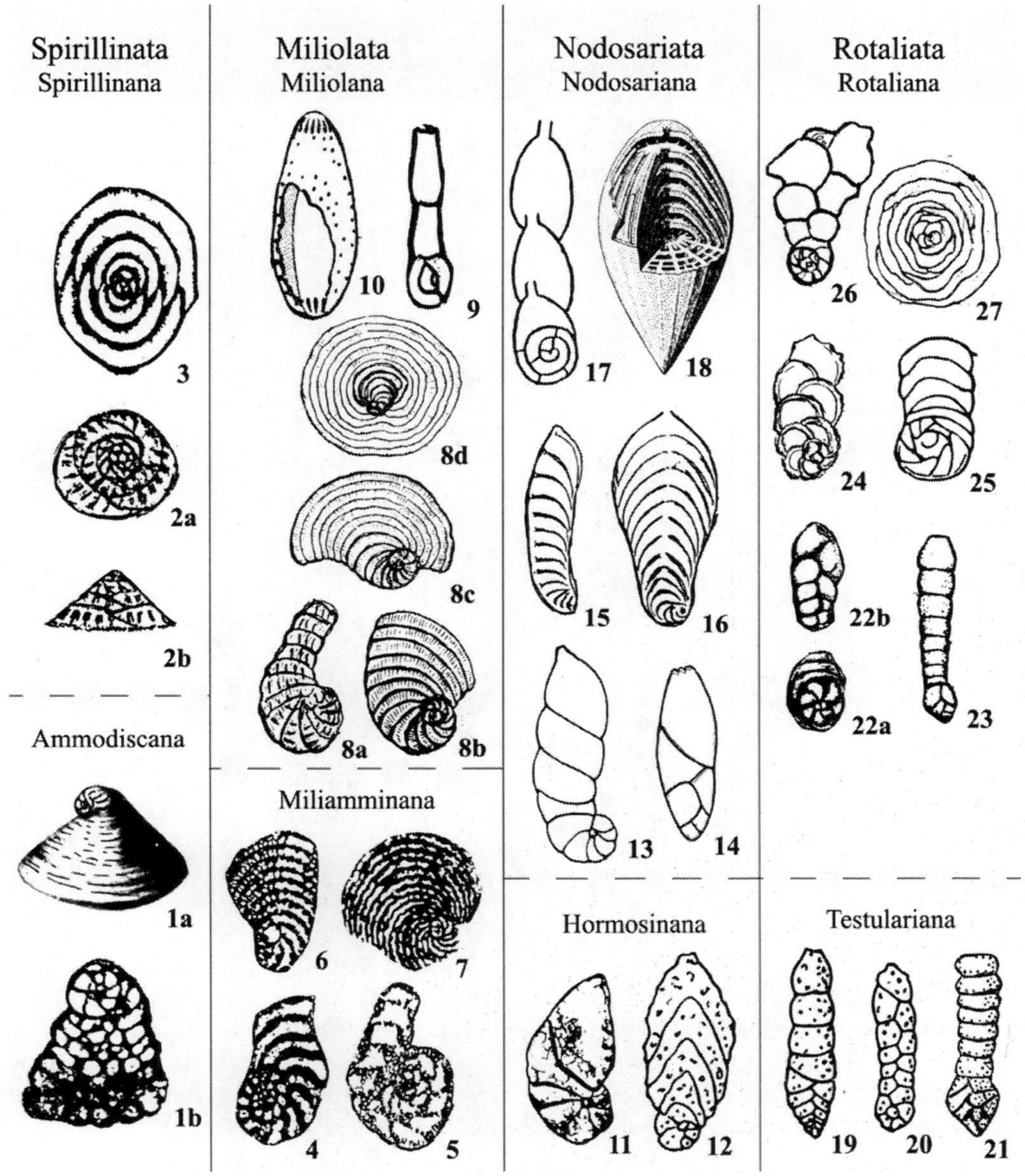

Plate 38. Oligomerizations of the chambers in the different foraminiferal classes.
1 – 3 – class Spirillinata (1 – subclass Ammodiscana, 2, 3 – subclass Spirillinana):1a,b – *Coleiconus*, b – section (early stage trochospiral, then becomes uniserial), 2a,b – *Patellinella*, 3 – *Paleopatellina*, five chambers in the initial whorl, two chambers in the last whorls; 4 – 10 - class Miliolata (4 – 7 – subclass Miliamminana, 8 – 10 – subclass Miliolana): 4 – *Feurtillia*, 5 – *Pseudocyclammina*, 6 – *Alzonella*, 7 – *Flabellocyclolina*, 8a,b,c,d – different species of *Peneroplis*, 9 – *Articularia*, 10 – *Idalina*; 11 – 18 - class Nodosariata(11, 12 – subclass Hormosinana, 13 – 18 – subclass Nodosariana): 11 – *Ammomarginulina*, 12 – *Flabellammina*, 13 – *Vaginulinopsis*, 14 – *Pyrulina*, 15 – *Astacolus*, 16 – *Palmula*, 17 - *Dimorphina*, 18 – *Colaniella*; 19 – 27 – class Rotaliata (19 – 21 – subclass Textulariana, 22 – 27 – subclass Rotaliana): 19 - *Bigenerina*, 20 - *Bolivinopsis*, 21 – *Clavulina*, 22a,b – *Caucasina*, a – view of the initial trochospiral part, side view with the biserial last chambers, 23 – *Rectobolivina*, 24 – *Siphonides*, 25 – *Rectoeponides*, 26 – *Dyocibicides*, 27 – *Cyclocibicides*.

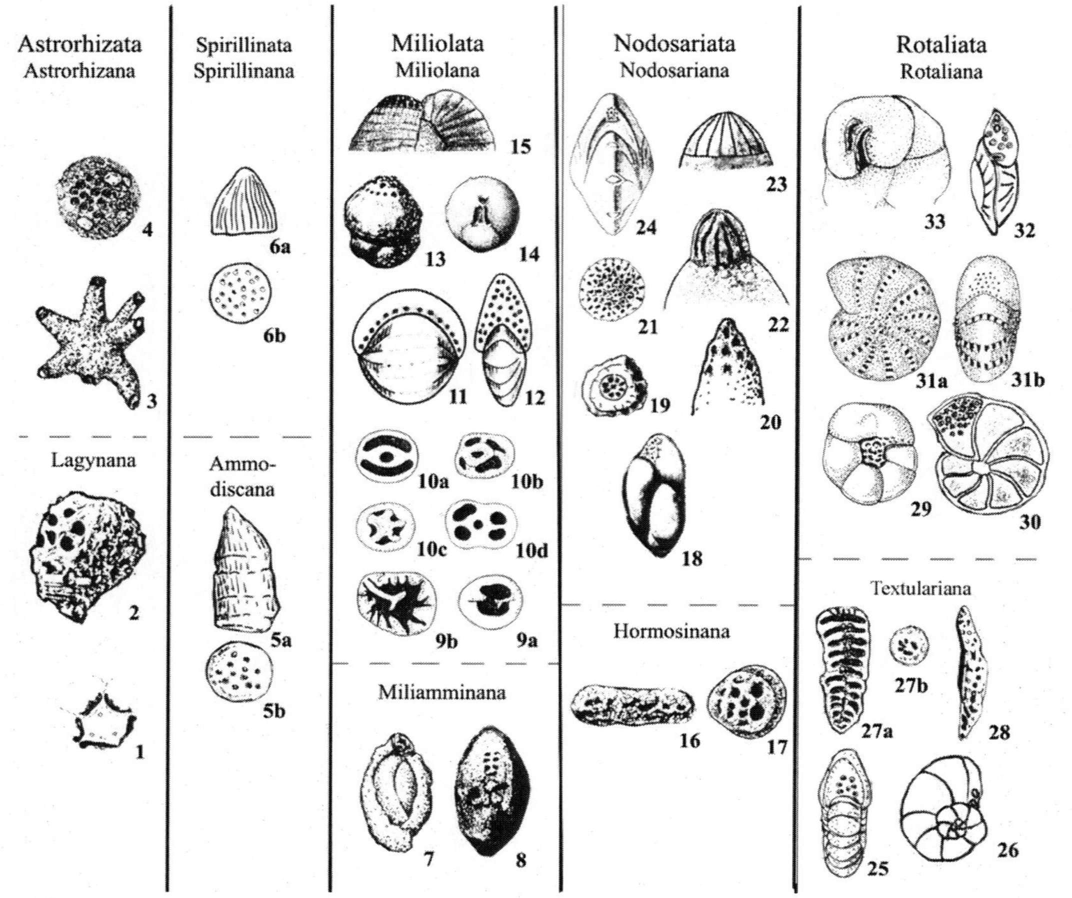

Plate 39. Polymerization of apertures in the different Foraminiferal classes.
1 – 4 – class Astrorhizata (1,2 – subclass Lagynana, 3,4 – subclass Astrorhizana), primary polymerized aperture: 1 – *Microcometes*, 2 – *Guanduella*, 3 – *Astrorhiza*, 4 – *Orbulinelloides* (later transferred with other Saccamminids to Hormosinana); 5, 6 – class Spirillinata (5 – subclass Ammodiscana, 6 – subclass Spirillinana): 5a,b – *Haurania*, 6a,b – *Ferayina*; 7 – 15 - class Miliolata (7,8 – subclass Miliamminana, 9 – 15 - subclass Miliolana): 7 – *Ammomassilina*, 8 – *Lituola*, 9a,b – polymerization of the inner teeth in *Dentostomina*, 10a,b,c,d – multiple aperture of Miliolids as a result of teeth fusion, 11 – *Ovalveolina*, 12 – *Pseudedomia*, 13 – *Rhapydionina*, 14 – *Pyrgoella*, 15 – *Pseudohauerina*; 16 – 23 - class Nodosariata (16,17 – subclass Hormosinana, 18 – 23 – subclass Nodosariana): 16 – *Polychasmina*, 17 – *Cribratina*, 18 – *Delosina*, 19 - *Amphimorphina*, 20 – *Cribropleurostomella*, 21 – *Cribrolenticulina*, 22 - *Laevidentalina*, 23 - *Pandaglandulina*, 24 – *Cribrorobulina*; 25 – 33 – class Rotaliata (25 - 28 – subclass Textulariana, 29 – 33 – subclass Rotaliana): 25 – *Cyclammina*, 26 – *Entzia*, 27a,b – *Cribrogoesella*, a – section, b – apertural view, 28 – *Poritextularia*, 29 – *Sestronophora*, 30 – *Anomalinella*, 31a,b – *Elphidium*, 32 – *Cribroparrella*, 33 – *Anticleina* (Cassidulinidae).

The observed and analysed material allows the conclusion that polymerization of the chambers which results in the rise of multi-chamberedness took place during the foraminiferal evolutionary development in separate classes in parallel several times (Mikhalevich, 1981, 1992, 2005). The first attempt to transfer to multi-chamberedness was undertaken even among the more primitive groups possessing an organic shell: *Belaria* with internal septum, *Phthanotrochus* and *Periptigma* with their unstable chambers, obviously *Resigella*). The data of molecular analysis (Bowser et al., 2006) also led to the conclusion that "the multi-chambered shell originated at least twice in foraminiferal evolution". The data of morphological analysis permit to make more wide conclusions.

A special case of primitive additional polymerization is represented by the pseudocolonial forms existing among the Astrorhizata (e.g., *Sorosphaera*, *Ammopemphix, Webbinelloides*) when the polymerized unilocular shells live together, but the chambers are not organized in regular order and the connections between these chambers are weak. The whole group in this case does not create an organism of higher level such as do true colonial forms. The biology of the foraminiferal pseudocolonial species is studied insufficiently.

VIII.5. OLIGOMERIZATION AND DIFFERENTIATION OF CHAMBERS IN THE FORAMINIFERA

After the structures consisting of numerous elemental units are already preformed the second evolutionary stage - the process of differentiation of these units becomes possible (Plate **37**). The elemental units of the foraminiferal shell are represented by their chambers. The chambers are separated from each other by septa. Both chambers and septa create the compartmentalization of the shell space and of the cytoplasm in these chambers. In the simplest case the polymerized chambers are similar, equal in size and form (Plate **35**, figure - s 1, C, 2, A, B; Plate **37**, figure - s 3, 4, 12 - 14). But the multiple structures provide further opportunities—these multiple chambers could be differentiated. The process of differentiation is the second step in evolutionary development in general. Thus, in Metazoa cells are differentiated into different tissues. In Foraminifera chambers may be differentiated in size (the simplest case) (Plate **35**, figure- s 1 A,B, D – H, 2 C – M; Plate **37**, figure- s 5 - 25) and also in form (Plate **37**, figure- s 18 - 25). In some advanced genera the chambers of Foraminifera are differentiated in function (Plate **37**, figure– s 21 - 23, embryonal chambers; 24 - floating chamber; 25 - brood chamber). In the latter case the process of differentiation is tightly bound with the process of specialization. This process also took place in parallel in different classes and at different geological times. Embryonic chambers were formed in the higher representatives of the classes Spirillinata (Cretaceous orbitolins), Miliolata (Paleozoic fusulinids, Cretaceous alveolinids), and Rotaliata (Paleocene nummulitids) (Plate **37**, figure- s 21 - 23). Differentiation of the chamber form at the different stages of ontogenetic development occurred in each of the classes possessing multi-chambered forms. *Heterostegina* among Rotaliata, *Discospirina* among Miliolata, and *Orbitolina* among Spirillinata (subclass Ammodiscana) can be named as the best examples of such differentiation (Plate **37**, figure- s 18, 20, 23). Such functionally specialized chambers as the brood and floating chambers are known only within the more advanced foraminiferal classes (Plate **37**, figure- s 24, 25).

The process of differentiation is often combined with the process of oligomerization when similar multiple structures become diminished in number. This process also is possible only in structures that were already polymerized. This phenomenon in the multi-chambered shells is usually expressed in the decreasing number of chambers in the last whorls compared with the initial ones (Plate **35**, 2 G; Plate **38**, figure- s 1 - 27). Thus, in planispiral and trochospiral shells with several chambers per whorl in their initial part, these may be only 3 or more often 2 opposite chambers in the last whorls, or even one chamber (Plate **38**, figure 1). Another example of such oligomerizational process may be formation in the final whorls of the subcircular and annular chambers. In monoaxial forms with planispiral, biserial or trochoid initial part the number of chambers in the terminal part often decreases (Plate **38**, figure- s 19 – 26). In these cases the shell may become quadriserial, triserial, biserial or uniserial at the end. In some species all the successive changes are visible, reflecting the process that has taken place over geologic time. Oligomerization taking place step by step is sometimes not complete, for instance the flabelliform shell of some initially planispiral shells may or may not be transformed into an annular one (Plate **38**, figure 8) when only one chamber forms the whole whorl instead of several chambers per whorl in the initial planispiral part. The process of oligomerization leads as a rule to enlargement of the oligomerized structures in their size and to intensification of their functions. Thus the dimensions of the last chambers which became few or single in number exceed the dimensions of the more multiple chambers of the previous whorls by several times (Plate **38**, figure - s 2, 3, 8d, 12, 17, 18, 22 – 26). The processes of oligomerization of the chambers take place in parallel in all the phyletic lines of the multi-chambered representatives (classes Spirillinata, Miliolata, Nodosariata, Rotaliata) in their calcareous representatives, and often in the agglutinated forms as well. Thus, even in Spirillinata where cases of chamber oligomerization are comparatively rare (in connection with the weakly developed multi-chamberedness) such examples can be found: in *Patellina* after a short distance of the tubular chamber following the proloculus the 1–2 whorls containing up to 5 chambers per whorl, then – 3, and very soon – only 2, forming the major part of the shell. The oligomerization of the chambers per whorl from the initial 6 to the terminal 3 is also evident in *Paleopatellinella*, from 5 or 3 to 2 – in *Patellinella* (Plate **38**, figure- s 2, 3). Another example of oligomerization of the chamber number in this class is represented by *Coleiconus* (Plate **38**, figure 1), *Coskinolina* and similar forms. Some other examples of oligomerized chambers in the classes Miliolata, Nodosariata and Rotaliata (where such examples are represented more often) are also given in Plate **38**. The highest degree of the chamber oligomerization is achieved in the classes Miliolata (*Idalina*), Nodosariata (*Colaniella*) and Rotaliata (Globigerinana – *Orbulina*) when the single last chamber embraces the whole shell (Plate **35**, figure- s 2, H, I, K, L, M; Plate **38**, figure- s 10, 18).

A special case of oligomerization is known in Nodosariata when septa between the two chambers are later resorbed and the shell becomes secondarily uniserial (e.g. *Bombulina* - Text figure 1). Some similar phenomena of the resorbtion of the wall of the earlier chambers occur in some Rotaliana (Pegidiidae) and Globigerinana (*Orbulina*).

As mentioned above, the oligomerized chambers are usually of greater dimensions than the preceding ones. In all of these cases the lesser amount of structures (chambers) fulfills the functions of the more numerous ones and hence intensification of functions also exists. The differentiation leads to decrease of the portability of the chambers, to their submission to the whole organism but at the same time – to higher complexity of the organism structure. The process of oligomerization also takes place in many multicellular groups, for instance, the fusion of the breast segments in Insecta and the reduction of the number of parapodia in Polychaeta. This process is also one of the fundamental evolutionary processes taking place in different branches of the unicellular eukaryotes and other phylogenetic organisms tree, having nevertheless its own peculiarities in each of the groups (Mikhalevich, 1981,1992, 2000, 2005).

VIII.6. Polymerization, Differentiation and Oligomerization of Apertural Structures in the Foraminifera

Among other shell structures, the functionally important aperture and its structural elements can also be polymerized (or oligomerized). This polymerization also goes in parallel in different phylogenetic lineages. Thus polymerization of the apertural openings on the apertural face can be observed in the agglutinated and calcareous Nodosariata (Plate **39**, figure - s 16–24) (*Polychasmina, Cribratina* – Hormosinana, in *Laevidentalina, Amphimorphina, Cribrolenticulina, Hemirobulina, Vaginulinopsis, Saracenella, Delosina* – Nodosariana), Rotaliata (Plate **39**, figure - s 25–33) (*Cyclammina, Entzia, Tawitawia, Cribrogenerina, Colomita, Cribrogoesella* – Textulariana, *Duostomina, Candorbulina, Acostina, Bifarinella, Anticleina, Rugidia, Eponides* and *Poroeponides, Hofkerina, Neocribrella, Sestronophora, Cribroelphidium* and *Elphidium,* and some others – Rotaliana) and especially in Miliolata (Plate **39**, figure- s 7 – 15). In the last class the numerous agglutinated (*Debarina, Mayncina, Stomatostoecha, Kolchidina, Lituola, Acruliammina, Acupeina, Bulbophragmium, Cyclolina, Choffatella, Spirocyclina* and others – Miliamminana) and calcareous genera with secondarily multiple aperture (*Cribrolinoides, Inaequalina, Hauerina* and other forms with trematophore, *Cribropyrgo, Nevillina, Crenatella, Poroarticulina, Pyrgoella, Fabularia*, and others – Miliolana) is even difficult to enumerate. In Spirillinata such examples can be noted in the higher Spirillinana (in *Angotia, Chapmanina, Ferayina* – Plate **39**, figure 6), in agglutinated forms in *Haurania* (Plate **39**, figure 5) and related genera questionably referred to this class (Mikhalevich, 2004).

The inner structural apertural elements such as Miliolata teeth (when there are two or more in one aperture) and Nodosariata radial ribs also represent polymerized structures (Plate **39**, figure 9). The coalescence of the separate elements of the aperture into a united structure shows the process of oligomerization of these units (Plate **39**, figure- s 10, 19, 22) going in parallel in the different genera of both classes. In these cases, the oligomerization of the elements of the inner apertural structures results in polymerization of apertural openings thus demonstrating the interconnection of both processes. In the Rotaliata where the inner apertural elements (different tooth-plates and so on) are extremely strongly developed, their polymerization within the single aperture does not take place in a strict sense (although the different parts of the complex tooth-plate in the buliminids for example may be regarded to some extent as polymerized and at the same time differentiated).

But the formation in higher Rotaliata (Buliminida, Bolivinida) of integrative apertural systems (Mikhalevich & Debenay, 2001) extending from chamber to chamber exemplifies the appearance of apertural and organismic systems on a more advanced, new level of organization. Thus polymerization with the subsequent oligomerization of the structures of the apertural system in this case also takes place.

The multiple apertural openings in the unilocular tectinous or agglutinated Astrorhizata (*Microcometes*, *Guanduella* – Lagynana, *Astrorhiza*, and others – Astrorhizana) (Plate **39**, figure - s 1- 4) and Lagenids (*Cribrolagena* – Nodosariata) are regarded here as primary multiple (see Mikhalevich, 1995). Circular or more often irregular openings without a definite position obviously arose primarily. The next step in the development of such forms results in the appearance of a single aperture of the definite position and usually of larger dimension

(*Apogromia*, *Paralieberkuehnia*, *Psammophaga*, *Saccammina*, *Lagena*). Their apertures can be regarded as oligomerized and secondarily single. Strict localization often takes place during the process of oligomerization.

The appearance of additional (called also as secondary or supplementary) apertures (sutural, umbilical, peripheral ones) (Plate **6**, figure - s 1 – 10) also represents the polymerization of the apertures and at the same time the rise of the new apertural structure of a new level of organisation. Additional apertures are mostly and widely developed in the agglutinated (*Norvanganina*, *Plotnikovina*, *Polystomammina*, *Rotaliammina*, *Toretammina*, *Trematophragmoides*, *Balticammina*) and especially in the calcareous Rotaliata (*Schwantzia*, *Pseudoeponides*, *Strebloides*, *Trochulina*, *Epistomaria*, *Cribroelphidium*, *Globigerinoides*, *Pseudobuliminella*, *Virgulinella*, and others). They arose in parallel in the calcareous representatives of the class Spirillinata (*Trocholinopsis*, *Hergottella*, *Lasiodiscus*, *Howchinia*). They are not developed in the classes Miliolata and Nodosariata where they appear only as an exception (*Polysegmentina* – class Miliolata, *Delosina* – class Nodosariata).

As was shown above, processes of the polymerization and oligomerization of the apertural openings occur in many foraminiferal phylogenetic lineages in parallel having in each of them their own peculiar features. In all cases of polymerization and oligomerization of the apertural structures discussed above the evolutionary changes have led to intensification of the functions of the aperture where cytoplasmic flows are most intense and need more space and support.

VIII.7. Polymerizational/Oligomerizational Processes in the Plane of Symmetry of the Foraminiferal Shell

The number of axes and planes of symmetry also is a subject of the processes of polymerization and oligomerization. Sphaerical forms have multiple and equal axes, the appearance of the planispiral and trochospiral forms with a smaller number of unequal axes may be considered as progressive. Sometimes a sphaerical form could be of secondary origin (*Orbulina*) thus repeating the initial form at the new level of organization. Elongated forms with longitudinal axes also could be of primary (*Rhabdammina, Hyperammina, Hormosinella, Syzrania, Nodosaria*) and secondary (*Dimorphina, Clavulina,Rectobolivina, Siphogenerina Pseudobuliminella*) origin (in these last cases originating from planispiral and trochospiral forms).

VIII.8. Processes of Integration in Foraminiferal Development

The super-multi-chambered, large shells need to organize themselves in a new way during their evolution to avoid many problems connected with large dimensions and the difficulty of communication of their different parts with each other and with the environment. The interactions of the cytoplasm of the inner and outer chambers become hard in this case. The surface area to volume ratio in the organism changes with increase of its dimensions and the power of functions related to its surface (such as sensitivity, respiration and so on) is diminished in this case. To withstand environmental competition the organism needs to undergo a morphological reconstruction. The equilibrium between organism and surroundings is broken and the organism comes to an unbalanced state (Prigogine & Stengers, 1986). In this case the rise of the new integrative system facilitating communication between disintegrated parts of the organism reinstates the lost equilibrium state of the organism. This third stage (integration) is represented in the foraminiferal shells by such integrative systems as foramens, tunnels, stolons, apertural integrative systems, and systems of canals (Plate **40**), and interlocular spaces (see Part V.1.3.). The latter are less studied.

The simplest of all of these systems is the system of the inner foramens between the chambers which are formed when the outer main aperture turns out inside the shell after the formation of a new chamber above the previous one during the growth process (Plate **40**, figure -s 1-6). Such foramens occur in all four classes with multi-chambered shells. They permit cytoplasmic flows from chamber to chamber and migration of the nuclei along the shell during reproductive processes (Plate **17**, figure 3). In Robertinacea formation of foramens goes in a more complex way: when part of the primary aperture becomes sealed and foramen is homologous with only part of the primary aperture. In some more advanced genera, in addition to these inner foramens new secondly formed systems of passages exist. These are tunnels and stolons.

Tunnels are known in Paleozoic fusulinids, which are regarded in the new system as belonging to the class Miliolata (Mikhalevich 2004, 2006, 2009 a, b, see below Part XI. 2. 2), and in pfenderinids (placed here in Spirillinata) and some nummulitids (Rotaliata) (see Part V. 1. 3). They represent the space between the inner skeletal structures [e.g. between chomata — the ridgelike secondary deposits of the shell material inside the chambers at their floor (Plate **8**, figure1; Plate **10**, figure5; Plate **40**, figure- s 7, 7a)]. Tunnels permit and direct in an easier way the cytoplasmic flows.

The stolon system is the system of passages between the chambers, studied in detail by Hottinger (1978) (Plate **8**, figure 4; Plate **40**, figure 8). Stolon system is formed in addition to foramens. It is known among representatives of the classes Miliolata, Spirillinata (e. g. orbitolinids) and Rotaliata (in Nummulitids, Lepidocyclinids and others). These three systems are rather primitive and do not have their own walls limiting their space.

Integrative apertural systems are developed in the classes Nodosariata and Rotaliata. They represent the inner apertural structures connected from chamber to chamber into a single joined system and going inside the chamber space. In Nodosariata they are formed by the closed entosolenian tubes (*Ellipsoidina, Ellipsoglandulina, Pleurostomella*) (Plate **5**, figure- s e, f; Plate **40**, figure- s 9, 10). In Rotaliata they can be represented also by closed tubes (although of a character different from that in Nodosariata) or more often by half-closed tubes, grooves, tooth plates, or complexes of tooth plates of variable form (Plate **5**, figure- s c, d; Plate **40**, figure- s 11 - 14; Plate **47**, figure– s 9 - 21) (Mikhalevich & Debenay, 2001, Plate **5**, figure- s 1 - 6). These integrative inner apertural systems partly or fully isolate cytoplasmic flows and serve as supporting structures inside the apertures where cytoplasmic flows are more intense. They provide full or partial differentiation of the cytoplasm inside the cell and also help to canalize cytoplasmic streaming.

The most complex integrative system in Foraminifera is the system of canals having its own walls and going not inside but between the chambers (having exogenous origin) (see above, Part V.1 3). These systems were studied and demonstrated in the detailed works of Hottinger (1976, 1978,1979, 2001, 2006) and Hottinger with co-authors (Hottinger & Dreher,

1974; Billman et al., 1980), and also in some other works (Muller-Merz, 1980; Hansen, 1999; Hansen & Reiss, 1971) (Plate **8**, figure 3; Plate **40**, figure- s 17 - 19). Complex canal systems are developed in many higher representatives of the class Rotaliata when there are thin canal tubes between the chambers in spiral and radial directions often ramifying into the complex system of multiple very thin tubes (Plate **40**, figure 19). Rotaliida and Nummulitida possessing canals are known from Cretaceous till Recent, some of them (Elphidiidae) managed to adapt themselves to life in cold water environments.

Canal systems of the higher representatives of the class Spirillinata (described by Hottinger, 1976, Mantzurova & Gorbatchik, 1982, Azbel, 1986) are less complex and differ from those of the higher Rotaliata: their tubes go between the whorls of one tubular chamber rather than between multiple chambers as in Rotaliata; their branching is not so complex, they may be also represented by the system of fissures (Plate **1a**, figure 10; Plate **40**, figure 15). Their thin structure, elements, composing these systems are not yet studied in such detail as in Rotaliata shells. The beginning of the primitive canal system is found in Paleozoic Lasiodiscids, Involutinids (in the Hottingerellida extinct in Cretaceous), in the Jurassic Spirillinids (in *Spirotrocholina*). It is characteristic that in all these cases this system arose in the trochoid shell type which creates additional structural opportunities. Some more Spirillinan representatives with the canal system may obviously be discovered in the future.

In the Rotaliata themselves some more primitive “canal” systems exist, formed by the system of radial and intraseptal fissures and system of interlocular spaces between the chambers, facilitating their communication with the environment as they are opened into the wide and partly opened umbilical area (*Ammonia, Lockhartia* - Plate **40**, figure 16).The space of these passages is separated by secondary lamellae or pillars rather than by the closed tubes of the more advanced Rotaliana representatives. The early chambers of such shells often do not even have such an integrative system of passages and are more strongly isolated. In Nodosariata a specific canal system arose only once as an exception in one genus *Delosina* (Revets, 1989) and is represented by a system of thin tubes going under the sutures and opening along them. Short tubes piersing the central thickening beneath the preseptal space in the shells of some advanced miliolins (*Rhapydionina*) may be questionably considered as primitive canals and needs further study.

As was mentioned earlier, Hottinger (1978) showed that the ectoplasm circulates in canals supplying the cell with environmental elements, including oxygen. Thus, forming the complex ramifying system of thin and thinnest hard calcareous tubes, it plays the role of the integrative blood system in higher multicellular animals and its complexity is striking for a unicellular organism. All of the last types of foraminiferal integrative systems (tunnels, stolons, apertural integrative systems, interlocular spaces and canals) represent systems of the new organism level in one cell, thus giving at the same time an example of the multiplication (polymerization) of organism systems. In some genera the two types of integrative systems are be combined in one shell (for instance, the stolon system coexists with the canal system in *Nummulites* and in some other nummulitid genera). The primitive system of foramens between the chambers also always occurs in all shells possessing more complex integrative systems.

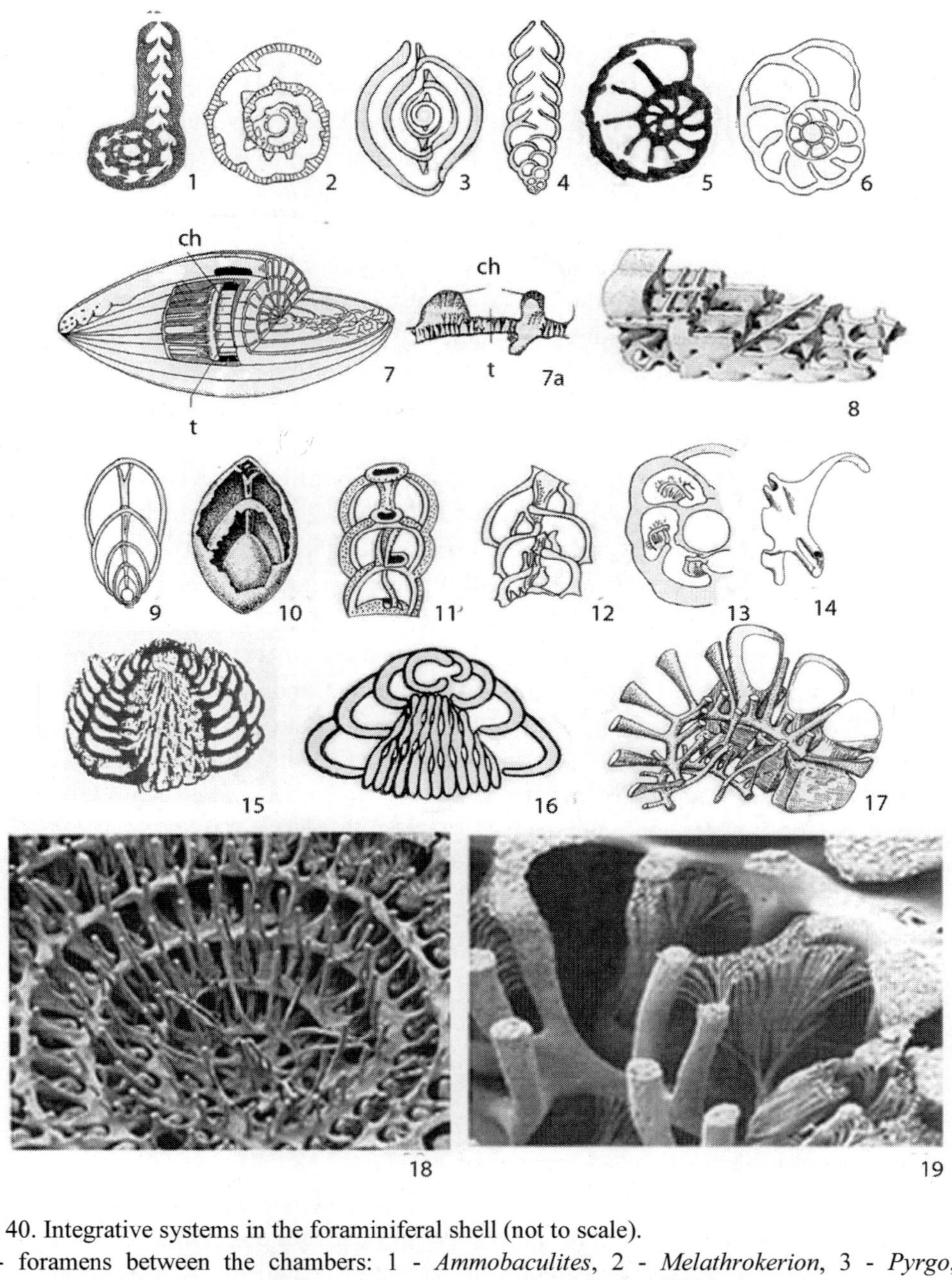

Plate 40. Integrative systems in the foraminiferal shell (not to scale).
1-6 - foramens between the chambers: 1 - *Ammobaculites*, 2 - *Melathrokerion*, 3 - *Pyrgo*, 4 - *Bigenerina*, 5 - *Lenticulina*, 6 – *Cibicides*, 7 - chomata (ch) and tunnels (t) in Fusulinoida (7 - the whole shell, 7a - the detail); 8 - system of stolons of *Marginopora*; 9 - 14 - inner apertural integrative systems: 9 - *Ellipsoglandulina*, 10 - *Ellipsoidina*, 11 - *Siphogenerinoides*, 12 - *Eouvigerina*, 13, 14 - *Reincholdella* (13 - chambers with intercameral plates, 14 - intercameral plate, enlarged) (9, 10 - Nodosariata; 11-14 - Rotaliata); 15 -19 - canal systems: 15 - *Lasiodiscus*, system of fissures in the center of the shell, 16 - *Lockhartia*, system of fissures between the pillars in the center of the shell, 17 - *Pseudorotalia*, system of tubes going between the chambers, 18 - *Elphidium*, 19 - *Pararotalia* (in 18 and 19 chambers are dissolved, tubes of the canals clearly seen), (15 - Spirillinata; 16-19 Rotaliata. (8 - after Hottinger, 1978; 9 -14 - after Mikhalevich, 2000, modified; 15 - after Rauzer-Chernousova and Fursenko, 1959; 16 - after Sigal, 1956; 17-19 - after Hottinger, 1979).

Each integrative system has its own material object. The material objects of tunnels are represented by chomata or system of pillar and other inner deposits of shell material, stolons—by a rather complex system of alternating openings, and canals—by the system of fissures limited by the secondary lamellae in the less advanced forms and by the closed tubes in the highly advanced genera. In the case of canals their inner ectoplasmic content also represents their special material object. Similarly in the metazoan blood system the vessel's walls are composed of specialized cells and their content consists of different types of blood cells. Of course the blood system is a much more complex plural-component and more deeply differentiated system than the foraminiferal canal system developed within one cell. But all such integrative systems permit the direction and canalization of special flows providing their special functions, thus facilitating the communication of different parts of organisms and increasing their wholeness. They provide the interaction of the disconnected parts and the strengthening and intensification of such interaction.

The progressive increase of organism size plays a significant role in the evolution of unicellular eukaryotes. Nevertheless the biggest individuals – the real giants in the protists world (up to 1 and even 18 cm) (e.g. *Bathysiphon*, *Cornuspiroides*) - exist among the primitive unilocular and pseudo-two-chambered forms, not only in the super-multi-chambered ones (such as fusulinids and nummulitids). The growth of multi-chambered shells is restrained because of the disconnection of its multiple chambers. The integrative systems mentioned above remove the contradiction between the multi-chamberedness and the integrity of the whole organism. The resorbtion of the septa between the earlier chambers of the shell preventing cytoplasm interaction between the chambers (the peculiar case of oligomerization mentioned above) represents another way of overcoming such a contradiction when the shell becomes secondarily unilocular.

Analysis of the foraminiferal integrative systems performed above shows that the most advanced foraminiferal groups tried to develop such systems in parallel achieving different degree of complexity and perfection in different systematic groups (Mikhalevich, 1981, 1992, 2000, 2005).

VIII.9. Aromorphoses in Foraminiferal Evolutionary Development

With the appearance of such a complex integrative system as the canal system the foraminiferal organism achieved a new higher level of organization, giving it new and more effective possibilities of surviving and competing with other organisms. As was written above (see part VIII.1.) such significant progressive promising evolutionary changes were named by Severtzov (1925, 1939) "aromorphoses"[1]. Later A. Panov (personal communication) called them "phase transfer," moving the term nearer to physical notions.

In foraminiferal evolutionary development the rise of multi-chamberedness, which took place in parallel in several phylogenetic lineages, and of the complex integrative canal system developed in the most advanced foraminiferal groups, are considered as such aromorphoses (Mikhalevich, 1981, 2000). In parallel to the rise of the more complex skeletal structures the

[1] Term "a key advance" in English-language literature obviously corresponds to "aromorphose" fixed in the Russian-language scientific literature.

nuclear apparatus also achieved in the two groups (in Miliolata and Rotaliata - as far as is known) the higher complexity being differentiated into somatic and generative nuclei. The rise of nuclear dimorphism may also be considered as aromorphosis.

The rise of the hard skeleton (even agglutinated) and then of fully secreted calcareous skeleton may also be regarded as aromorphoses arising in parallel in different classes during different geologic times (Mikhalevich, 1981, 1992, 2000, 2003). Among the types of secreted calcareous shell walls, microgranular and porcellaneous wall types are composed of disordered randomly oriented crystals (Text figure 3: 1, 2). In Spirillinata, Nodosariata, and Rotaliata their calcareous crystals are strictly oriented perpendicularly to the shell surface (Text figure 3: 3, 4). The well-ordered crystal disposition comparing the previous disoriented one represents aromorphosis in the development of the ultrastructure of calcariously secreted shell wall. Such crystal disposition makes the shells look hyaline and often translucent. But only in the latter of these three classes are these perpendicularly organized crystals situated in two lamellae with the middle organic layer (POM) between just from the beginning of the shell formation (Text figure 3: 4). In this case the wall layer is doubled in the process of polymerization. The functions of these two layers (lamellae) also differ, which can be considered as a result of the process of differentiation. The ultrastructural disposition of the calcareous crystals in the foraminiferal shell wall serves as the material object of the shell construction. It is possible to regard the rise of the bilamellar character of the calcareous secreted shell wall of the higher Rotaliata also as aromorphosis. This type of shell ultrastructure provides a less heavy and thin wall possessing new constructive possibilities. As a result Rotaliata have three additional aromorphoses in their development compared with the class Astrorhizata and two additional aromorphoses compared with other classes. This gave them the opportunity to develop during their geological history the most complex shell structures with a great degree of functional differentiation and specialization of their different parts (e.g., floating and brood chambers, canals). All this in turn provided them with a new level of organization, new complexity and wholeness of the organism, higher evolutionary rates of development and new possibilities in their competition with other groups in the environment, adaptive radiation (see Text figure 6), and expansion into new ecological niches. Rotaliata is the only foraminiferal group which could transform from the benthic to the pelagic mode of life by really getting up from the bottom and thus demonstrating once more aromorphosis in its development. And the evolutionary rate of planktonic foraminifera is 50 to 100 times faster than any other foraminiferal lineages being on average 4 subst/site/10^9years and sometimes exceeding 7 subst/site/10^9, thus, surpassing all rates reported earlier for SSU rDNA sequences (Pawlowski, 2000). This fact is in full concordance with general evolutionary genetics concluding that younger genes tend to show accelerated evolutionary rates with respect to older genes (examples given in mammalian genomes – Toll-Riera et al., 2008).

Aromorphoses provide organisms with a transition to a new level of organization which usually goes through the stage of a non equilibrium, unbalanced state (Prigogine & Stengers, 1986) (see above, Part VIII. 7). The creation of the new integrative system permits the organism to pass such an unbalanced state successfully and to preserve and enlarge its wholeness. The level of organization increases. According to evolutionary theory such results are possible only when there already exist appropriate structures and conditions precluding attaining a new equilibrium state. Thus the emergence of life itself was the most significant aromorphosis in evolutionary development of matter and it became possible only after the availability of accumulated polymerized molecular structures and precluding conditions. Aromorphoses of rising Procaryota and Eukaryota were the next steps in this development. Each of such transfers gives birth to a new form of activity and complexity of matter, a new level of organization of the system. Foraminifera acquire in their development their own aromorphic stages of complication obtaining a hard skeleton, multi-chamberedness, complex integrative systems of multi-chambered organism, perfection of the skeletal wall with the rise of its bilamellar constitution, complex nuclear apparatus.

As described above, the processes of polymerization (with subsequent differentiation) and integration, and the aromorphoses happened in different classes of Foraminifera independently and in parallel, having in each of them similar tendencies and at the same time their own specific features (Mikhalevich, 1981,1992, 2000, 2005).

VIII.10. Parallelism and Convergence in Foraminiferal Evolutionary Development and Peculiar Features of the Different Foraminiferal Classes

As can be clearly seen from the above discussion, foraminifera demonstrate a striking number of cases of parallelism in the evolutionary development of their phyletic lineages (classes): rise of the hard agglutinated skeleton, then of fully calcareously secreted skeleton, rise of multi-chamberedness with subsequent differentiation and oligomerization of the chamber number, polymerization of the apertures, rise of different types of integrative systems. All of these progressive changes happened in their historical development several times in parallel (Mikhalevich, 1981, 1992, 2000). Parallelism in the shell form and structure will be discussed below (this Part). All this may be explained as a result of their wide distribution during geological times and at present, of variable conditions of their habitation and the high variability of their forms. It is possible to say that they used nearly all variants of combinations of possible features among the set of features theoretically possible in nature and composing the universal space of possibilities. Knowing the combination of their different structural features and the tendencies of their development it is possible in some cases to predict the discovery of new forms with definite structural features. Thus, for example, we know in Miliolana the presence of the different apertural types: apertural flap, simple or complex apertural tooth, multiple apertures. All these apertural types are combined variably with definite modes of coiling (quinqueloculine, triloculine, biloculine, spiroloculine, sigmoidal) (Plate **41**). For a long time the sigmoid shells with multiple apertures and spiroloculine shells with complex tooth were not described although theoretically thinking they could exist. Later such forms were discovered (*Longiapertina* Seiglie, 1979; *Bidentina* Mikhalevich, 1988). Another example – I hoped to find the agglutinated analogue of *Lenticulina* (with planispiral shell and terminal aperture at the external angle of the chamber) and such genus *Agardhella* was described by Nagy and Basov in 1998.

The greater number of parallelisms within the bounds of Foraminifera embraces the first steps of the complication of their cell structure and skeleton (Mikhalevich, 1981, 1992, 1999, 2000): the rise of multinuclearity, the rise of multi-chamberedness, the modes of formation of the new chambers, the more simple cases of complication of the apertures. These tendencies are developing until the divergence will not exhaust all the possibilities of previous structures (thus tunnels are replaced by canals, the agglutinated wall – with the fully calcareous one). Agglutinated shell wall was replaced by fully secreted calcareous shell wall in the classes Spirillinata, Miliolata, Nodosariata and Rotaliata in parallel (Mikhalevich, 1981, 1992, 1998,1999, 2000, 2003), although the ultrastructure of their calcareous wall in each of these classes has its peculiar features. As it is possible to see from the above chapters and from the comparison of the variability and complexity of the genera within the two subclasses in each class, the agglutinated type of the shell wall possesses lesser structural potentials. Under the calcareously secreted wall the considerable number of structural prohibitions obviously turned out to be eliminated.

All this once more provides the basis to regard the rise of the calcareous skeleton and especially of the bilamellar calcareous wall as an aromorphoses, which permit to change substantively the structural plan of the organism, as well as its basic properties.

Although the analysis performed had shown that changes in the apertural structures, integrative systems and wall types may take place in parallel in shells of different structure, nevertheless in each of the examined phylogenetic lineages the tendencies of their development are realized to an unequal degree, achieve the different level of perfection and have their peculiar features in each of them.

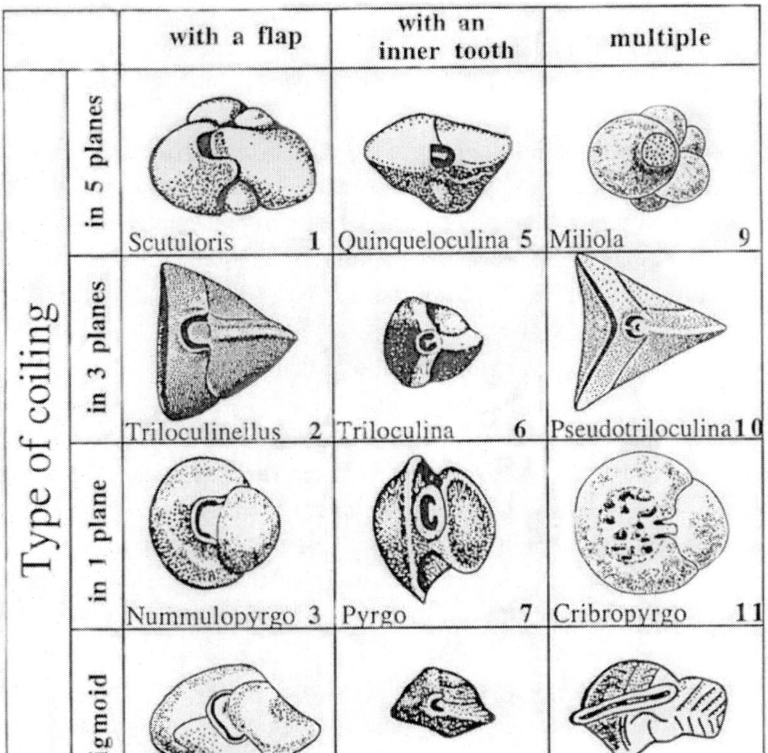

Plate 41. Apertural structures in Miliolana with the different types of chamber arrangement. Types of coiling: chambers situated in 5 crossing planes, in 3 crossing planes, in 2 planes, sigmoidal. Types of tooth structure: outer plate-like tooth (apertural flap), inner tooth, multiple aperture (several to many openings).

The qualitative originality of evolution is most pure in the more primitive subclasses Lagynana and Astrorhizana with their nearly exclusively unilocular (or rarely pseudo-multi-chambered – pseudocolonial) shells often of irregular form, simple aperture and tectinous or agglutinated, rarely microgranular wall. Their agamonts stay rather long time mononuclear, they lack nuclear dimorphism, their cytoplasm is not differentiated into endo- and ectoplasm. It is possible to suppose that after detailed cytological and molecular studies some of them would be replaced into other protistan groups as for instance in the case of *Pelosina* which after examination of its cytology (Mikhalevich & Voronova, 1999) was transferred to Xenophyophoria. [Despite the close relationship of Foraminifera and Xenophyophoria discovered in molecular studies (Pawlowski et al., 2003) the profound cytological differences of the latter (plasmodium instead of the cell, the presence of linellae, granellare, stercomare, and intracellular barite crystals and some other features of their organization and physiology seem to provide a basis to place them somewhat separately from the main bounds of typical Foraminifera].

Homokariosity is also characteristic of the other ancient peculiar foraminiferal branch – Spirillinata possessing a pseudo-two-chambered shell with a long tubular second chamber (or derived from such forms), with a simple aperture, but possessing well-ordered radial ultrastructure of the wall in their advanced forms. Their ancient forms combine a set of plesiomorphic features (the presence of the tubular chamber in many of them, at least at the earlier stage, pseudochamberedness, primitive septa, simple aperture, glomerate coiling) with some advanced apomorphic structures (the presence of tunnels and canals, additional apertures, trochospiral shells).

In miliolats and nodosariats the shell types are rather diverse with the predominance of a specific coiling, chamber form and apertural structures in each of them, with rather purely developed integrative systems. The shell wall in the representatives of their calcareous subclasses is monolamellar, with peculiar packing of crystals in each of them. Nuclear apparatus of miliolats acquires nuclear dimorphism (Plate **19**, Plate **34**, figure - s 8, 9), data on Nodosariata are absent.

Whichever evolutionary trends examined above that we take – nuclear apparatus, shell, apertural and integrative structures, wall ultrastructure – all of them have the most development, diversity and complexity in the Rotaliana group. As was already mentioned, the rise of the light bilamellar shell wall permitting their shell to be strong with a minimum thickness gave them the possibility of aromorphic changes that resulted in a sharp rise of their evolutionary rate, rise of the number of taxa both of the higher and lower rank, adaptive radiation and the advent of pelagic forms. The origin of the latter from the benthic rotaliats (close to some discorbids, nonionids, bolivinids) may be supported with their morphological resemblance and with the presence of the benthic stage in their life cycle. Such a close relationship was recently conclusively and brightly demonstrated in the work of Darling et al., 2009 showing the genetic and morphologic identity of *Bolivina/ Streptochilus* species living usually as benthic forms but also surviving and growing as plankton. (This work also demonstrates the remarkable coincidence of the results of morphological and molecular studies). The authors introduced the term tychopelagic for such a mode of life. The mechanism of the transition of the benthic inhabitants to the pelagic mode of life through the turbulent waters and wind driven upwelling was earnestly shown by Darling & Leckie (2010). Obviously such a double lifestyle is not so rare a phenomenon among the Rotaliana and possibly may also occur in some other foraminiferal taxa, providing the wider distribution of the species. In the process of studying the zoogeographic distribution of several foraminiferal taxa a wide climatic and bathymetric distribution was noticed for a set of genera among the former Rotaliina (Mikhalevich, 1978) It was supposed that such wider spreading was connected with the higher lability of these organisms at the moment of reproduction or at some other stage of their life cycle:" It is possible that during part of their life cycle they live as pelagic forms. Their natural habitats resemble the wide natural habitats of the planktonic foraminifers of the families Globigerinidae and Globorotaliidae. The outer experience of some of the widely distributed species, which till now are regarded as benthic forms, more likely shows their adaptation to soaring up in the water mass: such are the umbrella-like stepped form of many of bulimins, of uvigerins with their broaded upper part, diverse keels and outgrowths of lagens. The possibility of such a supposition may be confirmed by the findings of pelagic stages in *Iridia* and *Tretomphalus* (Loeblich & Tappan, 1964), as well as in the record of the finding of *Bolivina* sp. in the upper water layer (Bermudez & Seiglie, 1963). It is possible to presume that reception of new data on the foraminiferal mode of life would provide a basis to diminish the number of widely spread benthic species in which the pelagic stage is unknown" (Mikhalevich, 1978, translated here from Russian) (the orders were taken in the understanding of Loeblich & Tappan, 1964, where at that time lagens were included into Rotaliina). Now this supposition is brilliantly studied and documented in the Darling et al., (2009) publication. The more specialized planktonic forms living the most part of their life as pelagic forms and acquiring special adaptive features changing significantly their shell morphology give reasons to separate them as an independent taxon within the class Rotaliana. They belong to the subclass Globigerinana.

The origin of Miliolata is according to Arnold, 1978 from the ancestral forms similar to *Periptigma* (Lagynana). He was the first to offer the idea of emergence of Miliolata and Rotaliata from the soft-shelled Lagynana (Arnold, 1978, 1979, 1984). The supposition of the possible origin of rotaliats from *Phthanotrochus arcanus* (Lagynana) (Loeblich & Tappan, 1987, plate 10, figure - s 7, 9) is confirmed by data on their life cycle and nuclear apparatus as *Phthanotrochus arcanus* is the only Lagynana representative possibly having nuclear dualism (Arnold, 1978). Salami (1976) also supposed the origin of trochamminids and rotaliids to be from the unicellular soft walled foraminifera (studying the biology and reproduction of *Trochammina* cf. *T. quadriloba* he wrote:" Ancestry from one of the non-agglutinating, round or subsphaerical allogromiids is suggested. The basis for such deductions is resemblance in the organization of the proloculus and nuclei and the absence of prolocular agglutinate"). It also is most likely that other foraminiferal classes had also originated from some ancient Lagynana representatives and these events obviously happened earlier than in the Cambrian, (Text figure 2 of this study) (Mikhalevich, 1980, 1992, 1999, 2000). The primitive prototypes of the morphologic forms of the higher testate subclasses could be found among the naked or soft-walled unilocular predecessors. The roots of Spirillinata and Nodosariata obviously could also be discovered among the Recent Lagynana or their ancient forerunners. Miliolids branching among unilocular lineages was later shown by Flakowski et al., 2005. There is a conjecture of the heterogeneity of the Lagynana group, which still is the least studied foraminiferal taxon. (The finding of rich allogromiid fauna in Antarctic waters (Habura et al., 2004) had showed that even the composition of this group is mostly unknown as yet.) The supposition of traditional taxonomists that ancestral forms of the later rose skeletal foraminiferal groups (representing separate phyletic lineages – classes) existed among the unilocular soft-walled forms representing distinct phyletic lines long ago was later supported by data of molecular analysis (Habura et al., 2004).

Discussion. The question concerning which of the cases of morphological similarity of organisms arose as a product of their parallel development and which ones – as a result of convergent development is a key-question for right understanding of the evolution and classification of the different groups, and of the correct definition of the taxa.

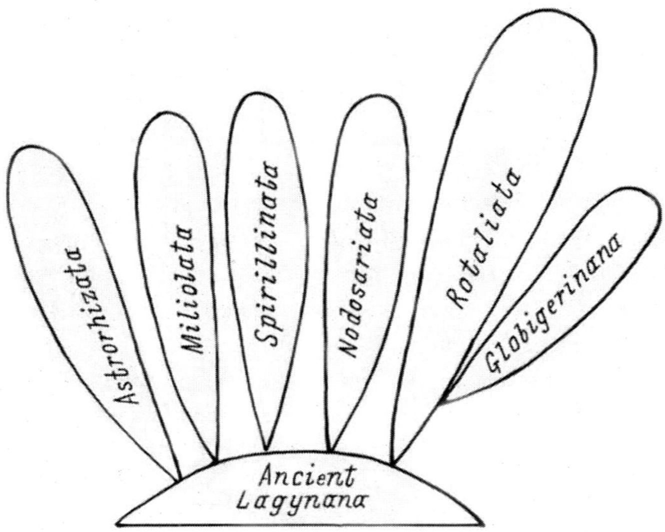

Text Figure 2. Scheme of the origin of the different Foraminiferal classes (after Mikhalevich, 1992).

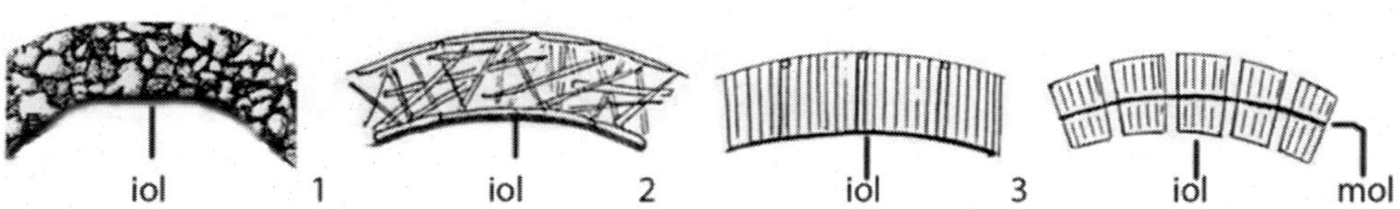

Text Figure 3. Ultrastructure of the calcareous shell wall (scheme).
1 – microgranular wall with the randomly situated grains of variable size occurring in the lower representatives of all the foraminiferal classes; 2 - porcellaneous wall with randomly oriented needle crystals, characteristic of Miliolata; 3 - hyaline wall with crystals situated perpendicularly to the shell surface, characteristic of Nodosariata; 4 - hyaline wall with perpendicularly organized crystals situated in two layers and separated by the additional organic layer characteristic, for the Rotaliata; (iol - inner organic lining, mol -middle organic lining (mol = POM) between two lamellae of the Rotaliata wall (after Mikhalevich 2000, modified).

In the foraminiferal classifications, widely adapted during the last two centuries and composed under the approach giving predominant significance to shell wall composition and structure, one meets a huge amount of pairs of agglutinated shells which are strikingly isomorphic analogues with one or another kind of the forms having the secreted wall. They were for a long time regarded as convergent forms. Only some clear examples are demonstrated here (Plate **42**):

1. Examples begin with the simplest unilocular *Lagenammina* and *Pygmaeoseistron*, *Saccammina* (a very interesting Antarctic species described by Mikhalevich, Pronina and Nestell (Mikhalevich et al., 2000) possessing a needle-like projection at its base) and *Lagena* with a similar projection, up to the uniserial pseudochambered *Hormosinella* and *Grigelis*, *Polychasmina* and *Geinitzina,* morphotypes of the agglutinated *Pseudopalmula* with unusual V-shape chambers and of the calcareous *Kyphopyxa*, and lastly among forms having a peculiar high-spire: the agglutinated *Nouria* and polymorphinids *Abdulevia* and *Paleopolymorphina* (Plate **42**, figure - s 1 – 7).
2. These examples continue among the planispiral and trochospiral pseudo-two-chambered forms with a long tubular second chamber (*Ammodiscus – Spirillina, Arenoturrispirillina – Trocholina)* and advanced multi-chambered forms obviously of the same line(*Alpillina –Ferayina* and some others) (Plate **42**, figure - s 8 – 11).
3. Such examples of isomorphism are very demonstrative among the shells with few (usually 2) tubular chambers per whorl both in agglutinated and calcareous secreted forms with different miliolid types of coiling [the irregular (*Schlumbergerina – Miliola*) and regular glomerate (*Dentostomina – Quinqueloculina*), sigmoid [*Sigmoilopsis – Neosigmella (=Sigmella*)], planispiral (*Psamminopelta – Spiroloculina*) and combined miliolid types of coiling (*Pseudoflintina,* early stage quinqueloculine with two chambers per whorl, later planispiral with 3 chambers per whorl – *Flintina,* early stage triloculine with two chambers per whorl, later planispiral with 3 chambers per whorl)] as well as in Miliolid forms with wide and more numeral chambers per whorl (*Trochamminoides* - *Hechtina*) (Plate **42**, figure - s 13 - 16).
4. Similar agglutinated and calcareous multilocular analogues are well known among the trochoid genera (Trochamminidae – Rosalinidae (Plate **42**, figure 12; Plate **43**, figure- s 11b, 12 a, b), planispiral forms (Haplophragmoididae – Nonionidae (Plate **la**, figure- s Textulariana -1, Rotaliana -1) and the monoaxial forms derived from them (Plate **la**, figure- s Textulariana - 2, 3, 4, Rotaliana - 6, 7, 4) having in principal a different plan of the shell structure than previous examples.

The list of such examples can be multiplied many times over. Such isomorphic repetitive forms were considered to have evolved as a product of evolutionary convergence. The phenomenon of convergence is known in the evolution of different groups of organisms but, perhaps, in no other group is it encountered so often as in Foraminifera. It is as if we have two nearly mirror reflections differing only in their wall character. It is unlikely that such full coincidence of their minute details in such pairs of genera having such a complex shell structure as *Tiphotrocha – Neoconorbina*, *Asterotrochammina – Discorbis*, *Spiroplectinata – Euvigerina*, *Minouxia – Sporobulimina*, *Toretammina – Schwantzia* (Plate **1a**, V – Rotaliata, Plate **43**, figure - s 12 a – 12 b) and in many other cases could arise so many times convergently. Why are they so strikingly similar? The isomorphic calcareous forms were subdivided into several taxa long ago. Is it right to unite all their agglutinated analogues into a single taxon, including all the known foraminiferal morphotypes in the former Textulariina as was previously done?

Is it possible to give them distinctive diagnoses when their only common feature is the agglutinated character of the wall? [(With the separation of the orders Astrorhizida, Lituolida and Textulariida (Loeblich & Tappan, 1992) it became possible to award the distinctive diagnosis to the first one, but two others are still too broadly defined)].

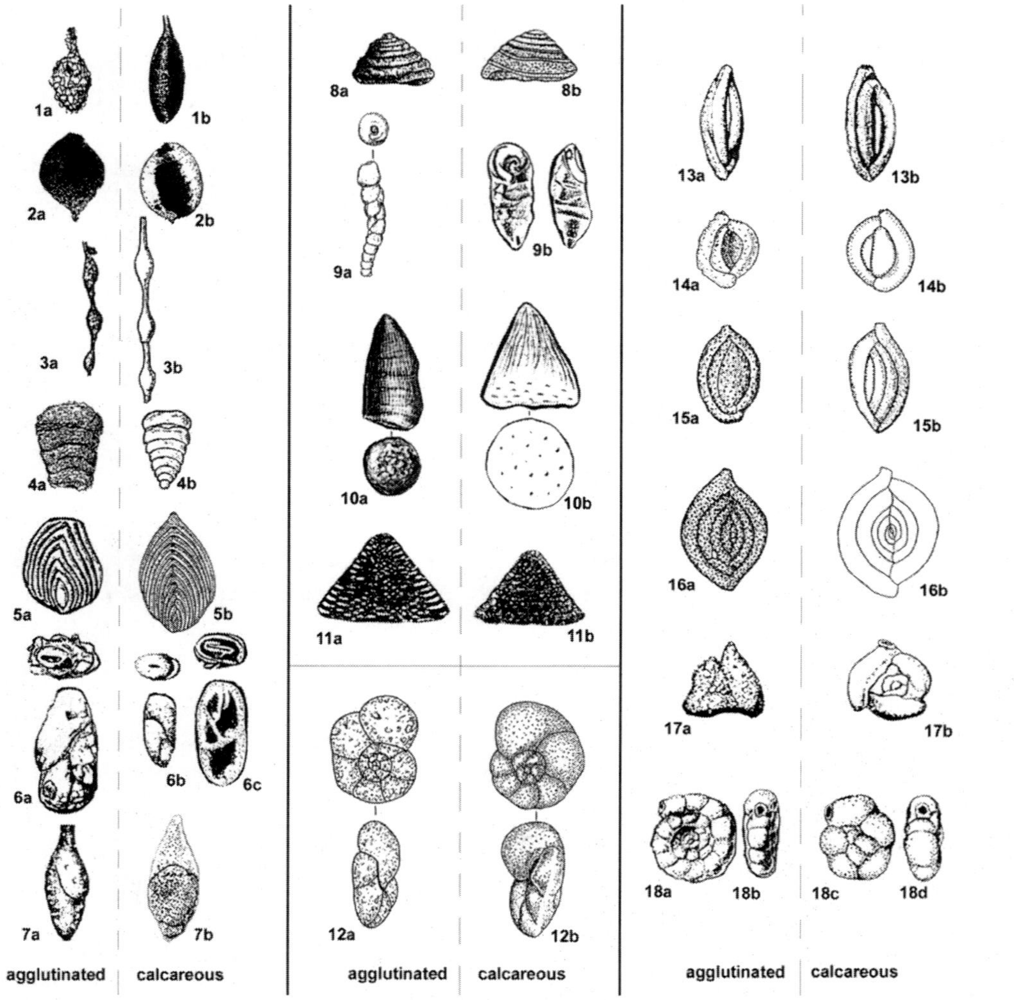

Plate 42. Examples of homeomorphism among the agglutinated and calcareous secreted forms (I, II, III, IV).

I. Class Nodosariata - 1 – 7: 1a - *Lagenammina*, 1b - *Pygmaeoseistron*; 2a - *Saccammina*, 2b - *Lagena*; 3a - *Hormosinella*, 3b - *Grigelis*; 4a - *Polychasmina*, 4b - *Geinitzina*; 5a - *Pseudopalmula*, 5b - *Kyphopyxa*; 6a - *Nouria*, 6b - *Obliquilingulina*, 6c - *Entomorphinoides*; 7a - *Abdulevia*, 7b - *Paleopolymorphina*; **II**. Class Spirillinata - 8 – 11: 8a -. *Arenoturrispirillina*, 8b - *Trocholina*; 9 - each whorl formed by one chamber (9a - *Plagioraphe*, 9b - *Ungulatella)*, 10a - *Haurania*, 10b - *Ferayina*, 11a - *Alpillina*, 11b – *Chapmanina*; **III**. 12. Class Rotaliata: 12a - *Trochammina*, 12b - *Discorbis*; **IV**. Class Miliolata - 13 – 18: 13a - *Schlumbergerina*, 13b - *Miliola*; 14a - *Dentostomina*, 14b - *Quinqueloculina*; 15a - *Sigmoilopsis*, 15b – *Neosigmella (= Sigmella)*; 16a - *Psamminopelta*, 16b - *Spiroloculina*; 17a - *Pseudoflintina*, 17b - *Flintina*; 18a,b - *Trochamminoides*, 18c,d - *Hechtina*.

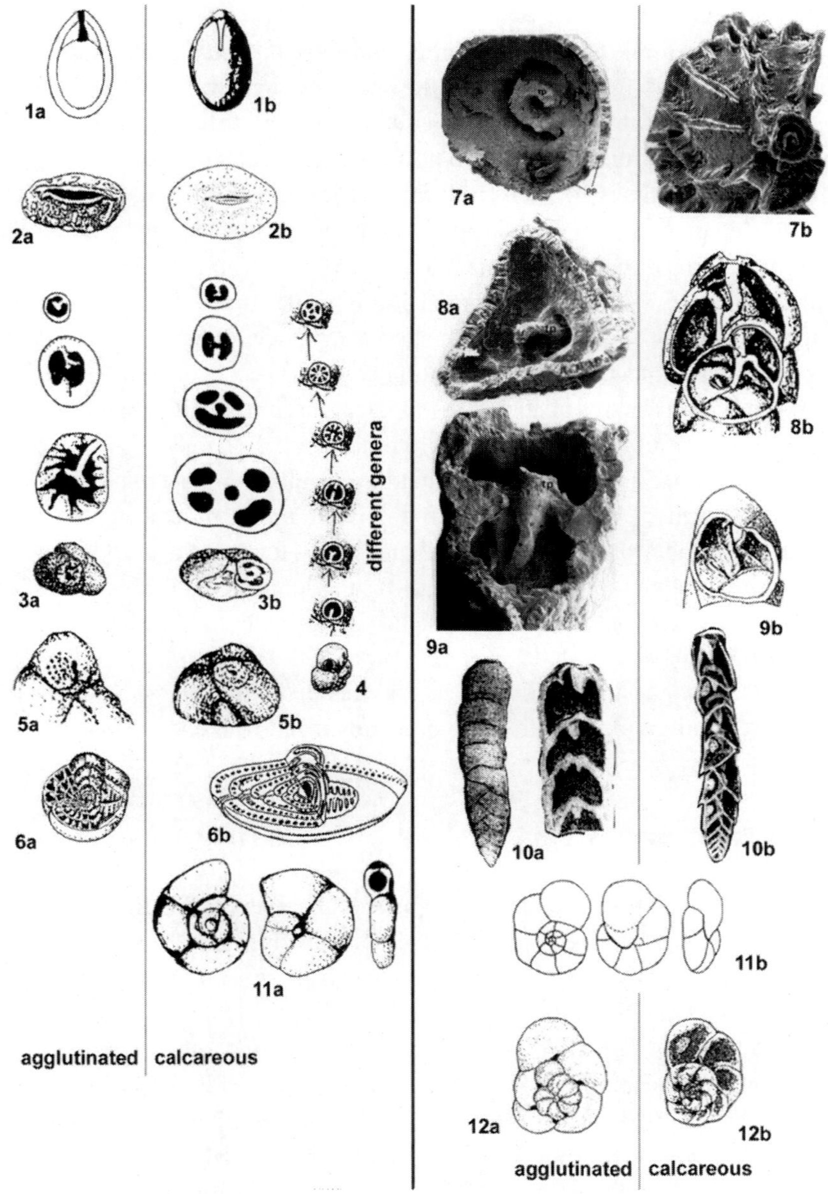

Plate 43. Examples of similarity of position and the structure of the main and additional apertures in the homeomorphic agglutinated and calcareous forms (A – agglutinated shell, B – calcareous shells).
1 – 11 – main aperture, 1a - *Ovammina*, 1b - *Oolina*, 2 a – *Cuneata*, 2 b - *Fissurina*, 3 - scheme of the stages of formation of the inner tooth apparatus during the individual growth in: 3a - *Dentostomia*, 3b - *Tortonella*, 4 - schematic changes of the development of the inner tooth apparatus in phylogeny of Miliolid line: 5 a - *Schlumbergerina*, 5b - *Miliola*, 6a - *Reticulinella*, 6b - *Alveolina*, 7 –10 - Inner apertural structures in some Textulariana and Rotaliana: 7a, 8a, 9a - *Clavulina* (after Hottinger et al., 1993), 7b - *Neouvigerina* (after Hottinger et al., 1993), 8b - *Bulimina* (after Hofker, 1947), 9b - *Neobulimina* (after Hoeglund, 1947), 10 a, a' - *Siphobigenerina*, 10 b - *Euloxostomum*, 11 - the different position of the main aperture in trochospiral forms of different origin: 11a - *Fisherinella* type of aperture (terminal) (Miliolata line), 11 b - *Trochammina-Discorbis* type of aperture (basal or interiomarginal) (Rotaliata line), 12 - additional apertures: 12 a - *Toretammina*, 12 b - *Schwantzia*.

In an effort to understand the whole Foraminiferal group I undertook a comparative morphological analysis of all the main Foraminiferal morphotypes, evaluating the taxonomic weight of the main morphological features of the shell and their wall (Mikhalevich, 1980 a, b; 1981). The initial attempt to compose the system strictly following theprinciple of the predominant significance of the shell wall composition gave rather absurd scheme. As a result of the aforementioned analysis, it transpired that basing mostly on the shell morphology it was possible to separate the main well-delimited morphotaxonomic groups regarded as classes: Astrorhizata, Spirillinata, Miliolata, Nodosariata and Rotaliata. Each of the four latter classes in turn contains within their composition, aside from their calcareous (higher) subclasses, also subclasses containing agglutinated forms (see above), and the most primitive class Astrorhizata contains tectinous and agglutinated forms.

In spite of the high diversity of the forms in each class, it's still possible to outline their main features using the following criteria[2]: 1. the number of chambers, 2. the characteristic chamber form, 3. pathways of chamber formation resulting in the character of chamber conjunction, 4. the form of the shell and predominant mode of coiling, 5. the form and position of the main aperture, 6. the development of inner apertural structures, 7. the development of integrative apertural systems, 8. the presence or absence of additional apertures, 9. the presence of additional skeletal plates, 10. the presence or absence of integrative systems of the whole shell.

Let us review the expression of all of these features in the following classes, each of them including subclasses with more primitive and more advanced wall character as well.

VIII.10.1. The Number of Chambers and Characteristic Chamber Form

In Astrorhizata – shells unilocular, rarely pseudochambered, chambers mostly circular, tubular or irregular in form. Some representatives of this class may actually belong to other protistan groups, as was shown earlier for *Syringammina* (Tendal, 1979), *Pelosina* (Mikhalevich & Voronova, 1999) and for some other genera;

In Spirillinata - shells pseudo-two-chambered with a long tubular second chamber or derived from them multi-chambered, often with a preserved initial tubular part (in multi-chambered forms the chambers are isomorphic, sensu Hottinger, 2000);

In Miliolata – shells are predominantly multi-chambered; unilocular ones are an exception, pseudo-two-chambered forms also exist, in multi-chambered shells the chambers are mostly tubular, a few (2 - 3) per whorl, in some shells having wide chambers and in higher number per whorl but nevertheless mostly bearing some tubular features, an initial tubular flexostyle is often present;

In Nodosariata – shells may be unilocular (in lagenids, which I don't consider as secondarily unilocular, as well as in their agglutinated analogues the saccamminids), but

[2] It would be worth emphasizing that namely the use of the complex of all these criteria enables to define and characterize each class rather than the use of the apertural character only as it was understood by some authors after the publication on the main morphological trends in the development of foraminiferal aperture (Mikhalevich & Debenay, 2001). This article was the first publication in the widely known English-language journal describing the new approach to the foraminiferal taxonomy with the apertural structures taken as the most demonstrative example but with consideration of the other morphological features as well.

mostly multi-chambered, chambers subsphaerical, isomorphic, often of a special V-shaped form not met with in any other phyla;

In Rotaliata –shells exclusively multi-chambered, chambers mostly isomorphic, no tubular chambers present.

Thus among the multi-chambered shells in Spirillinata and Miliolata we can see the predominance of tubular chambers, in Nodosariata – the presence of special V-shaped chambers, in Rotaliata – the predominance of isomorphic chambers. These features characteristic of each class are expressed in both subclasses of every class under consideration.

VIII.10.2. Pathways of Chamber Formation

There are two possibilities: a. pseudochambers - formed under continuous growth when the previously formed chamber lumen is subdivided; b. true chambers – newly formed units as a result of additional growth (see Part VIII. 4.). These two modes of chamber formation give us the different chamber conjunction character (Mikhalevich, 1981) (Plate 35, figure - s A – H). In the Astrorhizata only the first mode is present, and pseudochambers only are sometimes formed. Among the Spirillinata – both modes are present, but the first mode predominates. In the Miliolata – also both modes are present, but the second mode predominates. In the Nodosariata – the first mode is very rare, and we mostly observe the second one. In the Rotaliata – exclusively the second mode is present and the genera form true chambers, the first mode is only present as an additional mode forming chamberlets within chambers.

VIII.10.3. Predominant Mode of Coiling

The coiling mode of pseudo-two-chambered (= pseudobilocular) and multi-chambered shells results in the form of the shell. In the Spirillinata (Plate **1a**, II) – the shell coiling mode may be irregular glomerate, planispiral (often combined with the glomerate mode), or trochospiral. In the Miliolata (Plate **1a**, III) – coiling may be predominantly irregular and regular glomerate and planispiral, often combined. The trochospiral coiling mode is only found as an exception (e.g., *Fischerinella* - (Plate **1a**, III, figure 8; Plate **43**, figure 11a)). Such type of coiling is so unusual for the Miliolata, that I checked the ultrastructure of its wall to determine whether or not it really belongs in the miliolids (Mikhalevich et al., 1988, figure – s 6, 2 a, 2 b)). In Nodosariata (Plate **1a**, IV) – the predominant shell types are elongated uniserial and monoaxial, sometimes biserial, and planispiral. Additionally, special polymorphinoid and plectofrondicularioid types occur. The trochoid coiling modes are practically absent. Among the Rotaliata (Plate **1a**, V) – nearly exclusively plani- and trochospiral forms, as well as monoaxial forms derived from both of them (mostly biserial and triserial). The predominance of one or another type of coiling is similar among pairs of subclasses within their respective class.

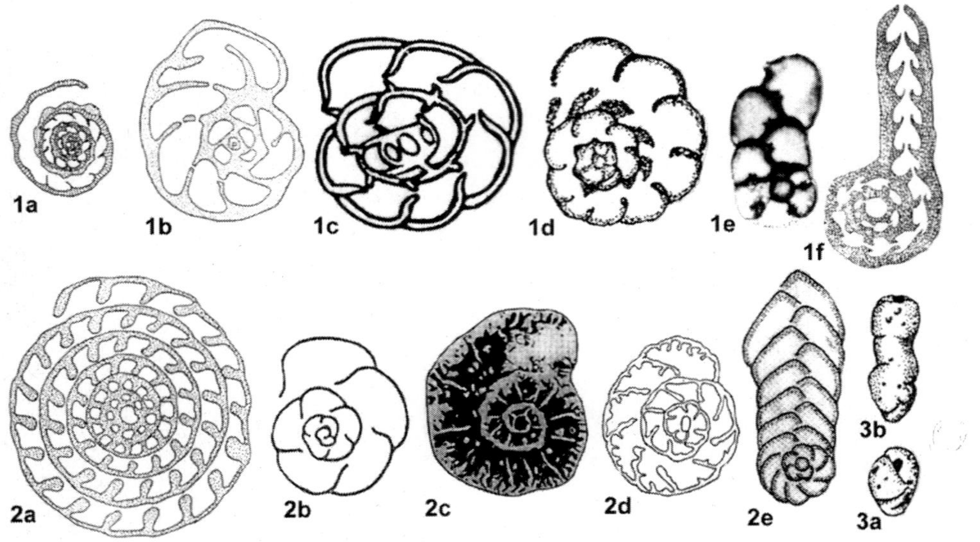

Plate 44. Position of the main aperture in some agglutinated and microgranular genera (1 and 2 – in sections or (1e) – in transmitted light) (scheme).
1 - areal and interio-areal aperture: 1a - *Melathrokerion praesigali* - areal (after Arnaud-Vanneu, 1980), 1b - *Lituola strogguloides* - areal (after Arnaud-Vanneu, 1980), 1c - *Recurvoides laevigatum* – interio-areal (after Hoeglund, 1947), 1d - *Paraendothyra nalivkini* – areal (after Fursenko, Rauser-Chernousova, 1959), 1e - *Orectostomina camposi* – areal (after Brönnimann and Beurlen, 1977), 1f - *Ammobaculites agglutinans* – areal (after Hoeglund, 1947); 2 - basal (interiomarginal) aperture: 2 a - *Nautiloculina cretacea* (after Arnaud-Vanneu, 1980), 2b - *Haplophragmoides membranaceum* (after Hoeglund, 1947), 2c - *Hemicyclammina sigali* (after Maync, 1953), 2d - *Remaneica helgolandica* (after Hoeglund, 1947), 2e - *Spiroplectammina variabilis* (after Subbotina *et al.*, 1981); 3 - change of the apertural position during ontogenesis of the shell: *Goesella flintii,* 3 a – young stage, 3 b – adult stage (after Mikhalevich, 1983).

VIII.10.4. Form and Position of the Main Aperture

The form and position of the main (initially terminal or areal or initially basal (Plate **44**)) and additional apertures, the development of inner apertural structures and of integrative apertural systems is the most demonstrative and conclusive character for the description of foraminiferal classes and may serve as one of key elements to distinguish them (Mikhalevich, 1980 a, b; 1981, 1992; Mikhalevich & Debenay, 2001).

In Astrorhizana (Mikhalevich, 1995, Mikhalevich & Debenay, 2001; Plate **1a**, I; Plate **2**, figure- s 1, 2; Plate **36**, figure- s 1, 2, 4, 5; Plate **37**, figure, 1; Plate **38**, figure- s 1 – 4; Plate 45, I and II of this study) – the aperture may be absent, may be of irregular position and form, or may be primary multiple. The single aperture in elongated forms is terminal (Plate **45**, figure- s 8, 12). Exactly the same character of the aperture together with the similarity of their shells in representatives possessing the tectinous wall of the subclass Lagynana (= Allogromiina) gives me reason to unite both groups into a single taxon – the class Astrorhizata (Mikhalevich, 1999, Mikhalevich, 2000). This is consistent with the opinion of Loeblich & Tappan (1987) and my own previous position, in which both taxa were considered as having equal rank. The artificial character of the separation of these two groups into separate rather remote taxa was previously taken as artificial by some other authors (Lipps, 1973, Bowser et al., 1995). The joining up of the isomorphous tectinous and agglutinated shells in a single taxon was supported by molecular data (Pawlowski, 2000, Pawlowski et al., 2003).

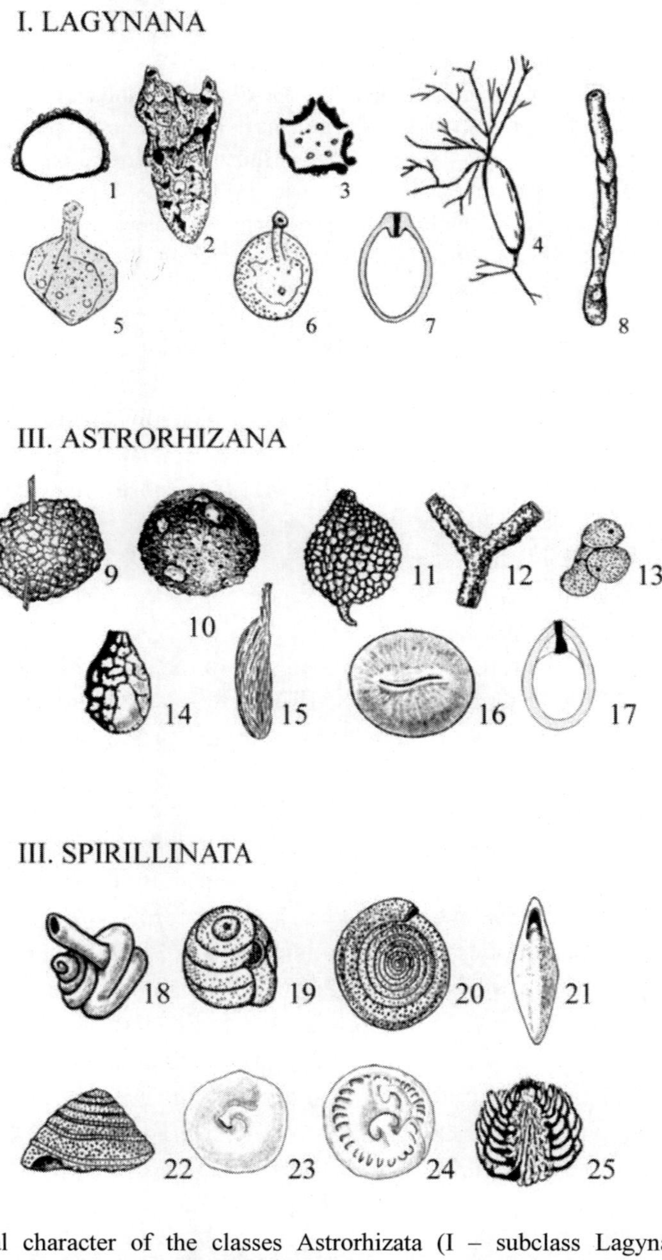

Plate 45. Apertural character of the classes Astrorhizata (I – subclass Lagynana, II – subclass Astrorhizana) and Spirillinata (after Mikhalevich & Debenay, 2001, modified).

I. 1 - *Myxotheca* - section, aperture not developed; 2 - *Cribratina* - shell with primary-multiple aperture; 3 - *Microcometes* - primary-multiple aperture; 4 - *Phainogullmia* - shell with two apertures; 5 - *Heterogromia* – aperture at the end of a neck; 6 - *Paralieberkuehnia* - aperture at the end of a neck, with a small border; 7. *Allogromia* – section, aperture with entosolenian tube; 8 –*Xenothekella* – aperture at the end of the tubular chamber.

II. 9 - *Psammosphaera* - shell without a definite aperture; 10 - *Orbulinelloides* - shell with primary-multiple aperture; 11 - *Armorella* - shell with two apertures; 12 - *Rhabdammina* – apertures at the open ends of the tubes; 13 - *Sorostomasphaera*– aperture as a circular opening on the shell surface; 14 - *Lagenammina* – aperture at the narrowing end of the shell; 15 - *Technitella* – aperture at the end of the elongated neck; 16 - *Pilulina* – aperture as a curved fissure with a small lip; 17 - *Ovammina* – section, aperture with entosolenian tube (later transferred to the subclass Hormosinana, order Saccamminidae, subfamily Ovammininae Mikhalevich, 2004).
III. 18 - *Ammovertellina (*Ammodiscana) – terminal aperture; 19 - *Repmanina* (Ammodiscana) - aperture at the end of the tubular pseudochamber; 20 - *Spirillina* (Spirillinana) – aperture at the end of the tubular pseudochamber; 21 - *Angulodiscus* (Spirillinana) - aperture on the peripheral margin, at the end of the tubular pseudo-chamber; 22 - *Trocholina* (Spirillinana) - aperture on the peripheral margin, at the end of the tubular pseudo-chamber, slightly displaced to the peripheral side; 23 - *Mychostomina* (Spirillinana) - aperture at the end of the tubular pseudo-chamber; displaced to the umbilical area and opening towards umbilicus; 24 - *Patellina* (Spirillinana) - aperture in the umbilical area, covered by a T-shaped plate; 25 - *Lasiotrochus* (Spirillinana) – canal system.

In Spirillinata (Mikhalevich, 1981, 1992, 1995, Mikhalevich & Debenay, 2001; Plate **1a**, II; Plate **45**, III of this study) – the aperture is terminal, at the end of the tube, although in trochoid, and especially in advanced multi-chambered calcareous shells it shows a tendency to be on the umbilical side (e, g. *Mychostomina* - Plate 45, III – figure 23). In the more primitive agglutinated forms (subclass Ammodiscana) it is more often terminal in position (Mikhalevich & Debenay, 2001; Plate **45**, figure - s 18, 19 of this study). With the inclusion in this class of textularielloids and chapmaninids (see Part XI.4.2) it would be possible to document the presence of the aperture as a numerous pores on a terminal face (e.g. *Coskinolina*) or as a multiple openings of the tube pillars (*Chapmanina*), patellinids also have a special character of their aperture on the umbilical side of the shell.

In the Miliolata (Mikhalevich, 1981, 1988, 1992, 1999; Mikhalevich & Debenay, 2001; Plate **1a**, III; Plate **2**, figure 8 a - c, Plate **4**; Plate **5**, figure- s 2 c; and Plate **4**, 7; Plate 11, figure- s 4, 5; Plate **19**, figure 1; Plate **22**, figure- s 1, 5, 6; Plate **36**, figure- s 14 – 16, 18 – 20, 23, 24; Plate **38**, figure- s 4 – 10; Plate **39**, figure- s 7 – 15; Plate **40**, figure- s 1 -3; Plate **41**; Plate **43**, figure- s 1 a, b, 2 a, b and Plates **46**, **52**, **53** of this study) – the aperture is always terminal in both subclasses Miliamminana and Miliolana. It is terminal and not connected with the base of the apertural face even among such coiled forms as *Fischerinella* (Plate **43**, figure 11a), and in the peneroplids, and alveolinids (Plate **51**, figure 8), (Mikhalevich & Debenay, 2001), as well as in fusulinoids (see Part XI.2.2.) the openings are always elevated above the base of the apertural face. Note the difference in the position of the aperture among the trochospiral forms of Miliolana (*Fischerinella*) and Textulariana (*Trochammina*) – Plate **43**, figure- s 11a and 11b). This difference is also demonstrative if we compare the Miliolata planispiral forms having multiple wide chambers in a whorl (agglutinated *Trochamminoides*, calcareous *Hechtina* (Plate **42**, figure- s 18 a-d, Plate **53**, figure- s 7, 8) with the planispiral agglutinated *Haplophragmoides* and calcareous *Nonion* in Rotaliata (Plate **48**, figure- s 1, 2).

In Nodosariata (Mikhalevich, 1980, 1981, 1992, 1995, Mikhalevich & Debenay, 2001; Plate **1a**, IV; Plate **2**, figure- s 4 – 6, 12 A, 14 A, B; Plate **5**, figure- s 2a – 2d, 4c, 4f; Plate **15**, figure- s F – J; Plate **22**, figure- s 1, 5, 6; Plate **36**, figure- s 25 – 35; Plate **38**, figure- s 11 – 18; Plate **39**, figure- s 16 – 24; Plate **43**, figure- s 1 a, b, 2 a, b; and Plate **47** of this study) – the aperture is always terminal in both subclasses Hormosinana and Nodosariana and often situated at the peripheral angle of the last chamber. It is terminal even among their coiled forms (Plate **15**, figure F, Plate **36**, figure. 34; Plate **39**, figure 24, Plate **40**, figure 5, Plate **47**, figure- s 25, 29). This feature facilitates the distinction between e.g., Lenticulininae and some involutely coiled Rotaliina with a basal aperture (Plate **40**, figure 6; Plate **48**, figure- s 1, 2, 17 – 19, 23). Accordingly, this is why it is necessary to move the agglutinated planispiral forms with a terminal aperture positioned at an angle of the shell from the Textulariana to other corresponding taxa, namely to the Hormosinana.

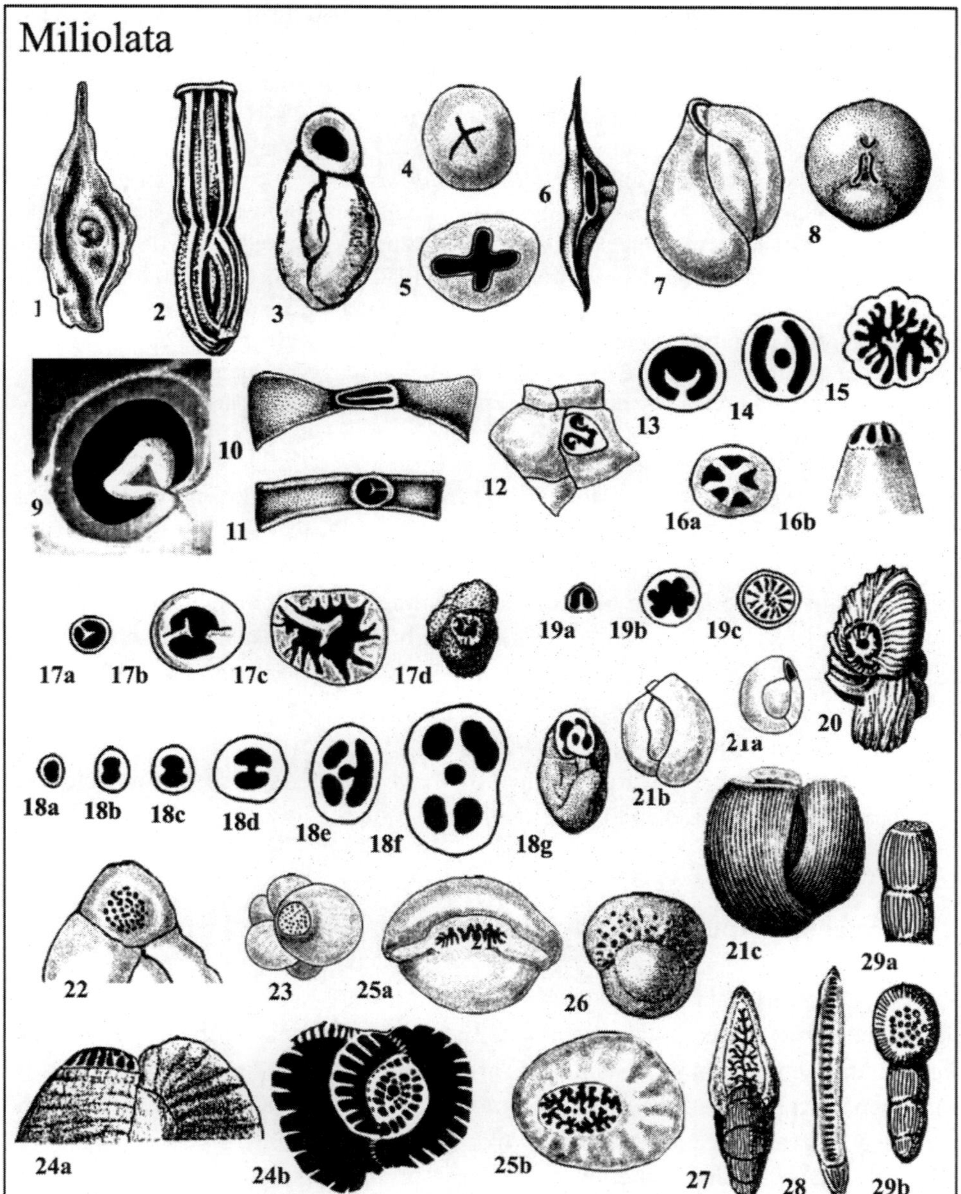

Plate 46. Apertural character of the class Miliolata (after Mikhalevich & Debenay, 2001).

1 - *Palaeomiliolina* (Miliolana) - simple aperture at the end of an elongated neck; 2 - *Articulina* (Miliolana) - simple aperture with a broad border; 3 - *Wiesnerella* (Miliolana) - simple aperture with a broad everted lip; 4 - *Nodophtalmidium* (Miliolana) - terminal cruciform aperture; 5 - *Rupertianella*

(Miliolana) - simple aperture with a lip; 6 - *Edentostomina* (Miliolana) - simple aperture; 7 - *Miliolinella* (Miliolana) - aperture with a flap; 8 - *Pyrgoella* (Miliolana) - aperture with a flap; 9 - *Pyrgo laevis* (Miliolana) - EM (x1000), structure of the inner tooth (from Mikhalevich 1988); 10 - *Spiroloculina* (Miliolana) - aperture with simple tooth; 11 - *Spiroloculina* (Miliolana) - aperture with bifid tooth; 12 - *Podolia* (Miliolana) - complex tooth gives lyre-shape appearance to the aperture, in the early stages the tooth is simple; 13, 14, 15 - bifid tooth (13) and complex teeth (14, 15) of different Miliolata; 16 a, b - *Nevillina* (Miliolana) - multiple radial teeth fuse, forming a multiple aperture divided by teeth (a - apertural view, b - edge view); 17 a-d - *Dentostomina* (Schlumbergerinana) - a, b, c - apertural structure from different age stages, d - apertural view of the shell (from Mikhalevich 1983); 18 a-g - *Tortonella* (Miliolana) - a-f - apertural structure from different age stages, g - apertural view of the shell (from Didkovskij 1957); 19 a-c - *Idalina* (Miliolana) - apertural apparatus from different age stages (from Munier-Chalmas & Schlumberger 1885); 20 - *Cribrolinoides* (Miliolana) - aperture with a complex annular tooth; 21 a, b, c - *Flintina* (Miliolana) - apertural apparatus from different age stages (from Mikhalevich 1983); 22 - *Schlumbergerina* (Schlumbergerinana) - multiple aperture with trematophore; 23 - *Miliola* (Miliolina) - multiple aperture with trematophore; 24 a, b - *Pseudohauerina* (Miliolana) - multiple aperture with trematophore (a - edge view, b - section, pillars supporting trematophore are clearly seen (from Ponder 1972); 25 a, b - *Periloculina* (Miliolana) - multiple aperture with trematophore through ontogeny (a - at the biloculine stage, b - at the adult stage with the embracing last chamber) (from Munier-Chalmas & Schlumberger 1885); 26 - *Involvohauerina* (Miliolana) - areal multiple aperture; 27 - *Dendritina* (Miliolana) - dendritic aperture; 28 - *Peneroplis* (Miliolana) - aperture as a row of openings; 29 - *Coscinospira* (Miliolana) - areal multiple aperture (a - side view, b - top view).

In the Rotaliata ((Mikhalevich, 1980, 1981, 1992, 1995, Mikhalevich & Debenay, 2001; Plate **1a**, V; Plate **3**; Plate **16**, figure 1; Plate **35**, figure- s D, E; Plate **40**, figure 6; Plate **42**, figure 12; Plate **43**, figure 11b; Plate **44**, figure - s 2a – 2e; 3a and Plate **48** of this study) – the aperture is basal in position, at the base of the apertural face in both subclasses Textulariana and Rotaliana, at least initially, becoming secondarily terminal usually in some monoaxial forms (Mikhalevich & Debenay, 2001; and Plate **5**, figure- s 4 c, 4 d; Plate **10**, figure 2 a; Plate **38**, figure- s 19 – 23; Plate **40**, figure- s 4, 11, 12; Plate **43**, figure- s 7 – 10; Plate **48**, figure- s 12 – 15 of this study).

The change in apertural position in the Rotaliata lineage is strictly connected with a profound change in the whole plan of their shell architecture.

VIII.10.5. Special Types of Apertures and Inner Apertural Structures

Among the Astrorhizata – inner apertural structures are not developed. The only inner apertural structure mentioned in this group was an entosolenian tube in the agglutinated genus *Ovammina*, but based on this fact (such tubes are characteristic of the Nodosariata) as well as on the morphological similarities of saccaminids and lagenids – saccamminids were moved from the Astrorhizana to the Hormosinana (Mikhalevich et al., 2000).

In Spirillinata – no inner apertural structures resembling some of such structures in Rotaliana or Nodosariana are developed in either subclass, although the tubular pillars of some calcareous spirillinans (Chapmaninidae) opening at the apertural surface represent a special apertural type absent in the other classes. In the advanced calcareous multi-chambered forms of this class external plates of a special S-shape or T- shape are rarely present (Mikhalevich & Debenay, 2001; Plate **1a**, II, figure 14 b; Plate **45**, figure- s 18 –25 of this study). The external plate of the aperture although of less developed form is present in *Ungulatella* (Plate **42**, figure9 b) removed here to Spirillinana. Its similarity with the Patellina's plate was noted earlier (Loeblich & Tappan, 1987, p. 555). Agglutinated *Haurania, Coskinolina* or calcareous *Ferayina* (Plate **42**, figure- s 10 a, 10 b) with their minute multiple openings at their apertural surface at the base of their high trochoid shell possibly may be referred to this class on the base of their shell structure, thus representing one more apertural type structure of Spirillinata representatives.

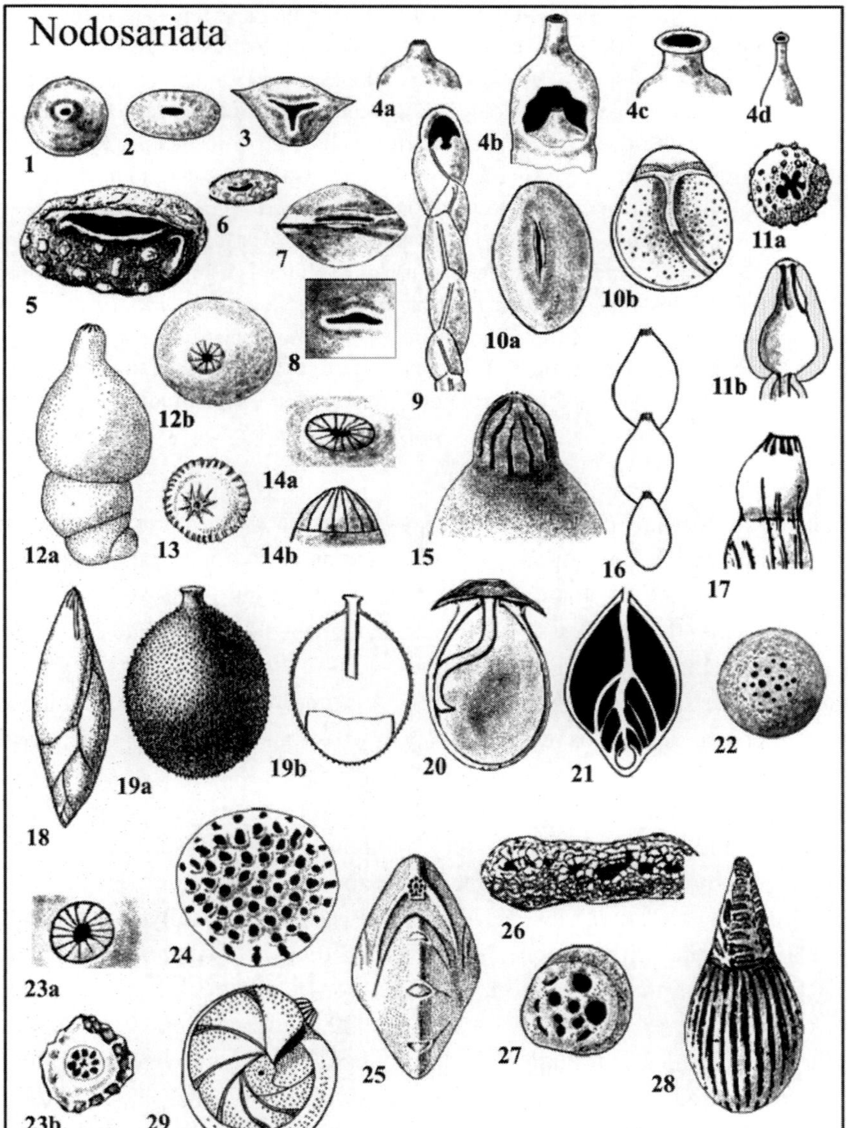

Plate 47. Apertural character of the class Nodosariata (after Mikhalevich & Debenay, 2001).

1 - *Lagenoglandulina* (Nodosariana) - radial aperture, slightly elevated; 2 - *Lingulonodosaria* (Nodosariana) - oval aperture; 3 - *Tollmania* (Nodosariana) - triradial aperture; 4 a-d - Nodosariate aperture (Hormosinana) (a - on a short neck; b - at the end of an elongated neck; c, d - on a neck with a

collar); 5 - *Cuneata* (Hormosinana) - asymmetrically fissured aperture; 6 - *Psammolingulina* (Hormosinana) - fissured aperture; 7 - *Lingulina* (Nodosariana) - fissured aperture; 8 - *Parafissurina* (Nodosariana) - hoodlike aperture (entosolenian tube is not seen); 9 - *Pleurostomella* (Nodosariana) - hoodlike aperture with entosolenian tube; 10 - *Fissurina* (Nodosariana) - asymmetrically fissured aperture with entosolenian tube ("asymmetrically fissured entosolenian aperture"); 11 a, b - *Nodosinum* (Hormosinana) - radial aperture (a - top view, b - longitudinal section, inner radial ribs are seen; 12 a, b - *Hemirobulina* (Nodosariana) - radial aperture (a - edge view, b - top view); 13 - *Pseudonodosaria* (Nodosariana) - aperture with external radial ribs; 14 a, b - scheme of the radial aperture (a - top view, b - edge view); 15: *Laevidentalina* (Nodosariana) - radial aperture closed at the apex (multiple radial aperture); 16 - *Nodosaria* (Nodosariana) -longitudinal section, radial apertures are seen; 17 - *Dentalina* (Nodosariana) - radial aperture; 18 - *Polymorphyna* (Nodosariana) - radial aperture with entosolenian tube (radial entosolenian aperture); 19 a, b - *Bombulina* (Nodosariana) - simple-entosolenian aperture (a - shell with the bordered elongated neck, b - longitudinal section) (from Mikhalevich 1983); 20 - *Irenita* (Nodosariana) - view from the dorsal side: asymmetrically fissured entosolenian aperture with broad asymmetrically developed lips; 21 - *Ellipsopolymorphina* (Nodosariana) - longitudinal section showing interconnected entosolenian tubes (from Silvestri 1904); 22 - *Cribrolagena* (Nodosariana) - primarily multiple aperture; 23 a, b - *Amphimorphina* (Nodosariana) - aperture at different stages of ontogeny (a - radial aperture of early stage, b - radial multiple aperture of adult test) (from Loeblich & Tappan 1964); 24 - *Cribrolenticulina* (Nodosariana) - radial multiple aperture; 25 - *Cribrorobulina* (Nodosariana) - radial multiple aperture; 26: *Polychasmia* (Hormosinana) - multiple aperture - a row of openings of irregular form; 27 - *Cribratina* (Hormosinana) - cribrate aperture; 28 - *Pytine* (Nodosariana) - rounded aperture at the end of the elongated neck, surrounded by a system of longitudinal and transverse ribs and by pillars going perpendicularly to the inner wall; 29 - *Lenticulina* (Nodosariana) - radial terminal aperture.

In Miliolata – in advanced forms the unique inner tooth and complex tooth-structures derived from it are well developed in both subclasses Miliamminana and Miliolana (Mikhalevich & Debenay, 2001; Plate **2**, figure 8 c; Plate **5**, figure- s 3 c – 3 m; Plate **7**; Plate **11**, figure- s 4, 5; Plate **30**, figure- s 2, 3; Plate **31 b**, figure- s 6 – 10; Plate **35**, figure- s H, I; Plate **39**, figure- s 7 – 15; Plate **41**; Plate **43**, figure- s 4 -5; Plate **46** of this study), although in the calcareous subclass Miliolana these structures are more developed and complex. Such a type of inner apertural apparatus does not occur in other classes or other agglutinated subclasses.

In Nodosariata – special types of radial apertures (sometimes with inner radial ribs), an asymmetrically-fissured aperture, an aperture with an inner entosolenian tube and with additional ampula are present – the first three of the above-mentioned types occur also within Hormosinana (radial with inner ribs in *Nodosinum,* assymmetrically-fissured aperture in *Cuneata,* and an aperture with an inner entosolenian tube in *Ovammina*) (Mikhalevich & Debenay, 2001; Plate **5**, figure- s 2 a – 2 d, 3 a, 3 b, 4 c, 4f ; Plate **35**, figure- s B, E, K; Plate **36**, figure- s 34, 35; Plate **39**, figure- s 16 – 24; Plate **40**, figure- s 5, 9, 10; Plate **42**, figure- s 1 - 7 and figure- s 1, 3: 1 - 2 and Plate **47** of this study). Here again we can see the profound similarity of agglutinated and calcareous analogues within the same class. These apertural types are not encountered in any other group of agglutinated or calcareous foraminifera and distinctly separate the subclass Hormosinana from the lituolids in the subclass Miliamminana.

Among Rotaliata – very complex special inner apertural structures are often developed, the most complex comparing the other groups of the equal taxonomic rank. These structures include tooth-plates, valves, groves, tubes (of a different kind than nodosariate entosolenian tubes) in elongated and trochoid forms not encountered in any other class (Mikhalevich &

Debenay, 2001; Plate **1 a**, figure 11; Plate **5**, figure- s 4 a – d; Plate **40**, figure- s 11 – 14 and Plate **49**, figure– s 1 – 9 of this study). The perfect photos of *Clavulina*'s inner apertural plates similar to those of some buliminids shown by Hottinger et al., (1993); (Plate **43**, figure-s 7 – 10 of this study) and similar integrative system of apertural plates in the genus *Olgia* (Mikhalevich, 2011) (Text figure 13) do not leave any doubt as to the close affinity of the agglutinated and calcareous analogues of the Rotaliata.

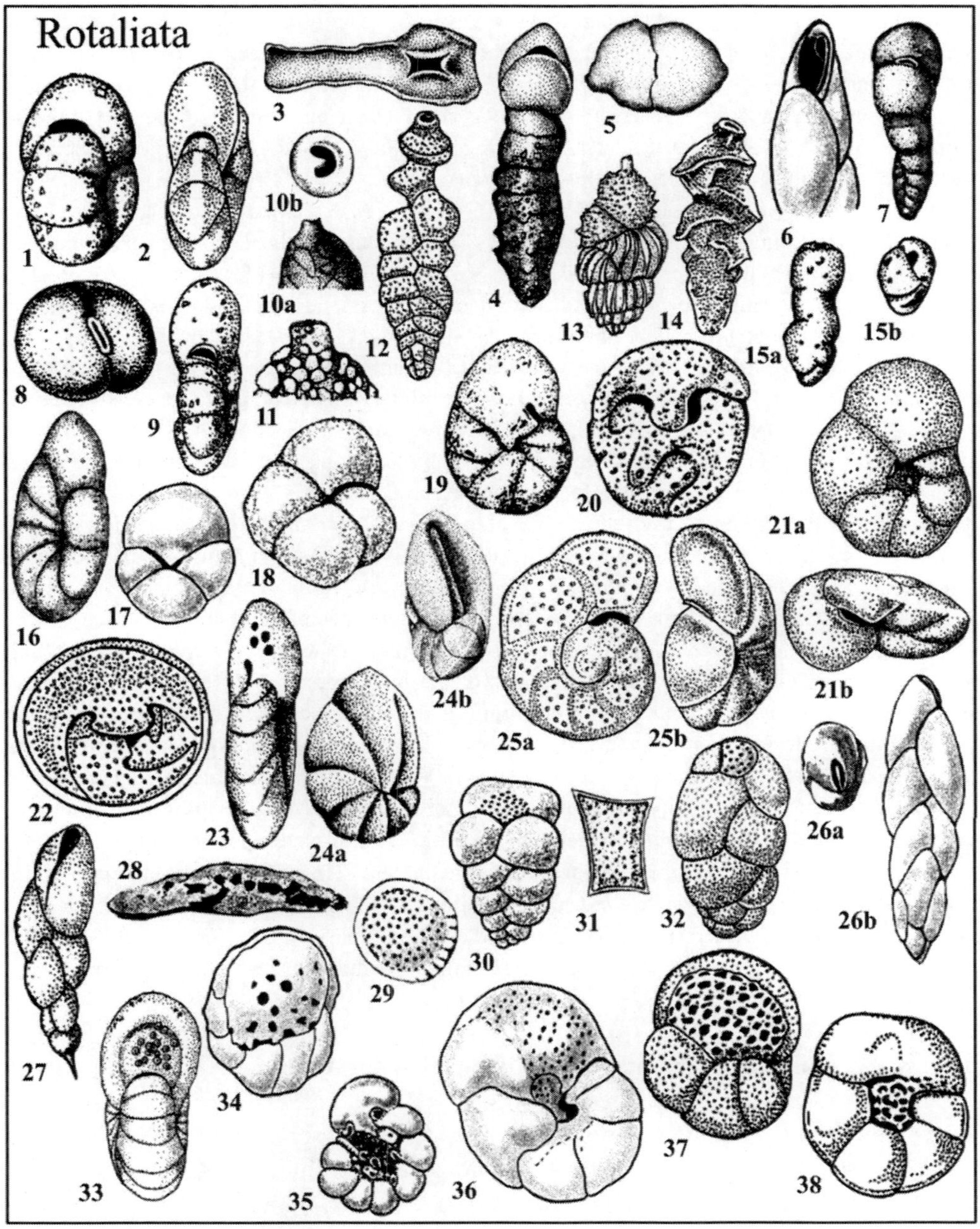

Plate 48. Apertural character of the class Rotaliata (after Mikhalevich & Debenay, 2001).

1 - *Haplophragmoides* (Textulariana) - basal arched aperture; 2 - *Nonion* (Rotaliana) - basal fissured aperture; 3 - *Tetragonostomina* (Textulariana) -basal rectangular aperture bordered by elevated lips; 4 -

Textilina (Textulariana) - basal arched aperture; 5 - *Norvanganina* (Textulariana) - basal fissured aperture; 6 - *Stainforthia* (Rotaliana) - loop-shaped aperture with narrow incurved lip at one margin and inner tooth plate (shell with inner apertural system of interconnected tooth plates); 7 - *Siphotextularia* (Textulariana) - oval bordered opening slightly elevated above the base (areal); 8 - *Karreriella* (Textulariana) - oval bordered opening slightly elevated above the base (areal); 9 - *Labrospira* (Textulariana) - areal aperture bordered by a lip; 10 a, b - *Cylindroclavulina* (Textulariana) - a - edge view, terminal aperture at the end of a short neck, b - top view, showing a tooth plate (shell with an inner apertural system of interconnected toothplates; early stages with a basal aperture); 11 - *Clavulina* (Textulariana) - terminal aperture at the end of a short neck (with an inner tooth plate; early stages with a basal aperture); 12 - *Spiroplectinata* (Textulariina) - rounded terminal aperture of the adult stage at the end of a short neck; 13 - *Euuvigerina* (Rotaliana) - terminal aperture of the adult stage at the end of a short neck (with inner apertural system); 14 - *Eouvigerina* (Rotaliana) - bordered aperture of the adult stage on a short neck (with a short inner tooth plate); 15 a, b - *Goesella* (Textulariana) - aperture at different stages of ontogeny (a - basal loop-shaped opening of a young stage, b - terminal opening of the adult stage) (from Mikhalevich 1983); 16-18 - *Trochammina* (Textulariana) - basal (interiomarginal) aperture of different species (16, 17 - on the umbilical side, entering peripheral margin, 18 - aperture wholly on the umbilical side); 19 - *Polystomammina* (Textulariana) - loop-shaped main aperture on the umbilical side and partly on the peripheral margin; 20 - *Tiphotrocha* (Textulariana) - basal aperture under umbilical chamber lobe; 21 - *Discorbis* (Rotaliana) - basal aperture on the umbilical side (with inner valves) (a - umbilical side; b - apertural view; 22 - *Neoconorbina* (Rotaliata) - basal aperture under umbilical chamber lobe; 23 - *Arenoparrella* (Textulariana) - loop-shaped aperture connected with the base of the apertural surface and areal multiple aperture; 24 - *Megastomella* (Rotaliana) - aperture as a long vertical fissure (connected with the base of the apertural surface) in deep fold of the apertural face; 25 - *Cibicides* (Rotaliana) - basal fissured aperture with a lip, going from the umbilical to the spiral side and entering the spiral suture (a - spiral side; b - umbilical side); 26 - *Fursenkoina* (Rotaliana) - elongated aperture connected with the base of the apertural face, free part of the toothplate projecting through the opening, opposite end attached to the previous foramen, forming inner apertural system (a - apertural view; b - edge view); 27 - *Fursenkoina* (Rotaliana) - comma-shaped aperture (lower part of the aperture closed, leaving only a suture to the base of the apertural face (toothplates of the inner apertural system are not seen); 28 - *Poritextularia* (Textulariana) - areal multiple aperture of the adult specimen (basal vertical elongated fissured aperture in the young stages); 29 - *Coskinolina* (Textulariana) - multiple aperture (obviously belonging to Spirillinata, Ammodiscana); 30 - *Minouxia* (Textulariana) - multiple sieve-plate aperture; 31 - *Chrysalidinella* (Rotaliana) - areal multiple aperture of adult shell (basal aperture with toothplate at the early stage); 32 - *Sporobulimina* (Rotaliana) - multiple sieve-plate aperture; 33 - *Cyclammina* (Textulariana) - basal fissured aperture and areal multiple one; 34 - *Arenagula* (Textulariana) - multiple sieve-plate aperture; 35 - ?*Dobrogelina* (?Textulariana) - multiple sieve-plate aperture; 36 - *Poroeponides* (Rotaliana) - basal fissured aperture and areal multiple one; 37 - *Neocribrella* (Rotaliana) - multiple sieve-plate aperture; 38 - *Sestronophora* (Rotaliana) - multiple sieve-plate aperture.

VIII.10.6. Presence or Absence of Additional Apertures

In the Spirillinata – sutural additional apertures along the spiral suture are found only in the calcareous subclass Spirillinana (Plate **6**, figure 2 and Text figure 4). Among the Miliolata (Plate **6**, figure 1 - *Polysegmentina*) and Nodosariata (Plate **39**, figure 18 - *Delosina*) (Revets, 1989) – sutural additional apertures occur only as exceptions within their calcareous subclasses. In the Rotaliata – sutural, peripheral and umbilical additional apertures are widely developed in both of the subclasses (Plate **6**, figure- s 3 – 10), often in a quite similar manner (e.g., *Toretammina – Schwantzia*, *Balticammina - Discanomalina* (Plate **43**, figure- s 12a, b)).

In the subclass Globigerinana adapted to the pelagic mode of life additional apertures are more strongly developed (Plate **2**, figure 10 c; Plate **6**, figure 10; Plate **35**, 2, figure M; Plate **49**, II, figure 17) (multiple additional apertures on the surface of enveloping chamber of *Orbulina* differ from its pores in significantly larger size).

VIII.10.7. Additional Skeletal Plates

These covering the sutural or umbilical cavities are well developed mostly in the Rotaliata in both subclasses, often representing mirror-image analogues (e.g., *Asterotrochammina – Discorbis, Astrononion* with umbilical plates (Plate **1a**, Textulariana, figure 6 b, Rotaliana, figure 9 b; Plate **15**, figure E), *Arenagula – Neocribrella, Lamellodiscorbis, Sestronophora* with apertural plates) (Plate **15**, figure D; Plate **48**, figure - s 34, 37, 38).

In the subclass Spirillinana, in lasiodiscids and involutinids possessing canal skeletal plates (secondary lamellar thickenings) above the umbilical area also occur, and in Hottingerellidae they possibly present above the sutures. These obviously have the similar function as in Rotaliata but rarely occur in this subclass.

In the subclass Miliolana in rhapydionins the central thickening bordered by subepidermal lamellae may play the similar role, but these structures do not received wide development in this class. Additional lamellae in the umbilical area of some other miliolats (e. g. *Vidalina*) obscuring the initial whorls do not cover any cavities resulting only in thickened umbo and thus differ in their function from those of Rotaliata and Spirillinata.

VIII.10.8. Integrative Apertural Systems (Only in Some Taxa with Multi-chambered Representatives Having Inner Apertural Structures)

In the Miliolata – integrative apertural systems are practically not developed. The only example of the attempt to develop such a system could be seen in Rhapydionidae where the prominent central thickening of the shell is pierced by so called "canals". They represent a special type of integrative system which needs further detailed study and is more primitive than true canals of Rotaliana or Spirillinana.

Among the Nodosariata – such systems are purely developed only in some representatives of the calcareous subclass Nodosariana (in *Pleurostomella, Ellipsopolymorphina* (Mikhalevich & Debenay, 2001; Plate **5**, figure - s 4 e, f; Plate **40**, figure – s 9, 10 and Plate **47**, figure - s 9, 21 of this study), although the very elongated apertural neck in the agglutinated *Ginesina* (Plate **5**, figure 2b; Plate **47**, figure 4 b) and elongated inner radial ribs in *Nodosinum* (Plate **47**, figure 11 b) may promote longitudinal cytoplasmic flow.

In the Spirillinata apertural tube pillars may rarely be joined from chamber to chamber and in this case they also can be regarded as integrative apertural system (e.g. – in chapmaninids). A special columella in the center of the shell may present in their agglutinated and calcareous representatives possibly serving as integrative system.

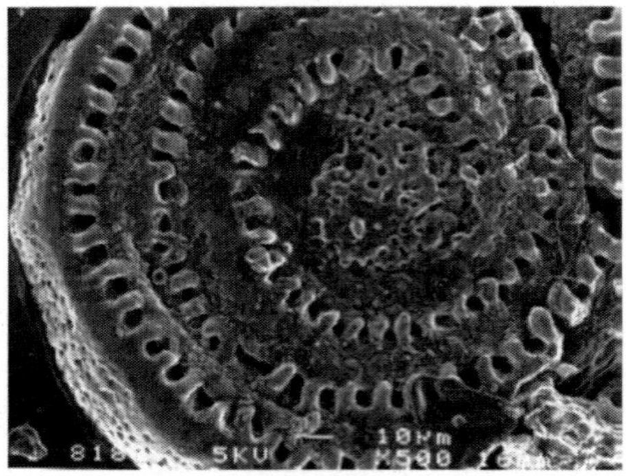

Text Figure 4. Outer view of the upper side of *Sejunctella* sp. (x 500).

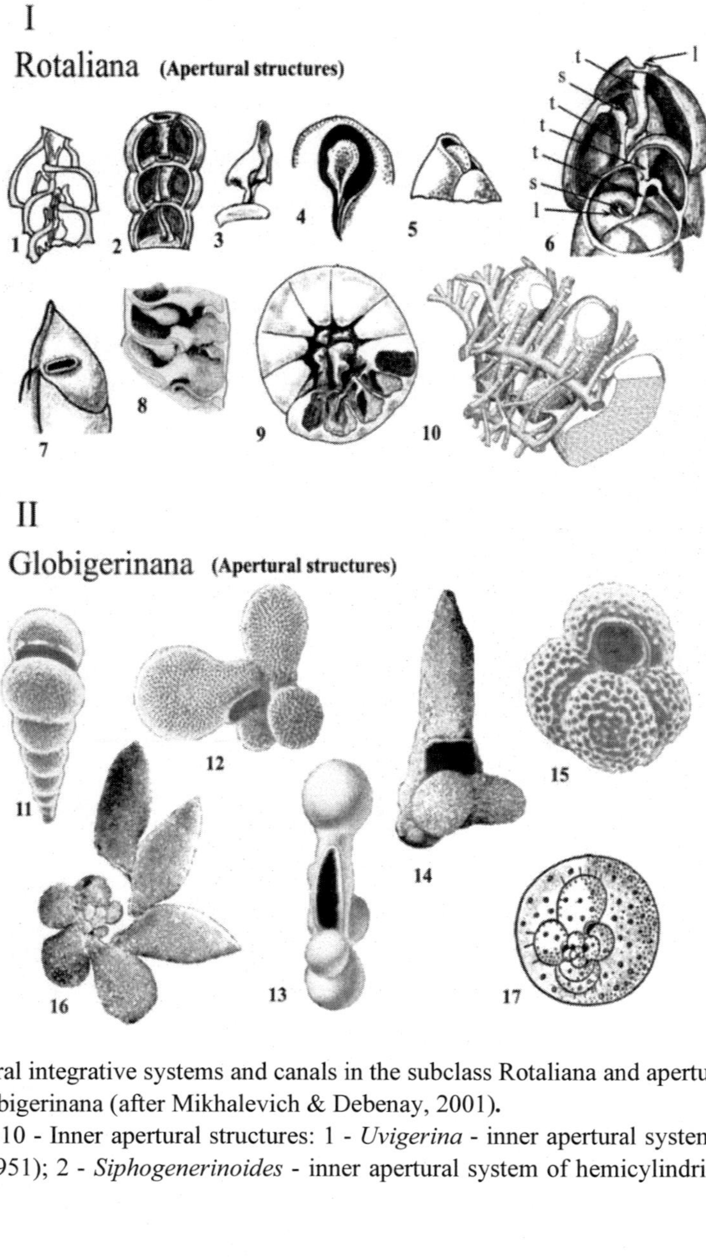

Plate 49. Apertural integrative systems and canals in the subclass Rotaliana and apertural character of the subclass Globigerinana (after Mikhalevich & Debenay, 2001).

I – Rotaliana. 1-10 - Inner apertural structures: 1 - *Uvigerina* - inner apertural system of groove-plates (from Hofker, 1951); 2 - *Siphogenerinoides* - inner apertural system of hemicylindrical groove- plates

(from Montanaro Gallitelli 1957); 3 - *Reussella* - inner toothplate (from Loeblich & Tappan 1964); 4 - *Globobulimina* - outer projecting fanlike part of the toothplate (inner parts forming inner apertural system) (from Höglund 1947); 5 - *Bolivina* - outer projecting broad short free part of the toothplate (inner parts forming inner apertural system); 6 - *Bulimina* - complex interconnected toothplates forming inner apertural system (l = apertural lip; s = free shank of the tooth plate; t = tooth plate) (from Höglund, 1947); 7 - *Osangularia* - outer view of the last chamber with areal aperture; 8 - *Osangularia* - inner part of the last chamber from the umbilical side showing bulging chamber ends, areal aperture, lip and indentation of apertural face runing to the apertural surface (from Reiss 1960); 9 - *Ammonia* - inner apertural valves forming inner apertural system (after Hottinger et al., 1993); 10 - *Rotalia* - fragment of the canal system (from Hottinger 1978).
II – Globigerinana. 11-17 - aperture of Globigerinana: 11-15: high arched gaping aperture (11 - *Laterostomella* , 12 - *Clavatorella*, 13 - *Clavigerinella*, 14 - *Beella*, 15 - *Globoturborotalia*); 16 - *Bolliella* - uncoiling spire of the last whorl with inner open ends of the chambers; 17 - *Orbulina* - multiple rounded apertures (larger than pores) on the surface of the last embracing chamber.

In the Rotaliata – integrative apertural systems are strongly developed and are widely present (Mikhalevich & Debenay, 2001; Plate **5**, figure - s 4 c, 4 d; Plate **43**, figure – s 7 – 10 and Plate **49**, figure - s 1 – 9 of this study) and although being more strongly developed in the calcareous subclass Rotaliana they nevertheless occur also in the agglutinated forms of the subclass Textulariana [e.g., in *Clavulina* (Hottinger et al., 1993 and Plate **43**, figure - s 7a, 8a, 9a of this study, in *Olgia (*Mikhalevich, 2011, Text. figure 13 of this study]. With further study, additional numerous examples among the Textulariana will possibly be found.

VIII.10.9. Development of Integrative Systems of the Whole Shell

Canals are developed in two classes: in pseudo-two-chambered forms with a long tubular second chamber (Spirillinata) and in multi-chambered forms (Rotaliata) among representatives of their higher subclasses (Plate **1 a**, II, Spirillinana, figure - s 7, 12 c, V, Rotaliana, figure 11; Plate **8**, figure – s 2, 3; Plate **40**, figure - s 15 – 19; Plate **49**, figure 10). In Nodosariata the peculiar canal system occurs only in a single genus *Delosina* (Revets, 1989). Stolons are met in both subclasses of the classes Miliolata, Spirillinata and Rotaliata (Plate **8**, figure 4; Plate **40**, figure 8). Those of agglutinated Spirillinata (subclass Ammodiscana) were studied in detail by Hottinger & Drobne, 1980. In ancient Fusulinids which are referred here to Miliolata (Mikhalevich, 2009), special system of tunnels is developed (Plate **8**, figure 1; Plate **40**, figure – s 7, 7a). According to Hottinger (see part V. 1. 3.) tunnels are present also in pfenderinids (included here into class Spirillinata) and in nummulitids (class Rotaliata). The representatives of the latter class may also possess systems of interlocular spaces. Inner integrative apetural systems are developed in Nodosariata and especially in Rotaliata (see Part V.1.3.). Central columella and system of connected tube pillars which present in some Spirillinata (see Part XI.4.2.) possibly may also be considered as a special integrative systems. The system of successive apertures in the uniserial uncoiled cylindrical part of rhapydionin's shells with their inner endoskeletal structures in the central area free of partitions may be considered as some special type of shell integrative system lacking (similar to tunnels and stolons) their own delimiting wall (see Part V.1.3.). The system of pre- and postseptal passages of alveolinids is also far from being perfect although more advanced than fusulinid integrative systems.

It is necessary to emphasise that all the taxonomic features in consideration are evolutionary and functionally significant. By combining the sets of all the above-mentioned features and also by using some additional less significant ones, it is possible to give a distinctive diagnosis to each of these lineages (classes) – whereas the previous Textulariina (sensu Loeblich & Tappan, 1987) included all the known foraminiferal morphotypes, and the only common feature for all of these were the presence of an agglutinated wall. This fact hindered the task of making a distinctive diagnosis for the group. The similarity of the subclasses comprising forms with agglutinated and calcareous walls within each class is stressed in this and previous (Mikhalevich, 1992, 1997, 1998, 2000) studies. There is also some additional evidence of the unity of agglutinated and calcareous analogues within these groups (see below – Part VIII.10.10. – VIII,10,14,).

VIII.10.10. The Similarity of Ontogenetic and Phylogenetic Development of Subclasses within the Classes

There are similar tendencies of ontogenetic and phylogenetic development within the representatives of the pairs of the subclasses in each class. For instance, the development of the complex inner tooth from the previous simple tooth in Miliolana and Miliamminana (Plate **43**, figure - s 3 - 5); the change in the position of the aperture from basal at the initial stage to terminal at the final stages in monoaxial forms of the subclasses Textulariana and Rotaliana (Plate **44**, figure 3). The tendencies of the development of the aperture in each phyletic line (class) from the simpliest forms to the more complex are demonstrated on Plate **50**.

VIII 10.11. Processes of Building New Chambers during Growth

The classes discussed above differ also in the mode of growth and formation of new chambers. Although these processes are studied insufficiently and unevenly in the different groups, a comparative analysis of previous work and my own observations on *Massilina secans* and *Miliolina* sp. (Miliolana) (Plate **30**, figure – s 1 – 6; Plate **31**, figure 6 a, b) (see also Part VII.1.7. and VII. 2.1. of this study) allow conclusion that this mechanism has profound distinctions in the classes under consideration.

Within the Rotaliata the situation is just the opposite – we can see the similarity in the process of the chamber addition in Textulariana (Angell, 1990 on *Trochammina inflata*, Bender, 1992 on *Textularia candeiana* with the mixed agglutinated-microgranular wall) and in Rotaliana (Angell, 1979 on *Rosalina floridana*). In brief, the main steps of chamber addition, based mainly on the works of Angell and Bender, are the following:

1. The exit of the veneer of the reticulopodia through the aperture of the last chamber forming a cyst around this last chamber or around the whole shell;
2. The formation inside the cyst in front of the aperture of the last chamber of a cavity having approximately the same size as the future chamber which defines the space for its construction (predetermining its future form at once);
3. Formation of the outer organic envelope (= plasmatic membrane) around the cytoplasmic projection;

4. The deposition of biomineralised calcareous material in the direction from the distal to the proximal part of the chamber wall, the oriented disposition of crystals forming the cylindrical cavities of the pore (or pseudopore) system if present (evidently there pre-exists a chemical "marking" of the plasmatic membrane into zones participating in the deposition of calcareous material and the "islands" in which such deposition is absent (the future pores) (Labas & Mikhalevich, 1987);
5. Formation of the organic lining of the pores (where present);
6. Formation of the thin inner organic layer of the chamber after the cyst is destroyed and then the increasing of this layer into the several layers during subsequent growth of the shell with the addition of the each new chamber.

The difference between the formation of new chambers in higher advanced hyaline Rotaliata (as opposed to the Miliolata) is that the calcareous crystal units are laid at the both sides of the primary organic membrane, which obviously corresponds to the "envelope" or the "plasmatic membrane" in *Textularia candeiana.* In both cases the new chamber is outlined at once and the organic and calcareous secretions alternate sequence and are separated in time, opposite to the process of the new chamber formation in the representatives having the porcellaneous wall. In tubular forms of Miliolana the new chamber grows piece by piece (according to Angell, 1980 who studied *Spiroloculina hyalina* and to our own observations in *Massilina secans* (Mikhalevich, 1999*,* Mikhalevich, Debenay, Stauff, in preparation). In brief (see part VI.1.5.), the process begins from the formation of a cyst around the whole shell, then the exit of a small amount of cytoplasm through the aperture. The chamber wall is built with the participation of this cytoplasm and the reticulopodia not at once but piece by piece in the direction from the previous aperture. The cytoplasm is transporting to the place of building two kinds of the vacuoles – with organic material and with secreted needle-crystals formed in the Golgi apparatus (Hemleben et al., 1986). The calcareous needles are embedded into the organic layer of the future wall (not laid upon it like in Rotaliata). The organic and calcareous secretions take place simultaneously. Each subsequently build new wall segments are then underlined by a thick inner membrane differing in its composition and tightness from the organic material in which the needles are embedded. The cytoplasmic projection and the envelope are not formed in tubular Miliolata (in the process of building the wide chambers of *Peneroplis* (Wetmore, 1999) the anlage was observed but the mechanism of laying secreted and organic material is supposedly the same as in the tubular Miliolata). The process of chamber formation in the agglutinated Miliolata and in hyaline radial Nodosariata type has not been studied.

The remaining hyaline radial Spirillinata type differs from other hyaline radial types in its mode of chamber growth - it proceeds (according to literature data – Loeblich & Tappan, 1987) as a gradual lateral accretion along the edge of the chamber wall, neither cyst nor envelope are formed. Thus the processes of the chamber formation differ in the majority of the classes under consideration.

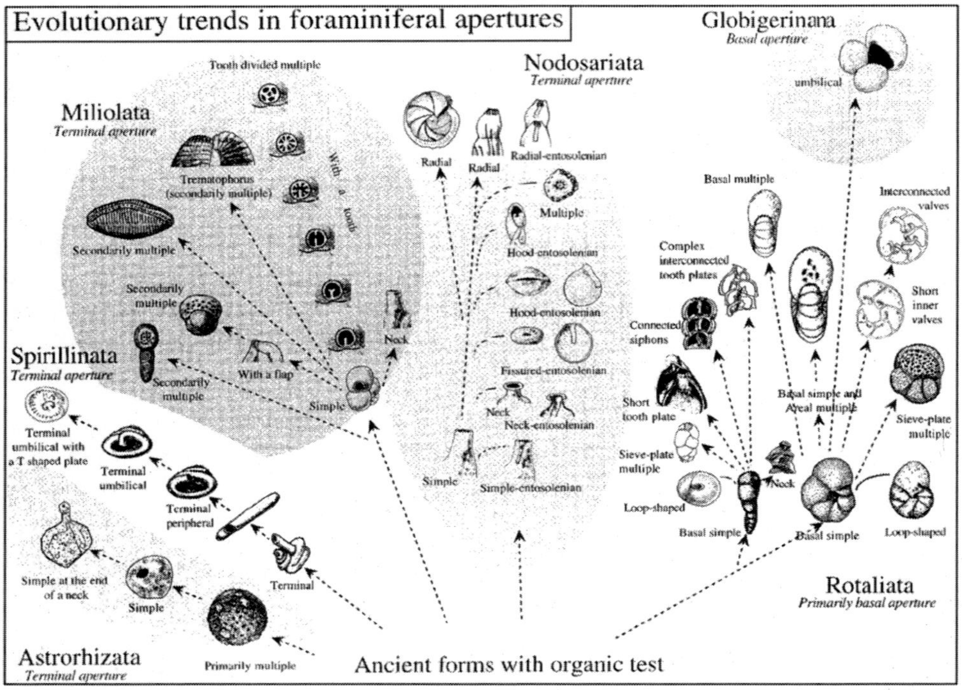

Plate 50. The scheme of evolutionary trends in the foraminiferal aperture (after Mikhalevich & Debenay, 2001).

VIII.10.12. The Gradual Transition from Agglutinated to Secreted Calcareous Walls

The gradual change from an agglutinated shell wall into the characteristic miliolid wall also demonstrates the close affinity between morphologically similar representatives of the subclasses Miliamminana and Miliolana (in *Dentostomina agglutinans* long calcareous needles lining the inner surface of the shell and forming the tooth are combined with a thick agglutinated external wall (Mikhalevich et al., 1986, Plate **13**, figure - s 1 a - d of this study). Such a combined shell wall with calcite needles forming a very thin inner layer in another agglutinated species of Miliamminana was observed also in *Siphonaperta compta* (Mikhalevich et al., 1986), in *Schlumbergerina alveoliniformis* and *Sigmoihauerina bradyi* (Hottinger et al., 1993).

The absence of a distinct boundary between the agglutinated forms and the microgranular calcitic forms among the other agglutinated groups, the variable amount of the cement composition (different types of organic matter) and the different ratio of the microgranular matter in their wall was also noted (Coleman, 1980; Plotnikova & Egorova, 1982; Toksvad & Hansen, 1983; Bender & Hemleben, 1988; Bender, 1992, 1995, Mikhalevich, 1981,1995, 2003). *Karreriella bradyi* wall (Textulariana) is fully microgranular and bears some features of the shell wall of calcareous rotalians although it is not bilamellar Mikhalevich et Peryt, 2007, Plate **4**, figure – s 5, 6; Plate **5**, figure – s 1, 2; Text figure 15 (4) of the present book). The examples of the transition from an agglutinated wall into a hyaline radial one, although very rare, nevertheless exist: in *Akcaya* (=*Sabaudia*) (Text figure 5) and in some orbitolinids, the hyaline radial wall composes part of the agglutinated shell. The formation of pseudopore openings during the evolution of agglutinated foraminiferal representatives obviously also was taking place gradually. The variable kinds of pseudopores cannot be a stable feature. Some authors (Desai & Banner, 1987; Banner et al., 1991) have documented the iterative change of microgranular calcitic canaliculate forms from noncanaliculate ones. It is also possible that the presence of tight calcareous material in the wall with closely packed particles makes it necessary to communicate with the surrounding water through a pore system, while the organic cement of the wall permits the penetration of molecules without forming any special openings (irregular occasional openings between sand particles are present in such walls (Plate **66**, figure – s 1 – 7). Different species of the same genus often have a variable proportion of the sand/calcitic material in their shell wall as well as different kinds of organic cement (Bender, 1995) among the organically cemented genera. Thus, it is hardly reasonable to create new isomorphic pairs of genera among agglutinated groups such as *Eggerelloides – Eggerella, Karrerulina – Karreriella* and so on. The form of the skeleton in spite of all its variability is a very conservative feature providing a stable basis for foraminiferal taxonomy, and I have attempted to re-emphasise its significance (Mikhalevich, 1980, 1981, 1992, 2000, 2002). From the taxonomic point of view, a varying unstable feature can hardly be of great use. Moreover, the incomplete state of the investigated taxa – [only 13% of the known agglutinated genera according to Bender (1995), and among them *Pelosina* belonging to Xenophyophoria (Mikhalevich & Voronova, 1999) was included] – does not permit to take these features as the basis for taxonomic studies. For example, in *Dentostomina agglutinans,* which was considered to be a noncanaliculate species, some irregular pore-like openings transversing the shell wall, were unexpectedly discovered (Plate **66**, figure 4). In the Lituolida, although identified as noncanaliculate, the character of the wall has not been studied in the majority of the type species and this feature has been extrapolated on the basis of morphological similarity. In the process of electron-microscopic study of *Recurvoides contortus* from Antarctic area pseudopore openings in its solid wall were also discovered (Mikhalevich, 2011) (Plate **66**, figure 5 of the present study). In *Placopsilina*, canaliculi were described (Bender, 1995), in Coscinophragmatoidea they also were reported (Loeblich & Tappan, 1987, Hottinger et al., 1993 – in *Haddonia*). In the Prolixoplectidae, only for the three genera out of the nine composing the family, the shell wall has been verified as being noncanaliculate. The data of Loeblich & Tappan (1985) have shown that even in recent species of *Dorothia* the canaliculate character is not always clearly seen, and may be masked as a result of calcification. Among the ataxophragmiids having diagnoses as noncanaliculate, nevertheless *Coskinolina* and *Coleiconus* have "radial pores"; and noncanaliculate *Indomarssonella* and *Cylindroclavulina* were placed among canaliculate forms. The pseudopores of *Glaucoammina*, "scattered and irregular" can hardly be regarded as canaliculi and the genus should be moved into the uniserial Hormosinana according to its shell morphology (Mikhalevich, 1992, 2000, 2004). I observed somewhat similar pseudopore openings in *Ginesina* (Hormosinidae) (Plate **66**, figure 6). In *Tholosina, Saccammina, Rhabdammina* (Astrorhizana) and in some other agglutinated genera (Mikhalevich, 2011 and Plate **66**, figure – s 1 – 3 of this study) whose shell wall is designated as "solid," the shell wall is, nevertheless, also permeable and has some openings of irregular form and distribution. These features of the agglutinated wall and their diagnostic value require further investigations. Questions about the taxonomic significance of wall structure versus the morphology in larger agglutinated foraminifera have also been raised by Gooday & Smart

(2000), Gooday & Pawlowski (2004). The more and more authors come to conclusion that the taxonomic significance of many of the characters of the wall structure and composition was exaggerated in the traditional classifications. Pawlowski also united agglutinated and microgranular textularians (textulariids in his system) and calcareous bilamellar rotalians (rotaliids) in one taxon of high rank (Pawlowski in Adl et al., 2012) similarly to uniting of Textulariana and Rotaliana in the class Rotaliata (Mikhalevich, 1980). All worded above gives reason to designate character of the shell wall in the diagnoses of the lower subclasses as agglutinated or microgranular. The morphological structure of the shell, although also variable, serves as a more stable and conservative basis for taxonomic purposes (Mikhalevich, 1981 – 2000).

VIII.10.13. Data of Molecular Studies

The close affinity of each pair of subclasses within the classes Astrorhizata, Spirillinata, Miliolata and Rotaliata were also supported by molecular analysis performed by the group of specialists in the field (K.Darling, J. Fahrni, M. Holzmann, J. Pawlowski and others) (Darling et al., 1996, 1997, Fahrni et al., 1997, Pawlowski, 2000; Pawlowski et al., 1995, 1997, 2002, 2003, Ertan et al., 2004, Flakowski et al., 2005, Bowser et al., 2006, Pawlowski & Gooday, 2009, Pawlowski & Burki, 2009). The general structure of the foraminiferal phylogenies was confirmed not only by the SSU rDNA sequences but also by the ACT1 and ACT2 and several other actin-deviating proteins (Flakowski et al., 2005). Among the clades shown in the classification scheme of Pawlowski (2012) clade I including Rotaliida, Buliminida, Globigerinida as well as Textulariida, Trochamminida and some representatives of Lituolida (sensu Loeblich & Tappan, 1987) corresponds to the class Rotaliata; clade II including Triloculina, Sorites, Peneroplis as well as Miliammina – to the class Miliolata; clade including Patellina and Ammodiscus - to the class Spirillinata, that including Lenticulina and possibly Saccammina – to the class Nodosariata. Some allogromiids together with astrorhizids correspond to the class Astrorhizata. The other 1parts of allogromiid and astrorhizid representatives are situated in this scheme closer to lagenids, thus confirming the heterogenous nature of this unilocular foraminiferal group. Lituolids (in the understanding of previous authors) are split in Pawlowski's scheme among rotaliat, miliolat and spirillinat clades which proves the heterogenous character of the former lituolid group - its representatives after revision of the order (Mikhalevich, 2004 a, c) where placed into the four different subclasses (Textulariana, Miliamminana, Ammodiscana and Hormosinana) on the basis of the morphological analysis.

Thus, foraminifera exemplify the case where the results of the morphological and molecular studies coincide in its main lines. Molecular data confirm the independent and parallel rise of agglutinated, and later calcareously secreted shell wall in different foraminiferal classes proposed earlier in the principally new classification of the phylum Foraminifra (Mikhalevich, 1980 – 2000, Saidova, 1981). These data of molecular analyses also show that the studies based on the analyses of the morphological structures, on the features reflecting the whole building plan structure of the organism as well as the tendencies of their development still work permitting the separation of large classes representing the main phylogenetic foraminiferal lines (Mikhalevich, 1980, 1992, 2000, Saidova, 1981).

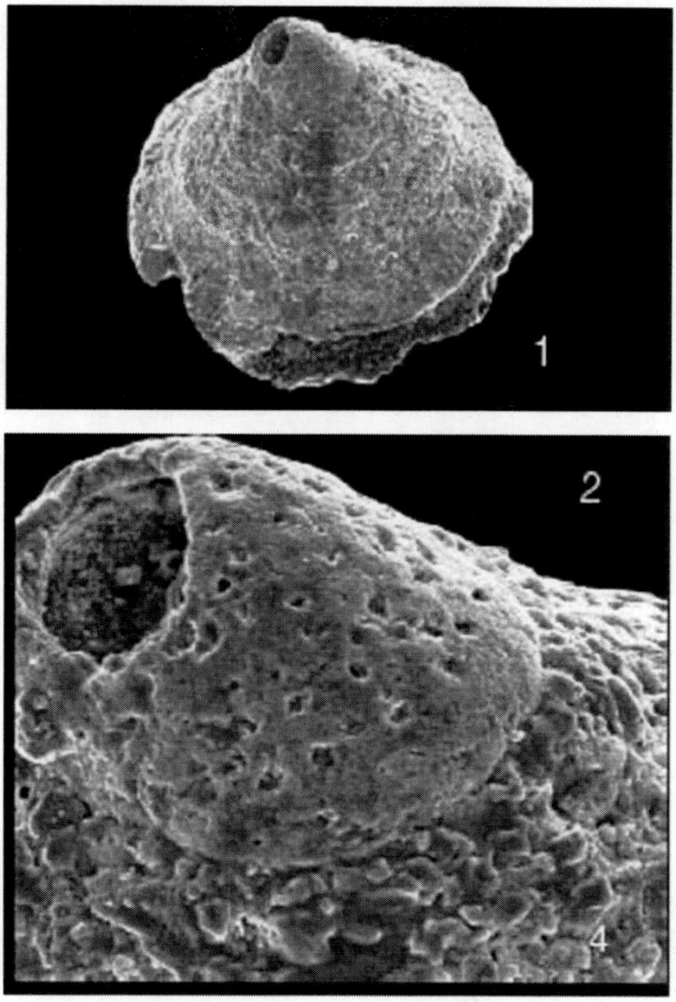

Text Figure 5. Outer view of *Akcaya* (=*Sabaudia*) sp.
1 - view of the whole shell with partly broken embryonal chamber (x 250); 2 - broken embryonal chamber – enlarged (x 2000) (photographed by G. Nestell).

The morphological methods didn't by far exhaust all their potentialities. These results were possible in the case of the study of the representatives with the complex morphology giving the possibilities of comparative morphological analysis. In the case of the more simple forms lacking the distinctive morphological features, especially of the representatives having naked or soft shell - molecular studies play the leading role being sometimes the only resulting methods.

VIII.10.14. The Biostratigraphic Record

The classification proposed here can also be supported by the biostratigraphic record: in each phylogenetic line under consideration the agglutinated forms were usually found earlier than their secreted analogues (Text figure 5). According to the previous data (mostly according to Loeblich & Tappan, 1987) the representatives of the subclasses Astrorhizana (class Astrorhizata) and Ammodiscana (class Spirillinata) were known from the Cambrian, of Hormosinana – from Ordovician. Agglutinated Rotaliata (Textulariana) are in existence from Carboniferous and may be somewhat earlier. In the case of Miliolata for the subclass Miliamminana, we do not have such evidence of an earlier appearance of their agglutinated forms supposedly due to several reasons. One of them might be that the agglutinated Miliolata (Miliamminana) were too fragile. Another reason may be difficulties in separating the initial planispiral and glomerate tubular forms from ancient Spirillinata as the primitive forms of both groups are not rich in distinguishable features and agglutinated planispiral tubular forms may be isomorphous to both: calcareous *Spirillina* and *Cornuspira*. The genus *Ammodiscus* may possibly be heterogenous. May be just in this case molecular methods may be of greater use. Fahrni et al., 1997 analysing molecular data had postulated a more ancient origin of Miliolata than previously recorded. I also hypothesised their pre-existence much earlier (Mikhalevich, 1992, 2000). I believe there is a need to search for their roots in the former Fusulinacea (in the understanding of Loeblich & Tappan, 1987), which I consider to be a polyphiletic group. With the inclusion of the fusulinoids into the class Miliolata (Mikhalevich, 2009) this class has now the more ancient origin as it was previously predicted. Some of the primitive representatives of the ancient fusulinids have agglutinated material in their shell wall. And we always need to keep in mind the incompleteness of the geological record. For the more and more species the new findings had widened their geological range as for instance it was for *Miliammina gerochi* Tyszka, 1997 - a rzehakinid which was found in the Middle Jurassic for the first time (Tyszka, 1997), or for some Globigerinids also recorded for the first time from Jurassic (Kuznetzova & Basov, 2001; Cetean et al., 2011). The number of such findings gradually increases. A thorough revision of Stratigraphic ranges of many of agglutinated genera through Phanerozoic (Kaminski et al., 2008) permitted to modify the ranges for more than two hundred of them. Gausher & Sprechmann (1999) reported the presence of agglutinated foraminifera in the lower Cambrian describing the oldest known fossil saccamminid genus *Titanotheka*. Bosak et al, (2011) discovered interesting unilocular tubular representatives possibly belonging to Astrorhizata as early as 716 – 635 Ma, much earlier than their known findings in Cambrian.

The skeleton reflects the plan of the structure of the organism and is significantly conservative in its qualities. The main types of the hard shells were supposedly formed in Cambrian – Carboniferous. Later the changes in their shell wall structure, inner shell structures and apertural complication took place. All this could be searched in details owing to their geological record demonstrating their phylogenetic history and giving us the detailed historical background that can be objectively controlled. But the molecular methods give much earlier data comparing the geological record data even for the ranges of forms having a hard skeleton (Fedonkin, 2009; Hengeveld & Fedonkin, 2004). Considering the early existence of foraminiferal forms with the soft organic shell envelope, not preserved in the geological record, the origin of the whole group according to molecular timescale was estimated initially as far as about 950 and 1350 (Pawlowski, 2000, Pawlowski & Gooday, 2009) to1500 millions years ago (Fedonkin, 2009, Sokolov, 2011). At present with the use of the Bayesian method, the age of the appearance of foraminifera is dated between 650 and 920 millions years during Neo-Proterozoic (Cryogenian) (Groussin et al. 2011), which is in accordance with the data of other authors on the early Eukaryotes origin in Neo-Proterozoic (Fedonkin, 2009). Foraminifera represent one of the earliest mitochondriate lineages among the ancient Eukaryotes. Namely in the group of soft walled unilocular representatives could be found the roots of all the other classes (Text-figure 6) and hence the reason the group is supposed to be polyphyletic (Mikhalevich 1992). The lack of morphological features in the group makes it the most urgent objective for the molecular studies.

According to Groussin et al., 2011, the results of Bayesian method are mostly in agreement with the fossil records of different foraminiferal phyletic lineages (classes) except the Miliolida (= class Miliolata) whose remains are known from Carboniferous (~ 350 Ma) but the estimation of their age by molecular analysis gives the time of their rise at 490 Ma - Devonian and Silurian). With the inclusion of fusulinoids into the composition of the class Miliolata, (Mikhalevich, 2004, 2009) the conflict seems to be resolved. This example shows the necessity and importance of work of specialists in the molecular methods in collaboration with the traditional taxonomists.

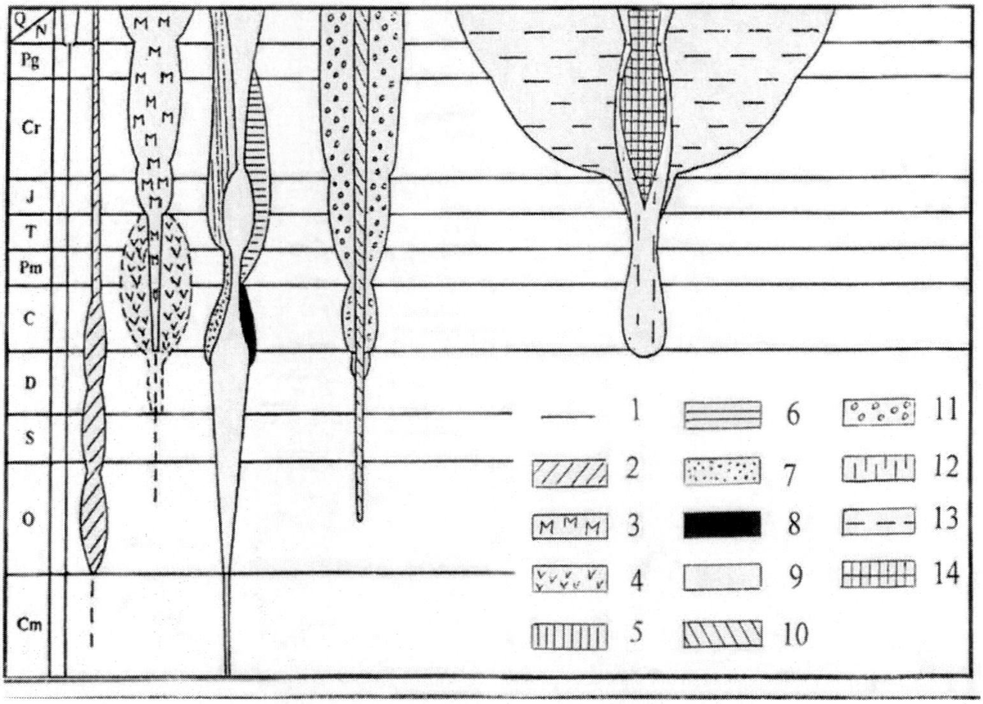

Text Figure 6. Scheme of the development of the different Foraminiferal classes (with the agglutinated subclasses) in geological history.

1- 2 - class Astrorhizata (1 - subclass Lagynana, 2 - subclass Astrorhizana); 3, 4 - class Miliolata (3 – Miliolata, 4 – Fusulinoida); 5 -9 - class Spirillinata: 5 –8. subclass Spirillinana (5 - superorder Spirillinoida, 6 – superorder Involutinoida, 7, 8 - superorder Archaediscoida (7 – order Lasiodiscida, 8 - order Archaediscida)), 9 - subclass Ammodiscana)); 10, 11 – class Nodosariata (10 - subclass Hormosinana, 11 - subclass Nodosariana); 12 – 14 - class Rotaliata: 12 - subclass Textulariana, 13 - subclass Rotaliana, 14 - subclass Globigerinana (after Mikhalevich 2000, modified).

The main evolutionary processes were going in each of the foraminiferal phyletic lines (classes) independently and in parallel at the same time having many specific features in each line (Mikhalevich, 1992, 2000).

It is my hope that I managed to demonstrate here the profound similarity of the basic shell structures, the unity of each of the classes, and the continuity of morphological features within each phyletic lineage (class) – from the agglutinated to the secreted calcareous groups, even if in the agglutinated groups the discussed features and especially the complex structures are usually less developed and expressed. They are not fully realized due to the lower constructive potentials of the agglutinated wall compared with the calcareous walls.

There are some further consequences resulting from the analysis performed (see Part IX).

Chapter IX

TAXONOMIC CONCLUSIONS OF THE COMPARATIVE MORPHOLOGIC ANALYSIS OF FORAMINIFERAL EVOLUTIONARY TRENDS

1. As the morphological regularities are the same for agglutinated and calcareous forms within a class and as they differ more significantly between the agglutinated forms from different classes than between agglutinated and calcareous analogues of a given class, I came to the conclusion to regard the different groups of the agglutinated foraminifera as the earlier step in the development of their calcareous analogues, and hence the latter ones may be their direct descendants (Mikhalevich, 1980 a, b)*. In some cases the agglutinated forms isomorphic to the calcareous ones may represent the sister groups of the latter rather than their ancestral groups. Nevertheless in both cases, they have the kindred relations within their classes. Therefore, the relations between the corresponding pairs of such isomorphic groups within each class are not convergent, but possess close affinity.And the subclasses with the agglutinated shell wall emerged in different phyletic lines (classes) in parallel but at different geologic time (Text figure 6) (Mikhalevich, 1981 – 2000) At the same time, agglutinated subclasses within appropriate classes compose distinctly outlined groups. The phylogenetic molecular analysis also shows that "… investigated calcitic foraminiferal species cluster together and are not interspersed by sequences of agglutinated taxa" (Ertan et al., 2004).
2. Therefore, different kinds of calcareous secreted wall obviously appeared in the evolutionary development of foraminifera independently, in parallel in different phyletic lineages and at the different times in geological history (the Spirillinata's calcareous type wall appeared during the Late Devonian, Nodosariata's – earlier in the Devonian, Rotaliata's – during the Triassic) (Mikhalevich, 1981,1992, 2005) (data on ranges of genera are taken from Loeblich & Tappan, 1987 but considering position of the genera in the new classification proposed).

---

* There are a congruous data of molecular studies: "The buliminids and rotaliids appear to be sister groups of the genera *Trochammina* and *Textularia*, and all together are separated from the lituolids – Ertan et al., 2004 - these conclusions correspond to the data based on comparative morphological analysis (Mikhalevich, 1980, 1992, 2004a, c) where lituolids and their closely related groups were separated from the other textulariins. .

3. The taxonomic significance of the composition and the ultrastructure of the foraminiferal shell wall has, although important, nevertheless subordinate significance compared with shell morphology. In the classification presented here this feature is characteristic of taxa at the subclass (not class) level (Mikhalevich, 1992, 2004, 2005, Mikhalevich & Debenay, 1998, 2001).
4. The former Textulariacea (later Textulariina) and the former Fusulinacea (later Fusulinina) represent heterogenous rather than a monophyletic groups (Mikhalevich, 1992, 2000, 2004, 2005, 2009). Their taxonomic composition was given in the mentioned publications.
5. The agglutinated foraminifera compose entirely the class Astrorhizata Saidova, 1981 (together with the tectinous ones) and the lower subclasses Ammodiscana Mikhalevich, 1980, Miliamminana Mikhalevich, 1980, Hormosinana Mikhalevich, 1992 and Textulariana Mikhalevich, 1980 of the classes Spirillinata Mikhalevich, 1992, Miliolata Saidova, 1981, Nodosariata Mikhalevich, 1992 and Rotaliata Mikhalevich, 1980 respectively (Mikhalevich, 1992, 2000) which represent several phyletic lines.
6. This new classification eliminates a significant number of cases of parallelism and convergence seen among the previous classifications, and comes nearer to the phylogenetic development of the group. This classification also provides possibilities to compose more distinctive diagnoses based on the first instance on shell morphology.

   Looking at the Plates **1a, 36, 38, 39** it is possible to take them as a tables comparable to some extent with the periodical table of elements where the increase of the level of organization goes along each morphological row (class) (as well as in the smaller lines within them). Although the raising of the level of organisation in the foraminiferal rows is not expressed so strictly as in chemistry, it, nevertheless, is regular and to high extent even predictable.
7. The classification proposed is composed hierarchically comparing the preceeding Loeblich & Tappan classification which was composed of the high rank taxa (superfamilies or later orders) of equal taxonomic rank when, for instance, the order Buliminida was equal in its rank to the order Astrorhizida or Spirillinida, while the two latter embrace the bulk of forms constituting the taxa of a class rank, and buliminids are only the order of the class Rotaliata. Hierarchical approach ought to reflect in classifications at least to some extent the phylogenetic relationship of the organisms classified. The wrong ranking reflects the wrong understanding of such relations. Moreover, classifying the whole foraminiferal group as an order or a class instead of a phylum limits consideration of phylogenetic relationships of the different taxa within the Foraminifera, as well as the position of Foraminifera among the other Eukaryotes.

   The accuracy of the classification depends of the right hyerachy of the diagnostic features. Thus the change of the understanding of hierarchy of the features used in foraminiferal taxonomy (shell structure/ shell wall versus shell wall/shell structure) fundamentally changed the structure of the foraminiferal classification.
8. The new approach to foraminiferal systematics together with a new understanding of their evolution requires a profound revision of many of the taxa (especially of agglutinated groups, of class Spirillinana, some major groups within Rotaliana) and

to some extent shows the pathway for such a revision (see Part XI.4.). In a set of articles (Mikhalevich, 1978 – 1998) and in the three books (Mikhalevich, 1983, 1999, 2000, in Russian, the last one with the English summary), I began to move in this direction (for instance separating the genuine lagenids (such as colaniellids, paleotextulariids) and some unilocular shells from the fusulinid group, separating from the former fusulinids also the groups of ancient Spirillinata). The taxonomy of the agglutinated Astrorhizata, agglutinated subclasses Schlumbergerinana, Ammodiscana, Hormosinana, Textulariana and calcareous subclass Miliolana were already elaborated down to the generic level (Mikhalevich, 1988, 1995, 2004 a, c, 2005, 2008 b, Mikhalevich & Kaminski, 2008), and the remaining subclasses - to the family level (Mikhalevich, 1992, 1998, 1999, 2000). One thing is certain – to elaborate the new up-to-date classification of all the Foraminifera, we will have to unite our efforts into a collective scientific effort. The set of monographs revising the major taxonomic groups at the new modern level of investigations are also necessary as well as the generalization of the new taxonomic data obtained during the quarter of the century passing since the monumental issue of Loeblich & Tappan, 1987.

Chapter X

UNIVERSAL CHARACTER OF THE EVOLUTIONARY REGULARITIES (FORAMINIFERA AS A PART OF THE UNIVERSE)

X.1. DIFFERENT LEVELS OF THE ORGANIZATION OF MATTER

This chapter does not regard the other important regularities as contributors to the evolutionary process, such as block-modular principles, feedback, nonlinearity, adaptations, increase of the rate of the evolutionary change, adaptive radiation, divergence, convergence, and parallelism, symbiotic evolution, and some others, but only the main stream ways—polymerization, differentiation, integration, and the phenomenon of aromorphoses. All of them give the organism a higher degree of wholeness, activation, and mobilization of its functions owing to the increased reciprocal activity of its functional parts and their better interaction.

As it was mentioned above, these rules were significantly elaborated by the Russian evolutionary school (Berg, 1922, 1977; Dogiel, 1929, 1954; Severtzov, 1925, 1939; Schmalchgausen, 1939, 1946; Beklemishev, 1964; Podlipaev et al., 1974; Naumov et al., 1977; Golubowski, 1994; Poljanskiy & Raikov, 1977, and others) mostly for the Metazoa. Here I try to demonstrate the manifestation of these developing processes using as an example one of the groups at the unicellular level. The continuum of the evolutionary processes of inorganic and living organic matter (biolevel) which represents the more complex level of organization and the general rules of the development for both levels are shown in Krylov & Libenson (2002). They put the "mark of equation" between the biogeochemical succession and the evolution of the biosphere. Wide character of the evolutionary regularities and the possibility of their application to the lower levels of organization of matter, beginning from the physical and chemical levels, are shown in a set of publications (Larin, 1977; Krylov & Libenson, 2002; Libenson & Przhibelskii, 2003; Krylov, 2005). Social life represents the most complex higher level of the biological form of life and at the same time the highest level of cosmic life. The processes of polymerization, differentiation, and integration along with the phenomenon of aromorphoses being the general universal evolutionary rules may be applicable to human social life as well.

X.2. PROCESS OF POLYMERIZATION, DIFFERENTIATION AND INTEGRATION IN THE DEVELOPMENT OF HUMAN SOCIETY

Let us have a brief look at the history of human society. It developed through the shoal to the genus, tribe, and then into the union of tribes to the state. The number of individuals increased in the human genera and tribes (polymerization), then more significantly in the union of tribes (new level of polymerization). Initial human population was initially 10 – 20 thousand with steady increase over time (about 200 million at the moment of Christ's birth, the first billiard in 1800, soon after Thomas Maltus publication, 3.5 billiards in 1968 and further increasing to 1 billiard every 13 years). This process still is continuing and according the newest data, mankind has now already reached a population of seven billion. According to the prognoses the human population will exceed 9 -10 billion at the middle of the 21st century and 30 billion in the year of 2300.

The enlargement of ancient human being society, the rise of more complex social relations and collective joint actions (e.g., hunting, defense against wild animals) resulted in the necessity for a communication system and the canalization of these new relations. It was a question of survival, of integration— and language emerged, the first social integrative system of human society. For Cro-Magnon man, not being very strong physically, this was of more importance than for other human species. The rise of language gave human beings an advantage during collective actions, quicker adaptation, and regulation. It provided the interaction between the disconnected parts and the strengthening and intensification of such interactions. Even in the early human stages of development from the shoal to the genus and tribe, some elements of differentiation could be marked. The most primitive of them was the division into age groups and division of labor between men and women. The further differentiation of labor in the tribe and unions of tribes caused the appearance of the three social groups: laborers, warriors, and priests. There were heralds to transfer the messages from tribe to tribe as the separate tribes needed communication between them just as the primitive multi-chambered foraminifera needed their foramens for communication between their chambers. The functions of such groups in ancient society differed. Accordingly, the functions were the creation of material products, security, and preservation of knowledge and information. Heralds provided communication between the tribes (prototype of the future postal service). With the advent of priests, the division into physical and mental work occurred—the most important event in the evolution of human social life (Achnazarov, 2002). Complex tasks in the process of joint work such as growing crops or building dams to create and enlarge the economic surplus (the necessary circumstance permitting a society to keep and feed priests and warriors), demanded more complex structures and relations in the society for it to survive.

When the amorphous tribal society became differentiated into different social groups a new integrative system emerged—a state and relations governed by a system of law and legal regulations. This is an example of centralized structures overcoming dissociation. The origin of the state could be considered at the same time as an event of aromorphic character. If, in the union of tribes, the chieftain was the supreme power (with the variants—sometimes he was simultaneously the leader of the bodyguard) while the other functions of management (collection of taxes, realization of justice, and many others) were not differentiated and were performed by the council of elders in the state, during the processes of centralization (i.e., integration). - the main part of the latter functions passed to the king, and the counts and barons preserved only the right to the possession of land. The functions of management became differentiated: special positions (and later special institutions) appeared for execution of special functions.

The following is also one of the evolutionary rules: the limitation of the functions of the subordinate structures in favor of a higher integrative structure (the subordination of the separate foraminiferal chambers to their unicellular organism as a whole, and of the metazoan cells to the centralized multicellular organism).

Although the comparison of the state with the human organism was mentioned earlier, it had a more artistic image based on domestic notions rather than on an analytical approach, based on scientific knowledge. Further it did not bring to light the mechanism of development of such a similarity. In such a likening the specific differences and complexity of each of these systems existing at the different levels of their organization were not considered.

State and law systems developed from the more primitive archaic eastern monarchy and medieval monarchy, to the early bourgeois society of the new time, to the industrial and postindustrial societies of the newest time and lastly to the information society.

In the medieval state the differentiation of the previously existing social groups was continued, mainly in relation to the different kinds of human activity (e.g., multiple guilds, corporations, etc.)—"the blooming complexity" (a term used by the well-known Russian philosopher. Leontiev (1996) of the Middle Ages. The partitions between these differentiated groups were strict: belonging to a definite group was defined by origin and usually could not be changed (as exclusion only).

With the rise of bourgeois society, centralization and integration increased. Gradually the medieval fragmentation and disunity was overcome and strict partitions between the different social groups were broken. Belonging to different social groups was not defined anymore by origin, but was mostly a result of personal possibilities and achievements, and differentiation of society was carried out on account of the more specific distinctions (educational, on the basis of property, etc.). The next very important stage of differentiation of the higher integrative functions of the state was the transition of the functions of the king's personal particular order to sectoral management by means of the institutions (the beginnings of ministries), and separation of the functions of the supreme power from the functions of management.

The disappearance in the bourgeois state of the estate partitions existing before may be compared with the disappearance of the inner chamber wall (dissolution of inner septa) in some foraminifera such as in *Bombulina* (Text figure 1) and the secondary reestablishment of the unilocular stage of the organism at the new level of organization. In both cases the elimination of the partitions made communication of the different parts easier and created in both systems a more dynamic state. The more effective social dynamics resulted in a more developed and effective economy, thus providing state advantages in international competition.

The transition from the feudal to the bourgeois state also had an aromorphic character as later did the process of transition to the recent postindustrial and information society and state. This process has not yet accomplished. This type of society consists of a big number of groups where people are differentiated and organized on the new principles mainly on the basis of their interests (not only material or industrial, but mostly wider and more variable: societies of fishermen, lovers of art, football fans, etc.). These groups are connected in the more complex and principally new systems of intercrossing and multiaspect connections resembling the human brain more with its neuron network than the blood system and forming the new “blooming complexity” at the higher level of organization.

The new process of recent integration is represented in international suprastate structures (e.g., the European Union, currently including 27 countries, international currency reserves, the World Bank), in the processes of globalization. These suprastate structures resemble the integrative stage of the unions of tribes but also represent a new higher and more complex level of organization. The integrative law systems of the different types of states mentioned above developed correspondingly from the primitive feudal law to the civil law after the Magna Charta and the Napoleonic Codex.

It is possible to compare early human society not differentiated into social groups with the primitive unilocular foraminifera not divided into chambers, and the law systems of the states of different levels of organization with the different types of the foraminiferal integrative systems in their different taxa with multi-chambered shells existing at various stages of development and complexity. Thus, feudal law can be compared with the more primitive integrative systems of the foraminiferal shell such as tunnels and stolons (Plate **40**, figure - s 7, 8), and the later civil law with the perfect canal system of the higher Rotaliata (Plate **40**, figure - s 17-19). The law systems during their further development diverged into two branches that can be outlined as codified common law (based on codex—German-Romanic branch) and based on court decision (Anglo-Saxon for all four branches of law). The latter can be compared with the most advanced Rotaliata canal system’; the first one with the variant of fissures (Plate **40**, figure - s 15, 16). A society composed of different groups cannot exist without a system of law in a similar way to how the super-multi-chambered foraminiferal shells cannot survive without their integrative systems providing easier cytoplasmic flows. Of course, the structures of societies with their multiple and interconnected variational components are many times more complex than the structures of one organism at the unicellular level. But the mechanisms of their development through the processes of polymerization, differentiation, and integration and via the nonequilibrium state and new aromorphic changes to the new organizational level are the same.

The law systems are not the only integrative systems of human society. Moral and religious standards also serve as such systems. The laws do not embrace all aspects of life in human society. Each of these standards acts as its own sphere (has its special functions). Economics and business are based on legal relations. Morals represent a more ancient system (compare this with the system of simple foramens in foraminifera - Plate **40**, figure - s 1-6), which appeals to the simpler and more natural spheres of being. Religion can serve as an integrative system inside the state and as a system of the new civilization level uniting several states (Christian civilization).

X.3. Process of Polymerization, Differentiation and Aromorphoses in the Development of the Material Objects of Social Integrative Systems

Like the foraminiferal canal system, the Vertabrata blood system and the brain for language, each social integrative system also has its own material object. For the state this includes bureaucratic institutions, and for the legal system, law institutions (Raskin, 2001). Different cultural institutions serve as a base for different social groups of the postindustrial state. Communication service is the material object for several integrative systems at once. The evolutionary development of material objects (of different types of integrative systems) also passed through historical stages of polymerization, differentiation, integration, and aromorphoses. Whatever example is used all of them display the same regularities.

The first human production tool was a primitive stone cutting instrument. Later a multitude of instruments were used, more complex, multipurposed, and specialized (nomenclature of the Late Stone Age counts dozens of them). The advent of metal instruments (firstly made from copper and bronze and later from iron) may be regarded as aromorphosis. The hand-manufactured production in the guilds of medieval times underwent a similar process of multiplication and differentiation until it was substituted by machine production. This new mode of production also represented an aromorphic change. Polymerization of the bearers could be exemplified by the polymerization in variable technical branches—multibarrel arms (revolvers), multiengine airplanes, etc.

One of the most important methods among these material objects for the integration at the social human level is the communication service that serves as a material object for several integrative systems. The line of its stages can be seen from the first heralds and sound signals to the regular postal service (relay-race horse mail, written mail) and from communications based on the wire principle (telegraph, telephone), to wireless ones based on wave principles (radio, mobile phones). In parallel the rise and development of printing and an increase in book and press publications took place. It was not by chance that the dissemination of Protestantism coincided with widespread book printing. With the appearance of television and the Internet, and with the integration of all the communication services (television, Internet, fixed and mobile communication systems) into a unified whole, there was a unified information system. The Internet represents a new type of communication that is difficult to distinguish from material matter and the integrative system itself. The recent units of countries would not be possible without a new unified communication service system as its material objects. In the development of material objects the examples of aromorphic changes also happened several times (invention of firearms, wireless communication service). The most demonstrative example would be the transition from the multicore processor to the quantum computer.

With the increasing complexity of human society, the number of integrative systems increases. While in the foraminiferal organism, the maximum number exceeds only four, in the human organism it is about ten, and the number in human society increases dozens of times.

Conclusion

It was shown above (Part VIII) that foraminiferal multichamber anatomy and the different types of calcareously secreted walls of the integrative systems originated in different classes many times independently and in parallel, and not simultaneously (e.g., canals of Lasiodiscids in the class Spirillinata in the Paleozoic, and of Rotaliata much later, since the Upper Cretaceous).

The origin of the states had the same character, often with the existence of intermediate forms (between the feudal and bourgeois, or the early feudal and late feudal societies). The ancient primitive structures and the advanced ones usually coexist simultaneously (unilocular Foraminifera live in the Recent seas together with the Rotaliata, i.e., unicellular organisms coexist with multicellular ones; the feudal type of state coexists with bourgeois and industrial societies). Human tribes existing at the level of the Stone Age still exist in South America, where they are not acquainted even with making fire. And the newly formed structures also often contain the remains of the previous one: some remains of the feudal law in civil law, of the estate partitions in bourgeois society (the caste system in modern India and the House of Lords in the UK have been preserved, for example).

The transition from one organizational level to a new one occurs when the possibilities of the previous structures are exhausted. It happens as a result of aromorphosis and usually goes through a nonequilibrium stage. Thus, the eukaryotic Rotaliata cells of higher complexity represent the highest level of organization at the unicellular level. The mainstream evolution of life later went the way of multicellularity in a similar way. The previous Procaryota level was followed by the eukaryotic way. When the potentialities of the feudal-level state were exhausted, bourgeois societies arose. The material objects of human activities exemplify the same regularities. Stone tools were replaced by metal tools, hand manufacturing by factory production, mechanical principles of arms and aircraft construction by the jet principle, etc.. In materials, bearers of foraminiferal integrative systems chomata and stolons were replaced by canals. One of the best examples is seen by the changes in Japan at the end of the nineteenth century (Maidzy revolution) when the elimination of the estate compartmentalization (of saoguns) resulted in equal rights and the advent of a new effective integrative systems of the western type (including the judiciary). These were enough to make the leap in development through several centuries at once.

The recent democratic rightful state having the most advanced integrative systems, with the most developed material objects of these systems, represents the highest level of social organization, which gives it the most effective possibilities for surviving in competition with other state systems. The foraminiferal higher class Rotaliata was not only widely distributed at the ocean floor, but owing to such a progressive material object of its shell as the bilamellar calcareously secreted wall, it was able to gain a new ecological niche and expand into pelagic waters. Similarly the first species of the genus Homo settled from the Africa to Asia and to the remaining remote continents. The information society of the new type possessing social systems with a high level of integration and also highly organized material objects gives mankind possibilities to expand into the cosmos.

Thus comparatively not numerous group of primates not only loosed itself from its initial native habitat into the global and cosmic space, but it also loosed itself from its biological nature forming absolutely new form of the organization of matter – social matter.

This short historical discourse was made to search for the mechanisms of evolutionary changes at different levels of structural organization of organic and social life and to show the continuum of these life levels and the general character of the main evolutionary regularities (polymerization, differentiation, integration, and aromorphic changes) through time. The same mechanisms could be applied and searched for within different organizational levels of varying complexity. There was no opportunity here to go into more detailed analysis of the peculiarities of the processes of polymerization and differentiation themselves, such as polymerization through the subdivision of homologous units or through their new formation and addition, and cases of differentiation after oligomerization (through the process of fusion or elimination of subunits, etc.). Only the main processes could be considered here. Knowledge of all of these regularities permits not only an understanding of the evolutionary processes of the past and present but also the ability to predict ways of development for the future.

Thus, the basic evolutionary regularities elaborated first for living organisms represent general evolutionary principles applied to all levels of organization of the matter of the universe (Mikhalevich, 2008). They promote the mechanism of transition to a new level of organization and the way of further progressive evolutionary changes through time.

Chapter XI

SYSTEMATIC PART

XI.1. POSITION OF FORAMINIFERA IN THE CLASSIFICATION OF THE UNICELLULAR EUKARYOTES

The former conception of Protista regarding them as a phylum changed cardinally since the development of the methods of electron microscopy. It became clear that it is an ecological rather than a systematic group (Krylov et al., 1980, Starobogarov & Mikhalevich, 1985). With the technical advance, molecular phylogenetic studies also demonstrated an extraordinary diversity of the eukaryotes (Nikolaev et al., 2004). These mostly unicellular organisms were united later into a group of variable number of phyla (from 8 to 45) (Krylov et al., 1980, Corliss, 1984, Cavalier-Smith, 1993). The tendency of the enlargement of the number of taxa of kingdom rank and uniting them into dominions was also outlined (Kusakin & Drozdov, 1998). Comparison of the Foraminifera with the nearest groups of the former Sarcodina (Starobogatov & Mikhalevich, 1985; Mikhalevich, 1980 a, b, 1999, 2000) had showed the profound differences between them. Between the Foraminifera and Radiolaria there are more differences (in their nuclear apparatus, pseudopodia, skeletal composition and structure, life cycles and living) than similarities (tubular mitochondrial crystae, biflagellate gametes and closed intranuclear pleuromitosis). If one will compare the nuclear apparatus of the Foraminifera with that one of Radiolaria (one highly polyploidal nucleus) and Acantharia (multinuclear, obviously without polyploidy), it would be turned out that Foraminifera had used in their development both possible evolutionary ways rather than one of them, contrary to both of the groups mentioned. The nuclear dimorphism among unicellular Eukaryotes occurs except of Foraminifera only in Infusoria. But the latter group more often has one contrary to the multiple Mi and differs also in life cycles. Obviously, the nuclear dimorphism of both groups emerged independently and in parallel. The usual way of the genome evolution begins from the initial polymerization of nuclei and of genomes inside nuclei (polyploidization), and then subsequent oligomerization of the nuclei. On the way of polymerization of genomes and oligomerization of the nuclei, Foraminifera are less advanced than Infusoria. The fate of its somatic nucleus differs of that one in the Radiolaria and Infusoria as it is not inherited by the daughter cells but degenerates, in the asexual stage also. Infusoria were regarded as a phylum long ago. All worded above led to the conclusion that Foraminifera ought to be regarded as a separate clearly outlined group of the phylum rank

(Mikhalevich, 1980 a, b). In the latter work it was regarded as subphylum Foraminifera d'Orbigny, 1826, and later as phylum (since Mikhalevich, 1992).

The similarity of the *Boderia turneri* flagella (Lagynana) (see above – Part V, 6) with the flagella of the Chrysomonadida allowed to some authors to regard both of these groups as being closely related (Loeblich & Tappan, 1987). Flagella possessing the mastigonemes were discovered in one more Foraminifera – in *Nummulites venosus* (Roetger et al., 1998) belonging to higher advanced Rotaliata. However the presence in the Chrysomonadida of the siliceous skeleton is in contradiction with previous understanding and prevents to consider some relations between these unicellular taxa (Mikhalevich, 1980). The ultrastructure of foraminiferal mastigonemes was not studied, meanwhile it has its peculiar features in different groups of algae and Mastigophora. The presence in the Foraminifera of the three-flagellate gametes in some of their genera (see above, Part VI. 6) and their capability to deposit calcareous matter earlier allowed to consider them more likely to be closer to Haptophyta (Prymnesiomonada) (Mikhalevich, 1980, 1999, 2000; Starobogatov & Mikhalevich, 1985).

The newest data of Nikolaev et al., 2004 place Foraminifera close to Haplosporidia and Gromiidae, and together with some other amoeboid eukaryotes include them into the supergroup Rhizaria (term of Cavalier-Smith, 2002). On the base of the protein data (Keeling, P.J., 2001) Nikolaev et al., 2004 consider Foraminifera to be close to the heterogenous assemblage Cercozoa (classification of the latter group was later given by Cavalier-Smith & Chao, 2003). Molecular data also give evidence to the close relations between *Gromia* and Foraminifera (Longet et al., 2004). The new insight into the evolution of Rhizaria was provided by Burki et al., (2010). With the progress achieved in understanding of high-level phylogenetic relationships within the Eukaryotes (Adl et al., 2012) the super-group Rhizaria was placed, in its turn, into the super-group SAR [acronym of Stramenopiles, Alveolata and Rhizaria (Byrki et al., 2007)]. Thus the former Rhizopoda including Foraminifera and Radiolaria (with the exclusion of amoeboid forms) received the new place and the new roofs. The sister relationship of Radiolaria and Foraminifera was supported by the data of multigene phylogenetic analyses (Ishitani et al., 2011).

At the same time it is necessary to consider all the difficulties, weakness and imperfectness of the new molecular methods and their sometimes contradictory results. The position of the Foraminifera on rDNA protist tree was misplaced several times. The data of Pawlowski et al., 2000 place them in the middle of this tree close to gromiids. Extraordinary difficulties in the molecular studies of Foraminifera (hard skeleton, the presence of symbionts and other contamination, high acceleration rate) resulted in the fact that this group was for a long time excluded from phylogenetic reconstructions based on molecular studies. Its position on the eukaryotes phylogenetic tree is still highly controversial (Pawlowski & Burki, 2009).

The peculiarities of the nuclear apparatus of Foraminifera, the features of its high specialization and advancement, as well as originality of the morpho-physiological trends of the evolution of their skeletons being capable to create the new unique integrative system of the organism level – the canal system, the peculiarities of the structure of their garnuloreticulopodia, gametes and flagella, along with the uncommonness of their life cycles (antithetic alternation of generations with the intermediate reduction), gives the base for scientists working in parallel to elevate their rank up to the phylum level (Margulis, 1974, Margulis & Schwartz, 1988, Margulis et al., 1990, Mikhalevich, 1980, 2000, Page, 1987).

The analysis of the DNA sequences in Foraminifera had shown their dissimilarity with other studied sequences among the known groups of Procaryota and Eukaryota and supported the originality and monophyly of this group (Langer et al., 1993; Pawlowski et al., 1997, Pawlowski, 2000).

XI.2. Systematic Notes and Major Taxonomic Changes

XI.2.1. Systematic Notes

The taxonomic classification presented here is treated unevenly. The elaboration of the classification is not yet complete, as large groups require thorough revision, such as, for instance, has been done for Stilostomellids and some other nodosariates by the scientific team of B. Hayward (Hayward et al., 2012). With the exception of the four agglutinated subclasses (Ammodiscana, Miliamminana, Hormosinana, Textulariana), agglutinated Astrorhizata (soft-walled genera of this class are waiting their classification) and group of superorders Miliolids in subclass Miliolana which are elaborated up to the generic level (Mikhalevich, 1988, 1995, 2004 a, c, 2005, Mikhalevich & Kaminski, 2008), the classification of the remaining subclasses is to some extent only outlined in the direction of future investigations. Nevertheless, I tried to show possible preliminary ways of future reviews. This explains why some categories are placed only tentatively (marked by a question mark). The former Textulariacea and Fusulinacea are in my opinion, the two largest polyphyletic groups that mostly needed to be revised. In the present work this revision is not fully completed. The separation of the five main phyletic foraminiferal lines (the classes Astrorhizata, Spirillinata, Miliolata, Nodosariata, Rotaliata) and the study of the regularities in the development of the main structures of their representatives (Mikhalevich, 1980, 1981, 1992, 1997, 1998, 2000, Mikhalevich &, Debenay, 1998, 2001), may help us bring to light their roots in the representatives of the subclass Lagynana (class Astrorhizata). It is easier to differentiate the complex multi-chambered forms with complex apertural structures. In the case of the unilocular or simple pseudo-two-chambered forms with a second tubular chamber nearly lacking morphological diagnostic features, it is nearly impossible to do so. For instance, as it was mentioned above, among the agglutinated ammodiscids the predecessors of both the miliolid cornuspira-like forms and of Spirillinata spirillina-like forms may possibly exist and hence the ammodiscid group may be considered as polyphyletic. It is just in such cases that we especially need the insight gained by molecular methods. It had been suggested earlier (Loeblich & Tappan, 1987) that the nodosariid, spirillinid and miliolid forms may have evolved from some extinct fusulinids. I believe that moreover some of the former fusulinacean groups previously included in fusulinoids are themselves the ancient representatives of these three phyletic lines concerning the heterogeneity of the former Fusulinacea. It is more clear in the case of the ancient nodosariids (= lagenids) and of the more advanced spirillinids, especially of those having a radial wall layer (Mikhalevich, 1992, 1998, 2000). That explains why some of the former Parathuramminacea, as well as Earlandiacea, Geinitzinacea, Colaniellacea, Palaeotextulariacea (all according Loeblich & Tappan, 1987), were transferred to the class Nodosariata (Mikhalevich, 1992), while Archaediscacea, lasiodiscids and Tetrataxacea as well as suborder Involutinina – to the class

Spirillinata (Mikhalevich, 1992, 2000). Thus it was considered the evolutionary continuity of the ancient nodosariate and spirillinate lines across the time boundaries to the later times up to the recent forms. All of these taxa in their turn need to be revised and some of them also may be heterogenous to some extent (but at the lower phylogenetic level of the smaller phylogenetic branches). It seems that their classification ought to be constructed mainly on their morphologic features rather than on their stratigraphic ranges. The evolutionary continuity of lagenids across Permian-Triassic boundary was also shown in Groves, 2000, Groves et al., 2004. Later Vachard et al., (2010) also regarded the Nodosariata line going from Earlandinida although these authors consider them to be closer to the Archaediscids and Involutinids (while I ascribe this two latter groups to the class Spirillinata). The specialists on the faunas of the geologically restricted periods sometimes composed the classification of the studied forms not considering their later development through the geologic time, and now many of the ancient groups need to be revised and united in a single system with their closely related forms of the later times.

On the basis of some characters it was possible to suppose that the typical fusulinids represent a special Miliolata line with a strong and peculiar direction of their inner structures in advanced forms. This preliminary hypothesis (Mikhalevich, 2004) was later regarded and confirmed by the detailed analysis performed (see below – Part XI. 2. 2). The view on the heterogeneity of the former Fusulinacea (Mikhalevich, 1992, 2000) is shared in the works of Pronina & Nestell, 1997, 2001, Nestell & Tolmacheva, 2004, Nestell & Nestell, 2006, Nestell et al., 2009, 2011, Karavaeva & Nestell, 2007, Vachard et al., 2010 and in some other studies.

The heterogeneity of the former suborder Textulariina (sensu Loeblich & Tappan, 1987) was shown earlier (Mikhalevich, 1980, 1992, 2000, 2004). In the further revision Conotrochamminidae were also excluded from this group on the basis of their inner shell and apertural structure (see below – Part XI.2.3). The position of Ataxophragmiids among the agglutinated forms is still unclear to me. I suppose this group (in the previous generally accepted understanding) is also polyphyletic, including genera belonging partly to subclass Textulariana (class Rotaliata) and partly to subclass Ammodiscana (class Spirillinata). High conical forms and those with the inner columella (which is not known in any other class) are likely belonging to Spirillinata and tentatively placed here in the latter class. All these complex forms need further detailed investigation and the thorough revision at the generic level.

The former suborders Nodosariina and Rotaliina are more homogenous, although the latter also includes some heterogenous groups.

Thus, the position of Stilostomellids having the uniserial shells and previously ascribed to Rotaliata was also revised on the basis of electron microscopic study of their shell wall (Mikhalevich, 2011, 2012, see also below, Part XI. 2. 4). The electron microscopic data supported the taxonomic conclusions made on the basis of the morphological characters of their shell thus showing once more the importance of the morphological methods in the taxonomic and evolutionary studies. Pleurostomellids were also transferred from Rotaliata to Nodosariata on the basis of their outer and inner apertural structure (Mikhalevich, 1993 a).

Some of the former Rotaliina (sensu Loeblich & Tappan, 1987) (such as Patellinidae and related forms and also Chapmaninidae) are transferred to Spirillinata (Mikhalevich, 2000). In the case of *Patellina* such transfer is supported with the ultrastructural studies of its shell wall which is monolamellar and looks like a single crystal (Text figure 11). In the case of Chapmaninids, the transfer is based on their resemblance with some representatives of

Textularielloidea, which are placed here (with some hesitation) into Ammodiscana (e.g. similarity of *Haurania* and *Ferayina* – Plate **42**, figure – s 10a and 10 b). The transition of the high conical shell structure (both with trochoid or planispiral initial part) into the uniserial part of a special character, the presence of the tube pillars, of the inner columella, some other shell and apertural features are not characteristic of the other Rotaliana.

The position of the Robertinids among Rotaliana also causes some questions. In general, in some cases it is difficult to define the border between some of Rotaliats and Spirillinats, and here many cases of the true evolutionary convergence may occur. Nevertheless, each of these classes represents a clear and independent phyletic lineage. The molecular data also confirm the separate existence of these two lines (Pawlowski et al., 2003, Bowser et al, 2006).

Significant difficulties may also exist when we need to make a distinction between multi-chambered agglutinated planispiral involute shells with a terminal aperture belonging to Miliolata (Miliamminana) and to Nodosariata (Hormosinana) (conventionally speaking to the "Peneropline" and "Lenticuline" types but with agglutinated wall and with a simple, non radial aperture) or between the planispiral forms with an uncoiled part. Additional morphological features, the molecular data, and perhaps in some cases the wall character can be useful in these studies.

Many of the genera need to be reinvestigated as their apertural structures and the position of the aperture at the initial stages of their development, which are important for the taxonomic conclusions, are not known yet in many of them.

The former Textulariacean forms included here in the subclasses Milamminana and Hormosinana are described as noncanaliculate. The forms isomorphic to higher Rotaliata and included here in the subclass Textulariana unite both canaliculate and noncanaliculate taxa, as the taxonomic value of canaliculated shell wall is not quite clear yet. As it was shown above (Part V.2.3. and VIII.10.12), the wall of noncanaliculate forms is not fully solid and is penetrated, although its openings are irregular in their form and position (Plate **66**). At least a greater number of forms of the subclass Textulariana are not studied yet concerning their shell wall structure, and among the families attributed to canaliculated groups, only 13% of genera were really investigated from this point of view. The problem needs further study.

Among the 5 phylogenetic lineages separated here as classes, Miliolata and Nodosariata may be considered as the most clear and compact groups. In spite that the kern (see Part III) of the classes Spirillinata and Rotaliata are well outlined, in some cases it is rather difficult to mark the border between the advanced multi-chambered Spirillinata and Rotaliata (see below, Part XI.5.) which permits to suppose, although with a high caution, some closer affinities between these groups or their ancestral forms, otherwise we need to state a high degree of evolutionary convergence. The most problematic is the most primitive class of unilocular Astrorhizata whose simple shells (if present) lack the morphological structures facilitating the taxonomic identification. Not by chance that some of the forms attributed as foraminiferans to this taxon were later transferred to the other remote groups of organisms (more often to algae). The naked or soft-walled allogromiids (subclass Lagynana) make the problem of their classification even more difficult. The border between these two subclasses (Astrorhizana and Lagynana) seems to be artificial. Their taxonomic separation only renders the tribute to the previous traditional classifications based on the prevailed significance of the shell wall structure (nevertheless, since 2000 year, they are already united in one higher taxon – Mikhalevich, 1999, 2000, Pawlowski, 2000). With the study of *Astrammina rara* and some other large agglutinated species showing the striking features of their biology (Delaca, 1982,

1986, Bowser et al., 1992, 1995, 2002), the boundary between these two groups seems unreal. With the fact that one of the astrorhizid species (the genus giving its name to the order and the class) (*Astrorhiza triangularis*) behave similary as *Astrammina rara* (cell exiting from its "shell"), the type species of the genus *Astrorhiza* (*A. limicola* Sandahl, 1858), ought to be reinvestigated concerning its mode of life.

Obviously, the system of this primitive class would be transformed but at present time the available data on the biology of the majority of its representatives are insufficient. Only the new data (including the biological observations and not only molecular but also cytological studies) will give the key to the right understanding of their phylogeny and classification.

XI.2.2. On the Systematic Position of Fusulinoids

XI.2.2.1. Introduction

Wedekind (1937) introduced the name Fusulinacea Wedekind in 1937 to designate a taxon comprising unusual Paleozoic forms. Fursenko (1958) proposed ranking this taxon as a separate order; this rank was subsequently accepted in *Fundamentals of Paleontology* (Rauser-Chernousova & Fursenko, 1959) and other works of Soviet micropaleontologists. The later rank of fusulinoids was raised to superorder rank, including the orders Endothyrida and Fusulinida Mikhalevich (1980). A new system of this superorder was developed by a team of Russian micropaleontologists headed by Rauser-Chernousova (Rauser-Chernousova et al., 1996) who additionally included the order Tournayellida. The same approach is accepted here. At the same time, in the best known and widely used classification developed by Loeblich & Tappan (1964, 1987), it was regarded as a suborder; only in 1992, its rank was raised to order (Loeblich & Tappan, 1992). Thus, in the studies listed, these ancient protists are regarded as a separate group, which is not connected with other large foraminiferal taxa of the same rank, such as the orders Miliolida, Nodosariida, Rotaliida, and others in the classification of Rauser-Chernousova & Fursenko (1959) or respective suborders in the classification of Loeblich & Tappan (1987). In the last study, fusulinoids are considered to comprise almost all ancient foraminifers, including taxa that, based on the shell and wall structure, should be assigned to other large taxa, for example, the classes Nodosariata and Spirillinata (Mikhalevich, 1980, 1992, 2000, 2004; Mikhalevich & Debenay, 2001).

XI.2.2.2. Morphological similarities in the shells of higher Fusulinoida and higher Miliolata

As the entire diversity of foraminiferal genera is investigated, attention is paid to the morphological similarity between higher fusulinoids (order Fusulinida) and higher Miliolata, primarily of the suborder Alveolinina Mikhalevich, 1980 (order Miliolida Delage et Herouard, 1896) (Plate **51**, figure - s 1–5, 7 – 9) and some higher members of the order Soritida Schultze, 1854 (Plate **51**, figure 10, *Fusarchaias*); both orders belong to the class Miliolata Saidova, 1981 the system of which is here following Mikhalevich (1988, 2005). This similarity is well known and usually regarded as a result of convergent development, i.e., independently acquired in remote groups that are not united by common origin. However, shells of the fusulinoid type are absent from other large foraminiferal taxa with multi-chambered shells, i.e., the classes Spirillinata, Nodosariata, and Rotaliata according to the classification of Mikhalevich (1992, 2000, 2004). In these fusulinoid shells, the last whorls are embracing planispiral, strongly extended along the coiling axis; the shells are frequently twisted from pole to pole. The initial whorls are glomerate (in alveolinids and some fusulinoids) or primarily planispiral (in some fusulinids and in *Fusarchaias* of the order Soritida). Chambers of the initial whorls are wide, with several chambers in a whorl, as in fusulinids and *Fusarchaias*, or elongated, tubular, with two chambers per whorl (in alveolinids). The inner structures of fusulinid and alveolinid shells differ to a greater extent. In particular, fusulinids differ from alveolinids in the undulating septa; the chambers incompletely divided into chamberlets; the communication between chambers through septal pores, tunnels, and, in some genera, through basal foramina, supplementary tunnels (which are widened basal foramina), and radial passages for the cytoplasm, which are open spaces in the upper part of chambers, formed as a result of incomplete septation. In alveolinids, the septa are straight, the chambers are completely divided into chamberlets and communicate through a series of foramina in all septa, the radial passages for the cytoplasm are provided by the preseptal and postseptal passages. These morphological distinctions provide different patterns of internal communications. In alveolinids, the system of internal communication is more advanced and effective. The septal pores of fusulinids and series of apertural foramina of alveolinids are considered to be homologous since the septal pores are also found in the outside septum of fusulinids (Leppig et al., 2005). However, in one group of the superorder Fusulinoida, namely, Verbeekinoidea (Neoschwagerinidae), the septa also are straight and the chambers are also completely divided into chamberlets, as in alveolinids; in this group, the inner structures and arrangement of the internal foramina are more similar to those of alveolinids rather than fusulinids, enabling more direct movement of both equatorial and radial flows of the cytoplasm, as in alveolinids. Leppig et al. (2005) remarked that this similarity between Paleozoic neoschwagerinids and later alveolinids was surprising. The similarity between these groups in both shell appearance and internal structure does not seem to be accidental; it supports the previous hypothesis of remote relationships between them (Mikhalevich, 2004).

Fusarchaias, which is recorded in the Oligocene and Miocene, is similar to fusulinids in the undeveloped preseptal and postseptal passages, as well as in the planispiral position of the initial chambers, these are wide and have several chambers in a whorl. The similarity between this genus and some ancient fusulinids in the structure of chambers and shells is even more profound (despite the fact that they have pillars inside the chambers). At the same time, despite certain distinctions, the sutural structures of ancient Bradyinoidea (former Bradyinidae, order Endothyrida) from the Lower Carboniferous (*Janischewskina*) (Plate **51**, figure 6) resembles the sutural structures (preseptal and postseptal passages) of Cretaceous alveolinids (for example, in *Bullalveolina* (Plate **51**, figure 11; Text figure 7). In both cases, the shell wall of the two neighboring chambers curves inside and forms an interseptal space between them. However, in *Janischewskina*, it is covered with a more convex cover the base of which is separated from the rest of the shell wall by small foramina. Judging from published figures (Loeblich & Tappan, 1987, pl. 250, figure 9; pl. 375, figure 5), the septal plates of bradyinids are detached to a greater extent. In addition, the picture depends on the position of sections. In any event, none of the other foraminiferal groups has similar structures at the sutures. The chomata, which are characteristic of fusulinids and absent from alveolinids, are not observed in some fusulinids (for example, in some Shubertellidae). These structures will be considered below in more detail.

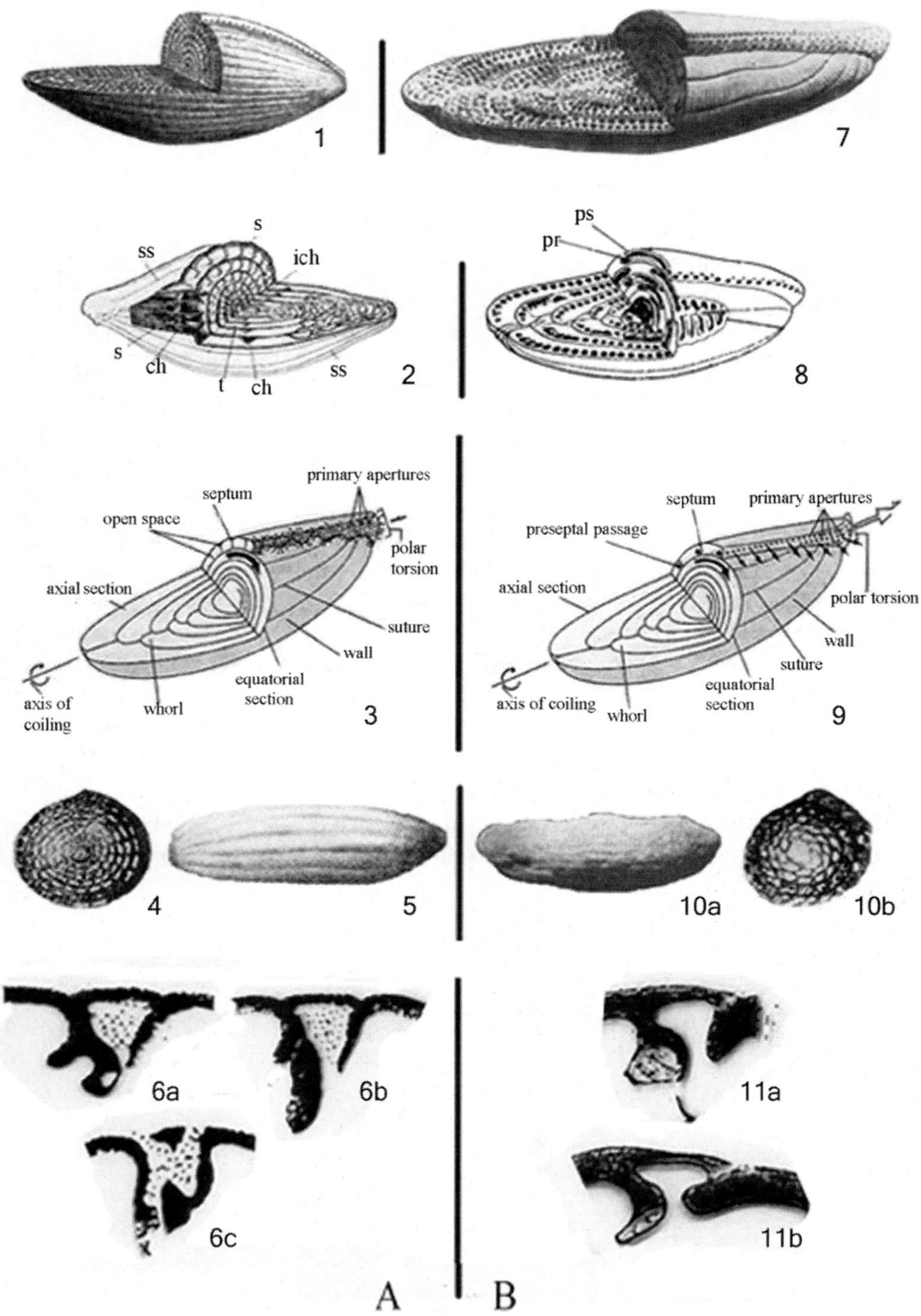

Plate 51. Scheme of the shell structure and sutural structures in representatives of the superorder Fusulinoida (A) and the class Miliolata (B) (order Miliolida, suborder Alveolinina, and order Soritida) (here and later the pictures of fusulinoids (A) and miliolats (B) are divided by a line) (after Mikhalevich, 2009 a).

1 - *Neoschwagerina craticulifera* (Schwager), Middle Permian (after Loeblich and Tappan, 1987); 2 - *Fusulinella*, Upper Carboniferous; designations:(ich) initial chamber, (s) septa, (ss) septal sutures, (t) tunnel, (ch) chomata; 3 - scheme of the shell structure in fusulinids and directions of cytoplasmic flows (indicated by arrows) (after Leppig et al., 2005); 4 - *Eoverbeeckina intermedia* J.S. Lee, Middle Permian, median section (scheme after Loeblich and Tappan, 1987); 5 - *Pseudofusulina prisca* (Ehrenberg), Upper Carboniferous, general appearance of the shell (scheme after Loeblich and Tappan, 1987); 6 - *Janischewskina typical* Mikhaylov, Lower Carboniferous, sutural structures of different chambers of one specimen; 7 - *Alveolinella quoli* (d'Orbigny), Holocene; 8 – *Alveolina* (*Fasciolites*), Upper Paleocene–Upper Eocene; (pr, ps) preseptal and postseptal passages; 9 - scheme of the shell structure in alveolinids and directions of cytoplasmic flows (indicated by arrows) (after Leppig et al., 2005); 10 - *Fusarchaias bermudezi* Reichel, Oligocene–Miocene: 10a - general appearance of the shell, 10b - median section; 11 - *Bullalveolina bulloides* (d'Orbigny), Middle Oligocene: 11a, 11b – sutural structures of two chambers of the same shell in the median section.

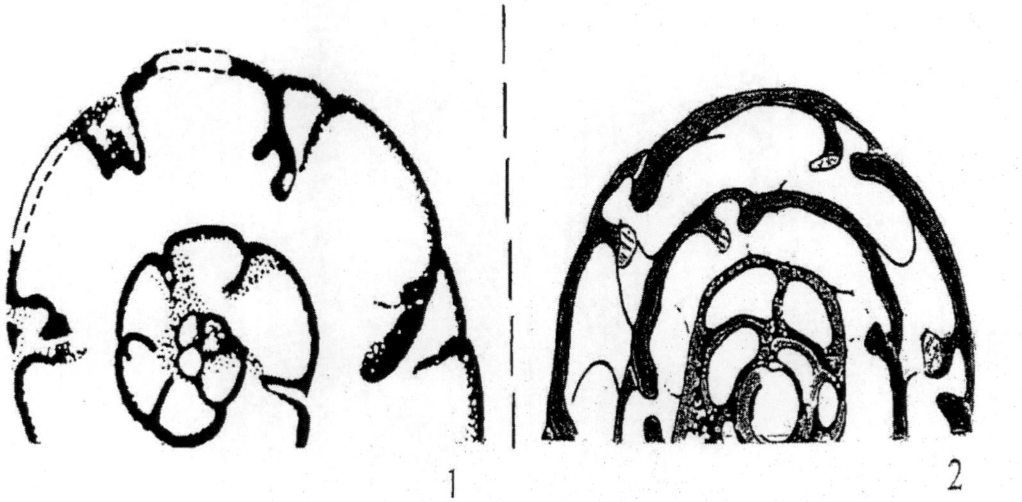

Text Figure 7. Schematic structure of the additional passages in the sutural area in the Fusulinoida and Miliolata (after Mikhalevich, 2009, b).
1 – *Janischewskina* (Bradyinidae, Fusulinoida), (sagittal section), 2 – *Bullalveolina* (Alveolinidae, Miliolata) (equatorial section).

XI.2.2.3. Morphological Similarity in Shells between the Lower Fusulinoida and Miliolata

It is well known that higher members of the two groups compared are rather similar in shell structure, which is complex and evolutionarily advanced (Loeblich & Tappan, 1964); however, little attention has been paid to the similarity between lower members of these groups, with simpler shells. Let us compare the structure of lower Fusulinoida and lower Miliolata in more detail, based on representatives of the orders Tournayellida and Endothyrida, on the one hand, and the orders Cornuspirida, Miliolida, the lower representatives of the order Soritida (Fischerinidae), and some representatives of the subclass Miliamminana of the class Miliolata on the other hand. The system of the subclass Miliolana follows Mikhalevich (1988), and that of the subclass Miliamminana follows Mikhalevich (2004).

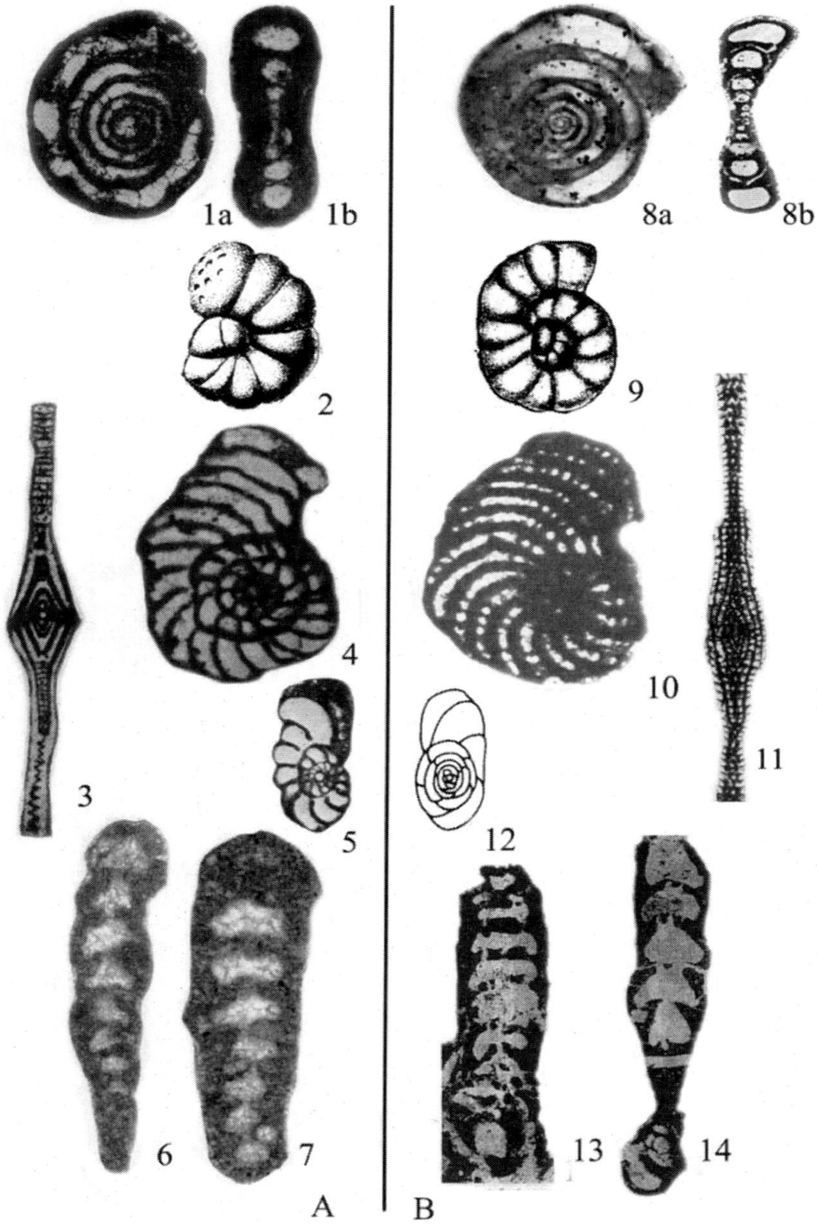

Plate 52. Scheme of the shell structure in the lower representatives of the superorder Fusulinoida (A) and the class Miliolata (B) (after Mikhalevich, 2009 a).

1 a, b - *Pohlia henbesti* (Skipp), Lower Carboniferous: 1 a - median and 1 b – axial sections; 2 - *Haplophragmella panderi* (von Moeller), Lower Carboniferous; 3 - *Parareichelina reticulata* K.V. Miklukho-Maklay, Upper Permian, axial section; 4 - *Rectomillerella texasensis* Liem, Upper Carboniferous, median section; 5 - *Hubiella simplex* J.X. Lin, Lower Permian, median section; 6, 7 - *Vachardella longiuscula* G.P. Nestell et M.K. Nestell, Middle Permian, axial sections of two specimens; 8 a, b - *Fischerina ? carinata* Peybernes, Upper Cretaceous: 8a - median and 8 b - axial sections; 9 - *Zoyaella trochamminoides* (Goes), Holocene; 10 - *Pseudorhipidionina casertana* De Castro, Upper Cretaceous, slightly inclined median section; 11 - *Spirapertolina almelai* Ciry, Upper Cretaceous, axial section; 12 – *Neomurciella butterlini* Fleuri et Fourcade, Paleocene, Eocene; 13, 14 - *Kayseriella decastroi* Sirel, Paleocene, axial sections of two specimens.

According to this system, shells with an agglutinated or combined microgranular with agglutination wall, isomorphic in the structure of secreted porcelain-like shells of the subclass Miliolana, are assigned to the subclass Miliamminana. Each of two large taxonomic groups considered (representatives of the superorder Fusulinoida and class Miliolata) evolved from pseudo two-chambered shells, which consisted of the proloculus and a long second undivided tubular chamber, with glomerate coiling (compare *Endochernella* with *Baisalina* – Plate **53**, figure - s 1a, 1b) or coiling within one plane (compare *Pseudoendothyra* with *Reticulogyra* (Plate **53**, figure - s 3a, 3b); *Tournayella* ((Plate **53**, figure - s 4a, 4b) and *Pohlia* (Plate **52**, figure 1 a, b) with *Fischerina* (Plate **52**, figure 8 a; b; Plate **53**, figure - s 6 a, 6 b). In both cases, the second tubular chamber is occasionally undivided, or it is divided into pseudochambers or even chambers (usually in the last whorls). The initial whorl or several first whorls of the tubular chamber are frequently undivided, for example, in *Tournayella* (Tournayellidae) (Plate **53**, figure - s 4 a - d), *Uviella* (Lituotubellinae), and *Chernyshinella* (Chernyshinellinae) from fusulinoids and in *Trochamminoides* – Plate **53**, figure - s 7 a, 7 b), *Paratrochamminoides* (Trochamminoidinae), *Baisalina* (Baisalinidae) (Plate **53**, figure 1 b), *Fischerina* (Plate **52**, figure 8 a, b; Plate **53**, figure - s 6 a, 6 b) (Fischerinidae), from Miliolata. The photographs of the series of *Paratrochamminoides* species in Kender et al., 2008 (Plate **7**, figure – s 1 – 6) and Neagu et al., 2011 (Plate **3**, figure – s 1 – 13) have even more similarity with some endotirid forms. A small piece of the tubular chamber embracing the proloculus in miliolids with two tubular chambers per whorl and in some soritids is usually named flexostyle.

In both groups considered, the initial streptospiral whorls are frequently replaced at later stages by planispiral whorls. This occurs in many Tournayellidae, such as *Costayella, Glomospiranella, Spinobrunsiina*, the endothyrids *Endothyra, Euxinita, Haplophragmella* (Plate **52**, figure 2), at early stages of some fusulinids (Fusulinoida), and in many representatives of Miliolata: *Trochamminoides* (Plate **53**, figure - s 7 a, 7 b), *Bosniella* (Plate **53**, figure 1 b) (somewhat irregularly coiling initial whorls) (subclass Miliamminana), in *Hechtina* (Plate **53**, figure - s 8 a, 8 b), *Zoyaella* (Plate 52, figure 1), *Kayseriella* (Plate **52**, figure - s 13, 14; Plate **53**, figure 10), *Massilina*, and many other miliolids, subclass Miliolana).

This similarity is particularly well pronounced in *Haplophragmella* (Plate **52**, figure 2) and *Zoyaella* (Plate **52**, figure 9), as well as in *Vachardella* (Plate **52**, figure – s 6, 7), and *Kayseriella* (Plate **52**, figure - s 13, 14; Plate **53**, figure 10). Thus, plates **52** (figure - s 1 – 14) and 53 (figure - s 1 – 10) show the isomorphic structure of shells, chambers, and their arrangement in many representatives of both groups. Such parallel series of similar forms can be easily continued.

In both groups, completely formed multiple chambers of later whorls often preserve more or less expressively a tubular character of their shape, for example, in *Endothyranella* (Plate **53**, figure - s 5 a, 5 b) and *Pseudoendothyra* (Plate **53**, figure 3 a) (Fusulinoida), *Hechtina* (Plate **53**, figure - s 8 a, 8 b) (particularly in early whorls) and *Fischerina* (Plate **53**, figure - s 6 a, 6 b) (Miliolata).

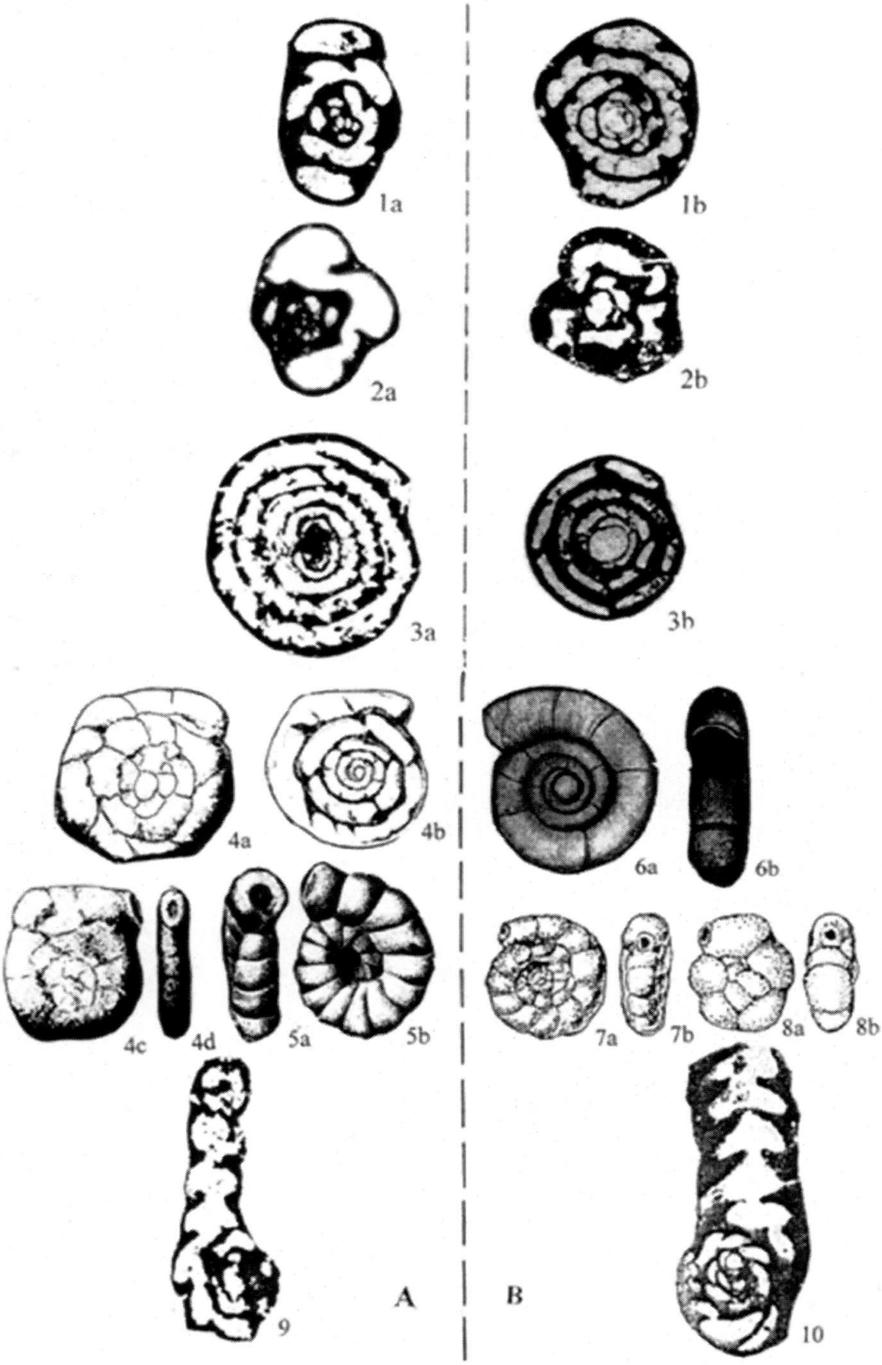

Plate 53. Schematic shell structure in the Fusulinoida (A) and Miliolata (B) (after Mikhalevich, 2009 b). 1a – *Endochernella*, 1b – *Baisalina* (compare the character of the deposits of the inner test matter); 2a – *Nodochernyshenella*, 2b – *Bosniella* (compare the character of the inner shell deposits and the chamber shape); 3a – *Pseudoendothyra*, 3b – *Reticulogyra*, 4a –d – *Tournayella*, 5a, b – *Endothy-ranella (Endothyra)*, 6a,b –*Fisherina,* 7a, b – *Trochamminoides*, 8a, b – *Hechtina*, 9 – *Chernyshinellina*, 10 – *Kayseriella* (compare the chamber form also).

All the genera listed are similar in the position of the aperture: it is always terminal. In the majority of genera, this is an opening at the distal end of the tubular chamber, which is occasionally more or less narrowed, although, in the case of a flattened tubular chamber and, consequently, wide and arched aperture, it is often described in the diagnosis as a basal aperture (for example, in *Tournayella*). Nevertheless, it differs essentially from a typical basal aperture of Rotaliata in its position and structure (Mikhalevich, 2009, Text figure 2). Even in the exceptional case that the shell of Miliolata has a trochoid arrangement of chambers (*Fischerinella* – Plate **1 a**, 8 c), the aperture is a terminal opening at the end of the last chamber, retaining the tubular pattern, and differs from the basal aperture of trochoid Rotaliata (Plate **43**, figure - s 11 a, 11 b); Mikhalevich, 2004, Text figure – s 3, 11 a, 11 c). Series of rounded foramina on the external apertural surface of alveolinids and small "septal pores" on the external apertural surface of fusulinids (Plate **51**, figure – s 1, 3, 7 - 9) are always raised above the base. These "septal pores" are external (and then internal) apertural foramina of fusulinids rather than wide horizontal foramina at the base of septa, which are often described as apertures. The last are never observed in complete shells on the external surface (Leppig et al., 2005), and they are formed as a result of subsequent resorbtion in the septa of internal chambers and should be named interchamber foramina as proposed by Leppig et al., 2005. In addition, there is one more unusual structural type of aperture, which is only recorded in the two groups considered, i.e., the aperture in adult shells in the shape of an elongated slit positioned perpendicular to the chamber base and several differently oriented slits and foramina of other shapes located above them: in *Glyphostomella* and *Pseudobradyina* (Bradyinidae: Endothyrida, superorder Fusulinoida) and *Involvohauerina* (Triloculinidae: Miliolida, class Miliolata) (Text figure 8).

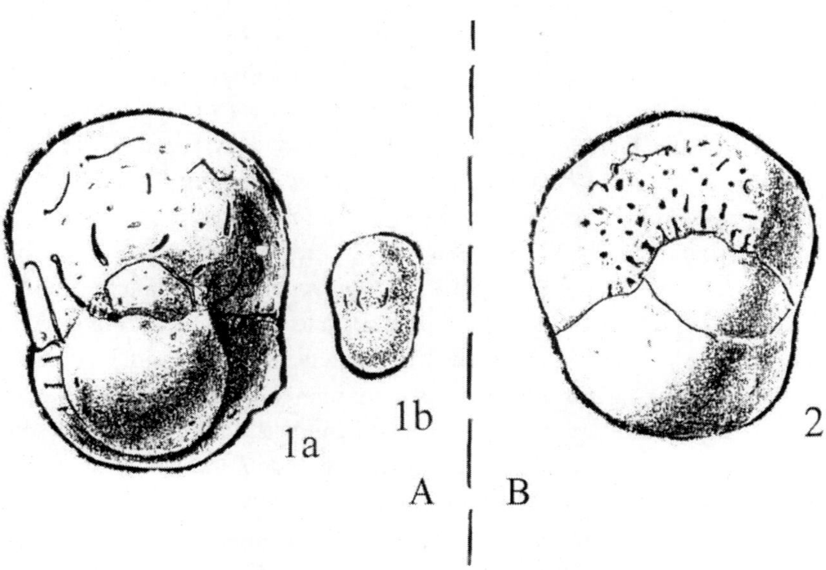

Text Figure 8. Structure of the aperture in some Fusulinoida (A) and Miliolata (B) (after Mikhalevich, 2009 b).

1a – *Glyphostomella*, 1b – *Pseudobradyina* (Bradyinidae, Fusulinoida), 2 – *Involvohauerina* (Triloculinidae, Milioloida).

In other foraminiferal classes with a multi-chambered shell (Spirillinata, Nodosariata, Rotaliata), the initially terminal aperture is only characteristic of the first two classes (Mikhalevich, 1992, 2000). However, the class Nodosariata essentially differs from Miliolata in the structure of shells, chambers, and apertural apparatus. Shells resembling lower Fusulinoida and Miliolata, with a long second tubular chamber with the terminal aperture, coiling in various directions are only observed in the class Spirillinata (for example, *Tournarchaediscus* of the order Archaediscida). However, the last group lacks deposits of shell material at the chamber floor, and the pseudochambers are formed exclusively by constrictions of the wall; in addition, they differ in the ultrastructure of the shell wall. None of the other phylogenetic foraminiferal lineages (classes) has this shell type (Mikhalevich, 1992, 2000; Plate **1 a**, Plate **38**).

XI.2.2.4. Apertural Structures and Deposition of Shell Material Inside Shells

An important taxonomic character of Fusulinoida is the presence of deposits of skeletal material inside the shell, usually at the bottom of chambers; supplementary deposits of the screen type, i.e., slightly projecting tubercles or pointed spines, projecting deep into the chamber, rostra, chomata, and parachomata. However, firstly, the chomata are occasionally poorly developed or absent, as in many endothyrids and some fusulinids, for example, in *Gallowayinella* (Shubertellidae). Secondly, many Miliolata also have deposits of skeletal material at the bottom of chambers. Examples are internal longitudinal endoskeletal ribs, which cross the bottom of chambers from one wall to another, as in *Periloculina* (Plate **54**, figure 22) (compare with the parachomata of *Misellina* - Plate **54**, figure 11) and *Pseudolacazina* (Loeblich & Tappan, 1987, pl. 306, figure 1). Let us compare the tubercles at the bottom of chambers of some representatives of the orders Tournayellida and Endothyrida with similar structures in various later species of the class Miliolata (Plate **54**). They are undoubtedly similar in morphology. Moreover, in Miliolata, these structures are expressed in both planispiral forms with wide chambers, for example, in *Nummoloculina* ((Plate **55**, figure 1 d; Text figure 9 (6)), *Danubiella* (Plate **55**, figure 4 f), and *Hechtina* (Plate **54**, figure 14; Plate **55**, figure 4 g), *Parasorites* (Plate **54**, figure 17), *Hottingerina* (Plate **54**, figure 18), and in shells with tubular chambers and various types of the regular miliolid coiling type, such as *Miliolinella* (Plate **54**, figure – s 12, 13; Plate **55**, figure – s 2 d - f), *Pyrgoella* (Plate **55**, figure 3 c)*, Nummulopyrgo* (Plate **55**, figure 3 d)*, Planispirinoides* (Plate **55**, figure – s 3 b – e)*, Miliolipora* (Plate **54**, figure 15), and *Kayseriella* (Plate **54**, figure 16; Plate **55**, figure 4 h). The genera *Danubiella* and *Hechtina* were investigated and described in detail by Neagu (1968, 1970, 1985). In miliolids, these structures are named teeth (internal teeth, or external apertural plates or flaps). As the structure and arrangement of tubercles and spines of the screening type of lower Fusulinoida (various species of *Endothyra*: (Plate **54**, figure – s 1, 3, 5; Text figure 9), *Endothyranopsis*: (Plate **54**, figure 4); *Omphalotis*: (Plate **54**, figure 2); and *Melatolla*: (Plate **54**, figure 7) are compared with well-developed flat teeth (flaps) in tubular chambers of *Miliolinella* ((Plate **54**, figure – s 12, 13; Plate **55**, figure – s 2 d - f)) or, which is even more evident, in wide chambers of *Nummoloculina* ((Plate **55**, figure 1 d; Text figure 9 (6)), the last planispiral whorl of which has several wide chambers the similarity between the two phylogenetic branches in these structures is even more demonstrative. The same is true in the comparison of deposits of the shell material in peneroploid shells of the two groups, i.e., chomata of *Reichelina* (Plate **54**, figure 6) and deposits in *Hottingerina* (Plate **54**, figure 18). In the diagnosis of the last genus, only septulae in the upper part of chambers are indicated;

however, oblique sections show deposits in the lower part of chambers (Plate **54**, figure 18; Loeblich & Tappan, 1987, pl. 395, figure 7). Tubercles and spines of the lower fusulinoids have been studied mostly in sections, as are the shell structure of all fusulinoids. In Miliolata, ordinary and fine sections have been studied less often. However, as available sections are compared, it is evident that the internal structures similar in morphology and function (in this case, apertural structures of the screen type) are named differently in different groups in connection with the fact that the majority of researchers deal with material restricted to a certain historical period. Figures 12 – 16 of the Plate **54** show that in the sections the teeth of Miliolata look like the chomata of Endothyridae (Plate **54**, figure – s 1 – 5). Both are similar in shape and position at the bottom of chambers, slightly posterior to the aperture, and screening a part of the chamber from the environment. The shape of these structures often depend on the position of the section; in particular, teeth of *Miliolinella subrotunda* (Montagu) are either flat triangular (Plate **54**, figure 12), as in *Endothyra* (Plate **54**, figure 3), or narrow and sharp (Plate **54**, figure 13), as in *Eoendothyranopsis* (= *Eomillerella* (?) (Plate **54**, figure 4). The same picture can be seen at the Plate **55** including some other endothyroid and milioloid genera where the same characters are shown in the sections of the whole shells. As was mentioned above, openings in the septa of fusulinids are similar to the interchamber foramina of other foraminifers (Leppig et al., 2005).

The Paleocene genus *Kayseriella* (Miliolata) (Sirel, 1999), with the initial whorls of one of the glomerate miliolid types and with planispiral subsequent whorls (with a greater number per whorl and wider chambers), passing into the untwisted part, has a tooth typical of miliolids, which is distinct in median (Plate **54**, figure 16) and axial sections (Plate **54**, figure - s 19 – 21; Plate **55**, figure h). The same teeth named in endothyrids the chomata, rostra, or tubercles are observed in various fusulinoids (for example, in the endothyrids *Novella* (Plate **54**, figure 8, *Pseudoendothyra* (Plate **54**, figure 9) and *Eoquasiendothyra* (Plate **54**, figure 10). The similarity between tubercles, rostra, and chomata of Tournayellidae and Endothyridae, on the one hand, and teeth of Miliolata, on the other hand, is also evident in their developmental trends. As the shell grows, it is sometimes possible to recognize that hardly discernible rounded tubercles or ridges of early chambers gradually increase in size and thickness and become long and blunt, pointed, or, sometimes, hooked spines in late chambers ((for example, in *Endothyra* and *Omphalotis* - Plate **55**, figure s 2a, 2b, c; Text figure 9 (1, 2, 5)). Similar growth changes are characteristic of miliolid shells: at early developmental stages their teeth are poorly developed and become larger at late stages, displaying the shape characteristic of particular species or genera (Text figure 9 (6)). In fusulinids, which undoubtedly evolved from endothyrids (Rauser- Chernousova & Fursenko, 1959; Leven, 1970; Rauser-Chernousova et al., 1996), as the shell becomes longer along the axis of coiling the chomata either become wider (e.g., in *Staffella, Pseudostaffella, Neostaffella*, and *Eowaeringella*) or increase in number (*Pseudofusulinella, Pseudodoliolina*, and *Cancellina*). The deposits of the shell material inside chambers, as those shown in Plates 54 and 55 and Text figure - 9, are absent from other foraminiferal groups, the inner structures of the rest groups differ in their character and arrangement. Thus, the features of the inner apertural structures characteristic of the class Miliolata are also observed in representatives of Fusulinoida.

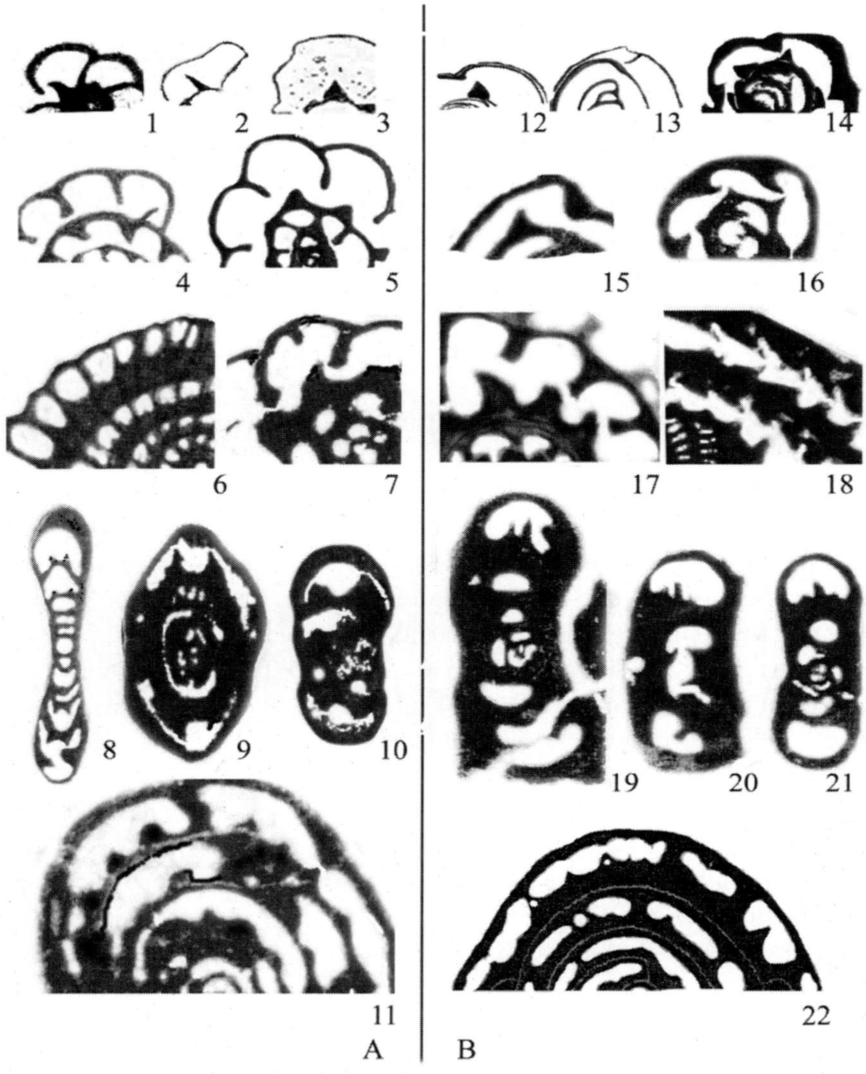

Plate 54. Scheme of the inner structure (deposition of shell material) in the shell sections of Fusulinoida (A) and Miliolata (B) (after Mikhalevich, 2009 a).

1 - *Endothyra* ex gr. *bowmani* Phillips, Lower Carboniferous; 2 - *Omphalotis frequentata* (Ganelina), Lower Carboniferous; 3 - *Endothyra* ex gr. *similis* Rauser-Chernousova et Reitlinger, Lower Carboniferous; 4 - *Eoendothyranopsis scitula* (Toomey), Lower Carboniferous; 5 – *Endothyra* sp., Upper Carboniferous; 6 - *Reichelina cribroseptata* Erk, Upper Permian; 7 - *Melatolla whitfieldensis* Strank, Lower Carboniferous; 8 - *Novella evoluta* Grozdilova et Lebedeva, Upper Carboniferous; 9 – *Pseudoendothyra simplex* (Vdovenko), Lower Carboniferous; 10 - *Eoquasiendothyra bella* (Chernysheva), Upper Devonian; 11 - *Misellina dyhrenfurhi* (Dutkevich in Likharev), Middle Permian; 12, 13 - *Miliolinella subrotunda* (Montagu), Holocene, section in the tooth crosses several planes: 12 - horizontal section, tooth looks triangular, 13 - longitudinal section of shell, the tooth of the last chamber looks like a long thorn (compare with figure - s 2, 4); 14 - *Hechtina antiqua* (Reuss), Lower Cretaceous; 15 - *Miliolipora cuvillieri* Bronnimann et Zaninetti, Upper Triassic; 16 - *Kayseriella decastroi* Sirel, Paleocene; 17 - *Parasorites orbitolitoides* (Hofker), Holocene; 18 - *Hottingerina lukasi* Drobne, Middle Paleocene; 19 – 21 - *Kayseriella decastroi* Sirel, Paleocene, longitudinal sections, showing teeth; 22 - *Periloculina zitteli* Munier-Chalmas et Schlumberger, Upper Cretaceous.

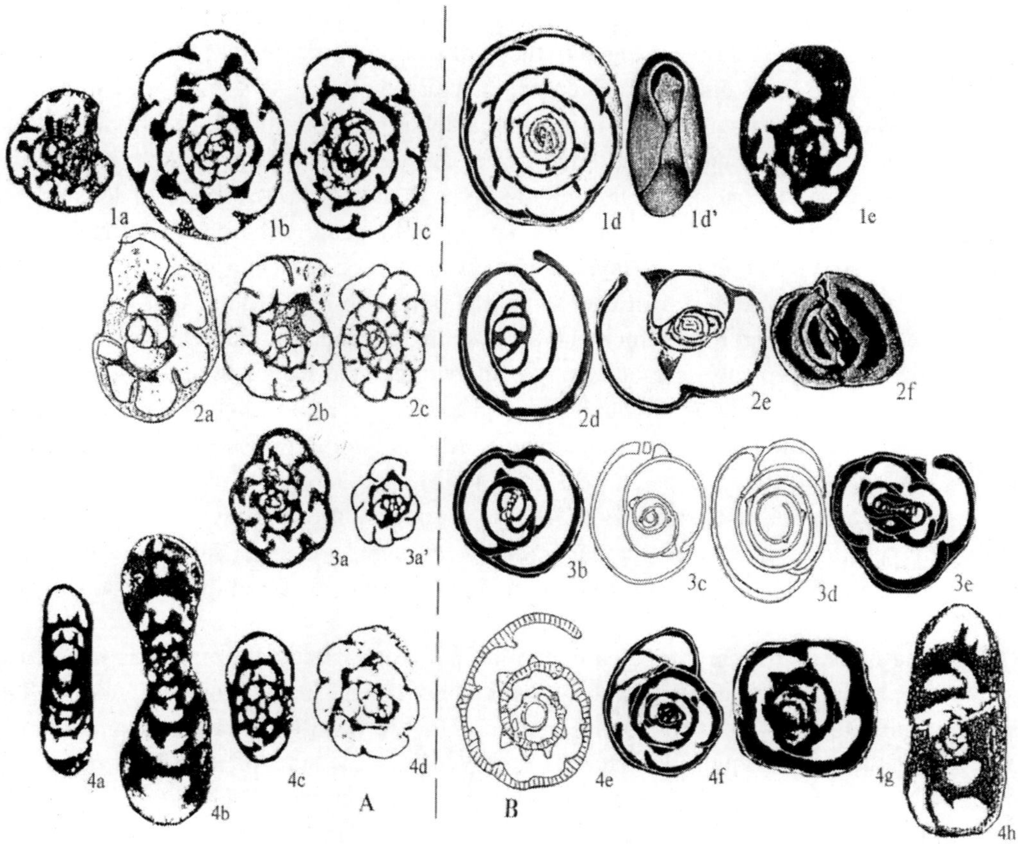

Plate 55. Scheme of the inner deposits of the shell material in the Fusulinoida (A) and Miliolata (B) (after Mikhalevich, 2009, b).

1a – *Tuberendothyra*, 1b – *Spinotournayella*, 1c – *Evlania*, 1d, 1d' – *Nummoloculina,* (1d – equatorial section, 1d' – outer view of the shell), 1e – *Kayseriella*, equatorial section, 2a – *Endothyra*, 2b, c – *Omphalotis*, 2d – f – *Miliolinella*, 3a – *Spinochernella*, 3a' – *Globoendothyra* (*Plectogyra*), 3b, e – *Planispirinoides bucculentus* (Antarctic), 3c – *Pyrgoella sphaera* (Antarctic), 3d – *Nummulopyrgo*, 4a – *Chomatomediocris*, 4b – *Quasiendothyra*, 4c – *Euxinita*, 4d – *Avesnella*, 4e – *Melathrokerion*, 4f – *Danubiella* (after Neagu 1985), 4g – *Hechtina* (after Neagu 1970*)*, 4h – *Kayseriella* (the teeth distinctly seen – compare with c 4a – 4c) (1a – c, 2d - f, 3a – c, 4a – c – scheme after Loeblich and Tappan 1987, 2a – c – scheme after Cozar 2000, 4f – scheme after Neagu 1985, 4g – scheme after Neagu 1970, 4h – scheme after Sirel 1999).

XI.2.2.5. Comparison of the Shell Wall of Fusulinoida and Miliolata

In the classification of Loeblich & Tappan (1987), in which the structure of the shell wall is regarded as a key taxonomic character, the wall of Fusulinacea is defined as calcareous, microgranular (Plate **56**, figure 1), and consisting of several (from one to four) layers. At the same time, the wall structure of many fusulinid groups remains uncertain. Even within the same genus, the wall occasionally ranges from double-layer (or even single-layer, in *Sumatrina* and *Pseudodoliolina*) to four-layer. In addition, in these ancient taxa it may be changed secondarily because of diagenesis. In the later foraminiferal groups (including Recent taxa) with a better preserved initial structure of the calcareous secreted wall, four ultrastructural types are known, including the types of Rotaliata, Nodosariata, Spirillinata, and

Miliolata (see Part V, 2, 2 in the present study, and Mikhalevich, 2000). In the first three, calcareous crystals are regularly arranged in the wall strictly perpendicular to the wall surface and extremely dense with only free pore spaces. Only in Rotaliata is this wall bilamellar (bifontinal), whereas, in other classes, it is monolamellar (monofontinal). In Miliolata (Plate **56**, figure 4 a, b), the internal and external layers are denser with the crystals arranged in parallel to the surface in one layer, the middle layer is the thickest, formed of loosely and randomly arranged crystals, which are usually strongly elongated, or slightly extended, ovate grains. The thin and dense external layer is not always preserved even in the recent forms. This is what makes the shells of Miliolata shiny and porcelain-like. If this layer is absent, the shell surface looks dull and rough. Incidentally, a shiny, polished looking wall is observed in some endothyrids. In typical representatives of higher secreted Rotaliata and Nodosariata, a completely or partially microgranular wall is recorded in some shells, for example, in *Syzrania*, the ancient nodosariates *Lunucammina, Multiseptida*, etc. Recent discorbids were originally divided into two groups and assigned to different remote taxa based on different microstructure of the wall (Reiss, 1963; Loeblich & Tappan, 1964). They were distinguished as bilamellar (bifontinal), with regular radially arranged crystals in one group, and monofontinal, microgranular (granular) in the other. The same occurred in the systematics of cassidulinids.

Later, the division of morphologically similar and closely related groups was rejected in both discorbids and cassidulinids, since it was shown that, in all Rotaliata, the wall is bilamellar (bifontinal), while the granular (microgranular) wall appeared secondarily. In *Globorotalia*, the microgranular structure is observed in some sites of the shell wall, while most of the wall retains an ultrastructure of radially arranged crystals (Maslakova, 1978). The microgranular wall type is widespread among Textulariana, a lower, les advanced subclass of Rotaliata. The porcelain-like wall, with irregularly arranged crystals in the middle layer is more easily transformed into the secondarily microgranular wall than in the case of dense radially arranged crystals.

Gubenko (1988, 1989) recorded a microgranular wall in Triassic *Ophthalmidium tricki* Janger and Permian *Agathammina pusilla* (Geinitz) (Plate **56**, figure – s 5, 6). The characteristic shell shape of the genus *Ophthalmidium*, which has not been recorded in other foraminifer classes, suggests that its ultrastructure changed with time. G.P. Pronina (personal communication) has recorded a microgranular wall in Permian foraminifers that are identical in morphology to Cornuspirida. In *Archaias angulatus* (Fichtel et Moll), the usual porcelain-like wall structure is replaced by the microgranular structure in some sites even including living individuals (Macintyre & Reid, 1998).

Similar changes are all the more probable in ancient shells. With proceeding from the above data, it is possible to propose that the shell wall of Paleozoic fusulinids resulted from recrystallization of a structure that is present among living foraminifers (most likely miliolids) rather than an essentially new type. This assumption is supported not only by the morphological similarity of shells in the two groups (see above), but also by the presence of the wall ultrastructure that is unusual for Miliolata in very large (up to 3 – 7 mm) shells of living individuals from the Antarctic Region. In particular, in the middle layer of the wall of several species of *Pyrgo*, in *Pyrgoella sphaera* d'Orbigny, *Planispirinoides bucculentus* (Brady) and some other miliolids, crystals are arranged in radially extended structures. According to Debenay et al., 1996, the miliolate elongated needle crystals represent secondary (second order) aggregates of tiny crystallites. The radially extended rows composed of the needles represent tertiary or third order aggregates – columns. These radially extended structures (columns) are separated by free spaces (rather than densely arranged), which form elongated, frequently branching passages (Plate **56**, figure – s 7, 8; Plate **57**, figure - s 4 – 6). These branching canals, which resemble pseudoporous canaliculi, open on the shell surface in irregular rounded foramina. This wall is comparable to the wall of fusulinids, which shows a radial fibrous pattern (Loeblich & Tappan, 1987, pl. 274 etc; Davydov et al., 1999, pl. 22; Leven & Davydov, 2001, pl. 4, and Plate **56**, figure – s 2, 3; Plate **57**, figure – s 1 – 3 of the present study). The branching passages (branching pseudopores) resemble ampullaceous branching pseudopores of keriotheca (compare with the fusulinoids *Sumatrina, Triticites*, and *Montiparus*).

In some cases, this thickened wall of living miliolids lacks pseudopores or they do not branch, and the thickened relatively homogeneous internal radial layer resembles the diaphanotheca. The thin, denser external and internal layers of a thick miliolid wall are apparently comparable to the internal and external tectoria or tectum of fusulinoids. Like the thin, dense, enveloping layers of the miliolid wall, the tectoria are changeable and frequently absent (Remizova, 2004). A comparison of the wall ultrastructure in *Pseudoschwagerina* sp. (Isakova & Filimonova, 2003) (Plate **57**, figure 1) and *Pyrgo sarsii* (Plate **57**, figure - s 4, 5) shows that the radially extending rows of crystals in the uppermost and lower parts of the wall are shorter, somewhat thinner, and denser in both taxa. Between these parts and the wider middle part, there is a weak boundary. A similar picture is seen in *Rugosofusulina* sp. (Plate **57**, figure 2). Figures 2 and 5 of Plate **57** show that the middle rows of grains are positioned at an angle to short upper rows. Another photograph (Plate **57**, figure 6), at higher magnification (x 7000), shows that the radial rows in the wall of *Pyrgo* are composed of elongated needles, which are characteristic of Miliolata.

In the photographs examined of the fusulinid wall, these rows are formed of microgranular grains (Plate **57**, figure - s 1 – 3). These data suggest that, as this miliolid wall changes secondarily, becoming microgranular, the similarity to the microgranular wall of fusulinids increases. The pseudoporous foramina of this miliolid wall are irregular or absent. The pseudopores of fusulinids are also changeable. It is known that, in the microgranular wall of lower fusulinoids, agglutinated material is occasionally present in varying quantities (Rauser-Chernousova et al., 1996). Varying amounts of agglutinated material are often present in the porcelain-like wall of Miliolata. In particular, it was shown that the lower internal layer of thick agglutinated wall of *Dentostomina agglutinans* (d'Orbigny) is formed by thick, slightly elongated, loosely arranged crystals; and the thin internal cover and teeth are formed by long needle-shaped crystals typical of a porcelain-like wall (Mikhalevich et al., 1986, Plate **13**, figure - s 1b, 1d of the present study). In the class Miliolata, the transition from the agglutinated wall to exclusively secreted calcareous wall is gradual; and it is impossible to indicate a distinct boundary between pure agglutinated and completely secreted forms (Mikhalevich, 2000, 2004). At the same time, within the classes Nodosariata, Rotaliata, and Spirillinata, a distinct boundary separates the agglutinated and agglutinated–microgranular taxa, which compose their lower subclasses, on the one hand, and secreted calcareous taxa of their higher subclasses with a wall ultrastructure characteristic of each taxon, on the other hand.

Another shared feature of the shell wall of Fusulinoida and Miliolata is the high magnesium content of their calcite (Loeblich & Tappan, 1964), which also distinguishes them from all the other foraminiferal classes.

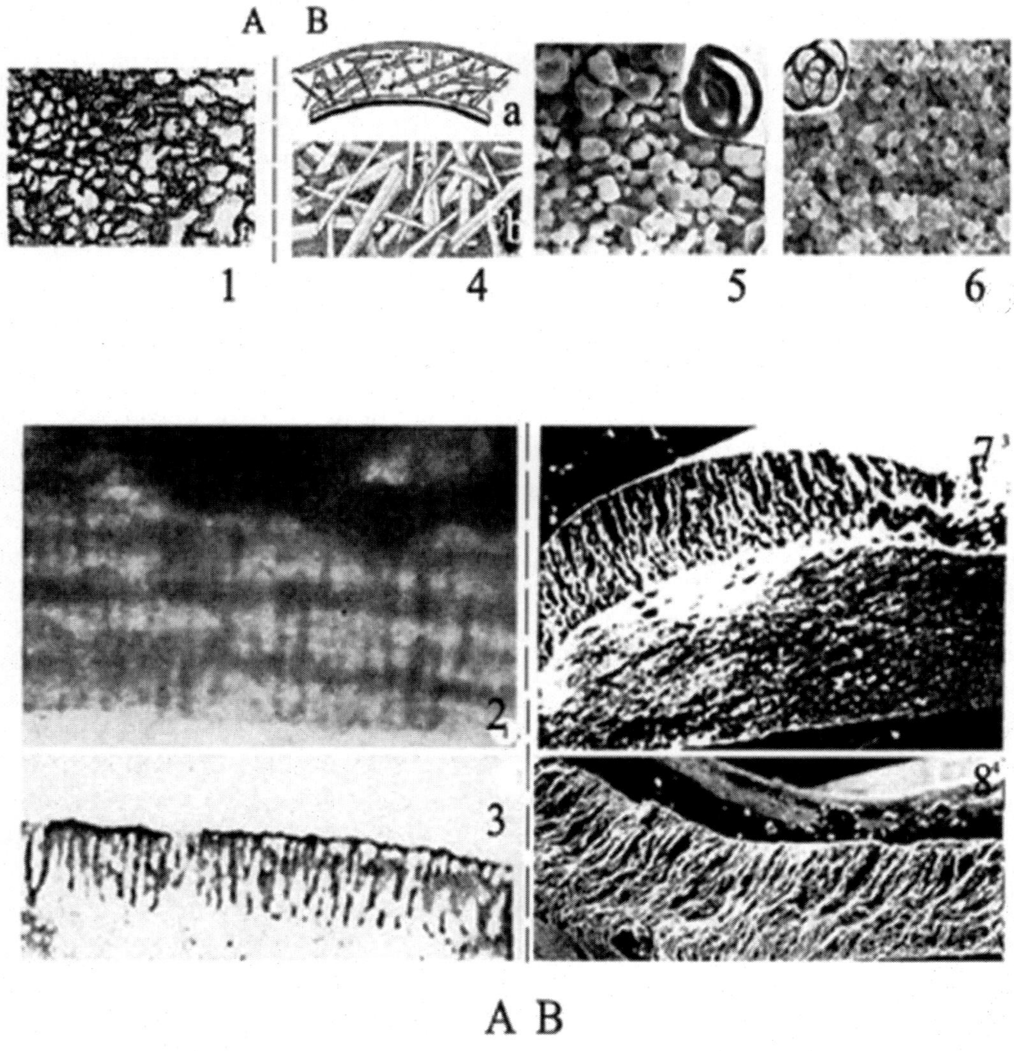

Plate 56. Wall ultrastructure of the fusulinids (A) and the fossil and recent (Antarctic) miliolids (B) (after Mikhalevich, 2009, b).

1 – microgranular wall of *Rugosofusulina*, 2 – fibrose wall of *Staffella elegantula*, 3 – fibrose wall of *Montiparus umbonoplicatus*, 4 – typical miliolid (porcelaneous) wall (4a – scheme), 4b - middle layer of *Pyrgo* wall, 5 – microgranular wall of *Ophthalmidium tricki* (after Gubenko 1988), 6 – microgranular wall of *Agathammina* (after Gubenko 1988), 7 – broken shell of *Pyrgoella sphaera* d'Orbigny 1839 (x 200), «Polarstern», 1996, station 5, sample 1, 8 – broken shell of *Pyrgo sarsii* Schlumberger 1891 (x 200), «Polarstern», 1996, station 2.

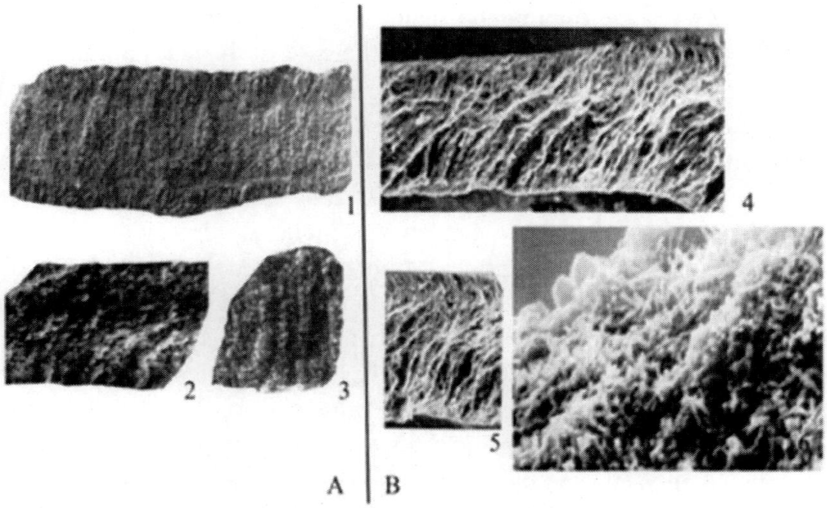

Plate 57. Wall ultrastructure in the representatives of the superorder Fusulinoida (A) and the class Miliolata (B) (after Mikhalevich, 2009 a).
1 - *Pseudoschwagerina* sp., x ?250, Lower Permian (after Isakova & Filimonova, 2003), note long columns (tertiary or third order aggregates) in the middle part of the wall and shorter columellae in the upper and lower parts of the wall; 2, 3 – columns of crystals in *Rugosofusulina* (2 – *Rugosofusulina* sp., x ?250, Southern Ural Mountains, Lower Permian (photograph is provided by T.N. Isakova), note the curve of the rows of crystals (collumellas); 3 - *Rugosofusulina* sp., x ?250, Lower Permian (after Isakova & Filimonova, 2003); 4 – 6 - *Pyrgo sarsii* Schlumberger, Holocene, Weddell Sea, collected by the vessel *Polarstern* in 1996, point 2: 4, 5 - various sites of a cross break of the wall, long columns (tertiary or third order aggregates) in the middle part of the wall and shorter columellae in the upper and lower parts of the wall, note the curve of the rows of crystals (collumellas) in the middle part of the wall, x ?200; 6 - fragment of fig. 5, (needles – secondary (or second order) aggregates, x 7000.

XI.2.2.6. Conclusion

Thus, it is shown that representatives of the superorders Fusulinoida and Milioloida and other lineages of the class Miliolata display profound similarity in the structural patterns of the shell, chambers, positions of the aperture and internal apertural structures (called rostra and chomata of Fusulinoida and teeth of Miliolata but similar in their structure, position and function). Note that both groups lack trochoid shells, which are widespread in the classes Spirillinata and Rotaliata (trochoid *Fischerinella* is the only exception among Miliolata; trochoid tetrataxids are transferred from fusulinoids to the subclass Ammodiscana of the class Spirillinata: Mikhalevich, 1992, 2000). The profound similarity in the external and internal shell morphology of the superorder Fusulinoida and the class Miliolata, similar trends in evolution and complication of the shell in each group, beginning from the pseudo-bilocular-chamber shell to complex multichamber shells of the fusulinoid type should be regarded as evidence of their relationship rather than accidental coincidence. These features are not observed in any other phylogenetic lineages of foraminifers (Mikhalevich, 1992, 2000, 2004). Based on this, it seems reasonable to include the superorder Fusulinoida in the class Miliolata. Some features of the composition and structure of the shell wall of the two groups considered suggest that the walls of Paleozoic Fusulinoida and later Miliolata differ to a lesser extent than was previously thought, although many details of the wall structure of fusulininoid's ancient Paleozoic taxon remain poorly understood. It is difficult to recognize a distinct boundary between the microgranular and porcelain-like walls, even when late and well-preserved shells are studied, the more so, in the case of the study of ancient specimens changed by diagenesis. Additional studies are required to determine whether a shell wall was initially microgranular or changed as a result of diagenesis. Possibly it has the same ultrastructure as the later Miliolata. Apparently, at the end of the Carboniferous and in the Permian (or earlier), certain primitive fusulinoids or their ancestors, possibly closely related to the Tournayellidae, gave rise to a new branch of Miliolata, with tubular chambers (two per whorl) (Vachard et al., 2010 also consider Miliolata derived from Tournayellida). In the Cretaceous, they reached a high degree of perfection and, in the advanced groups (suborder Alveolinina), produced super-multi-chambered shells of the fusulinoid type. Thus, at late developmental stages, they repeated the shell pattern implemented earlier by their predecessors. According to the block–module principle of evolution, high efficiency, peculiarity, and specific functionality of new systems are achieved through the combination of standard functional elements and blocks, which appeared during the previous evolutionary development period (Krylov, 2005). The Alveolinina provide a prominent example of this evolutionary pattern. However, at a later evolutionary stage, they lost a number of primitive ancestral features and acquired some new, more progressive characteristics (in particular, the system of their internal communications is more advanced than in fusulinids, see above). The second branch of calcareous Miliolata, the order Soritida (Cretaceous–Holocene), has the initially planispiral shell with several wide chambers in a whorl, as in endothyrids. One branch of this group also developed the fusulinoid shell type at the adult stage. This is the subfamily Fusarchaiasinae (Oligocene–Miocene). However, this trend is represented in Soritida by a few taxa. A few representatives of the order Miliolida, the initial part of which has a typical miliolid structure, acquired at the adult stage the planispiral whorls with several wide chambers in each whorl, resembling planispiral whorls of multi-chambered endothyrids (for example, *Danubiella, Derventina, Nummoloculina*). The presence in *Nummoloculina* of the apertural blade, accentuates this similarity as is seen in sections (Text figure 9). In endothyrids, the initial whorls are also frequently glomerate (plectogyroidal), but formed of wider chambers (among Recent taxa, *Zoyaella* is most similar to them). Thus, three lineages (Fusulinoida, Milioloida, and Soritida) of one large foraminiferal stem (class Miliolata) display similar features at different evolutionary stages. This similarity should be regarded as a result of parallel development in closely related lineages rather than convergent or accidental similarity. This reconstruction of the evolution of the two groups considered agrees with the results of molecular studies (Fahrni et al., 1997), which suggest earlier origin of Miliolata than it was previously accepted. With the inclusion of the Fusulinoid supergroup into the composition of the class Miliolata, the data of the geological record and of the molecular studies on the age of this class will be approximately the same.

XI.2.3. On the Systematic Position of Conotrochamminids

XI.2.3.1. Introduction

The systematic position of the *Conotrochammina antarctica* Saidova, 1975 was unclear. The inner structure of the shell and the ultrastructure of the shell wall of this species were not investigated previously. The data on the position and structure of its aperture were contradictive. Meanwhile this feature is a very important taxonomic feature in the definition of the systemic position of the genus. As a result, in some works it was attributed to different genera (e.g. – *Conotrochammina* sp. in Fillon, 1974, Tabl. 2, figure 4; *Trochammina conica* in Milam & Anderson, 1981, Tabl. 3, figure 5; and presumably *Paratrochammina scotiaensis* in Igarashi et al., 2001, Tabl. 4, figure 3 (photograph not quite clear). The species was described by Saidova (1975) from the shelf Antarctic waters in the Pacific sector of Antarctica near the Oates Coast, Northern part of the Victoria Land from the depth 695 meters ("Ob", station 336), and is widely distributed in the Antarctic area in sublittoral and upper bathyal subzone (Fillon, 1974, Saidova, 1975, Milam & Anderson, 1981, Igarashi et al., 2001, Mikhalevich, 2004a and many others).

XI.2.3.2. Material and methods

The material at our disposal was that one gathered by "Ob" during the years of 1956-1972 from the collections of the Zoological Institute of the Russian Academy of Sciences in St. Petersburg and from the collections of the Shirshov Institute of Oceanology of the Russian Academy of Sciences in Moscow. These latter was kindly presented for this study by Khadyzhat Magomedovna Saidova. The value of her material is of a special significance concerning the fact that it included the topotypes of her new species. The studied specimens were picked up from samples at the stations 336, 69° 36'S, 161°50'E, "Ob", 1958, 650 – 700 m depth.

The shell and wall structure were studied using the Scanning Electron Microscope under magnifications ranging from 50 to 8000 times. The inner structure of the wall and apertures was studied on the breakings of the shell. Observations were carried out on the SEM Hitachi, model S – 570 with the assistance of the engineer T.K. Tsoegoev (Zoological Institute, Russian Academy of Sciences) and partly on Philips XL-20 (Institute of Paleobiology, Polish Academy of Sciences).

XI.2.3.3. Details of the shell and apertural structure

The thorough study of the external character of 15 exemplars of the species designated by Kh. M. Saidova as *Conotrochammina antarctica* in the light and electronic microscopes had unexpectedly showed that nearly half of them lack the outer aperture (Plate **58**, figures. 1 a, b, c; 2), while in the others the clear areal aperture could be discovered (Plate**58**, figures. 3 a, b; Plate **59**, figure1 a, b). In the shells without external aperture, the internal apertures were clearly seen in all the inner chambers of the broken shells (Plate. **59**, figure - s 2 - 7; Plate **60**, figure – s 4, 5). The series of successive breakings of the same shell (Plate **59**, figure – s 2 - 7) permit the revealing of the inner apertures of the shell up to the initial chamber (Plate **59**, figure 7). All of them have areal position and are represented by the oval opening surrounded externally with a rather high, very thin and delicate white collar (or neck) (Plate **59**, figure – s 2 – 7; Plate **60**, figure – s 1, 2; Plate **61**, figure – s 1, 6), constituted by very thin (pelite) fine sand material with no special characters when viewed from the inside of the chamber (Plate **59**, figure 6; Plate **60**, figure 5). This collar can be clearly seen only using SEM. When using the light microscope the aperture looks only flash with the surface. The photographs in SEM show that the collar of the inner aperture is higher and better developed than in the outer aperture of the last chamber when present. The distance of the apertural position from the base of the chamber varies strongly from being placed in the center of the chamber (Plate **59**, figure – s 2, 4, 6) to being situated more closely to its base (Plate **58**, figure 3 a, b; Plate **59**, figure 1a, b).

In the case when the aperture is very close to the sutural line it looks from some angles like an interiomarginal aperture and could be erroneously taken for such a base opening (Plate **58**, figure 3a).

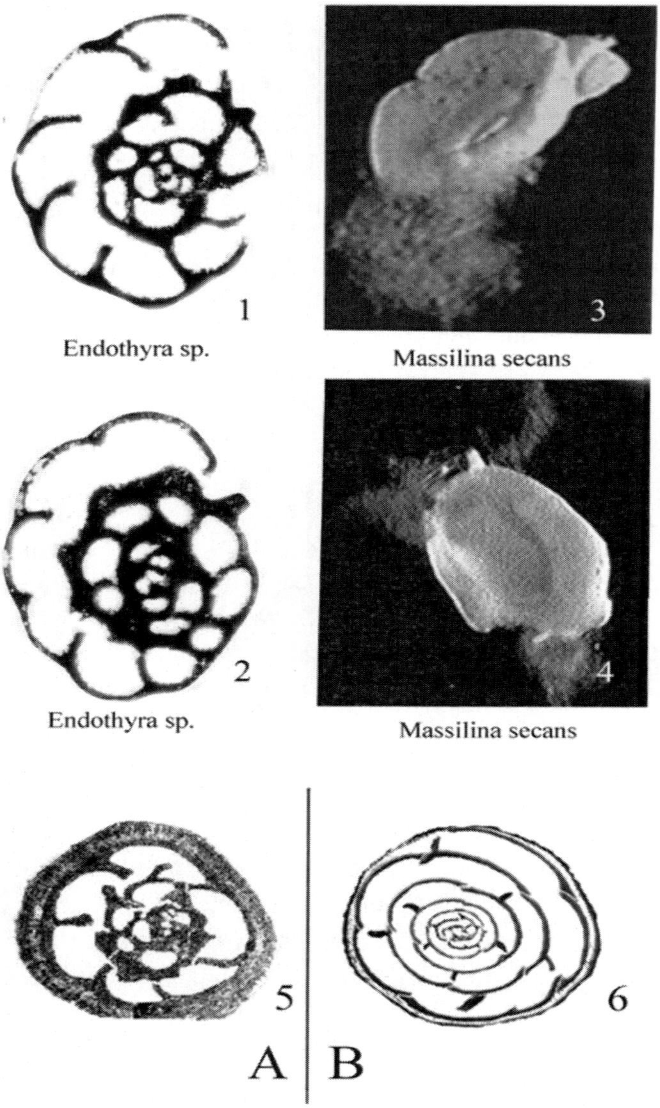

Text Figure 9. The scheme of median sections of shells of lower Fusulinoida (A), Miliolida (B) and photographs of the living *Massilina secans* in culture (after Mikhalevich 2009 a, 2009 b).
1, 2 - *Endothyra* sp. , median sections with the broken last chamber of different specimens, note rather wide chomata at the bottom of the chambers enlarging in size with growth up to the prominent tooth-like projection (2); 3, 4 – living *Massilina secans* in culture, after leaving the cyst, the remains of the cyst visible near the aboral end of the shell (3) and near the both ends of the shell (4) with the broken last chambers revealing the teeth (tooth of wide triangular form in specimen 3, straight and prominent like a nail at the upper and lower parts of the specimen 4; 5 – *Endothyra* ex gr. *bowmani* Phillips, Middle Carboniferous, chomata are visible at the bottom of chambers (scheme after Cozar, 2000); 6 - *Nummoloculina contraria* (d'Orbigny), Miocene, teeth are visible at the bottom of chambers.

The trochoid spire of the whole shell varies from rather low (Plate **59**, figure1) to rather high, with the shell height and breadth nearly equal (Plate **58**, figure 1 a, b; Plate **59**, figure 7). Sometimes the spire is more prominently elongated with the height slightly exciding the breadth of the shell (Plate **59**, figure 2). From three to four to rarely five globular chambers are visible from the apertural side (Plate **58**, figure – s 1a, b, 2, 3; Plate **59**, figure 1; Plate **61**, figure 7) and the fifth chamber may have less regular form (Plate **58**, figures. 2, 3a).

Sutures are deep and narrow, and in some places show the rows of the openings somewhat bigger than pseudopore openings of the wall. Such sutural openings can be seen from outside (Plate **59**, figure 1b) and inside the shell (Plate **61**, figure 7), but only under the bigger SEM magnifications. Most of the sutures show only a series of minute openings or lack even such minute spaces (Plate **60**, figure 7).

XI.2.3.4. Details of the shell wall ultrastructure

The shell wall of *Conotrochammina antarctica* was considered previously as noncanaliculate (Loeblich & Tappan, 1987), and much earlier such wall was called imperforate. In the light microscope, the outer shell wall of this species looks very smooth, nearly polished and solid, without any openings, and formed by the sand particles of middle size (Plate **61**, figure 8 a, b). In SEM, it also looks smooth and polished, encrusted by the sand particles of ordered orientation, having angular or irregular outline and turned to the outer surface by their flat sides, with very thin pelite material between them (Plate **58**, figure 1c). These pelite areas as well as the sutures are pierced by the small pseudopore openings of subcircular (in the middle part of pelite areas) or irregular (around the bigger sand particles) form. In some places such openings in the sutures form a row of bigger subcircular openings resembling the sutural additional apertures. In this species they are not regular for all the sutures (Plate **59**, figure 1 b; Plate **61**, figure 7). In the places where the outer flat sand particles were detached, the more coarse and regular pseudopore openings could be observed.

The inner shell surface opposite to the outer one is rough with the sand particles turned to the inner surface by their acute angles (Plate **59**, figure 6; Plate **60**, figure - s 3 – 7; Plate **61**, figure 7). This surface is pierced by the regular dense subcircular or irregular form openings (Plate **60**, figure 4; Plate **61**, figure 7). On the transverse breakings of the wall, occasionally the band of the upper flat sand grains can be seen (Plate **61**, figure - s 1, 5) somewhat resembling the dense roof layer of the miliolid wall or the flat surface layer of sand grains covering *Norvanganina* wall (Mikhalevich, 2012). Below it a layer of angular sand grains is situated although these rather big grains are broken off the wall in the process of fracturing or making slides, leaving the yawning cavities (Plate **61**, figure 3). Nevertheless, the clear spaces piercing the transverse wall section can be seen in many places (Plate **59**, figure - s 2, 4, 6, 7; Plate **60**, figure - s 3, 8; Plate **61**, figure - s 1, 2, 4 - 7) which occasionally look like they are going in the radial direction (Plate **59**, figure 4; Plate **60**, figure - s 3, 5; Plate **61**, figure - s 1, 2, 4, 7). In this case the breaking of the wall goes in another direction (not strictly perpendicular) and the dense subcircular openings can be seen resembling those of the sections of a canaliculate wall when the cutting goes perpendicularly to canaliculi (Plate **59**, figure 2; Plate **61**, figure 5 – the upper piece). The ultrastructure of the wall exhibits features similar to those found in canaliculated wall, although it is not of the true canaliculated type. These more regularly organized spaces are usually met in the areas of the finer material. The SEM photographs of the shell wall of the species studied resembles in many features (rough sand grains in the middle part of the transverse wall section easily broken off, the rough character of the inner chamber wall) a wall structure of *Saccamminella salsa* in the work of Bronnimann et al., 1992 (Tabl. XIX, figure - s. 1, 3, 7) differing from it in its smooth outer wall. Even some radially going spaces between sand particles were also met by the above authors (Bronnimann et. al., 1992, Tabl. XIX, figure 8, right lower part of the picture).

The very thin organic lining on the outer and inner wall surface can also be seen in the transverse breakings as a thin white line (Plate **60**, figure 3; Plate **61**, figure - s 2, 3, 5).

Thus in reality the wall of this species is not solid and is fully permeable. Under the higher magnifications up to x 2400 and 8000, the ultrathin structure of the collar of the inner aperture can also be seen. It consists of the minute white particles arranged in a radial direction (Plate **60**, figure - s 1, 2).

XI.2.3.5. Discussion

The type species of the genus *Conotrochammina* - *C. whangaia* Finlay, 1940 was described as having areal aperture. But Bronnimann et al., (1983) didn't find an apparent aperture in the topotypes and paratypes of the type species. This was the cause for these authors to suppose the aperture as being interiomarginal and obscured by the rough sand particles of the wall. Loeblich & Tappan (1987) confirmed the presence of areal aperture in this species. As it turned out in our material, nearly half of the exemplars of the species definitely lack aperture in their last chamber (Plate **58**, figure 1a – c). But in the broken specimens of such forms, inner areal aperture with a distinct collar was discovered through the series of breakings in all the previous chambers up to the initial one. The only possible suggestion in this case might be that the aperture is formed later in the fully performed last chamber, although this could be hardly imagined in the agglutinated forms. The indirect confirmation of such supposition could be the fact that the apertural collar of the series of the inner apertures is higher and more strongly developed (Plate **58**, figure – s 2 – 7; Plate **60**, figure – s 1, 2) than the slightly low rim of the outer aperture of the last chamber when present (Plate **59**, figure - s 1 a, b). The terminal rather than the interiomarginal position of the aperture and may be its later formation after the last chamber of the shell is fully formed distinguishes this genus and species from the true trochamminids where the next chamber is formed simultaneously with its aperture (Angel, 1990). The latter is interiomarginal and positioned at the base of the last chamber. Loeblich & Tappan (1987) had moved the genus *Conotrochammina* from trochamminids to verneuilinids (to the former superfamily Verneuilinacea Cushman, 1911, later the order Verneuilinida Mikhalevich et Kaminski, 2003). But the typical verneuilinids have interiomarginal aperture at least initially. Their aperture becomes terminal at the later stages of shell development. In the species studied here the aperture is areal from the very beginning of the shell. The same situation is in the genera of the family Reophacellidae Mikhalevich et Kaminski, 2003. The conical shell and the character of the chambers also resemble more *Reophacella* or *Uvigerinammina* than true trochamminids, although in these two genera their terminal aperture is elongated into a more distinct neck. As these genera were moved from verneuilinids of the subclass Textulariana to the nouriids belonging to the subclass Hormosinana (Mikhalevich, 2004b), the taxonomic position of conotrochamminids was changed correspondingly.

This variable character of the aperture at the different stages of the onthogenetic development of the shell in *Conotrochammina antarctica* could explain why this widely distributed in the Antarctic waters species was often mentioned under the other names (see above – this Part).

XI.2.3.6. New systematic position

Phylum **Foraminifera** d'Orbigny, 1826

Class **Nodosariata** Mikhalevich, 1992

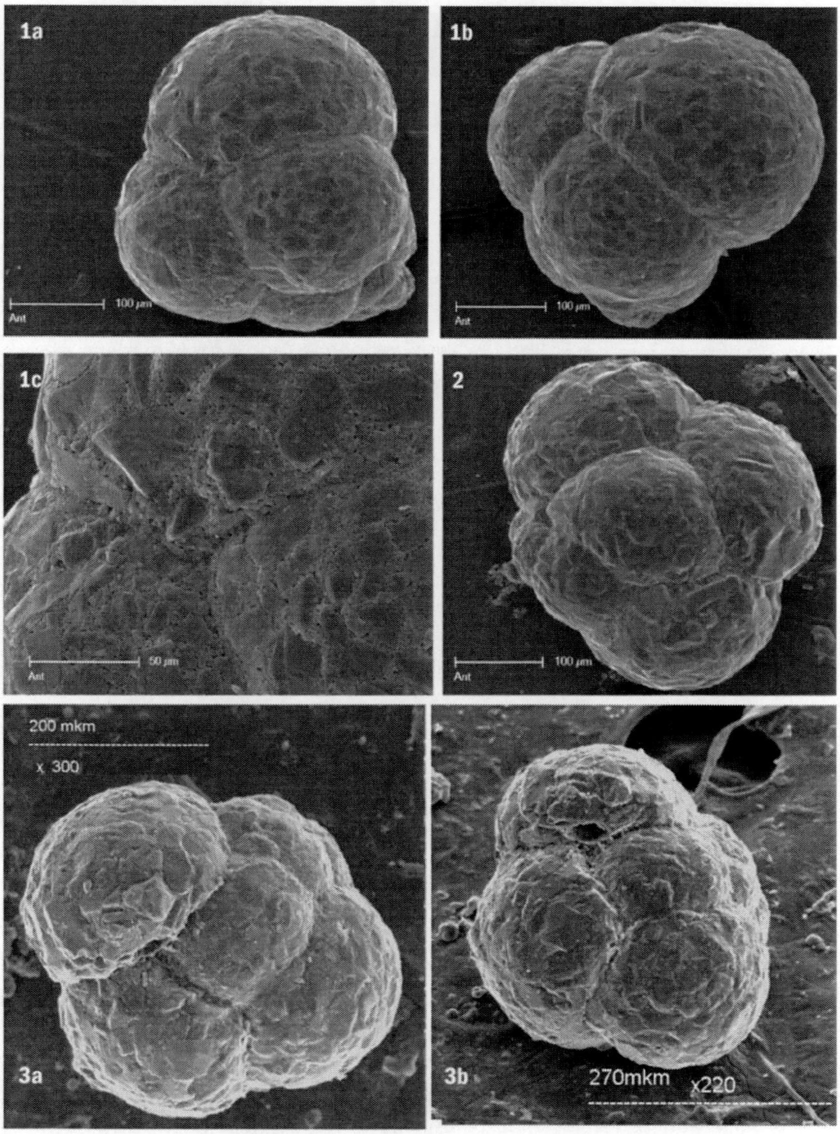

Plate 58. *Conotrochammina antarctica* Saidova, 1975.

1a, b, c – view of the shell from the different positions of the "apertural" side with the three chambers in the final whorl but aperture absent ("Ob" 1958, station 336), 1a,b – x 250, 1c – x 500; 2 – view of another specimen from the "apertural" side with the five final chambers visible, aperture absent ("Ob" 1958, station 336); 3a, b – view form the different positions of another specimen with five final chambers visible from the apertural side with the aperture present, 3a – from this position aperture looks as a narrow slit and seems to be placed at the base of apertural face – x 300, 3 b – from the position of the shell which was little changed, the aperture looks as an oval opening only slightly elevated above the suture ("Ob" 1958, station 336).

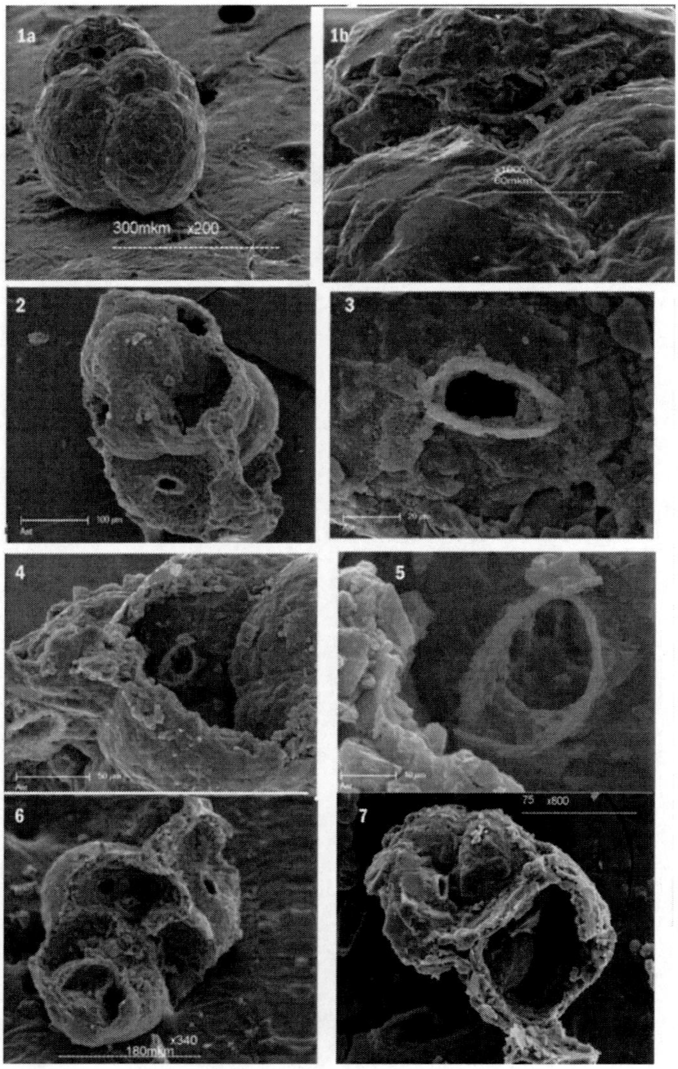

Plate 59. *Conotrochammina antarctica* Saidova, 1975.

1a, b – apertural side of the same specimen as Plate58, fig. 3a, b from the little changed positions: 1 a – x 200, 1 b – x 1000, mark the size of the sutural openings at the different sutures ("Ob" 1958, station 336); 2 – the broken specimen, mark two inner apertures with the collars from the outer side of the chambers, character of the inner shell wall, transverse breaking of the shell and character of the circular openings of the broken wall at the left side of the shell near the junction with the nearest chamber (openings resembling the ones of canaliculated wall when the cutting goes perpendicular to canaliculi) – x 250, ("Ob" 1958, station 336); 3 – part of the fig. 2 enlarged showing the inner aperture viewed from the outer side of the chamber – x 1250; 4 – the same shell as on the fig. 2 after the next breaking showing two inner apertures with the collars at the outer side of the chambers – x 500 (mark the character of the wall at transverse breaking showing the regular openings above the aperture; 5 – inner aperture from the outer chamber side from the fig. 4 enlarged, x 2500; 6 – another shell broken up to the initial chamber with inner apertures seen from the outer chamber surface and from the inside of the chamber – x 340 (mark the character of the transverse breakings of the wall and its openings) ("Ob" 1958, station 336); 7 – part of the successive breaking of the exemplar of the figure 6 revealing the aperture of the initial chamber (mark the character of the transverse breaking of the wall in the middle part of the picture) – x 800.

Subclass **Hormosinana** Mikhalevich, 1992
Order **Nouriida** Mikhalevich, 1980
Superfamily **Nourioidea** Chapman et Parr, 1936 (nom. corr. pro Nouriidoidea Mikhalevich, 2004)
Family **Conotrochamminidae** Saidova, 1981
Genus ***Conotrochammina*** Finlay, 1940
Type species: *Conotrochammina whangaia* Finlay, 1940, p. 448.
Syn: *Pseudotrochammina* Frerichs, 1969, type species *P. triloba* Frerichs, 1969, p. 1.
Trochammina conica Milam et Anderson, 1981, Table 3, figure 5.

Diagnosis. Shell with low conical spire of about two to seven coils, from a few (three to six) chambers per whorl, with chambers inflated to subsphaerical; aperture areal, wide slit to oval or circular opening, often with a distinct collar.

Paleocene to Holocene.

Remarks. Such features as the degree of the wall roughness, and correlated with it clearly visible or obscure sutures usually mentioned in the diagnosis of this genus, ought to be considered as having a specific rank. The main taxonomically important features of the genus *Conotrochammina* are the conical form of its shell, arrangement of chambers and areal aperture. The genus *Pseudotrochammina* described later does not principally differ in these features from the *Conotrochammina* and is regarded here as its synonym.

Conotrochammina antarctica Saidova, 1975

C. antarctica Saidova, 1975, p. 105, Table XCIX, figure 6.
Conotrochammina sp. Fillon, 1974, Table 2, figure 4.

Diagnosis. Shell very small (0.45 – 0. 5 mm), conical and slightly asymmetric, with five to seven coils of spiral and mostly three but rarely up to five chambers per whorl, chambers strongly inflated, subsphaerical, quickly enlarging in size as added, last chambers may be somewhat irregular in growth and position, sutures narrow, distinct, occasionally with pseudopore openings; wall externally smooth, mosaic, with flat middle sized sand grains of angular or irregular form and thin pelite material between them, areas of thin material rather densely pierced by pseudopores, inner wall surface rough, pierced by subcircular pseudopores, transverse sections of the wall shows flat "roof" sand particles with middle sized sand grains below alternating with areas of fine material and pierced through by the empty spaces everywhere (pseudopore passages going under the bigger flat particles as well); aperture areal, somewhat irregularly widely oval, may be slightly curved or subcircular, encircled by very thin white collar constituted by extremely thin grains, apertural opening situated variably – close to the suture or nearly at the center of the chamber's outer surface.

Holocene.

Remarks. The species is distributed all around the Antarctic at the shelf and upper subzone of the bathyal zone.

Among our material, nearly half of the specimens have no apparent aperture. The same was observed in another Conotrochamminid species - *C. whangaia* by Bronnimann et al., 1983.

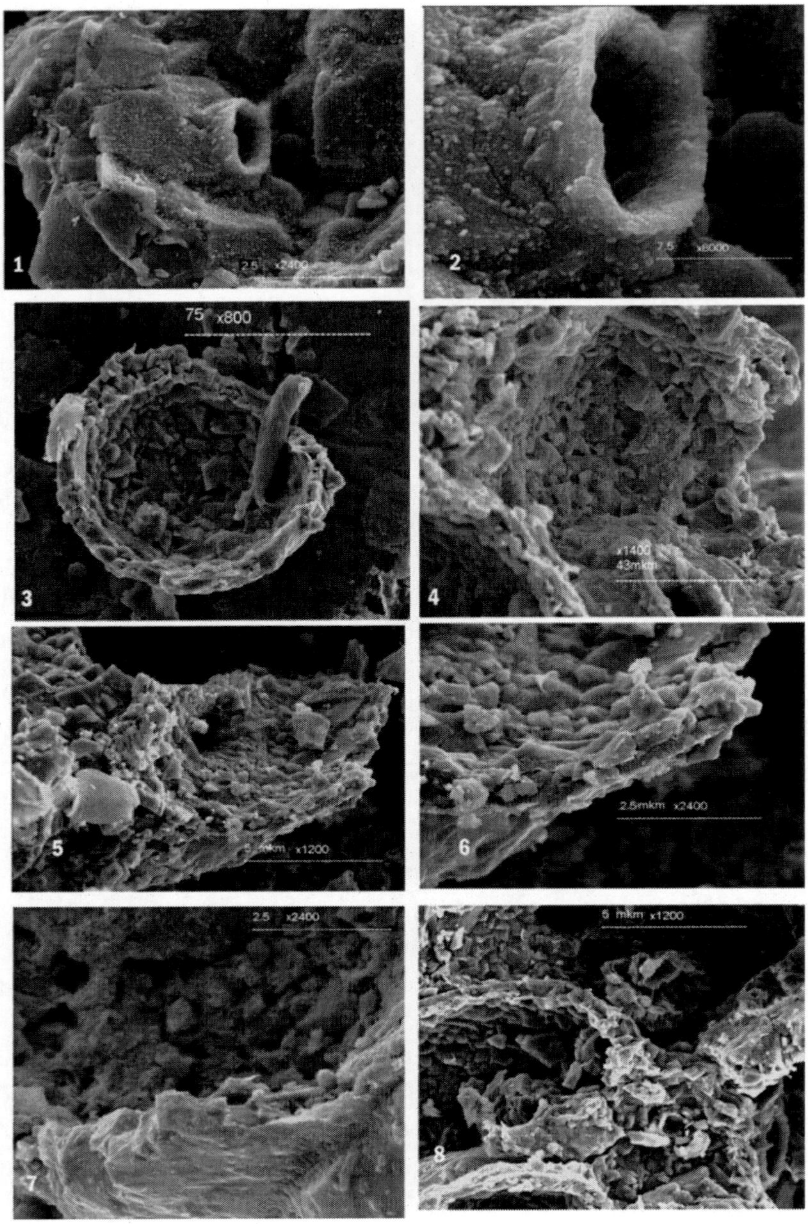

Plate 60. *Conotrochammina antarctica* Saidova, 1975.

1 - the outer collar of the inner aperture of the initial chamber of the fig. 7 (Plate 59) – x 2400; 2 – the same, enlarged – x 8000; 3 – a piece of the broken initial chamber showing the structure of the inner wall and transverse breaking of the wall (mark the character of the openings at the upper part of the piece going in radial direction and white line of inner organic lining at the right side) – x 800; 4 - part of the fig. 6 (Plate 59) showing the inner aperture from the outer chamber side and the structure of the inner shell wall - x 1400; 5 – another piece of the broken shell of figure 6 (Plate 59) showing the inner aperture from the inside, the transverse section of the wall (lower part) with radial openings and outer organic lining (white) and rough inner wall of the chamber – x 1200; 6 – the same, enlarged – x 2400; 7 – the breaking of the same shell showing the structure of the inner and outer wall and the suture – x 2400; 8 – another place of breaking shows the structure of the transverse wall (central part of the picture) with the thick outer layer of the flat surface particles – x 1200.

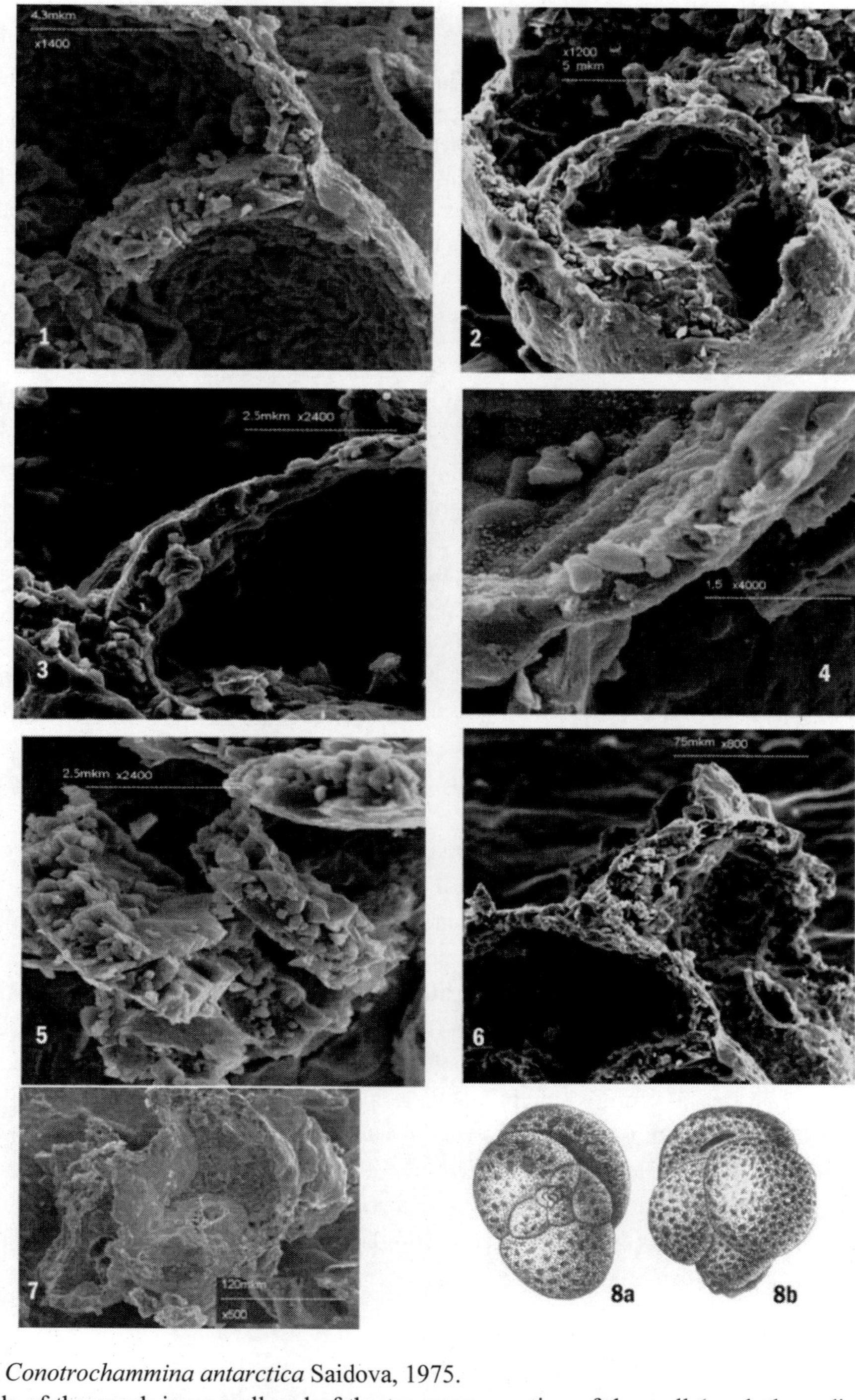

Plate 61. *Conotrochammina antarctica* Saidova, 1975.

1 – details of the rough inner wall and of the transverse section of the wall (mark the radial openings below the pierced outer wall of fine material at the very upper part of the picture and at the left part of the middle septa and thick roof of the flat sand particles at the lower right side of transverse section) – x 1400; 2 – details of the wall structure at the transverse breakings of the initial chambers (mark the thin inner organic lining, the radial fissures at the left and right parts of the transverse section and longitudinally going spaces in the septa dividing two lower chambers) – x 1200; 3 – details of the transverse breaking of the septa (mark thin white lines of the outer and inner organic lining and dark

spaces where the bigger sand particles are broken off) – x 2400; 4 – details of the transverse breaking of the septa with occasional radial openings – x 4000; 5 – details of the transverse breakings of the wall in the three small fragments (mark the outer organic lining in the lower and middle pieces, inner organic lining and roof layer in the middle piece, and openings in the upper piece resembling the transverse cutting of canaliculi in canaliculated wall) – x 2400; 6 – details of the transverse breakings of the wall (mark occasional radial openings) – x 800; 7 – initial part of the broken shell with a clear row of sutural openings (right side) and clear subcircular openings seen between the particles of the inner wall (mark also radial openings in the upper right part of the transverse breaking of the wall) – x 500; 8 a,b – holotype of *Conotrochammina antarctica* Saidova (after Saidova, 1975), "Ob", 1958, station 336 – x 40.

The successive series of breakings of such shells up to the very initial chambers show the presence of distinct areal apertures encircled by the thin fragile collar in all of them. This fact allows us to suppose that their aperture may be formed after the ultimate chamber is composed. At least the collar of the inner apertures is higher and more developed than the low rim around the external aperture (when present) of the last formed chamber. Rather rough sand grains in the middle part of the broken wall are often broken off. The wall could be considered as noncanaliculate as some occasional radial passages in the wall are not regular.

XI.2.3.7. Conclusion

The study of the wall of the Antarctic species *C. antarctica* in SEM revealed its ultrastructural character. Its agglutinated wall externally is smooth and mosaic, it consists of oriented flat sand particles of middle size with thin pelite material between them pierced by the pseudopore openings. From the inner side of the chamber the wall looks rough, inner pseudopore openings are also visible.

The thin layers of the inner and outer organic lining usually covering each type of the wall were discovered as well. In nearly half of the specimens studied the apparent external aperture was absent, presenting in all their previous chambers up to the initial one. The character and position of the aperture of this species and genus, on which there were some contradictory evidence in the previous literature, was clarified. The aperture of *Conotrochammina* is definitely areal and has a fragile collar surrounding it. The ultrastructure of this collar was also studied. All these structural features (including often absence of the external aperture on the last formed chamber) were also marked by Bronnimann et al., 1983 in the type species of the genus *Conotrochammina whangaia.* Belonging of the *C. antarctica* to this genus may be considered undoubted. Based on all the facts shown above, the taxonomic position of the genus concerning the structure of its aperture was changed: the genus was transferred from the order Verneuilinida (subclass Textulariana) to the order Nouriida (subclass Hormosinana).

XI.2.4. On the Systematic Position of Stilostomellids

XI.2.4.1. Introduction

The widely adapted classification (Loeblich & Tappan,1987) places Stilostomellids family (Stilostomellidae Finlay, 1947) as part of the superfamily Stilostomellacea (Finlay, 1947) which, together with Bulliminacea, enters suborder Rotaliina (Delage et Herouard, 1896).

The comparative morphological analysis performed above (Part VIII, 9) showed that the main characteristic shell structure in this class is planispiral and trochospiral or derived from them. Meanwhile Stilostomellid group of genera differs from the overwhelming majority of the other rotaliins in its uniserial shell character. Such shell form is characteristic of the big groups of Nodosariata. Placing these uniserial forms into Rotalians one needs to suppose the secondary origin of stilostomellids from the elongated trochospiral forms which theoretically reasoning could be possible. It seemed that the presence in some of stilostomellids of the inner apertural tooth also shows their affinity to rotalians. Their outer nodosariid appearance, however, looks like a special case among rotalians. In the rare cases of the uniserial rotalian shells (in some representatives of the families Siphogenerinoididae Saidova 1981, Eouvigerinidae Cushman, 1927) the small rudimentary and hardly observable initial trochoid part (triserial, loosely biserial) is nevertheless preserved at least in one of the generations. One of the main taxonomic features of this suborder (Loeblich & Tappan, 1987) was the presence of the bilamellar shell wall. It was supposed under this approach that the representatives of the family Stilostomellidae also possess such a bilamellar wall. All these facts show the necessity of investigation of *Stilostomellids* structure and ultrastructure in details in the Stereoscopic Electron Microscope to affirm or to change its taxonomic position.

XI.2.4.2. Details of the shell and apertural structures

The material at the author's disposal was not sufficient for such a study and additional material was kindly presented by Ch. Saidova from the samples collected from: (i) "Vitiaz" 22 trip in 1955; (ii) The P.P. Shirshov Institute of Oceanography, Russian Academy of Sciences, Moscow and (iii), Station 3513 (26° 22'08" N, 143°09'E) submerged 2368 to 2668 meters deep at the region of tropical islands in the Pacific Ocean, Pampo Island and Mariana Islands. Although stilostomellids are often found in the seafloor sediments because they used to be abundant in deep-sea sediment of Eocene faunas their extinction during the MPT has now been firmly documented in over 40 ODP sequences in a set of works, (Weinholz & Lutze, 1989, Hayward 2002, Hayward and Kawagata 2005, Hayward et al 2006, Hayward et al 2007, Kawagata et al., 2005, Kawagata et al., 2006). The summary of their extinction is given in Hayward et al., 2007.

In the material studied, it was possible to distinguish the form *Strictocostella matanzana* (Palmer et Bermudez, 1936) and *Strictocostella matanzana peritae* (Mikhalevich, 2011). The latter differs from the first one in its apertural structure with more elongated neck prominently sculptured, in ornamentation of its surface and even in the form of the elongated crystals forming the shell wall as it will be shown below. Both of the forms were represented by the microsphaeric and megalospheric generations (Plate **62**, Figures 1a, 1b; Plate **64**, Figures 2a, 2b; Plate **65**, Figures 8a, 8b). In both of the forms the microspheric shells were longer and more slender having acute base with smaller and narrower initial chambers and widening significantly to the apertural end, with the final chamber number of the whole shell from eight to ten (Plate **62**, Figure 1a; Plate **64**, Figure 2a). Their megalospheric shells are shorter with a big and rather wide initial chamber and wide rounded base, lower chamber number (four to five), the shell with parallel sides or only slightly widening towards the apertural face (Plate **62**, Figure 1b; Plate **64**, Figure 2b; Plate **65**, Figure 8a). The both forms show the similar trends in the changes of the shell form through the sexual and asexual generations.

Plate 62. *Strictocostella matanzana* (Palmer et Bermudez), 1936.
1 a, b – the outer view of the whole shells (a – microspheric, b – megalospheric forms) (x 65); 2 – the middle part of the shell (1a), enlarged (x 250), note the large circular holes presumably made by some predator; 3 – suture between the penultimate and antipenultimate chambers of specimen 1a, enlarged, note the rough character of the wall, and the more dense pore openings in the sutural area (x 500); 4 – surface of one of the lower chambers of specimen 1a, having a comparatively smoother outer wall than that of the latest chambers (x 1200); 5 – side view of the aperture of the last chamber of the specimen 1a, note rough surface of the last chamber and elongate crystals of the apertural border aligned in

parallel (x 250); 6 – inner aperture and the edge of the broken chamber around it (x 1200); 7 – left part of figure 6, enlarged, note the rough outer surface of the chamber and rough inner surface of the broken chamber with the prominent perpendicular crystals around the inner aperture (x 1200); 8 – central part of figure 6, enlarged, note the mass of sand grains and coccolithophorids in the central opening partially encircled by the organic lining (x 1200).

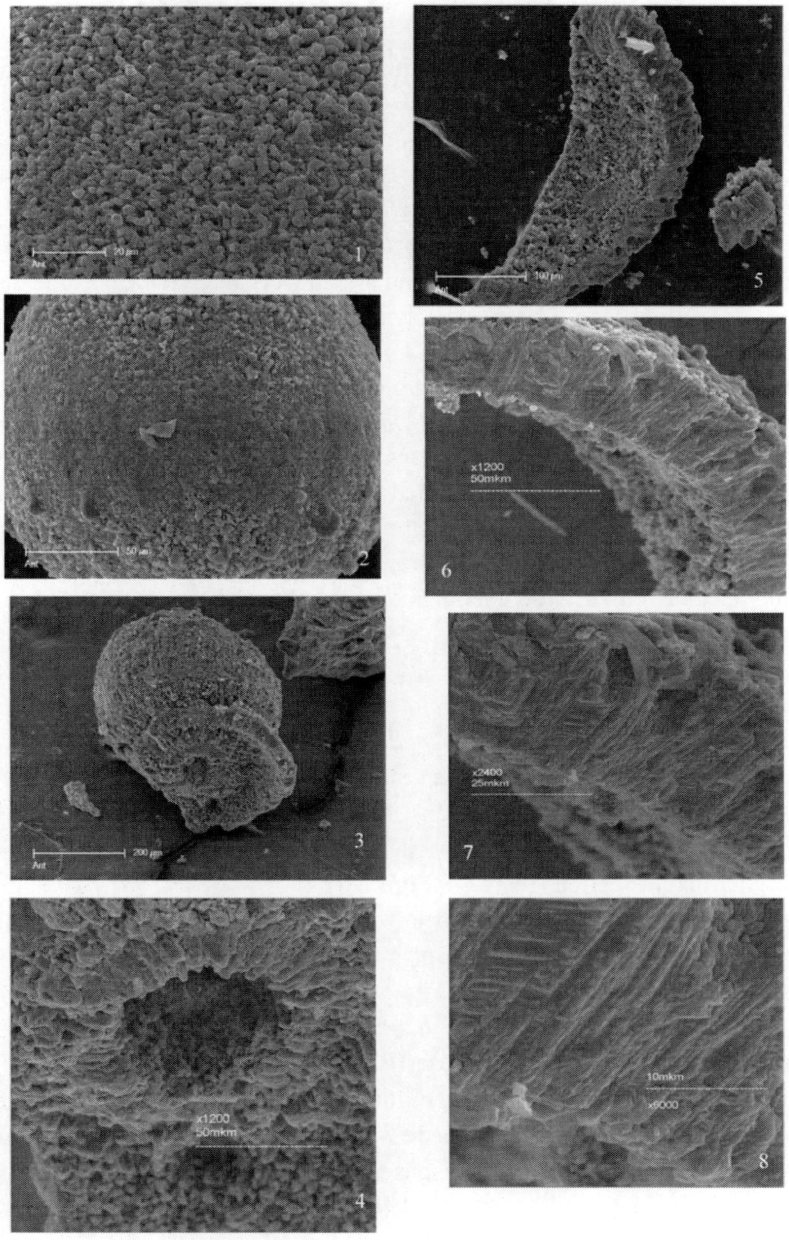

Plate 63. *Strictocostella matanzana* (Palmer et Bermudez), 1936.

1 – rough outer surface of exemplar 1a (Plate 62) with the pore opening between the tubercules (x 1200); 2 – surface of the fourth chamber of exemplar 1a (Plate 62) which is more smooth in its central part (x 600); 3 –the broken piece of another shell showing the inner aperture and transverse breakage of the encircling wall, note the radial disposition of the elongate crystals in the apertural border and

transverse wall section (x 130); 4 – the border of the inner aperture of the broken exemplar 3 (Plate 63), enlarged, note the inner pore openings in the rough surface of the inner wall (x 1200); 5 –the broken semilunar piece of the wall of another specimen of this species showing the radial direction of the elongated crystals of the chamber wall in its transverse breakage and the rough inner chamber surface with the prominent ends of the elongated crystals projecting inward as well as pore openings between them and along the transverse chip (x 250); 6 - breakage of figure 5, Plate 63, enlarged (x 1200); 7 – part of the figure 5, Plate 63, enlarged (x 2400); 8 - part of the figure 5, Plate 63, enlarged (x 6000).

In typical *Strictocostella matanzana* the apertural neck is rather short and funnel shaped, widening to its open end, without prominent collar (Plate **62**, Figures 1a, b, 5). Its outer surface is pustulose like in the chamber wall surface, its inner tooth-like projection is slightly developed and even hardly visible nevertheless resulting in the asymmetrical shape of the apertural opening which have irregular wide oval outline (Plate **62**, Figures 6, 7, 8; Plate **63**, Figures 3, 4). In the inner aperture of the broken shell it is possible to see the mass of coccolitophorids shells and sand grains bound by the organic matter inside the apertural opening (Plate **62**, Figures 7, 8). These possibly represent the residue of the cytoplasmic mass containing some inclusions.

Strictocostella matanzana peritae has a long cylindrical apertural neck slightly tapering to its end, its apertural opening is encircled by a wide and flat slightly everted smooth collar formed by wide flat segments with visible lines of division between them (Plate **64**, Figures 2a, 3, 4). The edges of this collar are going obliquely down. The surface of this neck represents rather wide band prominently sculptured by a closely set tubercles and bordered beneath with a line of scalloped plates (Plate **64**, Figures 3,
4). Between the pustules rather wide openings of the wall can be seen (Plate **64**, Figure – s 3, 4). In the apertural view (Plate **64**, Figures 5, 6; Plate **65**, Figure 8a) the apertural opening looks like a bean-shaped opening with slightly prominent tooth-like projection which nevertheless is developed better than in *Strictocostella matanzana* aperture. The apertural end of the megalospheric shells of *Strictocostella matanzana peritae* was broken. In the specimens studied in SEM it was clearly seen that the apertural structures are partly dissolved as it often happens in the reworked specimens. But the regular radial depressions at the inner margin of the apertural collar marking the borders of the denticles nevertheless can be clearly seen under a higher magnification (Plate **64**, Figure 6).

The outer shell wall of the chambers looks in the light microscope as smooth with slightly rough sutures in both the forms. But in SEM even under a low magnification it is possible to see some additional details. The wall of the typical form of *Strictocostella matanzana* has a rougher character (Plate **62**, Figures 2 – 6; Plate **64**, Figure 1 - 3). Its appearance varies from being fully pustulose (Plate **62**, Figures 3, 5; Plate **63**, Figure 1) to having more smooth central part (Plate **62**, Figure 2; Plate **63**, Figure 2). The outer pustules represent the prominent ends of calcite needles forming the wall (see below) and going in radial direction. The pore openings are disposed as evenly dispersed depressions between them being encircled by the needle ends (Plate **63**, Figure 1; Text figure 10). These needle ends form here and there the prominent rosettes (Text figure 10). Pore openings situated between the prominent ends of the needle crystals look nipple-like as being elevated on some papilla. The similar pore character was observed in some other nodosariid genera (Meyn & Vespermann, 1994, Plate 31, Figure 1; Plate 15, Figure 7; Plate 37, Figure 8). Sometimes the lower part of the chamber may have a rougher sculpture where several pustules coalesce into somewhat bigger tubercles or flattened particles (Plate **63**, Figure 4), or very rarely into a short spines. Usually the last chamber of the shell is fully covered by the pustules while the earlier chambers are preserving their partly smooth areas (Plate **62**, Figure – s 1a, 4). The sutures of this form are straight, horizontal, rather wide and sculptured with radially going thin elongated tubercles and the small pore openings between them (Plate **62**, Figures 2, 3). The junction of the radial rows of pustules of the adjacent chambers can be seen as a very thin, nearly dotted horizontal fissure (Plate **62**, Figure 3). The outer surface of the chamber wall of *Strictocostella matanzana peritae.* is mostly smooth, especially in its microspheric form, often with leakages of the shell matter (Plate **64**, Figures 2a, 7), the pore openings are rare, unevenly dispersed and hardly visible even in 1000 magnification (Plate **64**, Figures 7, 8). Sometimes very fine burrows visible only in SEM are going longitudinally along the chamber surface (Plate **64**, Figure 7). The sutures are horizontal, wide and rough like in *Strictocostella matanzana* but their sculpture is more prominent being composed of wider and higher rows of tubercles going longitudinally with large openings between them exceeding many times the size of the pore openings which with some caution may be considered as additional apertures (Plate **64**, Figures 7, 8). At least it is possible to suppose that their larger sizes compensate to some extent the insufficiency of communication with the environment which can provide small and rare pore openings of the rest of the chamber surface.

The shell wall of the both forms in some places is penetrated through with the deep holes subcircular in their outline (Plate **62**, Figures 1a, 2; Plate **64**, Figure 2a; Plate **65**, Figures 1, 2). These are supposedly made by some predators. In the wall of the chambers of Plate **65**, Figures 1, 2 these openings are of much smaller diameter and of regular disposition somewhat resembling the additional apertures.

It is necessary to note that the degree of sculptural details varies widely in the both forms studied and that the species as well as the entire genus and the majority of nodosariids is highly variable.

XI.2.4.3. Details of the shell wall ultrastructure

In the transverse breakings of the wall of the both species (Plate **63**, Figures 5 to 8; Plate **64**, Figure 1; Plate **65**, Figures 4 to 7) and in the surface of their inner aperture (Plate **62**, Figure 5 and Figure 7 – partially seen; Plate **63**, Figures 3, 4; Plate **65**, Figures 4, 5) distinct long crystals can be easily seen going in radial direction from the inner to the outer surface. In *Strictocostella matanzana* (Plate **63**, Figures 3, 4, 5 to 8; Plate **64**, Figure 1) they look somewhat flattened as compared with those in *S. matanzana perytae* (Plate **65**, Figures 4 to 7). The crystals of the latter look more voluminous and sub-prismatic while being wider at their bases (Plate **65**, Figure 6, 7). Under the higher magnifications exceeding 10 000 (Plate **64**, Figure 1) the smaller subunits composing these crystals could be discerned. In the broken elongated crystals of the Plate **65**, Figure 7, some dark centers looks as showing the empty space inside but these are most likely the artifacts. Unfortunately the higher magnification was limited in the microscopes used in this study. No transverse lamellae or primary organic lining were discovered in the both forms of *Strictocostella*. Secondary lamellae characteristic of many representatives of nodosariids were not observed either.

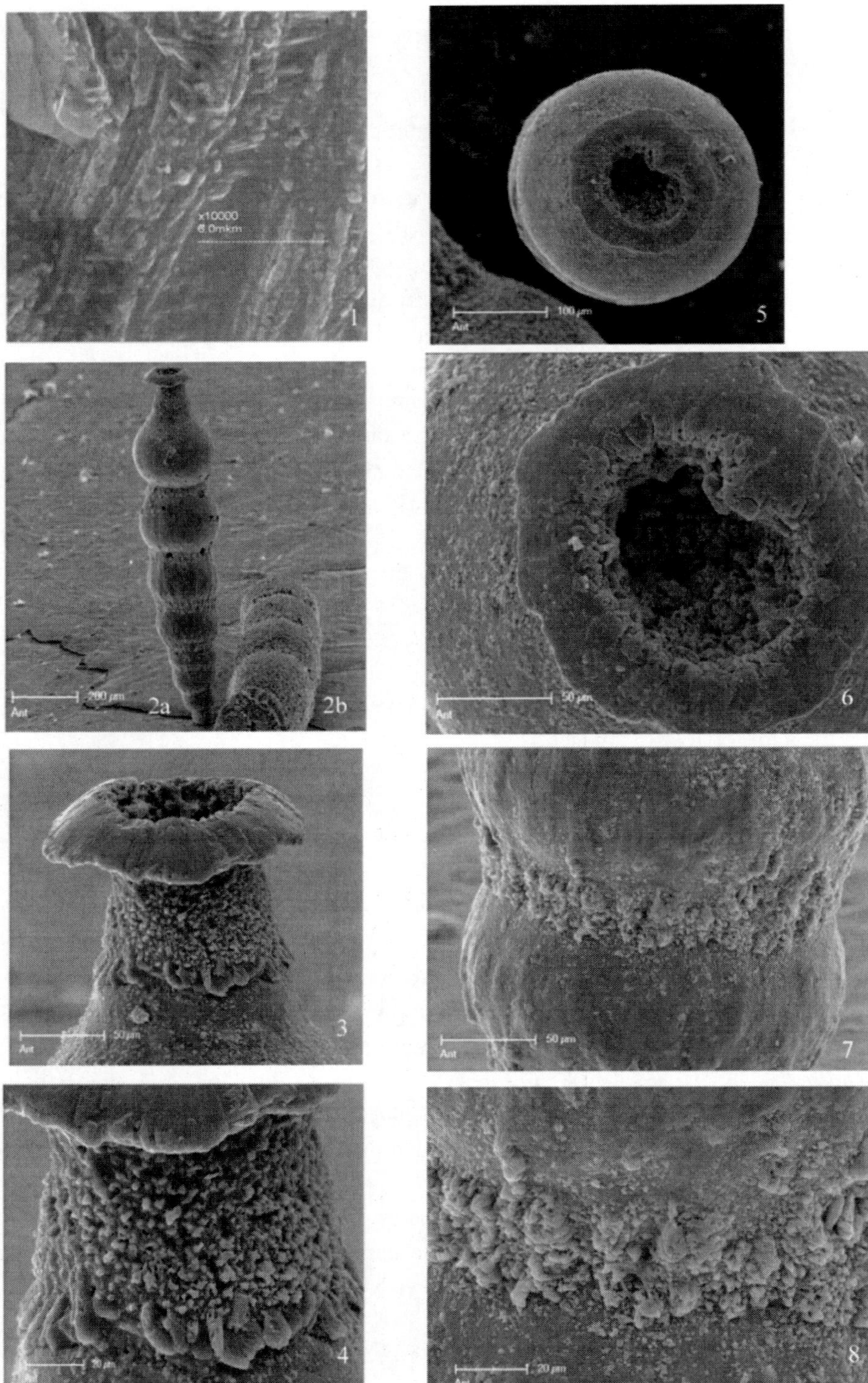

Plate 64. *Strictocostella matanzana* (Palmer et Bermudez), 1936 (figure 1) and *Strictocostella matanzana* (Palmer et Bermudez), 1936 *perytae* Mikhalevich, 2011 (figures 3 – 8).

1 - *Strictocostella matanzana,* part of the figure 8, (Plate 62), enlarged showing that elongated crystals are composed of rows of smaller subunits (x 10 000); 2 a, b – outer view of the microspheric (a) and the

broken megalospheric (b) shells of *Strictocostella matanzana perytae*; megalospheric shell with the dark holes circular in outline made supposedly by some predators, and area of the regularly placed smaller circular openings above the last suture (additional apertures ?) (x 90); 3 – aperture and a sculptured apertural neck of the exemplar 2a, Plate 64, note pore openings of the wall visible inside the aperture and the bigger openings within the sculptured part of the apertural neck (x 500); 4 – the sculptured apertural neck of the exemplar 2a, Plate 64, enlarged (x 1 000); 5 – apertural view of the exemplar 2a (Plate 64), (x 250); 6 - apertural view of the exemplar 2a, Plate 64, enlarged, note the smooth character of the surface of the apertural border and rough surface of the partly visible inner wall with the pore openings between the ends of the elevated tubercles (x 600); 7 – the suture between the fifth and the sixth chambers of the microspheric shell (of figure 2a, Plate 64), note the sutural sculpture, openings between the sculptural grains and longitudinal striae along the smooth chamber surface of the sixth chamber (x 500); 8 - the suture between the fifth and the sixth chambers of the microspheric shell (Plate 64, 2a), enlarged (x 1200).

The pore spaces are going between these long crystals along them and opening at the inner and outer shell surfaces between the ends of these crystals (Plate **62**, Figures 3, 4, 7; Plate **63**, Figures 1, 4, 6; Plate **64**, Figures 3, 4; Plate **65**, Figures 2 to 6; Text figure 10). At that they could be situated in the depressions between the crystals or may be elevated and disposed between the prominent elevated ends of several crystals. The pore character of the outer wall of the both forms was described above. The difference of this character in each of them was also marked where the most part of the outer wall surface in *Strictocostella matanzana perytae* is more smoothly finished.

Text Figure 10. Rough outer surface and pore openings of *Strictocostella matanzana* (Palmer et Bermudez) 1936, exemplar 1a (Plate 62) with the pore opening between the tubercles (enlarged view of the right upper part of figure 1, Plate 63), (x 5000).

The inner shell surface of the *Strictocostella matanzana* (Plate **62**, Figures 4, 7; Plate **63**, Figures 3, 4, 5, 8) and of *S. matanzana perytae* (Plate **65**, Figures 4 to 7) has principally the same character: it is rough with the small subcircular or somewhat angular pore openings in the depressions between the elongated crystals or between the elevated ends of the crystals. The inner surface of the both species looks as tuberculous and knobby.

1 – suture beneath the last chamber of the microspheric exemplar 2a, (Plate 64), note horizontal rows of openings at the sutural line between the rough pustules and above this line on the smooth chamber surface (x 500); 2 – right part of the figure 1, (Plate 65), enlarged, note the row of larger openings at the lower chamber part and very small and rare openings on the smooth middle part of the shell (x 1200); 3 – the initial chamber and sutural area of the megalospheric shell 2a of Plate 64, note the more expressed bigger rough pustules in the sutural area and the bigger openings along the very sutural line compared to the smaller pustules and pore openings of the chamber surface (x 500); 4 – inner aperture of the broken specimen (Plate 64, 2b), note the large elongated crystals in the transverse breakage of the wall situated in one layer and in the apertural border ((with the partially disposed second layer in the area of the slight inner apertural projection (tooth) protruding nearly to the center of the apertural opening); the inner ends of the elongated crystals forming the wall prominently protrude into the inner chamber cavity, pore openings being situated between them in the depressions or between their elevated ends (x 250); 5 – part of the Plate 65,(figure 4, Plate 65),enlarged (x 500); 6 - part of the Plate 65, (figure 4, Plate 65), enlarged (x 1200); 7 – elongate crystals in the transverse breakage of the wall: part of the figure 6, Plate 65, enlarged (x 2400); 8 a, b – megalospheric (a) and microspheric (b) specimens, the same of the figure 2, Plate 64, from the other side view; note rough postulate sculpture of the sutures pierced by the larger openings in the megalospheric form.

XI.2.4.4. New Taxonomic Position of the Genus Strictocostella and Family Stilostomellidae and of Their Agglutinated Analogues

As no lamellae or primary organic lining were discovered in *Strictocostella* wall its wall ought to be considered as monolamellar (termed monofontinal in Russian literature in Mikhalevich, 2000). This ultrastructural type of the shell wall is similar to that one described for Jurassic (Lias) *Saracenaria pyramidata* (Reuss), 1863 in Meyn & Vespermann, 1994 (Plate 37, Figure 4, 7) and for the Recent *Laevidentalina frobisherensis* (Loeblich et Tappan, 1953) described by Eiland & Gudmudsson, 2004 (Plate 5, Figures C, D) although secondary lamination formed in the process of growth in our specimens was not observed. This is not surprising as lamination pattern in nodosariids varies within and among species and genera. The similar ultrastructural type of the shell wall was observed in *Saracenaria pyramidata* (Reuss), 1863 by Bartenstein, 1994 (Plate 37, Figure – s 1, 4, 7). Such kind of hyaline calcareous radial shell wall which was observed here in *Strictocostella* is characteristic of the representatives of the class Nodosariata. The forms studied here are the typical representatives of the genus *Strictocostella* and hence the genus ought to be replaced from the class Rotaliata to nodosariats, to the superfamily Stilostomelloidea Finlay, 1947 (= suborder Stilostomellina Saidova, 1981). The type species of the genus *Ellipsonodosaria modesta* var. *prolata* Cushman and Bermúdez, 1937 (*Strictocostella modesta*) was not at the author's disposal but the thorough electron-microscopic investigations of the stilostomellids made by B. Hayward and his group in their monograph (Hayward et al., 2012) surely show the belonging of *Strictocostella matanzana* to this genus

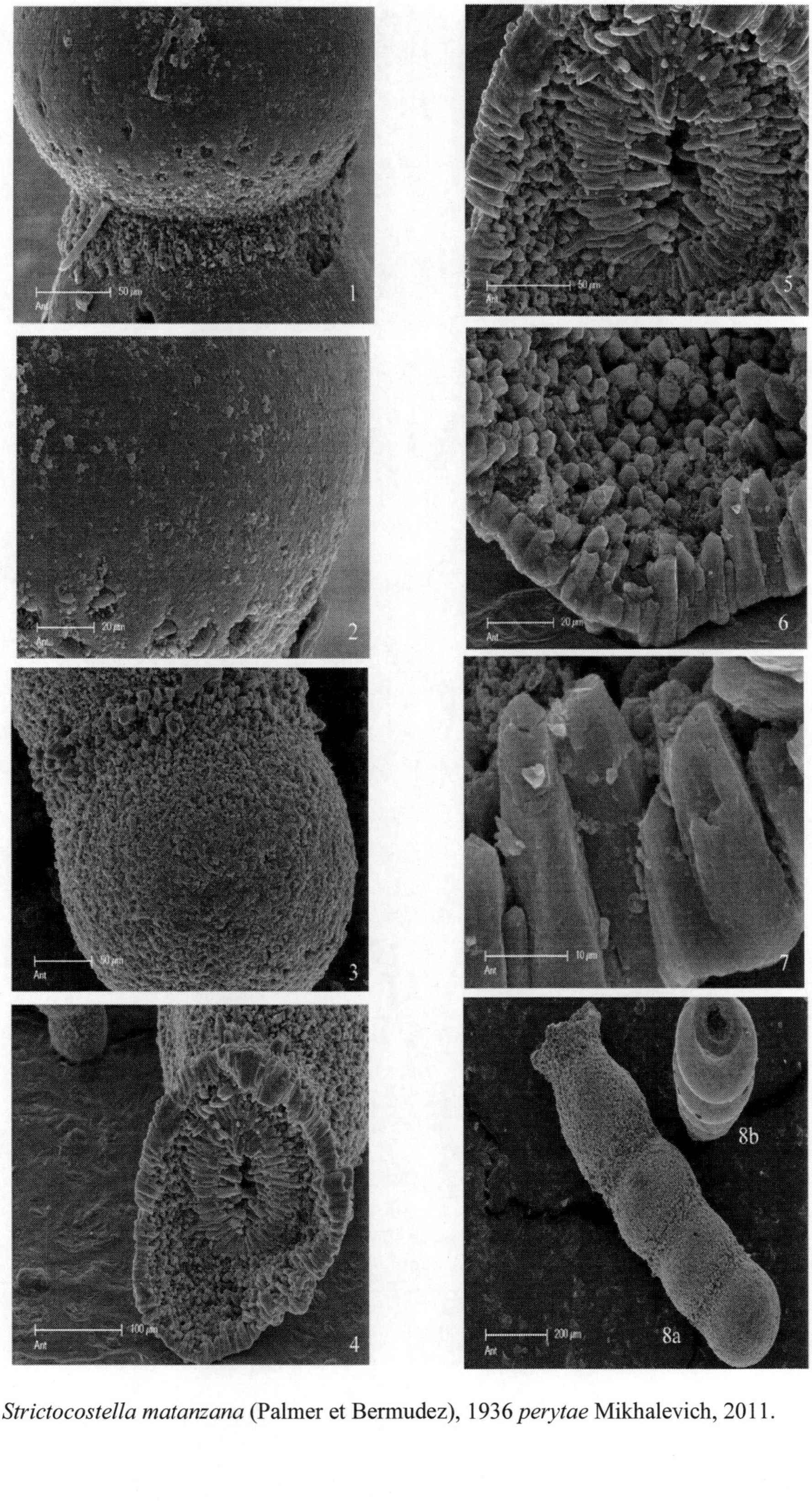

Plate 65. *Strictocostella matanzana* (Palmer et Bermudez), 1936 *perytae* Mikhalevich, 2011.

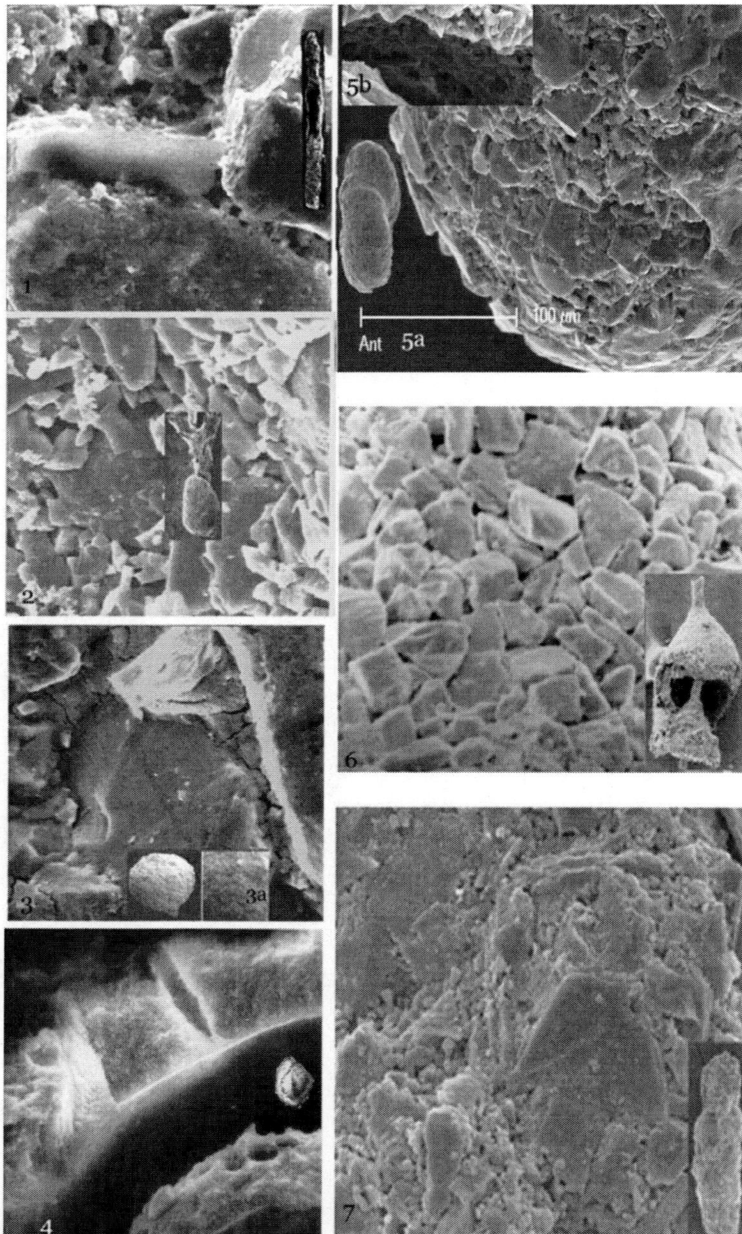

Plate 66. Shell wall ultrastructure of different agglutinated genera.

1 - *Rhabdammina antarctica* Saidova, 1975, (x 400), Polarstern, 1996, station 18; 2 - *Tholosina vesicularis* Brady, 1879, (x 1500), Polarstern, 1996, station 24; 3 - *Saccammina basispiculata* Mikhalevich et al., 1997, (x 1500), Polarstern, 1996, station 26 (3 – inner wall, 3a – outer wall); 4 - *Dentostomina agglutinans* (d'Orbigny, 1839) (SEM), (from Mikhalevich et al., (1988), part of the longitudinal section showing thick agglutinated outer layer of the wall, very thin white inner layer, pseudopore canal in the last chamber and circular pseudopore openings of the preceding chamber (x3000); 5 a, 5 b - *Recurvoides contortus* Earland, 1934, Ob, station 354, (5 a – (x 300), outer surface, 5 b - inner wall of the broken specimen showing its rough character and openings (x 1200); 6 - *Ginesina guttifera* Brady, 1881, (x 250); 7 - *Karrerulina apicularis* (Cushman), 1911 (x 1000), Viatiaz, station 2502.

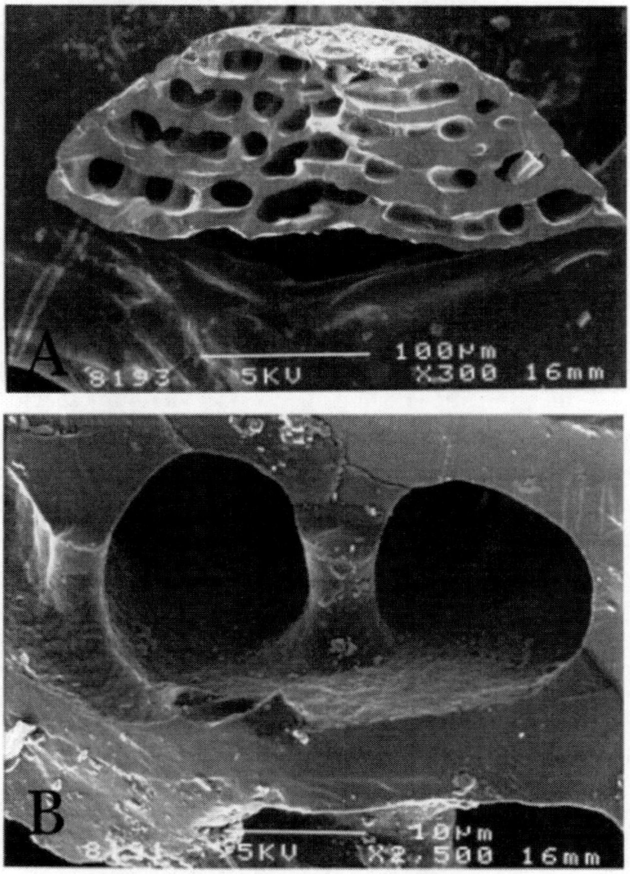

Text Figure 11. Transverse section of *Patellina corrugata* Williamson, 1858.
A – section of the whole shell (x 300); B - part of the broken shell (x 2500).

The agglutinated genus *Dusenburyina* Bermudez et Key, 1952 isomorphic to stilostomellids was previously included into the textulariin group (superfamily Textulariacea Ehernberg, 1838 in Loeblich & Tappan, 1987 and later suborder Textulariina Delage et Herouard, 1896 in Loeblich & Tappan, 1992, order Textulariida subclass Textulariana Mikhalevich, 1980 in Mikhalevich, 1998, 2000, Mikhalevich et al., 2000). The both genera have the wholly uniserial shell and terminal aperture with an inner tooth. Most likely this genus also belongs to the class Nodosariata, subclass Hormosinana Mikhalevich, 1992. Its strongly developed tooth permits to compare it with the genus *Siphonodosaria* A. Silvestri, 1924 among the representatives of the family Stilostomellidae Finlay, 1947. Thus the row of the agglutinated and calcareous forms in the class Nodosariata isomorphic in their shell and chamber form and in their apertural structure such as *Ovammina* Dahlgren, 1962 – *Oolina* d'Orbigny, 1839, *Saccammina* Carpenter, 1869 – *Lagena* Walker et Jacob in Kanmasher, 1798, *Hormosinella* Shchedrina, 1969 – *Grigelis* Mikhalevich, 1981, *Polychasmina* Loeblich et Tappan, 1946 – *Geinitzina* Spandel, 1901, *Warrenita* Loeblich et Tappan, 1984 – *Lingulonodosaria* A. Silvestri, 1903, *Nodosinum* Hofker, 1930 – *Nodosaria* Lamarck ,1812 and many others multi-chambered forms (Mikhalevich, 2004) is replenished on account of one more structural isomorphic type. The asymmetrical hooked aperture of the uniserial agglutinated *Cuneata* is of the same type as the one of the monothalamous calcareous *Fissurina* (Plate **47**, Figure - s 5, 10 a). Such type of aperture as well as the radiate nodosariid-nodosinellid aperture and the aperture having the entosolenian tube (like in *Ovammina* and *Oolina*) is characteristic of the class Nodosariata and is not present in the other foraminiferal classes. This circumstance confirms the uniting of the agglutinated Hormosinana and calcareous Nodosariana genera in one class Nodosariata and permits to suppose the common origin of the both groups from the same root. The agglutinated analogues of the calcareous nodosariats may be their preceding ancestors or their sister group (Mikhalevich, 1992, 2000, 2004). Thus the definition of the taxonomic position of *Strictocostella* and obviously of the other related genera of the family Stilostomellidae helps to understand the true taxonomic position of its agglutinated analogue.

The definition of the specific belonging to the definite species of its different representatives of this genus is difficult because of the extremely high variability of its forms. This feature was also noted by Cushman (1948) for the genus *Stilostomella* who wrote: "The genera *Stilostomella* and *Daucina*, although sometimes placed here, are difficult to determine with full accuracy until type material shall have been more carefully studied than has yet been done."

The presence in our material of the shells of the both generations of the studied species permits to give their differentiated diagnoses to the both forms and to mark the common trends in the change of their shell outlines studied in both the generations. Judging from the wide range of photographs published it is possible to suppose that some of the micro- or megalospheric stilostomellid forms were sometimes described as a separate species.

Thus Scanning Electron Microscopic study of the two *Strictocostella* representatives permitted to reveal the details of the ultrastructural character of its shell surface, some details of the apertural character and the ultrastrusture of their shell wall. The latter is formed by one layer of crystals going in radial direction. Such ultrastructure is characteristic of the monolamellar calcareous shell wall. Inner and outer pore openings are disposed between the ends of these elongated crystals. The inner shell wall was rough in all the specimens studied, the outer shell wall may be of the same or even much rougher character as the inner one or may be smoothly finished. The taxonomic position of the genus *Strictocostella* is changed according to the character of the ultrastructure of its wall: it is transferred from the class Rotaliata to the class Nodosariata which is characterized as possessing the monolamellar rather than bilamellar wall. Genus *Strictocostella* is placed here into the family Stilostomellidae of the superfamily Stilostomelloidea as the suborder of Nodosariata class.

Class **Nodosariata** Mikhalevich 1992
Subclass **Nodosariana** Mikhalevich 1992
Suborder **Nodosariina** Calkins. 1926
Superfamily **Stilostomelloidea** Finlay, 1947 (=Stilostomellina Saidova 1981, p. 59)
Family **Stilostomellidae** Finlay 1947
Genus ***Strictocostella*** Patterson, 1987

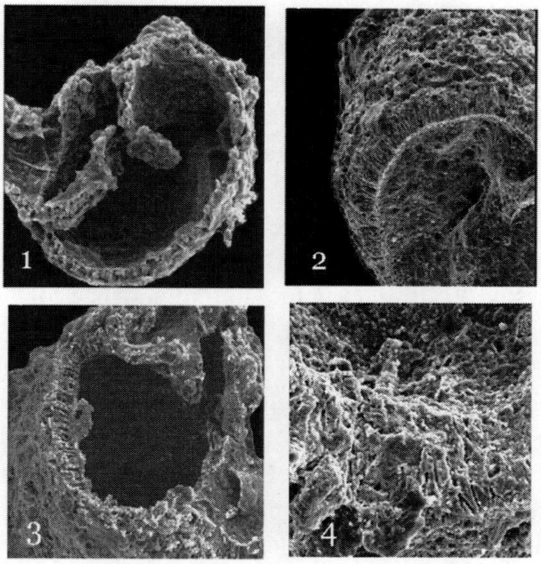

Text Figure 12. Ultrastructure of agglutinated shell wall.
1 - *Eggerelloides scaber* (Williamson), 1858, "Kovalevski", 1957 – 1959, (Black Sea), transverse section of the wall of the last chamber (x 350), note the inner and outer organic coating (white) at the right part of the picture and inner organic lining inside the chamber underlying its inner surface and falling off at the upper right part; 2 - *Gaudryina triangularis* Cushman, 1911, "Pelamida", 1961, station 6 (Tonkinsky Bay), transverse breaking of the shell (x200), note the radial disposition of canaliculi ; 3 - *Gaudryina atlantica* (Bailey), 1851, "Belogorsk", trip II, 1965, station 13 (Caribbean Sea, near Tobago Isl.), transverse breaking of the shell, (x 200), note the canaliculi at the left corner of the picture; 4 - *Martinottiella (Clavulina) communis* d'Orbigny, 1826, "Ob", 1958, station 467 (Antarctic), a piece of the transverse breaking of the wall with the anastomosing canaliculi, (x 350), note the pseudopores of the inner chamber wall.

XI.3. Principles of the Construction of the Foraminiferal Macrosystem

Separating the taxa I based myself on the complex of the following features: the number of chambers (1, 2 or many), the structure of the shell and mode of coiling, the shell wall ultrastructure, the inner and outer additional skeletal structures, the structure and position of the aperture, the presence or absence of additional apertures, the integrative systems, sometimes chamber form and some other minor features as well as the tendencies of development of all these features (Mikhalevich, 1980, 1981, 1992, 1998, 2000). Cytological data were also concerned where possible. Among the enumerated features, plan of the structure of the shell reflecting the plan of the organism in connection with its functions (including number of chambers, mode of their arrangement in connection with the apertural position and structure) I consider to be basic (but not the aperture only as was understood by some authors). In the pair of features: shell structure – shell wall ultrastructure, priority is given here to the first of them. Thus with the change of the hierarchy of the basic taxonomic features the understanding and constitution of the whole foraminiferal group fundamentally changed. Such an approach was always characteristic of the school of Rauser-Chernousova and Fursenko (Rauser-Chernousova & Fursenko, 1959; Rauser-Chernousova, 1990, Rauser-Chernousova et al., 1996, Fursenko, 1978, Solovieva, 1978, Jakovleva, 1979, Saidova, 1981, Subbotina et al., 1981, Grigelis, 1978, 1995, Basov, 1990, Maslakova, 1990, Vdovenko et al., 1993 and some other). And this approach to the foraminiferal taxonomy which became a posterior result of the comparative morphological analysis of the most typical representatives of all the foraminiferal groups is in contrast with the classical view in evaluating the taxonomic value of the shell wall composition and structure. Molecular analysis results also coincide with this new approach based mostly on the shell structure. Pawlowski et al., 1997 also formulated the necessity "…to reexamine the importance of wall structure and composition in the classification of foraminifera".

At the same time many of the peculiarities of the complication of the shell structure, the inner chamber structure, of their inner wall as well as of the apertural structure are regarded here as arising in parallel in the different shell types having a different plan of their structure (Mikhalevich, 1981, 2000, 2005, Parts VIII- X of this study) (plan of the shell structure is considered as the disposition of its parts in connection with their function). That is why plan of the shell structure is considered the basis for the separation of the different classes. In the classes Spirillinata and Nodosariata the agglutinated forms isomorphic in their shell structure to their advanced calcareous analogues are regarded as more primitive subclasses Ammodiscana and Hormosinana respectively, similarly as was previously done in the classes Rotaliata (Mikhalevich, 1980 a) and Miliolata (Saidova, 1981). In the subclasses Lagynana and Astrorhizana only representatives with the more primitive wall types (tectinous and agglutinated, or rarely microgranular) are known.

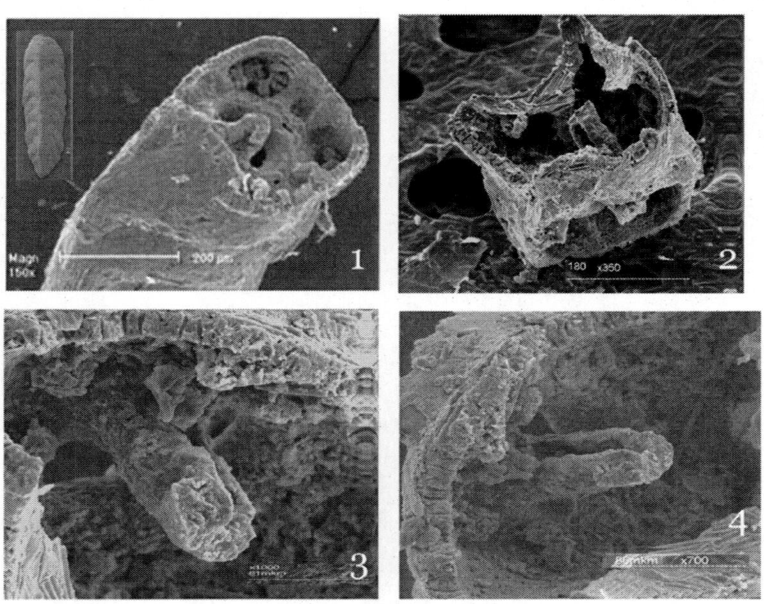

Text Figure 13. Inner structure of *Olgia pacifica* (Cushman), 1924, station 384 of the Soviet Antarctic Expeditions (SAE) of "Ob" (collected in 1958), off New Zealand.

1 – broken shell showing the inner tooth plate and partitions at the bottom of the chamber (x 150); 2 - the broken part of the shell of another exemplar showing the tooth plate going through two chambers (x 350); 3 – the inner grooved tooth plate of the broken shell (x 1000); 4 - the grooved tooth plate of the same break from another angle of view (grove of the tooth plate is seen better) (x 700).

The taxa of equal taxonomic rank were separated using categories of the features of the single level (for instance the superfamilies were separated on the basis of the complication of their chamber lumen and inner chamber wall). In cases when some morphological feature was not yet settled and was unstable and highly variable, its taxonomic value was considered as having less taxonomic weight than in the case of its stabilization (for instance mode of coiling is highly variable in some primitive forms with long tubular second chamber, or shell type at earlier stages of the occurrence of multi-chamberedness is not clearly expressed).

Paleozoic and later forms are viewed in a unified system with the recent ones. This permitted to eliminate a significant number of parallelisms of the previous systems. Large fully extinct taxa and taxa with tectinous shell (subclass Lagynana) are regarded more briefly. Taxa whose position was significantly changed comparing their place in the previous systems and especially in the monumental and so far most complete classification of Loeblich & Tappan, 1987 (with full synonymy of all the taxa) are asterisked. In the diagnosis of the subordinate taxa the features of their higher taxa, are not repeated. The new genera described after the Loeblich & Tappan Treatise (1987) and the new agglutinated genera published after Mikhalevich (2004) system of agglutinated forms are added here wherever possible when the relevant literary sources were available to me.

The endings of the superfamily names are accepted as –oidea instead of -acea. The latter ones are more commonly used in botanical nomenclature.

It is noteworthy to stress the significance of the hierarchical composition of classification. Ranking of taxa is not a formal technique only. The natural system of organisms reflects in its ideal variant the phylogeny of the group. Indeed, life itself has a hierarchical structure. The hierarchical relations of the taxa of the lower taxonomic level (from species up to the order) in the traditional foraminiferal classifications were usually adequately ranked. But hierarchy at the higher taxonomic level (superorder, class, phylum) was mostly neglected. Thus, taking all the bulk of foraminifera belonging to the number of superfamilies (later progressively elevated to the rank of order) all being of the equal rank disfigures their real relations. One may come to wrong conclusions comparing as an equal taxa for instance fusulinids (superorder Fusulinoida) with all rotaliats (= subclass Rotaliana) (while the first is only a component of the class Miliolata, and the second – a separate class Rotaliata). Wouldn't it be the same as comparison of one of the groups of Mammalia, - the order Rodentia (gnawing animals) with some other class (for instance Reptilia)? Similarly, when we consider the orders Buliminida and Nonionida united in one taxon (entering subclass Rotaliana) the data on their similarity received in molecular studies (Schweizer et al., 2008) would not seem so striking as in the case when we consider them to be separate unrelated orders. Placing of the subclasses Globigerinana and Rotaliana in the same foraminiferal class Rotaliata shows their close relationship, while placing them within Foraminifera as separate orders equal in their rank to the other foraminiferal orders (Astrorhizida, Fusulinida, Spirillinida, Nodosariida… so on) does not show their affinities. Meanwhile the relationships of the subclass Globigerinana, uniting the ends of several lineages of the smaller lineages of the lowest ranked Rotaliana and including the forms whose adaptation to the pelagic mode of life results in some profound changes of their morphology and in acquiring by them the new level of organization, can be clearly seen when they are placed as Rotaliana branch under the hierarchical composition of the taxonomic system.

Underestimation and depreciation of the rank of the whole foraminiferal group (regarding them as an order or a class) also prevents reflection on the subordination and phylogenetic relations of different taxa within the phylum Foraminifera.

Complete mutual understanding between specialists using different methods of investigation is also of high importance. The discussion on the results obtained by the methods of molecular analysis, using the term "textulariids" which is still in common use, loses its taxonomic sense without preliminary agreement on the volume and understanding of this term - whether is it used in the understanding of Loeblich & Tappan of 1964, or 1987 or 1992 years? Or in the understanding of the classification offered here, or of some other classification? The same situation exists with the term "fusulinids" - precise definition of their volume (with or without ancient Nodosariata and Spirillinata) is necessary for the appropriate understanding.

In the composition of the diagnoses of the high rank taxa I try to relay on the features of their skeleton structure which are of evolutionary significance, concerning also some cytological data.

Comparative morphological analysis permitted to elucidate some cases of parallelism and evolutionary convergence, mostly between classes, permitting to outline their limits more clearly. The elucidation of such cases reflecting phylogenetic development within classes and subclasses and revision of the smaller subordinate groups are touched on only partially and may be regarded as the next goal for the future taxonomic studies.

Keeping in mind all these circumstances, the outline of the new classification of the Foraminifera up to the family level is given below.

XI.4. Review of Taxa

PHYLUM FORAMINIFERA d'ORBIGNY, 1826
nomen translat. Mikhalevich, 1980

XI.4.1. CLASS ASTRORHIZATA Saidova, 1981

(= subclass Astrorhizicae Saidova, 1981, part, = class Monothalamids Pawlowski et al., 2003, part)

Shells unilocular, sometimes pseudocolonial or pseudochambered, of subspherical, elongated, tubular, stellate, branching or other irregular form; wall tectinous, agglutinated or microgranular, sometimes pseudopores are visible, may be simple or with incomplete partitions, may be thickened from the inside (spongy, labyrinthic); aperture may be entirely absent, may be primarily multiple (openings often irregular in their form and position or situated at the end of the tubes), or single in more advanced forms (usually of the definite form and position; may be slightly elevated on the neck, may have a small lip); free-living or (very often) attached; gametes biflagellate or amoeboid; agamonts uninuclear, nuclear dualism unknown.

Cambrian – Holocene.[1]

Remarks. Fresh water forms are also considered (see above, Part II), the data of Lejzerowicz et al., 2010 showed that foraminifera of the early foraminiferal lineages may also be a component of soil communities, Two genera (*Rhynchosaccus* and *Thalamophaga*) are parasitic, while *Globidendrina* is notable as boring species.

Astrorhizata is one of the most ancient foraminiferal groups, their shells are known from the Cambrian to the Recent with with isolated instances of findings even earlier (Gausher & Sprechmann, 1999, Bosak et al, 2011); the age of the naked forms is dated between 650 and 920 millions years during Neo-Proterozoic (Groussin et al. 2011). It is surprising that some families and subfamilies in this class have persisted for a very long period judging from the findings of their fossil remains (Diffusilinidae and Stegnammininae: Ordovician – Holocene; Jaculellinae and Hyperammininae: Silurian – Holocene) and even some genera (*Astrorhiza*: Ordovician - Holocene; *Bathysiphon*: Triassic - Holocene).

The taxa of the order and family level are separated on the basis of their shell form, the presence or absence of the aperture, number of the apertures and their position, mode of life (free, attached, pseudocolonial), and on the basis of the complication of their shell wall and chamber lumen, as well as on the combination of these features. Thus the forms having an aperture or lacking it, or singular and vice-versa pseudocolonial shells, which were united previously in one family are separated here into two different families. The multiple apertures of some astrorhizats, usually not strictly fixed in their position, are regarded here as primitive, primarily multiple, contrary to the secondarily multiple apertures of advanced representatives of the other classes, which are formed in addition to a single aperture.

The Astrorhizata is a primitive group with many species having the shell irregular, branching or attached; some forms are pseudocolonial. It seems that some species and genera of this group may belong to the initial forms of the other foraminiferal classes or even to some other groups of unicellular eukaryotes, as it was, for instance, the case with *Pelosina* [earlier attributed to Astrorhizats, it is transferred to Xenophyophorea (Mikhalevich & Voronova, 1999) on the basis of its structure (multiple, crossing, soft, hyaline organic tubes piercing the amorphous sand-grained mass)]. Although the molecular data shows the close relationship of Xenophyophorea to Foraminifera (Pawlowski et al., 2003), nevertheless they are not included here in the latter phylum on the basis of their structural (granellare plasmodium and stercomare rather than usual cell structure) and some physiological distinctions from the Foraminifera. The Komokiids are included here only tentatively, as their shell structure and the presence of stercomata (Shires et al., 1994) distinguishes them from the other foraminifera. They are studied insufficiently, inspite of intensive investigations of last decades devoted to xenophyophorean, komokiacean and soft-walled foraminiferan groups (Kamenskaya, 1987, Kamenskaya et al., 2012, Kamenskaya & Saidova, 1998, Gooday, 1986, 2002, 2012, Gooday et al., 1995, 1996, 1997, 2007 a, b, 2008 a, b, 2010 a. b, Gooday & Smart, 2000, Gooday & Bowser, 2005, Leckroq et al., 2011), for the most part, the bulk of these species remains unidentified.

The class includes many transitional forms whose features permit them to be included in Astrorhizana or with some provisional remarks into other subclasses: thus, the Vanhoeffenellidae, Causiinae and Amphifenestrellinae are transitional from the subclass

[1] Here and further the geological range is given following the literary data and for the taxa of the higher rank only (from the class and up to the order level).

Lagynana, Ammovoluminidae – to subclass Ammodiscana, Hippocrepinidae – to the subclass Hormosinana. Obviously, the whole order Hippocrepinida ought to be placed in the subclass Hormosinana but it would be initially preferable to make a thorough revision of all its taxa before such a step. Finding of the genus *Jugimurammina* Zheng, 2001 strongly supports such transfer. Some of the microgranular forms included previously in the lower Fusulinacea are transferred to Astrorhizana as the boundary between the agglutinated and microgranular wall is often transitional. Indeed, it is often not possible to separate the forms only on the basis of this feature (see Part VIII.10.12.) Middle Devonian and Low Carboniferous *Chrysothurammina* is marked as having "hyaline rather than microgranular wall" (Loeblich & Tappan, 1987). The most ancient typical Astrorhizana coexisted with their microgranular and tectinous analogues from the Cambrian. The known range of many of them has been significantly widened owing to the new data (Tyszka, 1997, Kaminski, 2004, 2012 and many others). The classification of all these groups is difficult and open to question because of the few distinct features of their shell structure.

With the study of *Astrammina rara* and some other large agglutinated species showing the striking features of their biology (Delaca, 1982, 1986, Bowser et al., 1992, 1995, 2002), the boundary between the subclasses of this class does not appear to reflect biological reality. Together with the fact that one of the astrorhizid species (the genus giving its name to the order and the class) (*Astrorhiza triangularis*) behaves similarly as *Astrammina rara* (cell exiting from its "shell"), the type species of the genus *Astrorhiza* (*A. limicola* Sandahl, 1858) ought to be reinvestigated concerning its mode of life. The biological observations on some large Antarctic unilocular agglutinated forms (*Astrammina rara, Notodendrodes antarcticos, Astrorhiza triangularis*) (Delaca, 1982, 1986, Delaca et al., 2002, Bowser et al., 1992, 1995, 2002) revealed some peculiar features of their structure and life behaviour ((the empty space between the cell body and its "shell", and the free exit of the cell out of its "shell" (theca)). On the basis of these features, Bowser et al., (2002), concluded that revision of the group was necessary. Judging from the comparison of such "theca" and *Massilina secans* and *Miliolina* sp. cysts observed here (see above Parts VII. 1. 3, VII. 2. 1 and Plate **26**, figure – s 1 – 9, Plate **31 a**, figure –s 12 a, b), it seems more reasonable to regard the cells of *Astrammina* as naked representatives of Lagynana (= allogromiids in Western literature), and their empty theca found here and there on the surface of the sediment, as abandoned cysts. The small houses built from the different material or foreign particles are known in several groups of other naked protists (e.g. some Infusoria, Euglenophyta, Chrysophyta), and this process underwent in the evolution of protists repeatedly. The main question concerning *A. rara* would be if the shell of the type species of the genus *Astrorhiza* represents a true shell or a kind of cyst. In the first case, the naked and soft walled forms living part of their life in cysts ought to be excluded from the astrorhizids and transferred to Lagynana (= Allogromiina of some authors). In the second case, the identification of the bulk of astrorhizids would be changed, and its other forms with the true shells would be united in a separate taxon with a new name. Until the solution of this problem, a profound reconstruction of the system of this class would be premature.

Obviously, all the system of this primitive class would be transformed but at the present time the available data on the biology and structure of the majority of its representatives are insufficient. More over, it is obvious now, that the bulk of these organisms remains to be discovered. At present state of our knowledge it is impossible to give their appropriate classification. It should be created anew on the basis of the future data and with the use of molecular and cytological methods. That is why traditional subdivision into hard-walled and soft-walled groups is here retained, in spite of conventionality of the border between these subclasses (but, nevertheless, within the frame of the single class).

Subclass *LAGYNANA Mikhalevich, 1980

Shells with one of the astrorhizats types of the shell and aperture; shell wall tectinous, often strongly ferruginous, of the characteristic brown or rusty-reddish colour, usually simple; reticulopodia branching or thread-like; besides the usual life cycles the budding, cytotomy and simple division into the two parts are often observed, gametes bilaflagellate or amoeboid; free living or attached, benthic forms often met in the empty foraminiferal shells and worm's tubes; mostly sea inhabitants, but freshwater genera also occur, two genera (*Rhynchosaccus* and *Thalamophaga*) became parasitic.

Cambrian – Holocene.

Remarks. This insufficiently studied subclass includes the more primitive foraminiferal representatives. Some peculiarities of their shell and cytoplasmic structure shows in this group the first stages of their formation. In many of them, the shell is only partially filled with the cytoplasm, its significant part being outside. They usually have one nucleus (usually rather large – 40 – 79 μm), with rather hard nuclear envelope. In many representatives of their genera there are also known the forms with several small nuclei (from 10 to 150) possibly undergoing the reproductive stage. The cytoplasm of Lagynana representatives is not differentiated into the endo- and ectoplasm. In *Allogromia* the proteinaceous shell base is formed by the endoplasm on the cytoplasmic membrane rather than by the endoplasm of reticulopodia. In *Shepheardella,* a thin cytoplasmic layer covers the shell from the outside (like it is in the advanced foraminiferal representatives), in the other Lagynana's representatives such outer cytoplasmic layer was not observed. The wall consisting of the two layers is noted in the *Saedeleeria.* The representatives of one of the Lagynana genera (*Rhumblerinella*) secrete the calcareous needles. In many of their forms, the transition to the pseudo- and multi-chamberedness was observed. Such forms are transferred here from the order Allogromiida to the polyphyletic order Ammoscalariida which needs further investigation and revision. The subclass Lagynana itself is obviously polyphyletic to a great extent but owing to the paucity of the morphological features of their primitive shells, the problem of their more adequate classification needs the use of the cytological and molecular methods (the present study is focused on the hard shelled foraminiferal taxa).

Some findings of such organic envelopes were usually considered as remains of multi-chambered calcareous shells. However, while the abundance of such linings in Callovian-Oxfordian limestones (Misik & Sotak, 1998) permitted their definition as belonging to more than 30 foraminiferal taxa having multi-chambered arrangement, this also gives reason to assume the existence of some multi-chambered shells having similarly an organic wall, as was earlier known (*Psammophaga, Periptigma, Resigella, Nodellum*). Unfortunately, the molecular studies of multi-chambered soft-walled species have not been performed yet. The results of such studies might confirm the supposition of the origin of all the main phyletic lines of hard-shelled foraminifera from the organic-walled forms as it was earlier supposed by morphologists. In this case, in each of the main foraminiferal lines (classes) the taxa with the organic wall of corresponding shell and chamber arrangement should be placed as the root taxa into the corresponding classes.

The heterogeneity of lagynans supposed by morphologists (Rauser-Chernousova & Fursenko, 1959, Arnold, 1978 a, b, Mikhalevich, 1992, 2000) was later proved by molecular data (Pawlowski et al., 2003). According to molecular data classification schemes showing the position of the different taxa not lagynans only but even the genus *Allogromia* turned out to be polyphyletic. During intensive studies of the last decades more than ten new soft walled genera and species were discovered, their classification needs further investigations; the new independent *Conqueria* lineage among the unilocular (monothalamous) representatives was discovered (Gooday & Pawlowski, 2004).

Significant number of the soft-walled representatives are fresh water forms, including the *Reticulomyxa* transferred to foraminifera from Athalamida (Pawlowski et al., 1999).

Order LAGYNIDA Mikhalevich, 1980

Shell of circular, subcircular, oval, elongated or irregular form; wall nude or with rare agglutinated particles scattered on the shell surface, may be with additional inner incomplete septa; aperture of one or several openings or may be lacking, in some genera aperture at the end of the elongated neck; gametes biflagellate.

Holocene. The much earlier representatives, which surely existed since Precambrian times were not preserved as they lack the hard shell.

Remarks. *Rhumblerinella* has calcareous needles on its surface which are tightly packed but not bound with the cement.

Families: Lagynidae Schultze, 1854 (with the subfamily Belariinae Mikhalevich, 1999 including representatives whose shells have the inner septa, genera *Belaria* and *Cystophrys*), Maylisoriidae E.V. Bykova, 1961, Blysmasphaeridae Bronnimann, 1988.

Order ALLOGROMIIDA Fursenko, 1958

Shell of circular, subcircular, oval, elongated, rod-shaped form; wall nude or may be covered fully or at some spots with the sand grains, at this preserving its elasticity; aperture of one or several openings or may be lacking, may be of subcircular or irregular form, in some genera aperture at the end of the elongated neck; lip or entosolenian tube may be present; gametes (if known) are amoeboid.

Ordovician - Holocene.

Remarks. A number of new soft-walled genera of allogromiids were described during the last years (*Niveus* of Altin et al., 2009, *Edaphoallogromia* of Pawlowski et Holzmann, 2002, *Xiphophaga* and *Allominuta* Goldstein et al., 2010 and some others) but a lot of them remains unidentified and often can be distinguished by the molecular methods only as they lack distinctive morphological features. Species with entosolenian apertural tube obviously represent the roots of agglutinated and calcareous unilocular forms of the class Nodosariata. The genus *Rhynchosaccus* may be a parasite on the more large foraminiferal shells.

Families: Allogromiidae Rhumbler, 1904, Shepheardellidae Loeblich et Tappan, 1948, Argillotubidae Avnimelech, 1952, Labyrinthochitinidae Mikhalevich, 1999 (wall labyrinthic, may be with inner subpartitions, including *Labyrinthochitinia*, *Chitinolagena* and probably some new genera).

Order AMMOSCALARIIDA Mikhalevich, 1980

Shell of elongated, cup-like, planispiral or primitive trochospiral or miliolid form, in the attached forms may be irregular and branching; in the Ammoscalariidae and Hospitellidae shell pseudo-multi-chambered, in Phthanotrochidae – of unstable multi-chambered character; wall nude or having agglutinated sand grains which may form the entire and rather hard cover but the inner septa stay tectinous in these cases also; aperture one, terminal; gametes (if known – in *Phthanotrochus*) – amoeboid.

Eocene – Holocene.

Remarks. Polyphiletic group including forms of different shell structure (uniserial – Hospitellidae, trochoid – *Phthanotrochus,* milioloid – *Periptigma*, planispiral – Ammoscalariidae). Such distinctions are regarded in the higher advanced Foraminifera as characteristic of the orders. However, as the formation of the multi-chamberedness is here at its very beginning (e.g. in *Periptigma* the chambers are not stable, often fall off, the trochoid spire in *Phthanotrochus* may erect itself, may become of irregular form), and, furthermore, because of the poor knowledge of all these forms, they are, for the time being, kept in this taxon, despite the relative artificiality of this classification. If it were the case that it could be confirmed that the *Periptigma* is the ancestor form of miliolats, and *Phthanotrochus* – of rotaliats, their systematic position would be changed.

Genus *Thalamophaga* (Hospitellidae) is known as a parasite of the calcareous multi-chambered Foraminifera.

Families: Phthanotrochidae Arnold, 1978 (including *Periptigma*), Ammoscalariidae Mikhalevich, 1982 (with *Starobogatovella* Mikhalevich, 1994 and possibly *Involutarena* Zheng, 2001), Hospitellidae Loeblich, Tappan 1984 (including free-living uniserial forms such as *Nodellum* in a special subfamily).

Subclass ASTRORHIZANA Saidova, 1981

Shells with agglutinated or microgranular wall.

Cambrian – Holocene.

Remarks. The system of this subclass is given here up to subfamily level, in Mikhalevich, 1995, 2004 - up to the genera level.

Order ASTRORHIZIDA Haeckel, 1894

Shell free, single, tubular or with tubes extending from central body; wall simple; apertures at open ends of tubes, two or more.

Middle Ordovician – Holocene.

Superfamily Astrorhizoidea Brady, 1881

Shell with the simple wall and chamber lumen not subdivided.

Remarks. The composition of the family Hippocrepinellidae is expanded by including the genera *Amphitremoida* and *Croneisella* transferred from Saccamminidae, *Crespinitella* Rauser-Chernousova et Reitlinger, 1993 transferred from Bathysiphoninae and the genus *Astrorhizinulla* transferred from the Rhabdamminidae, as well as the genus *Lakites* Nestell & Tolmacheva, 2004). The similarity of the apertural structure of the genus *Jugimurammina* by Zheng (2001) with the apertural structure of *Nodosinum* confirms the necessity of revision of this subclass and of moving part of its taxa to the subclass Hormosinana of the class Nodosariata.

Families: Astrorhizidae Brady, 1881 (including *Globidendrina* Plewes, Palmer et Haynes, 1993, genera *Clados* and *Linea* of Schroeder, Medioli et Scott, 1989 assigned by its authors to Astrorhizinae belongs more probably to Komokiida), Vanhoeffenellidae Saidova, 1981, Rhabdamminidae Brady, 1884 (with Rhabdammininae Brady, 1884, Bathysiphoninae Avnimelech, 1952), Hippocrepinellidae Loeblich et Tappan, 1984 (including the genus *Lakites* Nestell et Tolmacheva, 2004, the family differs from the closely related family Rhabdamminidae in the shorter shell, tapered apertural end, and structure of the aperture), Jugimuramminidae Zheng, 2001(with the type genus *Jugimurammina* Zheng, 2001).

Order DENDROPHRYIDA Haeckel, 1894, emend.

Shell attached or free, single or pseudocolonial, of irregular and often of fanciful form, strongly branched; wall simple or complicated by inner partitions or labyrinthic, with different amounts of cement, more often friable, flexible; apertures – open ends of ramifications or apertures not fixed, looking as perforations of wall.

Upper Cretaceous, Palaeocene – Holocene.

Remarks. The genus *Arborammina* Shires, Gooday et Jones, 1994 of the family Arboramminidae Shires, Gooday et Jones, 1994 belongs more possibly to Xenophiophorea as it has no apertural and chamber lumen and its body consists of organic fibrils, though *Luffammina* Kamenskaya, Bagirov & Simdianov, 2002 placed into this family (Kaminski, 2008) may belong to Astrorhizata – further studies are necessary; the genera *Saccodendron* and *Globodendrina* are transferred into the subfamily Dendrophryinae from the family Astrorhizidae as their representatives have strongly branching outgrowths; the genus *Radicula* is transferred from the family Astrorhizidae as its representatives have a branched shell buried in the ground; the genus *Atelikamara* is removed from the subfamily Hemisphaeramminae into the family Diffusilinidae as having an attached shell of irregular form with incomplete inner septa.

Superfamily Dendrophryoidea Haeckel, 1894

Shell wall simple, test lumen not subdivided.

Families: Dendrophryidae Haeckel, 1894 (with the subfamilies Dendrophryinae Haeckel, 1894, Spiculidendroninae Mikhalevich, 2004),*Notodendrodidae Delaca, Lipps et Hessler, 1980 (included earlier in Hippocrepinacea), Dryorhizopsidae Loeblich et Tappan, 1984, ?Arboramminidae Shires, Gooday et Jones, 1994 (with type genus *Arbor* Schroeder, Medioli et Scott, 1989, including *Luffammina* Kamenskaya, Bagirov & Simdianov, 2002), *Komokiidae Tendal et Hessler, 1977 (including *Rhizammina* according Cartwright, Gooday and Jones, 1989 and genera *Calos, Cactos, Cerebrum, Crambis, Reticulum, Tuber* Schroeder, Medioli et Scott, 1989 and *Skeletonia* Gooday, Kamenskaya & Cedhagen, 2007), *Baculellidae Tendal et Hessler, 1977(including *Neocatena* Özdikmen, 2009 pro *Catena* Schroder, Medioli & Scott, 1989 (non Richter, 1975), *Normaninidae Mikhalevich, 1994.

Superfamily Schizamminoidea Norvang, 1961

Shell with inner cavity complicated by inner subdivisions or with labyrintic wall.

Families Schizamminidae Norvang, 1961, Halyphysematidae Loeblich et Tappan, 1984, nom. corrected pro Halyphyseminae, Diffusilinidae Loeblich et Tappan, 1961.

Order PSAMMOSPHAERIDA Haeckel, 1894

Shell circular, oval, slightly elongated or hemispherical, free or attached, singular or pseudocolonial; wall simple or complicated (labyrintic, cellular, with inner septa); apertures numerous, single or absent.

Ordovician – Holocene.

Remarks. The genus *Amphifenestrella* is removed from the subfamily Vanhoeffenellinae into the subfamily Amphifenestrellinae Mikhalevich, 1995 on the basis of its shell morphology (without tubular extensions open at ends) and the absence of marked aperture. The genus *Blastammina* is excluded from the subfamily Stegnammininae on the basis of its wall character with only sparse sand grains; the genus *Raibosammina* is transferred from the subfamily Thurammininae (as well as some other genera from the subfamily Psammosphaerinae) into the subfamily Stegnammininae Moreman, 1930 as lacking the aperture and having a single (not pseudocolonial) shell. Although in the diagnosis of *Raibosammina,* the possibility of branching of its shell is indicated, the specimens of the type species have an oval shell without any branches (Loeblich, Tappan, 1988, pl. 20, figures. 6,7); only the genera with singular attached shell without aperture are left in the subfamily Hemisphaerammininae Loeblich et Tappan, 1961, the genus *Sorosphaerella* from Hemisphaerammininae needs further investigations, it is possible that it has pseudocolonial shell and hence it should be placed in Telamminidae; the genera *Colonammina* and *Jascottella* are excluded from the Hemispaerammininae and placed in the family Colonamminidae (subfamily Colonammininae) on the basis of the presence of distinct aperture at the apex of the shell; the genus *Nubeculariella* is removed from the family Halyphysematidae into Colonammininae, which is characterized by differing external structure of the shell and by subdivision of the inner lumen.. Contrary to the authors of the family Colonamminidae, I left in it only representatives with a single shell and one aperture, excluding *Tholosina, Iridia, Mesammina*, *Scyphocodon* (with multiple aperture) and *Webbinelloidea* (pseudocolonial), as well as *Saccamminis* whose representatives have pseudochambered shell with one common aperture.

Suborder STEGNAMMININA Mikhalevich, Pronina G.P., Nestell, M., 2000

Shell singular.

Superfamily Stegnamminoidea Moreman, 1930

Shell with simple wall.

Families: Stegnamminidae Moreman, 1930 (with subfamilies Amphifenestrellinae Mikhalevich, 1995, Stegnammininae Moreman,1930, Hemisphaerammininae Loeblich et Tappan, 1961), Colonamminidae Rauser-Chernousova et Reytlinger, 1993 in Vdovenko et al., 1993 (stat. elevated, with subfamilies Causiinae Mikhalevich, 1995, Colonammininae Rauser-Chernousova et Reytlinger, 1993 in Vdovenko et al., 1993, Tholosininae Mikhalevich, 1995).

Superfamily **Crithioninoidea** Goes, 1894

Shell with labyrintic wall or chamber lumen subdivided by inner septa.

Remarks. The genus *Masonella* is removed from Crithioninae into Oryctodermatidae.

Families: Crithionidae Goes, 1894 (differing from the Oryctodermatidae in the presence of inner septa subdividing the shell lumen and in the absence of the labyrintic wall) (with subfamilies Daitroninae Mikhalevich, 1995, Crithioninae Goes, 1894), Oryctodermatidae Saidova, 1981 [= Subfam. Masonellinae Saidova, 1981 (as "Marsonellinae"), part].

Suborder PSAMMOSPHAERINA Haeckel, 1894

Shell pseudocolonial or pseudo-multi-chambered.

Remarks. The genera with singular shell without aperture are transferred from the former subfamily Psammosphaerinae to the subfamily Stegnammininae as well as the genus *Luekatiella* Zhigulina, 1999; the subfamily Telammininae is removed from the Hormosinacea, as its representatives have no true apertures and represent pseudocolonial organism; the genus *Aggerostramen* is excluded from the subfamily Telammininae and placed here in family Lacustrinellidae as its representatives are characterized as pseudocolonial organisms having an aperture; the genus *Patellammina* is included into Telammininae conditionally as, according to its morphology, its pseudocolonial shells are the sooner detached from the surface, and the basal side with the "aperture" is supposedly the side of attachment (regarded as close to *Webbinelloidea)*; the subfamily Amphicervicinae is separated from Hemisphaeramminiinae on the basis of its pseudo-multi-chambered embracing shell, and presence of apertures; the family Lacustrinellidae is separated from the heterogenous subfamily Hemisphaeramminiinae, as Lacustrinellidae are pseudocolonial and have an aperture; the genus *Aggerostramen,* which was previously included in Hormosinacea, is transferred from the family Telamminidae; the former now includes only representatives with true multi-chambered shell.

Families: [Haeckel, 1894 (with subfamilies Psammosphaerinae Haeckel, 1894 ([including *Palmierina* Özdikmen, 2009 pro *Teichertina* Palmieri, 1994 (non Veevers, 1959)], Telammininae Loeblich et Tappan, 1985], Polysaccamminidae Loeblich et Tappan, 1984 (with subfamilies Polysaccammininae Loeblich et Tappan, 1984, Saccamminidinae Mikhalevich, 1995, Amphicervicinae Mikhalevich, 1995); Lacustrinellidae Mikhalevich, 1995.

Order *PARATHURAMMINIDA Mikhalevich, 1980

Shell free or attached, single or pseudocolonial, of circular, elongated or irregular form; wall microgranular, simple or complicated, perforated or imperforated; aperture one, many or aperture absent.

Silurian – Permian.

Remarks. This order differs from the order Psammosphaerida including the suborder Stegnamminina only in its wall structure. As I regard the step of the formation of the microgranular wall following the agglutinated wall the order Parathuramminida is transferred here from the previous Fusulinacea into Astrorhizata. Their revision is not finished because of the poor preservation and unclear features of many of them. For the detailed diagnoses see Mikhalevich, 1995, the composition of genera in Mikhalevich, 2004). Some of these ancient forms obviously did not even develop the microgranular wall, having only organic one as is mentioned, for instance, for *Chrysothurammina* Neumann, Pozaryska et Vachard, 1975.

Superfamily Parathuramminoidea E.V.Bykova, 1955.

Shell with simple wall, inner chamber cavity not subdivided.

Families: Archaesphaeridae Malakhova, 1956 [with the subfamilies Archaesphaerinae Malakhova, 1956, Neoarchaesphaerinae Sabirov, 1987, Insolentithecinae Loeblich et Tappan, 1986, Usloniinae A.D. Miklukho-Maklay, 1963, Eoammosphaeroidinae Mikhalevich, 1995, Atjussellinae Zadorozhnyy, 1987 (subfamily reinstated by Vdovenko et al., 1993), without Pachysphaerininae Sabirov, 1987 regarded by Loeblich et Tappan, 1987, p. 728 as erroneously assigned to foraminifera], Parathuramminidae E.V.Bykova, 1955 [with the subfamilies Parathuramminінae E.V.Bykova, 1955, Irregularininae Zadorozhnyy et Juferev, 1984 (including *Cribrauroria* Vachard, 1994) and Chrysothuramminінae Loeblich et Tappan, 1986).

Superfamily Marginarioidea Loeblich et Tappan, 1986 (= Ivanovelloidea Chuvashov et Yuferev, 1984).

Shell with complicated wall (labyrinthic, or largely perforated, or pierced by radiate canaliculi leading outside the shell) or with subdivided shell lumen.

Families Marginaridae Loeblich et Tappan, 1986 (with the subfamilies Marginarinae Loeblich et Tappan, 1986, Uralinellinae Chuvashev et Yuferev, 1984), Tuberitinidae A.D.Miklukho-Maklay, 1958 [with the subfamilies Tuberitininae A.D. Miklukho-Maklay, 1958 (with Tubeporininae Zadorozhnyy et Juferev, 1987 as synonym), Hemithuramminінae Mikhalevich, 1995), Auroriidae Loeblich et Tappan, 1986, Eovolutinidae Loeblich et Tappan, 1986, Rauserinidae Sabirov, 1987, Cribrosphaeroididae Sabirov, 1987.

Order HIPPOCREPINIDA Saidova, 1981

Shell free or attached, pseudo-two-chambered, rarely pseudo-multi-chambered, commonly elongate; from the subspherical and distinct or scarcely marked proloculus, the second tubular pseudochamber extends upwards, may be straight, branched at end, curved or weakly coiled, or widened into a conus or funnel, and, may be undivided or subdivided by incomplete inner partitions, or may have external or inner constrictions; wall agglutinated or microgranular; aperture terminal (wide or tapering open end of tube), circular or oval; branched shells are accordingly with two apertures.

Ordovician – Holocene.

Superfamily Hippocrepinoidea Rhumbler, 1895

Shell with undivided inner cavity.

Remarks. The genera whose representatives have a widely open aperture are excluded from the subfamily Hippocrepininae. At the same time the genera *Hyperamminoides, Giraliarella* and *Pseudohyperammina* are transferred from Hyperamminoididae on the basis of similarity of their shell and apertural structure with those of *Hippocrepina.*

In the heterogenous family Hyperamminoididae there were previously included ancient forms (Devonian – Cretaceous) of different structure. Inclusion of forms of different age but of similar morphology in one subfamily appears to be justified as it is done in Hyperammininae (Silurian – Holocene) and in many other subfamilies. The subfamily Jaculellinae was separated from the Hippocrepininae on the basis of apertural structure. The genera *Kechenotiske* and *Sansabaina* were transferred into the subfamily Jaculellinae from the Hyperamminoididae. The position of the latter is not clear and needs further investigation. The genus *Sacchararena* was transferred from the Hyperamminoididae to the Hyperammininae, as it has distinct proloculus. The subfamily Saccorhizinae is restored as a separate taxon from the synonymy of Hippocrepinidae. The subfamily Saccorhininae differs from Saccorhizinae in the microgranular wall and subdivided shell cavity. The genus *Saccorhina* placed here in Saccorhininae (Paratikhinellidae, Syzranioidea), was previously included in Moravamminidae (which are now considered as algae - Vachard, 1994 and Vachard & Cozar, 2010). The family Ammovoluminidae is transferred here from the order Ammodiscida: it is intermediate between the two orders, but the irregular shell form, short and often only slightly curved tube more sharply differentiate it from the regular ammodiscid shells with numerous volutions.

Families: Hippocrepinidae Rhumbler, 1895 [(with the subfamilies Hippocrepininae Rhumbler, 1895 = Hyperamminoididae Loeblich et Tappan, 1984 (part) and Jaculellinae Mikhalevich, 1995 = Hyperamminoididae Loeblich et Tappan, 1984 (part)], Hyperamminidae Eimer et Fickert, 1899 ((with the subfamilies Hyperammininae Eimer et Fickert, 1899 = Hyperamminoididae Loeblich et Tappan, 1984 (part), Saccorhizinae Eimer et Fickert, 1899)), Ammovoluminidae Chernych, 1967.

Superfamily Botellinoidea Chapman et Parr, 1936
(= Caligelloidea Reytlinger, 1959)
Shell pseudo-multi-chambered, with pseudochambers formed by incomplete septa or wall constrictions.

Remarks. The family Botellinidae is restored from the synonymy of Hippocrepinidae, as its representatives differ from those of the Hippocrepinidae and Hyperamminidae (Loeblich, Tappan, 1988) in the peculiarities of their external and internal structure.

Family: Botellinidae Chapman et Parr, 1936.

XI.4.2. CLASS SPIRILLINATA Mikhalevich, 1992

(=Tubothalamea Pawlowski et al., 2012, part in Adl et al., 2012)

Shells pseudo-two-chambered, with a long tubular second chamber coiled around the proloculus, in higher taxa with the tendency of transition to multi-chamberedness, as a rule in the last coils, often with very limited chamber number (2 - 3 per a coil), at the same time undivided tubular second part around proloculus may be preserved during 1-3 volutions, types of coiling - glomerate, planispiral, trochospiral, the predominant type of shells - subsphaerical, lenticular, flat, patellinoid; high conical; wall monofontinal, insufficiently studied, agglutinated in their lower less advanced forms (Ammodiscana), usually with the ferrugenous cement or microgranular, fully secreted in higher ones (Spirillinana), in the ancient forms often two layers are marked - microgranular and radial, in Involutinids - usually radial, in recent Spirillinids – radial crystal units may act as a single crystal, shell wall may be pitted by regular pseudopores, may have inner and outer additional skeletal sculpture, pillars, typical layers of growth are not formed but secondary layers of thickening of a special character, often multilamellar are formed, especially in the umbilical area; aperture usually at the end of the tubular chamber, simple, in *Patellina* with special T- shaped plate, in advanced agglutinated and calcareous forms aperture may be multiple, additional apertures rarely present, in higher more advanced forms weakly developed canal system is rarely present; columella in the center of the shell and tube pillars are specific to this class; agamonts homokaryotic, their proloculus larger than that of the gamonts (the reverse correlation), gametes amoeboid; benthic, mostly free living forms.

Cambrian - Holocene.

Remarks. The process of chamber formation insufficiently studied, in *Patellina coruugata* neither cyst nor cytoplasmic projection (anlage) were observed, the process of calcification begins near the apperture and goes as accrition of already existing structure.

The author of the class is reinstated by G. Nestell et al., 2009 according to the ICZN. In many representatives of the subclass Spirillinana the aragonitic wall was documented. Spirillinata is one of the most ancient foraminiferal classes. During the process of evolution the aperture shows a tendency to be displaced from the peripheral to the umbilical position even in the forms with the tubular second chamber (compare *Trocholina* and *Mychostomina* (Spirillinana)- Plate **45**, figure – s 22, 23). Its position usually becomes umbilical in the multi-chambered forms. Glomerate forms with the radial wall represent a blind alley, trochospiral calcareous forms are the most advanced and even acquired a primitive canal system. Agglutinated forms from Cambrian till recent, microgranular ones died out in Carbon and Perm, there is a succession in the development of the forms with the radial wall: Archaediscida - Carbon, Lasiodiscida - Carbon - Perm, Involutinida - Perm - Cretaceous, Spirillinida - Trias – Holocene (Text figure 6). I regard them as consecutive steps in the development of one phylogenetic branch. This succession confirms the supposition that the microgranular layer of their wall is a result of the diagenetic changes of the primary fully radial shell wall.

Class Spirillinata is a special separate branch of the development of the foraminiferal shells, which is characterized by some primitive features of their structure: pseudo-two-chambered shell with long tubular second chamber curved in different planes around proloculus, mostly simple aperture at the end of the tube. They demonstrate neither the diversity of the shell types, nor the complication of the aperture (except that in Patellinidae and multiple aperture in the rare advanced forms). The recent forms have the nuclear apparatus more primitive than in Rotaliata (homokaryotic: Grell, 1979) and possess ameboid gametes. But some progressive features were developed in their more advanced groups: multi-chamberedness, the inner subdivisions of the chamber by septa and pillars, the primitive canal system (in the representatives of Involutinida and Spirillinida, and even in the Paleozoic Lasiodiscida, as well as in the Palaeocene—Miocene Chapmaninidae) (Hottinger, 1976, Mantzurova & Gorbatchik, 1982, Azbel, 1986). The development of this new system of organism level, being a feature of high specialization (along with retention of primitive features), testifies to evolutionary advancement of these groups. It is interesting to note that the rise of integrative systems in the calcareous spirillinats is mostly correlated with the presence of trochospiral shell in which a rather wide umbilical area is filled with the shell matter (tubercules, pillars, laminae) pierced by the fissures, more rarely spiral and transverse canals and bridges are formed. This type of shell creates the greatest phylogenetic opportunities for development of supplementary openings in the umbilical area which connect the cytoplasm with the environment (Mikhalevich, 1981). The formation of the canal system in this group takes place in the representatives possessing monolamellar shell wall and obviously the elements of the structure of these systems differ from those forming the canal system of Rotaliata. It would be extremely interesting to study their canal system formation in more detail at the ultrastructural level of investigations. In the higher more advanced representatives of the class a central columella and tube pillars may be present both in their agglutinated and calcareous forms. In the latter (Chapmaninidae, *Ferayina*) the ends of hollow pillars extending to the previous septum obviously form the inner integrative system.

The system of the whole group is strongly revised here, some agglutinated and calcareous groups are transferred from the class Rotaliata, from its both subclasses (see below).Their Paleozoic forms are separated out from the former Fusulinacea. Nevertheless, further studies are necessary to compose its appropriate system.

A set of works contributed significantly to the studies of this group (Hohenegger & Piller, 1975, 1977, Piller, 1978, Zaninetti et al., 1987, Arnau-Vanneau et al., 1988, Rettori et al., 1994, Neagu, 1994, 1997, 2000, Boudagher-Fadel, 2008 and some others).

Subclass AMMODISCANA Mikhalevich, 1980

Shells with agglutinated or microgranular shell wall.

Cambrian - Holocene.

Remarks. Subclasses Orbitolinana and Pfenderinana of Neagu, 2003 are regarded as entering subclass Ammodiscana. The system of this subclass is given here up to subfamily level, in Mikhalevich, 2004 - up to the genera level.

Order AMMODISCIDA Mikhalevich, 1980

Shell with the second tubular chamber coiled in planispiral or sometimes weakly trochospiral or streptospiral planes, sometimes may even form a high trochoid spire in free living forms, in attached forms – uncoiled or irregularly coiled.

Lower Cambrian – Holocene.

Remarks. Genus *Ammovertella* ought to be transferred from Ammodiscidae to Calcivertellinae (subclass Miliolana) and considered to be the synonym of *Calcitornella* [according to the new data of Vachard & Krainer (2001) showing the porcellaneous character of its shell wall and its identity with *Calcitornella*].

Families: Ammodiscidae Reuss, 1862 [with Ammodiscinae Reuss, 1862 (without *Agatamminoides*) and Paulbronnimanniinae Rettori et Zaninetti, 1993 (nom. corr. pro Paulbronnimanninae Rettori et Zaninetti, 1993) with the type genus *Paulbronnimannia* Rettori et Zaninetti, 1993], *Pseudoammodiscidae Conil et Lys, 1970 (from the Fusulinacea) (without *Warnantella*), Turritellellidae Saidova, 1981 (fam. reinstated and given stat. nov), ?Tolypamminidae Cushman, 1928 (including later described *Ammodiscellites* of Resig et Glenn, 1997,without *Ammolagena* transferred by Kaminski et al., 2009 to the subclass Hormosinana as having a second aperture that opens out of the proloculus opposite the tubular chamber), Turriglominidae Zaninetti, in Limongi et al., 1987.

Order AMMOVERTELLINIDA Mikhalevich, 1999

Shell with the second tubular chamber coiled glomerately at least at the early stage (Ammovertellinidae) in distinction to ammodiscids, or with a sharp change of the regular types of coiling (Usbekistaniidae).

Lower Cambrian – Holocene.

Remarks. Genus *Agatamminoides* is transferred into Glomospirellinae from the Ammodiscinae. In the Ammovertellininae were previously included forms with both the agglutinated and microgranular wall (*Gandiella, Pilammina*) although in the diagnosis of their higher taxon Ammodiscacea only the agglutinated character of the wall was marked. Indeed, often it is not possible to separate the forms only on the basis of this feature.

Families: Usbekistaniidae Vyalov, 1968 (without *Turritellella* and *Repmanina*), Ammovertellinidae Saidova, 1981 [with Ammovertellininae - having an uncoiled part, Glomospirellinae Ciarapica et Zaninetti, 1985 (without an uncoiled part, subfamily reinstated from Ammovertellininae of Loeblich & Tappan, 1987), and Pilammininae Urosevic, 1992], *Pseudolituotubidae Conil et Longerstaey, 1980 from the Fusulinacea.

Order PLAGIORAPHIDA Mikhalevich, 2003.

Shell with chambers in a high spire, following the proloculus, each chamber forming an entire whorl or more with strongly oblique sutures between the chambers; wall agglutinated; aperture circular, large, at the open end of the enrolled chamber.

Middle to Upper Triassic .

Family: Plagioraphidae Mikhalevich, 1995 (*Plagioraphe*).

Order *?ATAXOPHRAGMIIDA Fursenko, 1958

Only forms with the typical strongly asymmetrical ataxophragmiid spire usually with elongated vertical chambers inclined to the previous whorls or conical shells of *Satorina*, typical *Coskinolina*, *Coskinon* types also often with the initial part inclined to the last shell, triserial or more usually uniserial part are left in this order (excluding the other forms into the order Verneuilinida). The phyletic line included here is a special trochospiral line differing from the *Trochammina*-like or *Eggerella*-like trochospiral types in its mostly high conical form with noninflated chambers and flat sides, the development of the uniserial part and its special character, development of columella (*Pfenderina*) or other kind of the integrative inner central structure and passages in the central space of the shell [the low conical *Dobrogelina* and *Nezzazatinella* belong to the same line having the inner central structure of the same Pfenderinid type as was shown by Neagu (1979)]. This line strongly resembles the *Glomotrocholina* (although pseudo-two-chambered) – *Cylindrotrocholina* (Involutinidae) – *Tetrataxis* – *Patellina* – *Chapmanina* – *Ferayin*a line, also with conical shells, noninflated chambers, similar character of the uniserial part (if present) and of the inner structures (columella in *Patellina*, hollow inner pillars are known in *Pseudolituonella* (Ataxophragmiida) and in Chapmaninidae (*Angotia* and others). The taxonomic elaboration of this line needs further study but obviously they belong to a single phyletic line, and probably to class Spirillinata, subclass Ammodiscana. Some ataxophragmiids are isomorphic to calcareous Robertinids having, like involutinids, aragonitic wall. In the Ataxophragmiidae and Pernerinidae forms with a typical *Ataxophragmium*-like spire and with the usual trochospire reduced to a triserial one are separated here into different families.

Middle Triassic – Holocene.

Remarks. As this group with complex inner structure underwent a significant change in their classification and it still is not clear and needs further study to define the position of many of its taxa, the provisionally outlined generic composition is here given contrary to the absence of generic body in the other taxa of this book (the authors of the genera given only for those published after Loeblich & Tappan, 1987). Their complex inner structure, shell wall complication in many cases shows the features arising in parallel and convergent with those of the inner complications in the agglutinated representatives of the other agglutinated subclasses (Miliamminana, Textulariana) and hence cannot serve in this case as effective and reliable taxonomic features specific for this group. At the same time some features of their outer shell structure ((a special high conical form, inclined character of the first volutions, special character of the final uniserial part (when present)) as well as of some inner shell structures (the presence of columella, hollow inner pillars) are specific for this group. Placing with some hesitation this group into a new place would be of use to attract the attention of specialists and to resolve the problem of its phylogeny. Wide ranging of the thorough investigations of its member groups of the lower rank is necessary before reaching the final conclusion. Many of their genera include only a few species, nevertheless a phylogenetic lineage was established for the genus *Coskinolinella* (Cherchi, 1983). Revising the agglutinated foraminifera and basing his approach mainly on the shell wall structure Neagu (2003) separated the subclasses Orbitolinana and Pfenderinana which are considered here to be heterogenous and including the forms with principally differing shell morphology, part of these subclasses enters here subclass Ammodiscana, the other part – subclass Miliamminana (see Part XI, 3). Ataxophragmiida as outlined here needs further revision, the present work only gives a new sight and a new approach to the detail elaboration of its new taxonomy.

Superfamily Ataxophragmoidea Schwager, 1877

Shell with a simple wall and undivided chamber lumen.

Families: *Duotaxidae Mikhalevich, 2003 (differs from the representatives of Verneuilinidae where it was previously placed in the character of the shell and chamber shape) (type genus *Duotaxis* Kristan, 1957), Ataxophragmiidae Schwager, 1877 (with Ataxophragmiinae Schwager, 1877 - only genera having *Ataxophragmium*-like largely involute spiral coiling with invisible early whorls and with high chambers turned to the axis of coiling are retained here) (*Ataxophragmium, Arenobulimina, ?Ataxorbignyna,* without *Hagenowella* and *Sabulina* (both transferred to the Globotextulariidae) and without *Pityusina* and *Praechrysalidina*), Dobrogelinidae Mikhalevich, 1992 (*Dobrogelina*).

Superfamily Textularielloidea Groenhagen & Luterbacher, 1966

Shell with a subdivided chamber lumen or alveolar wall.

Remarks. The genus *Pseudomarssonella* was placed by Mikhalevich, 2004 tentatively into the subfamily Paravalvulininae of the family Chrysalidinidae (which was raised to the status of superfamily Chrysalidininoidea) following Banner et al., 1991. In the works of Tasli et al., 2008 and Tasli & Altiner (2010) it was brought out clearly that the genus belongs to the family Pfenderinidae, subfamily Paleopfenderininae. The belonging of the genera

Riyadhoides and *Redmondoides* to pfenderinids was also confirmed. Families Orbitolinidae and Parurgoninidae separated by Neagu (2003) into suborder Orbitolinina are included here into this superfamily as well as the family Pfenderinidae [separated by Neagu, 2003 into the special order Pfenderinida including also family Nautiloculinidae which in the present work is placed in the order Nautiloculinida of the subclass Textulariana (see Part XI, 5)].

The subfamily Dictyorbitolininae Schroeder, 1990 (in Schroeder et al., 1990) has "marginal apertures situated on the border of the central zone and form a circular ring. These apetures are aligned in a straight line from the apex and are directed towards the base. The first order chamberlets reflect the design traced by the apertures and show the same arrangement" (Scott, 2002, p. 44).

The interesting finding of the Aptian orbitholinid representative designated earlier as chlorophyte algae and probably belonging to the new genus was shown by Bassi et al., 2009. Some later works contributed to the new morphologic data of this group with a complex internal structure (Peybernes, 1976, Hottinger & Drobne, 1980, Canerot, 1984, Pelissie et al., 1988, Septfontaine, 1988 and some others).

Families: Pernerinidae Loeblich et Tappan, 1984 [with Pernerininae Loeblich et Tappan, 1984 (only genera having *Ataxophragmium*-like largely involute spiral coiling with invisible early whorls and with high chambers and intercameral sutures inclined to the axis of coiling and shells lacking a uniserial part are retained here) (*Pernerina, Orbignyna, Voloshinoides, ?Opertum*, *?Anatoliella,* without *Agglutisolenia, Coprolithina* and *Kaeveria* transferred to the subclass Hormosinana and without *Crenaverneuilina, Hagenowina* and *Voloshinovella*) and Voloshinovellinae Mikhalevich, 2003 (differing from the Pernerininae in its small coiled part, with most of the shell formed by the uniserial rectilinear widening part giving it a subconical appearance, and from the Coskinolininae in its endoskeletal features: more simple and situated in the marginal part of the chambers (type genus *Voloshinovella* Loeblich et Tappan, 1964*, Trochamijiella* Athersuch, Banner & Simmons, 1992 (originally put into the Biokovinidae but differs in its early trochoid coil)], Textulariellidae Groenhagen et Luterbacher, 1966 [*Textulariella, Cuneolinella, Guppyella,* (without *Alveovalvulina, Hagenowinoides* transferred to the *Crenaverneuilinidae*)], ?Syrianidae Kaminski, 2004 (*Syriana* Fourcade and Mouti 1995 - if initial part really trochospiral (unclear, broken), final part uniserial and compressed), Cuneolinidae Saidova, 1981 [(with Cuneolininae Saidova, 1981 *(Cuneolina* (in the understanding of Neagu, 2000), *Palaeolituonella, Pseudotextulariella, ?Vercorsella*), Scythiolininae Neagu, 2004 in Kaminski, 2004 (*Histerolina* Neagu, 2000*, Scythiolina* Neagu, 2000 – these two last genera with an initial scarce planispiral part according to Neagu, 2000) and Montsaleviinae Zaninetti, Salvini-Bonnard, Charollais et Decrouez, 1987 ((type genus *Montsalevia* described by Zaninetti, Salvini-Bonnard, Charollais et Decrouez, 1987)], Dicyclinidae Loeblich et Tappan, 1964 *(Dicyclina* and *Nakkadyia* described by Ismail, Hussein-Kamel, Boukhary & Ghadour, 2009*),* Pfenderinidae Smout et Sugden, 1962 [(with Pfenderininae Smout et Sugden, 1962 (*Pfenderina, Drevennia, Pfenderinella*, without *Accordiella* transferred to the Chrysalidininae), Pseudopfenderininae Septfontaine, 1988 (without tunnels, subfamily reinstated here from the synonymy of Pfenderшnidae in Loeblich & Tappan, 1987)(*Pseudopfenderina, Pseudoeggerella*), Paleopfenderininae Septfontaine, 1988 (*Paleopfenderina* Septfontaine in Kaminski, 2000, *Conicopfenderina, Chablasia* (transferred from the Biokovinidae), *Satorina, Sanderella, ?Steinkella* (chambers subdivided), and

Kurnubiinae Redmond, 1964 (*Kurnubia, Gyroconulina, Praekurnubia, ?Conicokurnubia*)]; Coskinolinidae Moullade, 1965 [(*Coskinolina, Coleiconus, Coskinon, Lituonelloides, Pseudolituonella*); Hauraniidae Septfontaine, 1988 (*Haurania, ?Cymbriaella* Fugagnoli, 1999, *Gutnicella, Meyendorfina, ?Platyhaurania* Bassoullet et Boutakiout, 1996, *Bostia* Bassoullet, 1998 and?*Socotraina* Banner, Whittaker, Boudagher-Fadel et Samuel, 1997, *Spiraloconulus* (from Spirocyclinidae), (without Amijiellinae)]; Parurgoninidae Septfontaine, 1988 (*Parurgonina*), Orbitolinidae Martin, 1980 [(with Orbitolininae Martin, 1980 (*Orbitolina, Alpillina, Conicorbitolina, Eopalorbitolina, Eygalierina, Mesorbitolina, Montseciella Cherchi et Schroeder 1999, Naupliella, Neoiraqia, Neorbitolinopsis, Palorbitolina, Palorbitolinoides, Praeorbitolina (*genus *Praeorbitolinoides* Matsumaru, 2005 is regarded by Cherchi & Schroeder, 2009 as a synonym of *Praeorbitolina* on the basis of the similarity of their embryonic apparatus), *Rectodictyoconus, Valserina*), Dictyoconinae Moullade, 1965 (*Dictyoconus, Abrardia, Calveziconus, Campanellula, Carinoconus, Coskinolinoides, Cribellopsis, Cushmania, Daviesiconus, Dictioconella, Fallotella, Falsurgonina, Heterocoskinolina, Iraqia, Karsella* Sirel, 1997*, Orbitolinella, Orbitolinopsis, Paleodictyoconus, Pseudorbitolina, Simplorbitolina, Urgonina, Valdanchella, Verseyella* and *Barattolites* Vecchio et Hottinger, 2007), Praedictyorbitolininae Schroeder, 1990 (*Praedictyorbitolina* Schroeder, 1990 in Scroeder, Clavel & Charollais., 1990, *Paracoskinolina*) and Dictyorbitolininae Schroeder, 1990 in Schroeder et al., 1990 (*Dictyorbitolina*].

INCERTAE SEDIS:

Families: Pyreninidae Mikhalevich, 1999 (*Pyrenina*) and Nezzazatidae Hamaoui et Saint-Marc, 1970 with Nezzazatinae Hamaoui et Saint-Marc, 1970 (*Nezzazata, Trochospira, Nezzazatinella*) and Coixitinae Hamaoui et Saint-Marc, 1970 (*Coixites, ?Antalyna, Rabanitina*). The shell wall of this group is fully secreted and perhaps diagenetically changed, Nezzazatidae were supposed to belong possibly to pfenderinid line (as *Nezzazatinella* (in Neagu, 1979), the initial part of *Nezzazata* shell with irregular coiling was considered by some authors as the evidence of its miliolid character, but such coiling may also present in spirillinat forms. The morphology of the shell looks more like some rotaliat forms, additional study is necessary. Pyreninidae is considered by some authors as belonging to Miliolata. Obviously this Insertae Sedis group is heterogenous and needs further study.

Genus *Lupertosinnia* Farinacci, 1996 is not included here as unclear although it seems to belong to Spirillinata line.

Subclass SPIRILLINANA Mikhalevich, 1992

Shells with lime secreted wall, of radial ultrastructure throughout or (in the fossil forms) in one of the layers (more often in the outer one, but in Tetrataxidae – in the inner one), with rich additional skeletal structures on the outer surface, in the umbilical area, sometimes with the subdivision of the chamber lumen into the chamberlets, with the inner pillars, in the advanced forms the additional apertures and the primitive canal system may be present.

Devon - Holocene.

Remarks. In spite of a set of works of the last decades (of Neagu, Blau and Blau with co-authors, Di Bari, Rigaud with co-authors – most of them unfortunately unavailable for me) contributing significantly to the composition and structure of some of the Spirillinana representatives, the minute forms of this group are still insifficiently studied, the features of the pores and additional apertures in the majority of the cases remain unclear. The canal system known at present in a few of them may possibly be discovered in some more of their representatives which will give the new basis for the revision of this important group and for construction its new classification.

The succession in the geologic time mentioned above (Archaediscida - Carbon, Lasiodiscida - Carbon - Perm, Involutinida - Perm - Cretaceous, Spirillinida - Trias – Holocene, Patellinida – Holocene (Text figure 6)), showing the unity of this phyletic lineage and its development, at the same time shows taxonomic unreliability of such features as the presence of the microgranular layer in the wall possessing also the radial layer. The classification of the subclass offered below in spite of significant changes nevertheless still follows the mainstream of its traditional classification according to the geologic periodization. In my opinion it needs to be thoroughly revised and reconstructed according to the development of their morphologic features and stages of morphologic complication (from pseudo-two-chambered to multi-chambered shells, from glomerate to fully planispiral, trochoid and high conical forms, and at last to the forms possessing the primitive canal system). The systems founded on the basic morphologic features of their shells showed their reliability in the other foraminiferal lineages and in most of the cases were later confirmed by the molecular data.

Superorder ARCHAEDISCOIDA Pojarkov et Skvortsov, 1979

Shell pseudo-woc-hambered (the second tubular chamber may be subdivided) or multi-chambered,; wall usually of the two calcareous layers microgranular and radial one (which more often is the outer one, but in Tetrataxidae –the inner one); pore openings not described. Lower Carboniferous – Upper Permian.

Remarks. The microgranular layer of the wall in some genera is absent, its presence is more likely the result of the processes of diagenesis of the primary radial shell wall. I regard this feature of the wall in such ancient forms of low (if any) taxonomic significance.

Order ARCHAEDISCIDA Pojarkov et Skvortsov, 1979

Shell oval or lenticular, with the second tubular half-cylindrical chamber variably coiled (whorls of coils may be irregular glomerate or slightly sigmoidal, the later whorls may be planispiral, rarely entirely planispiral), involute or evolute; chamber lumen may be subdivided into the pseudochambers in the advanced forms, entire shell may be covered with the mighty secondary layers of thickening, developed especially in the umbilical area.

Lower and Upper Carboniferous.

Remarks. Two superfamilies of this order are separated here on the basis of the difference of their chamber lumen.

Superfamily Archaediscidoidea Cushman, 1928

Shells with chamber lumen lacking any subdivisions.

Remarks. Subfamily Eosigmoilininae Brazhnikova et Vdovenko, 1977 is reinstated from the synonymy of Archaediscinae (Loeblich & Tappan, 1987) according to Ponomareva, 2012.

Families: Archaediscidae Cushman, 1928 [with Archaediscinae Cushman, 1928, Eosigmoilininae Brazhnikova et Vdovenko, 1977 including two genera – *Eosigmoilina* Ganelin, 1956 and *Brenkleina* Zaninetti et Altiner, 1979 previously entering

Asteroarchaediscidae), Ammarchaediscinae Conil et Pirlet, 1974 in Pirlet & Conil, 1974 including *Donodiscus* Vdovenko, 1988), Uralodiscinae Grozdilova, 1993 in Vdovenko et al., 1993, Kasachstanodiscinae Marfenkova, 1983 – the three latter subfamilies reinstaited in Vdovenko et al., 1993], Asteroarchaediscidae A.D. Miklukho-Maklay, 1957, Plano-archaediscidae Gubenko, 1989, Tubispirodiscidae Mikhalevich, 1996 (fully planispirally and evolutely enrolled).

Superfamily Glomodiscidoidea Mikhalevich, 1998

Shells with the second tubular chamber subdivided by the inner skeletal formations.

Family: Glomodiscidae Mikhalevich, 1998 (*Glomodiscus, Uralodiscus, Tournarch-aediscus*, *Permodiscus).*

Order *TETRATAXIDA Mikhalevich, 1981

Shell trochospiral, multi-chambered, chamber lumen may be subdivided (Abadehellidae), canal system not developed.

Lower Carboniferous – Upper Permian.

Remarks. Transferred from the former Fusulinacea on the basis of its shell and wall structure. .

Families: Tetrataxidae Galloway, 1933, Pseudotaxidae Mamet, 1974,?Valvulinellidae Loeblich et Tappan, 1984 [(with Valvulinellinae Loeblich et Tappan, 1984, Abadehellinae Loeblich et Tappan, 1984 (chambers subdivided)], Akcayidae Özdikmen, 2009 pro Sabaudiidae Brönnimann, Decrouez et Zaninetti,1983 [the genus *Akcaya* Özdikmen, 2009 pro *Sabaudia* Charollais et Brönnimann, 1966 (non Ghigi, 1909)] transferred from Textularielloidea as having a hyaline radial layer of the wall in the embryonal stage), Endotaxidae Bogush et Brazhnikova, 1996 in Rauser-Chernousova et al., 1996.

Order LASIODISCIDA Mikhalevich, 1993

Shell from low trochoid, nearly planispiral to high conical, with the second tubular half cylindrical chamber coiled evolutely; umbilical area filled by radial fibrous mass with the tubercules and pillars, in the advanced genera this mass is pierced with the fissures of canals; the sutural apertures may be present situated along the spiral suture (perpendicular to it) on the outer side of the shell, in specialized forms these apertures posses tubular outgrowths.

Lower Carboniferous – Permian.

Remarks. Tubular outgrowths (bridges) lead into the spirally coiled tubular chamber through additional apertures, and together with the fissures of the umbilical area they form the primitive canal system. The order differs from the order Archaediscida in the shell form, regular evolute coiling of the whorls, the presence of the primitive canal system.

Families: Lasiodiscidae Reytlinger, 1956 Vissariotaxinae Reytlinger in Vdovenko et al., 1993), Monotaxinoididae Mikhalevich, 1999 (slightly trochoid shell lacking the additional apertures and bridges), Howchiniidae R. Martini et Zaninetti, 1988, Pseudovidalinidae Altiner, 1988.

Superorder INVOLUTINOIDA Hohenegger et Piller, 1977

Shell pseudo-two-chambered; shell wall of one layer of radial crystals, having pseudopore openings, aragonitic.

Lower Permian – Cretaceous.

Remarks. The shell wall often becomes fully microgranular as a result of the processes of recristallization.

Order INVOLUTINIDA Hohenegger et Piller, 1977

Shell without the canal system.

Lower Permian – Upper Cretaceous.

Superfamily Involutinoidea Butschli, 1880

Shells with the inner lumen of the tubular chamber without the inner formations.

Remarks. Family Triasinidae previously placed in this superfamily was transferred from it into a separate superfamily Triasinoidea (Mikhalevich, 2000). Loeblich & Tappan (1987) placed Trocholinidae into the synonymy of Involutinidae but may be such special gigantic trocholins as described by Pelissie et al., 1988 argue for the reinstatment of the Trocholinidae as a separate taxon.

Families: Involutinidae Butschli, 1880 (including the genus *Bancilina* of Neagu, 1995), Planispirillinidae Piller, 1978, Triadodiscidae Zaninetti, 1984 [with Triadodiscinae Zaninetti, 1984, Lamelliconinae Zaninetti, Ciarapica, Decrouez, 1987 (the latter subfamily was placed by Loeblich & Tappan, 1987 as synonym of Triadodiscinae, it includes now the genus *Ornatoconus* Di Bari, 1998)], Aulotortidae Zaninetti, 1984, [with Aulotortinae Zaninetti, 1984 (including the genus of Neagu, 1994 – *Andersenolina)*, Auloconinae Zaninetti, Ciarapica, Decrouez, 1987 (the latter subfamily placed by Loeblich & Tappan, 1987 into synonymy of Aulotortinae) and Parvalamellinae Rigaud, Martini et Rettori, 2012, with ball-like coiling and reduced lateral laminar extensions (Rigaud et al., 2012)], Hirsutospirellidae Zaninetti, Ciarapica, Cirilli et Cadet, 1985, ?Ventrolaminidae Weynschenck, 1950.

Superfamily Triasinoidea Loeblich et Tappan, 1986

Shell with the chamber lumen of the tubular chamber subdivided by pillars.

Family: Triasinidae Loeblich et Tappan, 1986.

Order HOTTINGERELLIDA Mikhalevich, 1993

Shell with the umbilical area filled by lamellar shell matter and pierced with the system of canals which includes spiral canal, going in parallel to the inner margin of the tubular chamber and system of the fissure-like passages connecting it with the chamber.

Jurassic – Lower Cretaceous.

Family: Hottingerellidae Mikhalevich, 1993.

Superorder SPIRILLINOIDA Hohenegger et Piller, 1975

Shell pseudo-two-chambered or multi-chambered, wall of one crystal or of mosaic of several crystals; pseudopores are filled with granular organic matter and have sieve plate with the micropores at their base.

Order SPIRILLINIDA Hohenegger et Piller, 1975

Shell pseudo-two-chambered, planispiral or trochospiral.

Upper Triassic – Holocene.

Suborder SPIRILLININA Hohenegger et Piller, 1975

Shells without the canal system.

Upper Triassic – Holocene.

Families: Spirillinidae Reuss et Fritsch, 1861, Conicospirillinidae Mikhalevich, 1995 (with the trochospiral shell and the umbilical area filled with secondary shell matter).

Suborder SPIROTROCHOLININA Mikhalevich, 1993

Shell with the lamellae in the umbilical area (forming a kind of pillar), lamellae going from one inner margin of the chamber to the other and with the spiral canal going beneath them.

Upper Jurassic.

Family: Spirotrocholinidae Mikhalevich, 1993.

Order PATELLINIDA Mikhalevich, 1992

Shell trochoid, with the early part of the tubular chamber composing part of the whorl or several whorls, sometimes tubular part absent, the later part of the shell multi-chambered, with the whorls formed by the two or rarely three narrow subcrescentiform chambers per whorl or by one cyclical chamber, chambers may be subdivided into the chamberlets by the additional septae or by radial tubes.

Lower Jurassic – Upper Cretaceous – Holocene.

Superfamily Hergottelloidea Loeblich et Tappan, 1984

Shell with simple chambers, not subdivided into chamberlets.

Families: Hergottellidae Loeblich et Tappan, 1984, Ungulatellidae Seiglie, 1964 (shell of *Ungulatella* – the type genus of the family is close in its structure to that one of the agglutinated genus *Plagioraphe*), Placentulinidae G.K. Kasimova, Poroshina, Geodakchan, 1980 [with Placentulininae G.K. Kasimova, Poroshina, Geodakchan, 1980, (obviously, including the genus *Metapatellina* from Ungulatellidae), Ashbrookininae Loeblich et Tappan, 1984], Pannellainidae Loeblich, Tappan, 1984 (possibly the genus *Heronallenita* close in its structure to this family also needs to be included here but the structure of its aperture is unclear).

Superfamily Patellinoidea Rhumbler, 1906

Shell with the chambers subdivided into chamberlets by the semiseptae or by the radial tubes or tube pillars (*Chapmaninidae).

Remarks. Families of this superfamily are transferred here from the former Rotaliacea. Tube pillars of the family Chapmaninidae represent inner apertural rather than canal system.

Families:Patellinidae Rhumbler, 1906 (including the new genus of Neagu et Cirnaru, 2001 b – *Rumanolina*), *Annulopatellinidae Loeblich et Tappan, 1964, Paleopatellinidae Mikhalevich, 1999 - shell similar in its structure to the patellinid shell, even the inner columella may be present, but the tubular undivided chamber enrolled around proloculus is absent (*Paleopatellina, Pseudopatellina, Patellinella, Subpatellinella,* ? *Paalzowella* from Ashbrookininae, ? *Pseudopatellinoides),* Patellinellidae Bugrova, 1990, *Chapmaninidae Thalmann, 1938, Dictioconoidessidae Bugrova, 1990.

Order *?CYMBALOPORIDA ord. nov.

Shell initially trochospiral, with a few or multiple small chambers (rarely two chambers per whorl) visible on the spiral side and only the chambers of the last whorl visible on umbilical side, later chambers in cyclic series in a single flat to conical layer (usually those of the successive cycles alternating in position) or in flaring tiers (Fabianiidae); umbilical area with symmetrically radial rather wide and incised sutures, either widely opened or may be obscured by the plate-like chamber extensions forming flange-like plate or a complex series of perforated plates (Cymbaloporettidae) or may be filled with lamellae and pillars (Halkyardiidae); wall perforated on the spiral side and imperforate on the umbilical side, often forming lamellae obscuring the spiral side (Cymbaloporidae) or filling the umbilicus (Halkyardiidae); apertures in the trochoid stage at the open umbilical ends of chambers, may also be on both sides of umbilical chambers in sutures as one or series of openings, which sometimes may be boarded with rim-like lips, lips of the apertures at the central ends of the umbilical chambers may form tubes, funnels or flange-like plates (Cymbaloporettidae), in tiered chambers – it consists of pore-like openings on the terminal face; in gamont generations in some genera the float and balloon chambers of different degree of complexity formed beneath the umbilical side are known, balloon chambers may have large rounded openings boarded by lips or may have branching channels (*Millettiana*).

Upper Cretaceous; Paleocine – Holocene.

Remarks. The representatives of the new order are transferred here from the former Rotaliina on the basis of their shell structure. Their unique pelagic mode of life forming rather complex balloon and float chambers is not met in other planctic foraminiferal group and may be regarded as a special parallel way to partial occupying of pelagic niche comparing the Globigerinana. Such peculiarities of their shell structure as right symmetrical rather wide and incised radial sutures with all the chambers of umbilical side having the equal dimensions, the hollow centres of the trochoid shells, the tendency towards the lamellar covering of the umbilical area characteristic of many ancient representatives of Sprillinana (archaediscids, involutinids), special character of flange-like plates and funnels in umbilical area of *Cymbaloporetta*, characteristic pore-like apertures with elevated lips and some other minor features of their appearance make it possible to put them closer to the other Spirillinana. The true bilamellar character of their shell wall was not observed. The new order differs from the order Patellinida also having trochoid shell in the structure of the umbilical side and aperture, and also in its unique adaptation to the pelagic mode of life in one of the generations. Further study is necessary to elucidate the resemblance of Cymbaloporids to the *Tretomphalus* – are their so similar floating chambers formed as convergent structures or are both groups in close relationship. The order is tentatively placed in this subclass. in hope that this transfer would provide the future attention and reinvestigation of the representatives of this group. Such a study ought to help revealing the problem wheather radially-symmetrical structure of the umbilical side of the representatives of the order is a feature characteristic of many Spirllinata taxa or a feature acquired by rotaliate form in its adaptation to the pelagic mode of life.

Superfamily Cymbaloporoidea Cushman, 1927, stat. elevated

Shell with chamber interior not subdivided.

Remarks. Belonging of the Cretaceous genus *Archaecyclus* to Cymbaloporidae is somewhat doubtful. The genus *Globigerinatella* conversely obviously belongs to Cymbaloporidae and ought to be transferred from globigerinids to this family.

Families: Cymbaloporidae Cushman, 1927, Cymbaloporettidae fam. nov. (shell similar to *Cymbalopora* and *Millettiana*, but lips of the apertures at the central ends of the umbilical chambers form tubes, funnels, flange-like plates or a complex series of such perforated plates over the center of umbilicus, type genus *Cymbaloporetta* Cushman, 1928), Halkyardiidae Kudo, 1931, stat. elevated.

Superfamily Fabianioidea Deloffre et Hamaoui, 1973, stat. elevated

Shell with chamber interior subdivided.

Familiy: Fabianiidae Deloffre et Hmaoui, 1973 stat. elevated.

Order ?SEABROOKIIDA Mikhalevich, 1980

Shell flattened, irregularly-oval in outline, at the early stage with the proloculus and two chambers on both sides of it, the later chambers one-half coil in length, fully embracing the previous ones, and alternate being disposed at an angle 180° to each other (the biloculine type of structure); aperture terminal, oval or fissure like, usually encircled with a lip and connected by a tube with the previous chamber.

Upper Cretaceous – Holocene.

Remarks. Shell structure, quite unusual for Rotaliata, and the terminal position of the aperture are the basis to transfer this order (although with some hesitation) from Rotaliata and for placing it in the Spirillinata. Monotypic taxon having some convergent similarity with the biloculinids (Miliolana), but differing in the shell wall ultrastructure.

Family: Seabrookiidae Cushman, 1927.

XI.4.3. CLASS MILIOLATA Saidova, 1981

(= class Miliolicea Saidova, 1981, = class Tubothalamea Pawlowski et al., 2012, part, in Adl et al., 2012)

Shells seldom unilocular, more often pseudobilocular, pseudo-multi-chambered in the more primitive forms, multi-chambered and super-multi-chambered in the advanced forms, the predominant types of coiling are irregular and regular glomerate, planispiral most often combined with glomerate coiling, trochospiral only as an exception, rarely biserial; shells of the higher, most advanced representatives, sometimes cyclical and fusulinoid in form; a flexostyle tubular chamber is most typical ; chambers of the higher Miliolana forms are predominantly tubular, usually two per whorl, in Soritida, and rarely in some Milioloida, chambers are broad and more numerous per whorl, in Paleozoic forms chambers multiple and mostly broad; chamber lumen sometimes subdivided by septa (the primary and the secondary septa may be formed), by pillars, or advanced forms may have integrative systems (stolons, tunnels), with inner apertural systems not developed, and canal system absent; shell wall monofontinal, agglutinated or agglutinated and microgranular in the lower forms (subclass Miliamminana), porcellaneous in the higher forms of subclass Miliolana , with irregular-shaped pseudopores between irregular randomly oriented needles of crystal units, of irregular character, not formed externally; chamber growth occurs by the gradual addition of the wall piece by piece, organic and calcareous secreted elements are added simultaneously; aperture terminal in position, simple in the most primitive forms, in more complex ones with an outer flap or inner tooth structure (the same apertural structures in Paleozoic forms, referred to differently), sometimes aperture secondarily multiple, additional apertures not developed (only as an exception); nuclear apparatus heterokaryotic in higher calcareously secreted forms, sometimes with the polymerization of somatic nuclei (*Sorites*); in *Triloculina* gametes were previously described as biflagellate (and having the axostyle?) but in recent investigations only biflagellate gamets were observed; benthic forms, mostly free living, more rarely attached.

Ordovician – Silurian – Permian – Carboniferous – Holocene.

Remarks. This ancient class is characterized by the shell structure of advanced forms of one of milioloid and fusulinoid types which are absent in any other foraminiferal classes, by the practical absence of trochoid shell types, the special character of its apertural structures, the unique ultrastructure of its calcareous shell wall in addition to the high magnesium content of the calcite (Loeblich & Tappan, 1964; Bender & Hemleben, 1988) opposite to the other phyletic lines with calcareous shells. In some forms the surface of the postembryonic chambers bears irregularly scattered pore openings of irregular form, which are unstable and seen mostly at the electron microscopic level. Some occasional pseudopore openings (or sometimes more densely dispersed) may be present in agglutinated (Plate **13**, figure 1 D; Plate **66**, figure 4) and calcareous forms (Text figure 14), pitted and tabulate inner or outer wall has been described earlier for many calcareous representatives (in Milioliporidae - Loeblich & Tappan, 1987, in *Nubeculina* – Clark, 1993, in *Peneroplis* – Hottinger et al., 1993, Langer et al., 2009). Clark (1993) even notes a lamination of the *Nubeculina*'s wall. Nevertheless this lamination originates from the mechanical sorting of wall crystals and their tighter or looser packing, rather than a result of physiological processes and formation of the middle organic lining (POM) in the Rotaliana wall. A true bilamellar wall is absent, as are regular pore canaliculi of definite form, characteristic of Nodosariata and Rotaliata. Pits on the shell wall provide the transparency of wall for light necessary for the symbionts, as has been shown for *Peneroplis* (Hohenegger, 2011). \

The system of the class is significantly elaborated here, advancing from the new concepts of the foraminiferal macrosystem published elsewhere (Mikhalevich, 1980 a, b, 1981, 1992, 1988, 2000, 2004). This new classification stresses the significance of a morphological approach, and also takes into consideration the wall structure and composition as having subordinate importance. Meanwhile the separation of the Paleozoic representatives of cornuspirid and ammodiscid forms is to some extent conventional because of the scanty features of their shells. All the main taxonomic characters are present both in the ancient Paleozoic and the later (Carboniferous – Holocene) branches of the Miliolata. Moreover, the two branches show similar trends in their evolutionary development (see below - Part XI. 2. 2). This is why the Paleozoic fusulinids (with the exclusion of some previously included groups belonging to ancient Nodosariata and Spirillinata) are included here in the class Miliolata, as its separate group of several superorders.

The Miliolata branch with tubular chambers (mostly two per whorl) may possibly have evolved from some ancient tournayellids or closely related forms during the Carboniferous and Permian (or perhaps earlier). They reached a high degree and developmental complexity in the advanced groups (suborder Alveolinina) by the Cretaceous (in parallel with fusulinids they have a super-multi-chambered fusulinid-like shell, but with a more perfect system of inner communication). Soritida - the second branch of later calcareous Miliolata (Cretaceous–Holocene) has an initially planispiral shell with several wide chambers in a whorl, as in the Paleozoic endothyrids and reichelinins (Plate **52**, figure – s 10, 11, 12, compare with 3, 4, 5). They may have evolved from a reichelinins-like ancestor or from other groups developing similar features in parallel which needs further investigation.

New data and work on soritids morphology (Gudmundsson, 1994) includs some peneroplid species into soritinids. Thus, three lineages (Fusulinoida, Milioloida, and Soritida) of one large foraminiferal stem (class Miliolata) display similar features at different evolutionary stages. A lot of work over the last decades (and some earlier) has contributed much to the knowledge of different Miliolata's groups (Bogdanovich, 1952, Subbotina et al., 1981, Seiglie, 1982, Hofker, 1983, Arnaud-Vanneau & Prestat., 1984, Andersen, 1984, Hamaoui, 1984, Reichel, 1984, Temirbekova & Antonova, 1985, Zaninetti et al., 1985, Calvez, 1988, Hottinger et al., 1989, Neagu, 1984, 1985, 1986, 1994, 1999, Clark, 1993, Sirel, 1999, Gawor-Biedowa, 1984, Septfontaine, 1984, 1985, 1988, Geroch & Nowak, 1988, Fugagnoli, 2000, Popescu & Crihan, 2002, Cherchi & Schroeder, 2005, Vicedo et al, 2011 a, b, Davydov et al., 1999, Bubík, 2000, Cozar, 2000, Kender et al., 2006, Cozar et al., 2007, Kobayashi & Altiner, 2007, Boudagher-Fadel, 2008, Zhang et al., 2010 and others). The system of the groups of Miliolid's superorders is now elaborated better, and the main phyletic lines are more evident. The system of the Fusulinoids group of superorders clearly should be constructed mainly on the shell structure rather than on the number and the character of the shell wall layers, which may vary depending of the ontogenetic stage of the shell and of the peculiarities of its preservation (see Part XI.2.2). The first steps in this direction were made in Rauser-Chernousova et al., 1996 revision of fusulinoids, but this work is not yet logically completed.

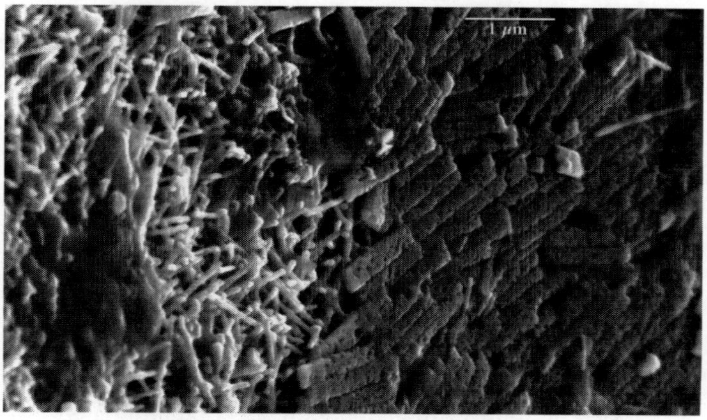

Text Figure 14. Shell wall ultrastructure of *Massilina secans* (d'Orbigny), 1826. Left side of the photograph – the upper parquet layer is destroyed exposing the needle-like crystals of the inner layer, right side – the upper parquet layer (x 10 000.).

Subclass MILIAMMINANA Mikhalevich, 1980
(= Rzehakinicae Saidova, 1981, = Schlumbergerinana Mikhalevich, 1992)

Shell with the wall agglutinated, microgranular or agglutinated partly with secreted needles.

Carboniferous – Holocene.

Remarks. In some ancient representatives the wall may be microgranular as a result of diagenesis and after future revision they may be transferred into the subclass Miliolana. The representatives of this subclass are often similar in form with some or another secreted Miliolata groups. Some groups were transferred from the former heterogenous Textulariacea (=Textulariina) on the basis of their shell and apertural structures Mikhalevich, 1992, 2004, 2005).

The system of this subclass is given here up to subfamily level, in Mikhalevich, 2004 - up to the genera level, the system of the order Lituolida and Schlumbergerinida is also given up to the genera level in a separate works (Mikhalevich, 2005 and Mikhalevich & Kaminski, 2008) .

Order LITUOTUBIDA Mikhalevich, 1992 (emend)

Shell where the second tubular chamber has irregular often alternating (glomerate, planispiral or combined) plan of coiling and undivided chamber lumen, with the following subdivision of the tubular chamber lumen into rather irregular pseudochambers, sometimes forming true chambers in the last volutions, usually of somewhat irregular form and dimensions, bearing some tubular features, the final part may be erected and irregularly rectilinear; wall agglutinated; aperture at the open end of the tubular chamber.

Cretaceous – Holocene.

Remarks. The representatives of this order are similar in their shell structure to some tournayelids but have an agglutinated rather than a microgranular shell wall, although microgranular tournayelids also often have agglutinated particles in their wall.

Families: Lituotubidae Loeblich et Tappan, 1984, Trochamminoididae Haynes et Nwabufo-Ene, 1998 (emend., stat. elevated.) - without an uncoiled part).

Order SCHLUMBERGERINIDA Mikhalevich, 1980

(= ord. Rzehakinida Saidova, 1981)

Shell multi-chambered, usually elongated, with one of the miliolid type of construction – glomerate at least at the early stage - quinqueloculine, sigmoidal, triloculine, may also be planispiral (mostly spiroloculine) or combined, more often with long tubular chambers, two per whorl, more rarely with wider chambers of three per whorl resulting in a broader shell; wall agglutinated on the organic base or agglutinated and combined with calcareous needles at the base of the wall; aperture simple or with a tooth or trematophore.

Jurassic – Holocene.

Remarks. The representatives of this subclass are often similar in form with some secreted Miliolata groups, and the families and subfamilies are separated here on the same principles as the calcareous miliolids. Some genera have an agglutinated or mixed shell wall and are removed from the subclass Miliolana. The finding of *Miliammina gerochi* Tyszka, 1997 in the Middle Jurassic (Tyszka, 1997) lengthens the geologic range of this group from Cretaceous to Jurassic.

Schlumbergerinids represent a peculiar clearly outlined foraminiferal group which is characterized by the predominance of several particular shell types (miliolid types) not occurring in any other foraminiferal phyletic line except the class Miliolata. They differ from their isomorphic forms in the subclass Miliolana in their agglutinated wall character, mainly on an organic base. Genera that have a thick agglutinated wall layer underlined by a thin calcareous layer, characteristic of the porcellaneous miliolids are also included here in the order Schlumbergerinida. They are regarded as transitional forms between these two wall types as was shown for some species of *Dentostomina* and *Siphonaperta* (Mikhalevich et al., 1986). Following the insights of Arnold (1978a, b, 1979), the primitive miliolids appear to have arose firstly as forms with a tectinous shell wall (*Periptygma* Arnold, 1978a). The next step in their evolution was represented by forms with an agglutinated wall on an organic base, and later on a secreted calcareous base. From this point of view the shell form is regarded as being a more conservative feature than the composition of the shell wall, and being preserved and repeated through geologic time. The gradual changes in the composition of the shell wall, from the organic, to the agglutinated and at last to the calcareous, were possibly going when the shell structure of definite type was already composed. Agglutinated isomorphic forms may possibly also appear as a sister group to porcellaneous ones.

The wall of all these forms as well as the wall of the porcellaneous miliolids was usually considered as imperforate. But nevertheless, as was mentioned above, occasional pseudopore openings penetrating the entire wall were observed in some species (Mikhalevich, 2004, 2010, Mikhalevich et al., 1986) (Plate **66**, figure 4).

Although the constructive opportunities of the agglutinated wall are generally less than those of the porcellaneous calcareous wall, and the variability and complexity of the apertural apparatus of such forms is usually more purely represented, nevertheless in the rare schlumbergerinid genera (*Schlumbergerina*) a complex trematofore is developed. The inner complication of the chamber lumen in these agglutinated forms is not developed contrary to their porcellaneous isomorphic forms. The system of this order is given here after Mikhalevich & Kaminski, 2008.

Families: Rzehakinidae Cushman, 1933 ((without *Rothina* which is considered to be a synonym of *Hormosinella,* representing its broken chamber (Bubik, 1997, Kaminski, 2004b, Kaminski et al, 1996), and without *Ammoflintina,, Spirolocammina, Spirosigmoilinella, Silicosigmoilina,Trilocularena, Silicomassilina* placed into another families according to their mode of coiling and shell structure, and *Miliammina, Birsteiniolla* transferred into Miliammininae) [with Rzehakininae Cushman, 1933, Psamminopeltinae Mikhalevich et Kaminski, 2008, (with evolute shell), including Spiroglutininae Mikhalevich, 1983 as synonym because the type species of the genus *Spiroglutina – Spiroloculina asperula* (Karrer, 1868) ought to be assigned to the genus *Psamminopelta*], Miliamminidae Saidova, 1981, stat. nov, (with Birsteiniollinae Mikhalevich et Kaminski, 2008, Miliammininae Saidova, 1981 ((=Siphonapertinae Saidova, 1975 (part), =Agglutinellinae El-Nakhal, 1985) and with *Rudoloculina* Guibault et Patterson, 1998 as synonym of *Siphonaperta* Vella, 1957 – see Mikhalevich & Kaminski, 2008], Triloculareniidae Mikhalevich et Kaminski, 2008 (including *Falsagglutinella* Loeblich et Tappan, 1994), Ammomassilinidae Mikhalevich et Kaminski, 2008 (with Silicomassilininae Mikhalevich et Kaminski, 2008, Ammomassilininae Mikhalevich et Kaminski, 2008), Sigmoilopsidae Vella, 1957 (with Spirolocammininae Mikhalevich et Kaminski, 2008, Sigmoilopsinae Vella, 1957), Schlumbergerinidae Mikhalevich, 1980, Ammoflintinidae Mikhalevich et Kaminski, 2008 (with Ammoflintininae Mikhalevich et Kaminski, 2008, Pseudoflintininae Mikhalevich et Kaminski, 2008).

Order HAPLOPHRAGMIIDA Loeblich & Tappan, 1989

Shell multi-chambered, with wide chambers streptospirally coiled at least initially (or in one of the generations), evolute or involute, later may be uncoiled, aperture terminal or elevated, single or multiple.

Middle Triassic – Holocene.

Superfamily Recurvoioidea Alekseychik-Mitskevich, 1973

Shell with a simple wall.

Remarks. Family Plectorecurvoididae stands apart from the rest of the Recurvoioidea as having biserially arranged enrolled chambers and possibly deserves to be elevated to the order rank. But the scrupulous work of Neagu & Platon, 1994 has shown the unstable mode of coiling in *Plectorecurvoides* ("both planispiral and streptospiral") and unstable biseriality in *Pokornyammina* ("biserial to larger uniserial"). Bubik (1995) has also shown the variable irregular coiling of some species of *Plectorecurvoides* (including the type species) as well as the terminal character of its aperture (judging from the inner tubes connecting the chambers seen in sections) and of the aperture of *Recurvoides* (Bubik, 2000). Moreover, some specimens of *Thalmannammina* (fam. Recurvoididae) appear occasionally as biserial (see Neagu & Platon, 1994). All of these genera have a terminal aperture although sometimes it appears externally as basal (interiomarginal) and further investigation of their variability is necessary for taxonomic conclusions. They appear to be closely related from a morphological perspective.

Families: Cystamminidae Mikhalevich, 2003 (differs from all others in Recurvoioidea in its strongly inflated shell and few number of chambers), Recurvoididae Alekseychik-Mitskevich, 1973, Ammobaculinidae Saidova, 1981 (with Ammobaculininae Saidova, 1981 possibly including *Carteriella* Haig et McCartain, 2010 – structure not clear enough),

Telatynellinae Gawor-Biedowa, 1987*)*, Acupeinidae Brönnimann et Zaninetti, 1984 (*Acupeina, Navarella* – transferred from the Ammobaculinidae as having multiple apertures), Plectorecurvoididae Loeblich & Tappan, 1964.

Superfamily Haplophragmioidea Eimer & Fichert, 1899
Shell with an alveolar wall.

Families: Haplophragmiidae Eimer & Fickert, 1899, *Mesoendothyridae Voloshinova, 1958 (transferred from the former Loftusiacea).

Superfamily Labyrinthidomatidoidea Loeblich & Tappan, 1987
Shell with the chamber lumen subdivided by inner septa.

Family: Labyrinthidomatidae Loeblich & Tappan, 1987 .

Order SPHAERAMMINIDA Mikhalevich et Kaminski, 2003 in Mikhalevich, 2003
Shell multi-chambered, planispirally coiled, involute, with a few strongly overlapping subsphaerical chambers; wall agglutinated, simple; aperture terminal, areal, with or without tooth .

Paleocene – Holocene.

Family: Sphaeramminidae Cushman, 1933 (with Sphaeramminae having a tooth (*Sphaerammina, Ammosphaerulina, Canepaia*) and Praesphaeramminae Kaminski & Mikhalevich, 2003 in Mikhalevich, 2003 – similar to Sphaeramminae but without a tooth).

Order LITUOLIDA Lankester, 1885
Shell multi-chambered, planispiral, at least initially, later may be uncoiled, coiled part involute (more often) or evolute, chambers from isomorphic nearly subtriangular in outline to broad and low (like in *Peneroplis*), with straight or arcuate sutures, wall agglutinated or microgranular-agglutinated, aperture terminal (areal to final), without a tooth, single or multiple.

Carboniferous – Holocene.

Remarks. Lituolids included here are to a strong degree isomorphic with some higher Miliolata (for instance *Charentia* to *Derventina; Stomatostoecha, Choffatella, Torinosuella* and many others – to different Peneroplids). In the composition of the order Lituolida (Loeblich & Tappan, 1994 and in some following classifications) genera having both agglutinated and microgranular walls and with or without canaliculi or some kind of pore openings were included, in spite of the diagnosis of the order defining it as agglutinated and lacking pores. It was indeed impossible to draw a sharp boundary in many cases. More often it is easier to distinguish the taxa based on shell form, and the position and the character of the aperture. The position of the initial aperture is in many genera unclear, and needs more detailed investigation. However in some cases the terminal character of the initial aperture is visible in thin sections, in spite that in their diagnosis it is marked as interiomarginal. As there is no sharp boundary between agglutinated and microgranular agglutinated wall, the details of the diagnoses of the genera concerning their wall structure have diagnostic significance only in some cases. Forms with the basal aperture, at least at the early stages, are excluded from lituolids and transferred to the class Rotaliata, subclass Textulariana (order Nautiloculinida Mikhalevich, 2003) (*Nautiloculina, Haplophragmoides* and some others). The uniserial forms as well as ammomarginulinids with the terminal aperture displaced to the peripheral angle and some others bearing the nodosariata's features - to the orders Hormosinida Mikhalevich, 1980, Ammomarginulinida Mikhalevich, 2002 and some other taxa of the subclass Hormosinana. Haurania-like forms, with a trochoid initial part, are transferred to Textularielloidea, trochospiral nezzazatids also are not included here. Pseudobilocular forms with long tubular second chamber are placed in the class Spirillinata (subclass Ammodiscana Mikhalevich, 1980, order Ammodiscida Mikhalevich, 1980). Lituotubida was also separated from lituolids (Mikhalevich, 1992, 2000, 2003). Some forms were placed within the same subclass, but in the other orders: tubular multi-chambered forms (Rzehakina-like) were placed in the order Schlumbergerinida Mikhalevich, 1980 (with Rzehakinida Saidova, 1981 as its junior synonym); forms streptospirally coiled were placed in the order Haplophragmoidida Mikhalevich, 1992; and some forms with stable cyclic and annular chambers in both generations were placed in the order Cyclolinida Mikhalevich, 1992. The system of the order is given after Mikhalevich, 2004a and Mikhalevich, 2005.

Superfamily Lituoloidea de Blainville, 1827

(Nom. corrected pro Lituolinoidea)

Shell free or attached, coiled part involute, wall simple.

Remarks. Subfamily Ammobaculitinae reinstated from the synonymy of Ammomarginulininae in Loeblich&Tappan, 1987. It differs from the latter in having a nonflattened shell, horizontal sutures and a central position of the aperture, but some of its genera may belong wholly or partially to the order Ammomarginulinida, and additional study is therefore necessary. Genera with a secreted wall and similar character of chambers occur both among higher nodosariats and higher miliolats, and it is sometimes difficult to distinguish such forms among their agglutinated ancestors. The genera *Adhaerentia* and *Flatschkofelia* which are described as having initial biserial parts are nevertheless kept here within Placopsilinids (one of them, *Flatschkofelia*, is transferred here to Adherentiinae), as the biseriality of these forms may be due to the irregularity of the attached forms. Two genera (*Ammocibicides* and *Ammocibicoides)* are not included into Placopsilinids as they have a trochoid shell.

Families: Lituolidae de Blainville, 1827 [with Ammobaculitinae Alekseychik-Mitskevich, in Subbotina et al.,1981and Lituolinae de Blainville, 1827 (as Ammobaculitinae, but with multiple apertures)], Buzasinidae Mikhalevich, 2003 [(differs from Haplophragmoididae where it was originally placed by the terminal character of the aperture (areal rather than basal), from the other families of this superfamily by the special character and position of its areal aperture, from the family Lituolidae in the absence of the uncoiled part of the shell)], Mayncinidae Loeblich&Tappan, 1985 (it is necessary in some genera to study the aperture at the initial stage, without *Deuterospira* with basal aperture at the initial stage) (with Mayncininae Loeblich&Tappan, 1985 and Comaliamminae Mikhalevich, 2004 b), Placopsilinidae Rhumbler, 1913 (attached) (with Placopsilininae Rhumbler, 1913, Acruliammininae Mikhalevich, 2004 b, Flatschkofeliinae Kaminski 2004 and Adhaerentiinae Loeblich&Tappan, 1986).

Superfamily Barkerinoidea Smout, 1956

Shell with chambers, internally subdivided, wall microgranular or microgranular-agglutinated.

Remarks. Character of the aperture of *Barkerina* is similar to that one in *Crenatella* Luczkowska, 1972 (Quinqueloculinidae, subfamily Scutulorinae Mikhalevich, 1987) – a low slit subdivided by a distinctly crenulate upper margin. Family Haymanellidae is placed here somewhat formally, as lituolid form with the inner complication of the wall. "Radiate aperture" described by Sirel (1999) in the type description gave reasons to compare this taxon with the agglutinated ammomarginulinids, with the aperture of *Nodosinum* (Hormosinana, Nodosariata). But discovery by Hottinger (2007) of a short flexostyle in *Haymanella hubery* lives now no doubt in its belonging to the class Miliolata. Some features of its apertural and inner structure are close to *Praerhapydionina*.

Families: Barkerinidae Smout, 1956, Biplanatidae Mikhalevich, 1992 (transferred from Nezzazatidae where it was originally separated as a subfamily), *?*Haymanellidae Mikhalevich, 2003.

Superfamily Biokovinoidea Gusic, 1977

Shell free, entirely planispirally coiled or with a later uncoiled part which may be straight or peneropliform and reniform, the planispiral part involute (sometimes semi-involute) or evolute, the proportions of the development of planispiral and uncoiled parts vary in different families, with agglutinated, microgranular thickened complicated wall, imperforate or perforated, may be with canaliculi and keriotheca-like and phrenotheca-like layers and additional endoskeletal elements (pillars, partitions, chomata-like structures), or may be alveolar (in Buccicrenatidae, Choffatellidae, Spirocyclinidae); aperture terminal, single or multiple.

Remarks. The very early part in many of the genera shows slight oscilation (see Mikhalevich, 2004 b). In some genera it is difficult to distinguish the border between biokovinid and some endothyrid genera (in Charentiidae triangular projections on the spiral septum of *Charentia* "superficially resembling fusulinacean chomata" (Loeblich&Tappan, 1987), they also resemble a miliolidean tooth, in *Melathrokerion* the whole character of the shell and its wall is close to that one of *Charentia,* in different species of *Evertycyclammina* such structures of the same position and appearance as endothyrid chomata can also be clearly seen in the equatorial and axial sections (see Banner & Whittaker, 1991, Plate 3, figure - s 2, 3, 6, Plate 4, figure – s 3, 6, 7, 10, 11); the presence of chomata-like structures at the bottom of the chambers is similar in Charentiidae and Buccicrenatidae, *Biokovina* (Biokovinidae) and *Bosniella* (Charentiidae) also have in their wall keriotheca-like layer. This similarity is not considered here as occasional or convergent, but as showing the close relations of these groups (Mikhalevich, 2003, 2009 a) and I regard this line as the single phyletic line of the planispiral Tournayellids and Endothyrids (see Loeblich & Tappan, 1987, pl. 79 (8) and pl. 233 (11); pl. 79 (11) and pl. 243 (16)), these representatives of Biokovinoidea differing from the latter in their agglutinated rather than special microgranular shell wall. Families Buccicrenatidae, Choffatellidae, Pseudochoffatellidae and Spirocyclinidae are transferred into this superfamily from the former Loftusiacea (the latter family without *Quataria, Saudia, Thomasella* Sirel, 1998, *Timidonella, Vania* having cyclical or annular chambers, without *Streptocyclammina* with the initial streptospiral coiling and without conical and initially trochospirally coiled forms (*Haurania, Spiraloconulus*)). Genus *Bosniella* is moved from

Biokovinidae into Charentiidae, *Chablaisia* – from Biokovinidae to pfenderinids, genera *Choffatella, Alveosepta* are transferred into Choffatellidae from the Hottingeritidae, genus *Ijdranella* is moved into Choffatellidae from Hauraniids (Amijiellinae), genus *Popovia* is transferred into Hottingeritidae (Popoviinae) from the Alveolophragmiinae on the basis of its terminal (even at the early stage) rather than basal aperture, the genera *Amijiella, Bramkampella* and *Rectocyclammina* are transferred from Choffatellidae to Ammomarginulinida Mikhalevich, 2002, the superfamily Flabellamminopsidoidea Mikhalevich, 2002 on the basis of their shell morphology. Genera *Balkhania* and *Montsechiana* are transferred from Pseudochoffatellidae to the order Cyclolinida Mikhalevich, 1992 (superfamily Vaniinoidea Mikhalevich, 2003) on the basis of the character of their shell with cyclic and annular chambers. The basis to all these changes was partially given in Mikhalevich, 2004 a, c.

Families: Charentiidae Loeblich &Tappan, 1985, Lituoliporidae Gusic et Velic, 1978, Biokovinidae Gusic, 1977 (without *Chablasia* and *Bosniella* but with *Dubrovnikella* Schlagintweit & Velić, 2011), Labyrinthinidae Septfontaine, 1988 [(with Labyrinthininae, Planiseptinae Septfontaine, 1988, nom. nudum (*Planisepta* Septfontaine in Kaminski, 2000) and ?Levantinellinae Fourcade, Mouty &Teherani, 1997 (the structure of the type genus is somewhat unclear)], Buccicrenatidae Loeblich & Tappan, 1985 (including *Evertycyclammina*),Choffatellidae Maync, 1958 (Banner & Whittaker, 1991 showed the planispiral character of its genus *Alveosepta* and its similarity with *Choffatella*), Hottingeritidae Loeblich et Tappan, 1985 [(with Hottingeritinae Loeblich et Tappan, 1985 (its planispiral character was also shown by Banner & Whittaker, 1991) and Popoviinae Mikhalevich, 2004 b)], Pseudochoffatellidae Loeblich & Tappan, 1946, Spirocyclinidae Munier-Chalmas, 1887.

? Superfamily Coscinophragmatoidea Thalmann, 1951

Shell attached, initial part may be coiled, later uncoiled, formed in curving irregular rows of chambers, or branching, chambers irregular in size and shape, wall complicated - canaliculate, may be alveolar, aperture terminal, single or multiple.

Remarks. Genus *Ammotrochoides* is not included into Haddoniidae as having trochoid shell.

Families: Haddoniidae Saidova, 1981 (*Haddonia, Stylolina* reinstated by Popescu et al., 1998 from the synonymy of Lituola), Coscinophragmatidae Thalmann, 1951.

Order CYCLOLINIDA Mikhalevich, 1992

Shell multi-chambered, initially planispiral or coiling oscillating, then peneropliform and later having cyclical or annular chambers; aperture multiple, rows of circular openings in the apertural face. Lower Jurassic – Eocene.

Superfamily Cyclolinoidea Loeblich et Tappan, 1964

Shell with a simple wall and undivided cavity. L.- U. Cretaceous.

Family: Cyclolinidae Loeblich et Tappan, 1964.

Superfamily Cyclopsinelloidea Loeblich & Tappan, 1984

Shell with the chamber lumen subdivided by pillars or septa, or both, advanced forms with stolons.

Families: Cyclopsinellidae Loeblich & Tappan, 1984 (with Cyclopsinellinae Loeblich & Tappan, 1984 and Levatinellinae Fourcade, Mouty & Teherani, 1997 with pillars and pilaroid structures, subepidermal partitions are absent), Ilerdorbidae Hottinger et Caus, 1982, Orbitopsellidae Hottinger et Caus, 1982.

Superfamily Vaniinoidea Mikhalevich, 2003

Shell in form as in other cyclolinids but with subepidermal network of beams and rafters, and often with pillars.

Remarks. The superfamily differs from the Cyclolinoidea in its complex wall and endoskeleton, and from the Cyclopsinelloidea in the character of the wall, having a subepidermal network. Family Vaniidae is separated from the former Loftusiacea (from the Spirocyclinidae and from Cyclamminidae (Pseudochoffatellinae)) as having annular and cyclical chambers in the adult stage (shell planispiral, with cyclical or annular chambers in the adult stage; wall with a subepidermal network of beams and rafters forming an alveolar wall and in some cases may form chamberlets, pillars often formed in the later stages; aperture multiple, rows of openings along the peripheral margin). Genus *Quataria* is transferred into Vaniinae from the Spirocyclinidae, *Balkhania?Montsechiana* - from the Pseudochoffatellinae, type genus of Thomasellinae *Thomasella* and another genus *Saudia is* moved from Spirocyclinidae.

Family: Vaniidae Mikhalevich, 2003 [with subfamilies Vaniinae Mikhalevich, 2003 (without pillars) and Thomasellinae Mikhalevich, 2003 (with pillars)].

Order LOFTUSIIDA Kaminski et Mikhalevich, 2003 in Mikhalevich, 2003

(= Loftusiida Kaminski et Mikhalevich, 2004 a)

Shell large, planispiral, axis of coiling elongated resulting in an alveolinid and fusulinid appearance, a short tubular chamber around proloculus may be preserved; chamber interior with exoskeletal beams and rafters and endoskeletal pillars, canal-like openings may be present; wall agglutinated and microgranular, with imperforate epidermis and an inner alveolar layer; aperture terminal in the initial stage, later multiple, with rows of rounded openings slightly above the base of the apertural face.

Lower - Upper Cretaceous.

Remarks. The affinity of these forms to the Miliolata is confirmed not only by the similarity of their outer (like Alveolinids) and inner shell structure (*Reticulinella* even has a preseptal canal) but also by the presence of a tubular chamber around the proloculus (*Praereticulinella*). The order was separated after our joint discussion on the foraminiferal classification in Buchurest University in 2002 and published in the IWAF 2004 edition. However, the diagnosis and the volume of the order, as well as its position in the foraminiferal system, differ much in the understanding of Kaminski, 2004 and Mikhalevich, 2004 a. In the latter work the separation of the order is based mainly on the shell morphology rather than on the shell wall structure, the order is considered to be monotypic, with the only family Loftusiidae, and is placed in the subclass Miliamminana. All the trochospiral forms are excluded from its composition.

Family: Loftusiidae Brady, 1884 (*Loftusia, Praereticulinella, Reticulinella*).

Subclass MILIOLANA Saidova, 1981

Shells pseudobilocular, pseudo-multi-chambered or multi-chambered (in one case unilocular - Squamulinida), glomerate and planispiral (more rarely uniserial, trochospiral as an exception only), usually with two tubular chambers per whorl and with flexostyle, more rarely number of chambers per whorl is greater; wall lime secreted, porcellaneous, monofontinal, with middle layer of elongated irregularly and randomly set crystal units (needle-like, rod-like, stick-like), paved from inside and usually outside by the facing layer of crystal units (elongated, plate-like or of the other form) packed horizontally and more tightly, going in parrallel with the outer wall surface, with occasional openings of an irregular form between them at the shell surface, pseudopore openings at the surface of the post embryonic chambers - also occasional and of irregular form, seen only in electron microscope, pore canaliculi of regular shape as well as pores characteristic of Nodosariana and Rotaliana, are absent; deposits of the shell material often formed at the bottom of chambers, being especially developed in fusulinids but may also be present in some miliolids (*Periloculina*), chamber growth occurs by the gradual addition of organic and calcareous secreted elements simultaneously; aperture simple or with a tooth of variable structure, sometimes with a complex tooth, or multiple teeth, additional apertures as a rule not developed; in advanced forms last chambers with a system of inner partitions and chamberlets, connected by foramens, stolons or tunnels, canal system absent.

Carboniferous - Holocene.

Remarks. Inner organic layer of the wall is formed by the fibrillar layer and the crossing tubular elements, as as shown by Leutenegger (1977) on living forms. Apertural structures in this subclass are well and variably developed compared to representatives of the subclass Miliamminana. The "canals" of Rhapydionidae, piercing the prominent central thickening of the shell, represent a special type of integrative system and more complex structure than in the other representatives of this subclass, nevertheless it does not represent a true canal system compared with that of one either Rotaliana or Spirillinana.

Superorder Fusulinoida is included here in Miliolana, as comparative morphological analysis (see above – Part **XI**, 2, 2) of the shell and apertural structures in lower more primitive and higher advanced members of this superorder, and of the representatives of the class Miliolata, has shown that they are similar in shell and apertural structures as well as in their developmental trends, beginning from the pseudo-two-chambered forms to complex multi-chambered and super-multi-chambered forms of the fusulinoid shape, not observed in any other class. As the series of sections were compared, supplementary inner deposits at the bottom of chambers of endothyrids (tubercules, rostra, nodes, chomata) showed homology with the teeth of Miliolata. It seems reasonable and acceptable to name the chomata of endothyrids as "tooth structures" of a different type (tuberculates, rostral, nodes-like, spine-like outgrowths), and for the massive wide or multiple chomata of fusulinids to preserve their previous name "chomata". Shell wall of the superorder Fusulinoida is supposed to be primarily of miliolid type, and later changing during the diagenetic processes (see above – Part **XI**, 2, 2), in recent miliolats it also may be easily changed to microgranular even whilst living (Macintyre & Reid, 1998). It is possible that in some ancient representatives the wall also could be primarily microgranular. Miliolid and a typical microgranular wall have much in common, being composed of irregularly disposed crystals and crystal units. The high magnesium composition of the shell wall, and some ultrastructural features shared by the two groups, support their affinity and suggest the assignment of the superorder Fusulinoida to the class Miliolata (subclass Miliolana) as an early phylogenetic lineage.

***Group of superorders FUSULINOIDS** Fursenko, 1958

Shell with a typical character of a microgranular wall, including forms of different shell form and structure (from pseudo-two-chambered up to multichambered and super-multi-chambered) with glomerate, planispiral or alternating mode of coiling and terminal aperture (single or multiple), advanced forms with deposits of shell material at the bottom of chambers, subdivided chambers and primitive integrative systems [(septal pores, tunnels, and, in some genera, basal foramina, supplementary tunnels (which are widened basal foramina), and radial passages].

Upper Devonian – Triassic.

Remarks. Fusulinoids microgranular shell wall is supposed to be secondarily altered, being primarily of the same character as a typical wall of the Miliolata (see above – Part **XI**, 2. 2). The classification of this taxon is given mainly according to Rauser-Chernousova et al. (1996) with the addition of the new taxa after Hance et al. (2012) and some other authors, and also with some changes (with the exclusion of the trochospiral forms which are transferred here into Spirillinata and of the group of ancient nodosariats). The collective of authors, with the participation of Rauser-Chernousova, attempted to compose a system of fusulinoids by giving paramount significance to shell morphology rather than wall structure, nearly all the taxa in this collective book were emended. This trend in the classification of the group was not logistically completed and needs further elaboration.

Superorder ENDOTHYROIDA Fursenko, 1958

Shell pseudo-two-chambered and multi-chambered, coiling entirely planispiral, glomerate or more often variable and combined, shell at the later stage may have erected uniserial or biserial part, periphery lobulate or even, segmentation from primitive constrictions and short septa forming the slightly expressed pseudochambers to full septa forming true chambers, wall of one to several layers, occasionally with some agglutinated material, aperture single or multiple.

Upper Devonian – Triassic.

Order TOURNAYELLIDA Hohenegger & Piller, 1973

(= order Tournayellida Dain, 1953 in Vdovenko et al., 1996)

Shell pseudo-two-chambered and seesaw multi-chambered at the end of the growth, the final stage may be erected, rectilinear or biserial; coiling glomerate, planispiral or combined, segmentation into chambers by the slight constrictions of the wall, pseudosepta or true septa (usually short) at the final stages of growth, pseudochambers usually short (wide) and inflated symmetrically or asymmetrically, microgranular shell wall sometimes weakly differentiated with the thin bordering upper layer, additional deposits of shell matter weakly developed or absent; aperture terminal, single or multiple.

Upper Devonian – Carboniferous.

Remarks. Although in some diagnoses of the genera the aperture is designated as "basal", nevertheless it really represents a terminal aperture which in some cases may be not evident because of the narrowing of its opening by the enlarged chamber wall around it.

Superfamily Moravamminacea Pokorny, 1951 is excluded here from the composition of the phylum Foraminifera after Vachard, (1994) and Vachard & Cozar, (2010) who showed that moravamminids and some other closely related genera belong to the class Algospongia G. Termier et al., 1977 (Kingdom and phylum indeterminate).

Superfamily Tournayelloidea Dain, 1953

Shell free, of rather regular form, with the chambers inflated symmetrically.

Remarks. Some of tournayellids with an agglutinated (or partly agglutinated) wall possibly ought to be included in the subclass Miliamminana, or their position may be interpreted as transitional between the two subclasses of the class Miliolata. Family Lituotubidae Loeblich et Tappan, 1984 is transferred to the order Lituotubida.

Families: Tournayellidae Dain, 1953 (with Tournayellinae Dain, 1953, Forschiinae Dain, 1953), Lituotubellidae A.D. Miklukho-Maklay, 1963 [with Lituotubellinae A.D. Miklukho-Maklay, 1963 (including *Alticonilites* Hance, Hou et Vachard, 2011), Septabrunsiininae Conil et Lys, 1977, Septaglomospiranellinae Reytlinger, 1996 in Rauser-Chernousova et al., 1996], Laxoendothyridae Hance, Hou et Vachard, 2011(including *Crassiseptella* Brenckle et Hance, 2005 and *Endolaxina* Hance, Hou et Vachard, 2011).

Superfamily Chernyshinelloidea Reytlinger, 1958

Shell free, of rather regular form, with the chambers inflated asymmetrically.

Families: Chernyshinellidae Reytlinger, 1958 (with Chernyshinellinae Reytlinger, 1958, Tournayellininae Reytlinger, 1996), Mstiniidae Lipina, 1989 with Mstiniinae Lipina, 1989, Eotextulariinae Hance, Hou et Vachard, 2011, Darjellinae Hance, Hou et Vachard, 2011), Palaeospiroplectamminidae Loeblich et Tappan, 1984 (with Palaeospiroplectammininae Loeblich et Tappan, 1984).

Order ENDOTHYRIDA Fursenko, 1958

Shell entirely multi-chambered, may be with enrolled erected part at the final stage; coiling plectogyroid to glomerate or planispiral; wall of one or several layers, rarely with fibrous layer, inner additional deposits of shell matter variable and rather strongly developed or absent; aperture single or multiple (sieve-like), rarely are additional sutural apertures of special structure present.

Upper Devonian – Triassic.

Remarks. The former Tetrataxacea and Paleotextulariacea are excluded from the order as having a shell of principally another type of structure.

Superfamily Endothyroidea Brady, 1884

Shell without additional apertures and simple septa.

Families: Endothyridae Brady, 1884 (with Endothyrinae Brady, 1884, Omphalotinae Vdovenko, 1996, Endospiroplectammininae Loeblich et Tappan, 1986), Haplophragmellidae

Reytlinger, 1959 [with Haplophragmellinae Reytlinger, 1959 (including *Spireitlina* Vachard in Vachard & Beckary, 1991, Vachard & Krainer (2001) consider *Pseudopaleospiroplectammina* of Mamet & Pinard, 1992 to be synonymous with *Spireitlina*), Paraendothyrinae Lipina, 1996], Endothyranopsidae Reytlinger, 1958 (with Endothyranopsinae Reytlinger, 1958, Eoendothyranopsinae Reytlinger, 1996, Neoendothyrinae Reytlinger, 1996), Globoendothyridae Reytlinger in Voloshinova et Reytlinger, 1959 [family reinstated by Hance, Hou et Vachard, 2011, with Globoendothyrinae Reytlinger in Voloshinova et Reytlinger, 1959 (includig genera *Carbotarima* described by Brenckle, 2004 and *Planogloboendothyra* described by Hance, Hou et Vachard, 2011 , Eblanaiinae Hance, Hou et Vachard, 2011], Dainellidae Cozar et Vachard, 2001 (including *Paralysella* Cozar et Vachard, 2001), Loeblichiidae Cummings, 1955 [with Loeblichiinae Cummings, 1955, Urbanellinae Hance, Hou et Vachard, 2011 (type genus *Urbanella* Malakhova in Anonyme, 1963, emend. Brenckle, 1997) , Mediocrinae Hance, Hou et Vachard, 2011 , Quasiendothyrinae Reytlinger, 1961], Endostaffellidae Loeblich et Tappan, 1984 (including *Praeeostaffellina* Cozar et al., 2008 and *Praeplectostaffella* Cozar et al., 2008), Endotebidae Vachard, Martini, Rettori, Zaninetti, 1994 (with *Endoteba* and *Endotebanella* of the same authors, Endotriadidae Vachard, Martini, Rettori, Zaninetti, 1994 with *Endotriada* and *Endotriadella* of the same authors).

Superfamily Bradyinoidea Reytlinger, 1996

Shell with additional apertures of different structure, in advanced genera thickened complex doubled septa forming interseptal spaces by the pre- and post-septal plates at the periphery of the sutures, and sometimes additional chamberlets in praeumbilical area.

Remarks. Vdovenko et al., (1996) term the structures forming additional apertures "preseptal apparatus". I cannot acccept the supposition by this group of authors that the bradyinids represent a transition to a pelagic mode of life: their shell is too heavy, especially in the cases when it includes agglutinated particles.

Families: Bradyinidae Reytlinger, 1950 (with Bradyininae Reytlinger, 1950, Glyphostomellinae A.D. Miklukho-Maklay, 1963), Janischewskinidae Reytlinger, 1996 (with the new genus *Parajanischewskina* described by Cozar and Somerville, 2006).

Superorder OZAWAINELLOIDA Solovieva, 1980 (nom. corrected pro Ozawainelloidea)

Shell nautiloid, discoidal, rhomboidal or in adult of irregular form, involute entirely or with the flat evolute later part, which may often be enrolled, sometimes flaring, with chambers quickly enlarging in size during the growth (somewhat peneroplis-like), periphery may be angular and even keeled, umbilical area may be flat, somewhat invaginated or elevated, coiling mostly planispiral, sometimes with a changing axis of coiling, chambers numerous, chomata and pseudochomata variably developed in different families; aperture single.

Lower Carboniferous – Upper Permian.

Remarks. This taxon was placed by Vdovenko et al. (1996) within Fusulinoidea, but is separated here on the basis of their shell structure, differing from the other Fusulinoidea, and to some extent isomorphic to later Meandropsinids (for instance *Pseudorhipidionina* and some other genera) from the order Soritida. Families of the suborder were separated by

Vdovenko et al., (1996) following their geologic succession. Following the principle of the predominant significance of the morphological features, the family Pseudostaffellidae differing in its shell structure is here moved to the order Staffellida.

Order OZAWAINELLIDA Solovieva, 1980

As of the superorder.

Middle Carboniferous – Upper Permian.

Remarks. In Rauser-Chernousova et al., 1996 the Incertae family (without the diagnosis) is presented as including the genera *Eostaffelloides, Kangvarella* and *Hubeiella.* It seems that the last two genera are closer in their shell structure to some Reichelinidae or Ozawainellidae, but *Eostaffelloides* differs from all of them and from all other fusulinoids by closer resembling the later miliolids of pyrgo-type, having a distinctive coiling, disposition of chambers and chomata also resembling pyrgo teeth (see Part **XI**, 2.2. 4). Possibly only this genus deserves to be separated into a special family Eostaffelloididae. Unfortunately Brenckle (2010), a publication on Eostaffellidae (2010), was not available.

Families: Eostaffellidae Mamet, 1970 (including *Bozorgniella* Cozar et Vachard, 2001), Ozawainellidae Thompson et Foster, 1937, Reichelinidae A.D. Miklukho-Maklay, 1959 stat. elevated (including *Quasireichelina* Ueno, 1992 originally placed by the author of the genus into Staffellidae), and Incertae family with the genera *Eostaffelloides, Kangvarella and Hubeiella*).

Superorder FUSULINOIDA Fursenko, 1958

Shell nautiloid, sphaerical, subquadrate, fusulinoid or cylindrical, mostly involute, rarely uncoiling in later stage, with periphery of the shell widely broad and convex or rarely broad and invaginated, often super-multi-chambered and with chambers partially or fully subdivided into chamberlets in the most advanced forms, early whorls often of endothyroid character, axis of coiling of the earlier whorls may be turned to the axis of later whorls at an angle of 90°, septa from straight to strongly flutted, chomata, parachomata and axial fillings often strongly developed, but may be weakly developed or absent, tunnels, cuniculi and foramina may be present; wall of one to four layers; aperture single or multiple as rows of openings above the base of apertural face.

Upper Devonian – Permian.

Remarks. Fusulinoida represent an ancient suborder with a complex and advanced shell structure of unique character. Some of the later miliolate forms such as fusarchaiasins and alveolinins developed the isomorphic fusiform shell, the latter also developed integrative systems (pre- and post-septal passages) which are less heavy and cumbersome and more effective than those of Fusulinoida [septal pores, tunnels, and, in some genera, basal foramina, supplementary tunnels (which are widened basal foramina), and radial passages, which are open spaces in the upper part of chambers, formed as a result of incomplete septation or partial resorbtion (see above – Part **XI**, 2. 1)]. Openings in the septa of Fusulinoida are similar to the interchamber foramina of other foraminifers (Leppig et al., 2005).

The classification of this superorder, with a highly complex structure in Vdovenko et al., (1996), is less distinctive than of the previous groups, the diagnoses of the orders significantly cover each other, and even based on the wide complex of features it is often difficult to define the correct position of some genera. In the diagnosis of schubertellids the authors wrote that their evolution “goes in the direction of the differentiation of the shell wall and the enlargement of the shell sizes” (Vdovenko et al., 1996, p. 85). It is possible to observe the same picture and trends in recent miliolids, when the shell wall was thin in small specimens (of typical miliolid structure, often without outer roof or inner thin layers which are easily destroyed), but in big specimens the inner layer of randomly oriented crystals becomes wider, with the more tight, oriented and differentiated disposition of crystals which could be taken as different additional wall layers (Plates **56, 57**). Moreover, such a wall structure, consisting of different layers as developed in adult chambers, in the earlier ones the wall is thinner and of usual miliolid character. The increase in number of layers in the endothyroid and fusulinoid wall with growth is marked in the diagnoses of many of their genera. The number of wall layers and their character in the shell wall of this group is correlated with shell size, which does not serve as a reliable taxonomic feature. In my opinion the system of the order needs to be fundamentally revised, and ought to be based mainly on the shell outer and inner structure. The precise translations of Vdovenko et al., 1996 diagnoses of the orders are given below except of the order Fusulinida.

Order STAFFELLIDA A.D. Miklukho-Maklay, 1949

Shell nautiloid, subsphaerical, lense-shaped, sometimes with a prominent umbilical area, in rare cases short fusiform or roller-like, involute, rarely evolute in the last volutions, single representatives may be erected at the final stage of growth; number of whorls from 4 to 14; shell wall either light-grey three-layered (tectum, luminotheca, inner tectorium), or more dark twolayered, with protheca, easily undergoing recrystallization, with simple pores; septa flat, straight or slightly convex, three-layered septa with luminotheca; aperture single, sometimes foramens are developed along the septa base; chomata and pseudochomata from weakly to moderately developed, sometimes parachomata are present in the outer whorls.

Lower Carboniferous – Upper Permian.

Remarks. Family Pseudostaffellidae is transferred to this order from the order Ozawainellida.

Families: Pseudoendothyridae Mamet, 1970, Nankinellidae A.D. Miklukho-Maklay, 1963, Cheniidae F. Kahler et G. Kahler, 1966, Staffellidae A.D. Miklukho-Maklay, 1949, Pisolinidae Rauser, 1985, Kahlerinidae Leven, 1963, Pseudostaffellidae Putrja, 1956.

Order SCHUBERTELLIDA Skinner, 1931

Shell of small or middle size, from spherical to subcylindrical, volutions usually few in number (6 – 8), coiling of the first whorls in the majority of genera is endothyroid; shell wall from monolamellar and undifferentiated, two-layered with protheca to three- and four-layered with the distinct diaphanotheca; septa from straight to intensively folded; additional deposits unstable.

Middle Carboniferous – Permian.

Families:Schubertellidae Skinner, 1931, Boultoniidae Skinner et Wilde, 1954 (including *Dilatofusulina* Wang et Ueno, 2009), Palaeofusulinidae A.D. Miklukho-Maklay, 1963, Yangchieniidae Leven, 1987.

Order FUSULINIDA Fursenko, 1958

Shell from subsphaerical to moderately elongated and ovoid, fusiform and subrhomboidal, ellipsoidal, involute, in the first and second whorls shell may be narrowly nautiloid or subsphaerical, with coiling in the first and second whorls sometimes endothyroid, axis of coiling of the earlier whorls may be turned to the later ones under the angle of 90°, number of whorls usually 4 – 5, rarely 8 – 9 and up to 11; shell wall from two-layered to four-layered, may be with the cleared layer in the upper part of primatheca; septa from straight to strongly contorted or undulated at the axial ends, may be arched near the axial ends, may be folded, with moderate number of septa; chomata distinct, subcircular, subtriangular or subquadrate in section, symmetrical or asymmetrical, becoming flatter to the poles, rarely chomata band-like or absent; additional deposits presented by pseudochomata, axial fillings or absent; aperture single.

Lower Carboniferous – Upper Permian.

Remarks. Diagnosis of the order not given in Vdovenko et al., 1996 (obviously occasionally.

Families: Profusulinellidae Solovieva, 1996, Aljutovellidae Solovieva, 1996, Fusulinidae Moeller, 1878 (with Fusulininae Moeller, 1878, Beedeininae Solovieva, 1996, Eofusulininae Rauser et Rosovskaya, 1959, Quasifusulininae Putrja, 1956), Hemifusulinidae Putrja, 1956, Fusulinellidae Staff et Wedekind, 1910 (with Fusulinellinae Staff et Wedekind, 1910, Pulchrellinae Solovieva, 1983), Wedekindellinidae F. Kahler et G. Kahler, 1966.

Order SCHWAGERINIDA Solovieva, 1985

Shell involute, from sphaerical to elongated cylindrical, rarely enrolled in the last volution, sizes from small to large; shell wall usually twolayered, consisting of theca and keritheca, in the primitive representatives sometimes the outer tectorium is also expressed; septa of different thickness, pierced with pores, from undulated at the axial ends to intensively and regularly folded along all the shell length, in the advanced forms pairs of folds touching each other subdivide chambers into chamberlets, cuniculi are not rare, sometimes frenotheca are present; additional deposits as chomata, parachomata and axialfillings; aperture single or multiple.

Upper Carboniferous – Upper Permian.

Families:Triticitidae Davydov, 1986, Rugosofusulinidae Davydov, 1980, Schwagerinidae Dunbar et Henbest, 1930 [with Schwagerininae Dunbar et Henbest, 1930 (= Biwaellinae Davydov, 1984), Pseudoschwagerininae Chang, 1963, Paraschwagerininae Bensh, 1996], Pseudofusulinidae Dutkevich, 1934 (with Pseudofusulininae Dutkevich, 1934, Chusenellinae F. Kahler et G. Kahler, 1966, Monodiexodininae Kanmera, Ishii et Toriyama, 1976), Polydiexodinidae A.D. Miklukho-Maklay, 1953 (with Parafusulininae Bensh, 1996, Polydiexodininae A.D. Miklukho-Maklay, 1953).

Order NEOSCHWAGERINIDA Minato et Honjo, 1966

Shell free, planispiral, involute, from slightly compresses along the axis of coiling to fusiform and cylindrical; axis of coiling of the initial coils is often turned to the axis of the next ones; shell wall consisting of tectum and finely porous keriotheca (?) and often of the inner thin and more tight layer corresponding to inner tectorium, sometimes shell wall becomes more compact and microstructure indistinguishable; septa flat; aperture multiple, divided by parachomata.

Lower Permian – Upper Permian.

Remarks. Genus *Thailandina* Toriyama et Kanmera, 1968 is considered in Vdovenko et al., 1996 as synonym of *Misellina* Schenk et Thompson, 1940, hence subfamily Thailandininae becomes synonym of Misellinidae.

Families: Misellinidae A.D. Miklukho-Maklay, 1958, Verbeekinidae Staff et Wedekind, 1910, Neoschwagerinidae Dunbar et Condra, 1927 (with Neoschwagerininae Dunbar et Condra, 1927, Lepidolininae A.D. Miklukho-Maklay, 1958, Gifuellinae Kobayashi, Ross C.F. et Ross J.R.P., 2010 based on the new genus *Gifuelloides*), Sumatrinidae Silvestri, 1933, Pseudodoliolinidae Leven, 1963.

Group of superorders MILIOLOIDS Delage et Herouard, 1896

Shell with typical miliolid shell wall. Carboniferous – Holocene.

Remarks. Classification is given after Mikhalevich 1986, 1988, 2005, with some changes. Many of the taxa, expecially those with the complex inner structure still need to be additionally studied to clarify the taxonomy and phylogenetic relations.

Superorder SQUAMULINOIDA Mikhalevich, 1988

Shell unilocular, free or attached.

Upper Cretaceous – Holocene.

Order SQUAMULINIDA Mikhalevich, 1988

As for the superorder.

Upper Cretaceous – Holocene.

Family: Squamulinidae Reuss et Fritsch, 1861 (with Brasiliellinae Loeblich et Tappan 1986 and Squamulininae Reuss et Fritsch, 1861).

Superorder CORNUSPIROIDA Jirovec, 1953

(= Order Cornuspirida Mikhalevich, 1980 (part), suborder Cyclogyrina Saidova, 1981)

Shell pseudo-two-chambered, without flexostyle. Carboniferous – Holocene.

Order CORNUSPIRIDA Jirovec, 1953

(=Order Cornuspirida Mikhalevich, 1980 (part), suborder Cyclogyrina Saidova, 1981)

Shell free or attached, pseudo-two-chambered, with a globular proloculus followed by a long tubular second chamber with planispiral, glomospiral, zigzagged or irregular coiling, final part may be uncoiled and rectilinear; aperture a simple terminal opening.

Carboniferous – Holocene.

Suborder CORNUSPIRINA Jirovec, 1953

Shell free or attached, second tubular chamber enrolled planispirally, at least initially, in the attached forms may be strongly changed, sometimes slightly branching.

Carboniferous – Holocene.

Families: Cornuspiridae Cushman, 1919 [with Cornuspirinae Schultze, 1854 (including *Barbuspira* Neagu, 1999), Cornuspiroidinae Saidova, 1981, Calcivertellinae Loeblich et Tappan, 1964, Pseudocornuspirinae Reytlinger, 1993 in Vdovenko et al., 1993, Glomospiroidinae Reytlinger, 1993 in Vdovenko et al., 1993, Falsotubinellinae Bogdanovich et Mikhalevich, 1983], Hoyenellidae Rettori, 1994 with the type genus *Hoyenella* Rettory, 1994 placed by the author in cornuspirids.

Suborder HEMIGORDIOPSINA Mikhalevich, 1986

Shell free or attached, the second tubular chamber with irregular glomerate coiling, at leastin the initial part.

Permian – Holocene.

Superfamily Hemigordiopsidea Nikitina, 1969

Shell with the simple shell wall.

Families: Hemigordiopsidae Nikitina, 1969 [with Hemigordiopsinae Nikitina, 1969 (including *Pseudohemigordius* Nestell et Nestell, 2006), Meandrospirinae Saidova, 1981 (including *Meandrovoluta* Fugagnoli et Rettori, 2003), Turriglominimae Zaninetti in Limongi et al., 1987, Orthovertellinae Mikhalevich, 1988, Kamuraninae Trifonova, 1984], Hemigordiidae Reytlinger, 1993 in Vdovenko et al., 1993.

Superfamily Shanitoidea Loeblich et Tappan, 1986 (nom corrected pro Shanitinidea Mikhalevich, 1988)

Shell with the inner outgrowths of the wall in the chamber lumen (pillars or thickenings of the chamber wall extending inward from the outer wall).

Families: Shanitidae Loeblich et Tappan, 1986 (= Shanitidae Mikhalevich, 1986), Baisalinidae Loeblich et Tappan, 1986 (with Baisalininae Loeblich et Tappan, 1986, Septagathammininae Mikhalevich, 1988), Neodiscidae Lin, 1984 – nomen translat. Gaillot & Vachard, 2007, pro subfamily, (after Vachard et al, 2008), with the new genera *Uralogordius* in Gaillot & Vachard, 2007, and *Glomomidiella* in Vachard, Rettori, Angiolini et Checconi, 2008 which is considered by these authors as an ancestor form for several genera).

Superorder COSTIFEROIDA Mikhalevich, 1988

Shell multi-chambered, without the flexostyle, with the chambers arrangement uniserial, in a straight or somewhat arched row; chambers irregularly pyriform or broadly oval; aperture simple, without the tooth, may be at the end of the elongated neck and may be with a flaring rim. Upper Triassic.

Order COSTIFERIDA Mikhalevich, 1988

As for the superorder.

Upper Triassic.

Family: Costiferidae Senowbary-Daryan et Zaninetti, 1986.

Superorder MILIOLOIDA Delage et Herouard, 1896

Shell multi-chambered, with a flexostyle; chambers primordially tubular, two, more rarely two and a half or three per whorl at least at the earlier chambers.

Lower Cretaceous – Holocene.

Remarks. In some genera the number of chambers in the last whorls may be more than three, and chambers become shorter and wider although nevertheless bearing some tubular features. Only in *Zoayella* are all the chambers (after streptospirally enrolled tubular part) short and multiple (up to 12 per whorl). Classification of the superorder is given after Mikhalevich, 1988, 2005, 2008.

Order NUBECULARIIDA Jones, 1875

(=order Cornuspirida Mikhalevich, 1980 (part), suborder Nubeculariina Saidova, 1981)

Shell free or attached, planispiral, evolute, chambers primordially tubular, two, rarely three per whorl, in advanced forms may become cyclical, in attached forms may become irregular in form and disposition; wall simple or with the inner septa incomplete; aperture simple, may be with a neck or a lip.

Middle Triassic – Holocene.

Superfamily NUBECULARIIDEA Jones, 1875

Shell with simple inner shell wall of the chambers.

Families: Nubeculariidae Jones, 1875, Ophtalmidiidae Wiesner, 1920, Meandroloculinidae Bogdanovich, 1981, Wiesnerellidae Saidova, 1981, Nodobaculariellidae Bogdanovich, 1981, Spiriamphorellidae Senowbary - Daryan et Zaninetti, 1986.

Superfamily DISCOSPIRINOIDEA Wiesner, 1931

Shell with the inner wall of the chambers complicated with the incomplete septa, final chambers cyclical.

Family: Discospirinidae Wiesner, 1931

Order MILIOLIDA Delage et Herouard, 1896

Shell in overwhelming majority of the cases free, with the primordially glomerate coiling usually of one of the regular glomerate miliolid types, often with the secondary transition to the planispiral coiling in some genera, whorls mostly involute, may have final evolute coils or rarely fully evolute (Spiroloculinids), often with the involute one or two of the final chambers, in the advanced groups (ALveolinids) of one of the fusiform types; chambers more often tubular, the last ones may be flattened and more wide (Hauerinidae) or nearly subsphaerical externally (in the case when the later chamber is embracing the whole shell); shell wall simple or with the inner incomplete septa, or may be with the complex system of inner partitions fully dividing chamber into chamberlets, with the additional system of the inner subdivisions, may be pitted externally or with deposits of skeletal matter at the bottom of the chambers, , rarely alveolar; aperture simple only in a more primitive forms, in the rest ones complicated with a flap (= lobe tooth - flat external tooth) or more often with the inner tooth of the variable degree of complexity, either cribrate or having a trematophore;

additional apertures known only as an exception (in the two genera of Hauerinidae; in advanced super-multi-chambered forms (Alveolinids) stolons and tunnels may present.

Lower Carboniferous – Holocene.

Suborder MILIOLINA Delage et Herouard, 1896

Shell mainly preserving one of the miliolide shell types, shell wall simple or externally complicated by pseudopores or inner incomplete septa.

Lower Carboniferous – Holocene.

Superfamily Quinqueloculinoidea Cushman, 1917

Shell with a simple wall, chamber lumen without subdivisions.

Remarks. Families are separated on the basis of their shell structure and mode of chamber coiling arranged as one of the milioloid types (quinqeloculine, cryptoquinqueloculine, rarely irregularly streptospiral, schlumbergerine, sigmoiline, spiroloculine, triloculine, biloculine) or a combination of these types, sometimes with the presence of an erected part, subfamilies are separated on the basis of the peculiarities of their apertural apparatus. Following these principles the genus *Ankaraella* Sirel, 1998 having initial quinqueloculine part and the later part trochospiral deserves to be separated into a special family Ankaraellidae if the presence of trochospiral part is not occasional as the irregular change of coiling pattern often occurs in miliolids. The full work of Sirel, 1998 is anavailable for me. If the presence of the trochospiral part is stable in many specimens, this genus represents one more finding of such chamber arrangement unusual in miliolids and represented in *Fischerinella* as an exclusion in this developmental line.

The family Zoyaellidae Saidova, 1981 is transferred to the order Soritida as the representatives of the genus Zoyaella Loeblich et Tappan, 1962 have planispiral coiling with broad multiple (up to twelve) chambers in their volutions, only the initial tubular chamber is coiled streptospirally. The entirely planispiral Fischerinidae Millett, 1899 and trochospiral Fischerinellidae Saidova, 1981 are also transferred to the Soritida. Most of the changes concern the former subfamilies Haueríninae and Miliolinellinae, whose composition was too wide, including forms with quite different shell and apertural structures.

Families: Hauerinidae Schwager 1876 [with Haueríninae Schwager, 1876 (including *Nurdanella* Ozgen, 2000, *Pseudolachlanella* Langer, 1992 and *Dervieuxina* Popescu et Crihan, 2002, firstly placed by the authors of the new genus into Miliolinellinae), Lorettaoidinae Mikhalevich, 1988, Flintininae Saidova, 1981, Nummoloculininae Saidova, 1981, Polysegmentininae Mikhalevich, 1986, Danubiellinae Mikhalevich, 2005, Hechtininae Mikhalevich, 2005, Flintinellinae Mikhalevich, 2005, Kayseriellinae Mikhalevich, 2008], Tubinellidae Rhumbler, 1906 (with Tubinellinae Rhumbler, 1906, Articulariinae Mikhalevich, 1987, Poroarticulininae Saidova, 1981), Pavoninoididae Saidova, 1981), Quinqueloculinidae Cushman, 1917 [with Quinqueloculininae Cushman, 1917 (including *Obliqueloculina* McNeil, 1997), Cribrolinoidinae Haynes, 1981, Scutulorinae Mikhalevich, 1987, Pseudoschlumbergerininae Mikhalevich, 2008, Labalininae Mikhalevich, 1988], Spiroloculinidae Wiesner, 1920 [with Kaloshinae Mikhalevich, 1988, Spiroloculininae Wiesner, 1920, Cribrospiroloculininae Mikhalevich, 1988, Stellarticulininae Mikhalevich, 1988, Parahauerinoidinae Mikhalevich, 2005, Elasigellinae Mikhalevich, 2008 (with type genus *Elasigella* Sirel, 1999))], Planispiroidinidae Saidova, 1981, Massilinidae Thalmann,

1941 (with Pseudomassilininae Mikhalevich, 1988, Rectomassilininae Mikhalevich, 1988, Massilininae Thalmann, 1941, Miliolinellinae Vella, 1957, Flintininae Saidova, 1981, Tortonellinae Mikhalevich, 1988), Sigmoilinitidae Luczkowska, 1974 [with Sigmoilinitinae Luczkowska, 1974 (including *Novosigmella* Özdikmen, 2009 pro *Sigmella* Azbel et Mikhalevich, 1983 (non Hebard, 1940), Mesosigmoilininae Mikhalevich, 1988, Sigmoinellinae Mikhalevich, 1987, Longiapertininae Mikhalevich, 1987], Glomulinidae Saidova, 1981 ((with Glomulininae Saidova, 1981, Orthellinae Mikhalevich, 2005 (transferred from the Hemigordiopsinae as having true chambers according to Azbel in Subbotina et al., (1981), although its chambers are rather primitive)), Triloculinidae Bogdanovich, 1981 (with Triloculininae Bogdanovich, 1981, Triloculinoidinae Mikhalevich, 1988, Triloculinellinae Mikhalevich, 1988, Involvohauerininae Mikhalevich, 1986), Pyrgoidae Mikhalevich, 1983 (with Pyrgoinae Mikhalevich, 1983, Biloculinellinae Mikhalevich, 1986, Cribropyrgoinae Mikhalevich, 1986, Idalininae Mikhalevich, 1988), Miliolechinidae Zaninetti, Ciarapica, Cirilli, Cadet, 1985.

Superfamily Milioloidea Ehrenberg, 1839

Shell with a complicated wall with multiple and regularly disposed pits and pseudopore openings on its surface, chamber lumen not subdivided.

Remarks. The occasional pseudopore openings in the miliolid wall are now known in many other genera of the order (see above, Part 4.2.2; 4, 2, 3) (Plate 56, figure – s 7, 8), though usually they are less in number and more rarely and irregularly scattered, not forming the pits on the shell surface. Nevertheless this taxonomic feature of the superfamily could hardly be considered as a reliable base for separation of the superfamily Milioloidea. It seems that a pitted character of the wall of its genera has only generic taxonomic value, the more so that usually these pseudopore openings extend up to half the thickness of the wall. Although in this work this taxon is kept with its previous diagnosis, I consider it would be more reasonable in future to place its genera into families of the present Quinqueloculinoidea, in accordance with their shell and apertural structure (type genus *Miliola* Lamarck, 1804 – to Cribrolinoidinae of Quinqueloculinidae with placing Cribrolinoidinae as junior synonym of Miliolinae Rhumbler, 1895, *Crenulostomina* - to Quinqueloculininae, *Picouina* and *Rupertianella* to Labalininae of Quinqueloculinidae, *Helentappanella* to Tortonellinae (Massilinidae), *Neaguites* - to Kaloshinae of Spiroloculinidae). In this case Quiqueloculinoidea becomes a junior synonym of Milioloidea Ehrenberg, 1839.

Families: Miliolidae Ehrenberg, 1839 (including *Pitella* Langer, 1992), Neaguitinidae Andersen, 1984.

Superfamily Riveroinoidea Saidova, 1981

[(= Riveroininea Saidova, 1981, p. 32 (nom.incorr.)]

Shell of one of miliolide types with the chambers partly subdivided by incomplete septa.

Family: Riveroididae Saidova, 1981.

Superfamily Austrotrillinoidea Loeblich et Tappan, October 9,1986

(= Austrotrillinoidea Mikhalevich, October 27, 1986)

Shell of one of the miliolide types with complex wall (with a subepidermal reticulum or alveolar).

Remarks. Superfamily need further revision.

Families: Austrotrillinidae Loeblich and Tappan, 1986, Brebinidae Mikhalevich, 1986, Pseudohauerinidae Mikhalevich, 1986, stat. elevated, [with Pseudohauerininae Mikhalevich, 1986 and *Malatyninae Hottinger, 2007 (type genus *Malatyna* Sirel et Acar, 1993 placed by the author of the genus tentatively into Riveroidae and later by Hottinger, 2007 into Alveolinidae).

Suborder ALVEOLININA Mikhalevich, 1980

(= Alveolinidae Loeblich, Tappan, 1964; Alveolinellinina Saidova, 1981)

Shell multi-chambered, often super-multi-chambered, subsphaerical or fusiform, initial part of the microsphaeric forms glomerate (of quinqueloculine or triloculine type), very small, in the macrosphaeric forms biloculine (only one chamber follows the proloculus), all the subsequent whorls of both generations are involute, planispiral in Fabulariidae and Alveolinidae, either glomerate in Keramosphaeridae; chambers fully subdivided into chamberlets with a complex system of partitions and pillars, septal and tangential partitions may divide chamber into two or more floors, when secondary spiral partitions do not reach the chamber septa, the preseptal and postseptal tunnels are formed, in some genera deposits of the shell material at the bottom of the chambers may be present; aperture multiple – cribrate or presented by the multiple openings on the apertural face but always above the base of the apertural face.

Lower Cretaceous – Holocene.

Remarks. Pores in the proloculus wall and wall of the first whorls resembling the pores of *Peneroplis* (Hofker, 1971) were observed in *Alveolinella.* The initial stages of growth in keramosphaerids are purely studied. Further investigations may show the heterogeneity of the group with several lineages developing from the different miliolid ancestors. Thus in *Keramosphaerinopsis haydeni* postembryonic, chambers were shown to be arranged in concentric layers (Cherchi & Schroeder, 2005).

Families: Fabulariidae Ehrenberg, 1839, Alveolinidae Ehrenberg, 1839 (including *Helenalveolina* described by Hottinger et al., 1989), *Keramosphaeridae Brady, 1884 (from Soritida).

Order SORITIDA Schultze, 1854 (= Orbitolinida Ehrenberg, 1839, Orbitolitida Wedekind, 1937)

Shell multi-chambered, with three or more usually wide chambers per whorl, in some representatives flexostyle is known in one of the generations; coiling planispiral (trochoid in one genus as an exception), more often evolute, more rarely involute, with a tendency to become fan-like and cyclical, more rarely with a rectilinear character of the later chambers; wall simple in the more primitive representatives, in the earlier chambers with pits or perforated, in the more advanced forms with a more complex inner arrangement consisting of partitions and pillars, with the sharply increasing chamber number, in the highly organized representatives stolons may present; aperture without a tooth, simple or complex, multiple, always above the base of the apertural face.

Upper Cretaceous – Holocene.

Remarks. The state of our knowledge on the structure and taxonomy of this group could be brightly demonstrated with the work of Hottinger (2007), who, giving his perfect investigation of the complex endoskeleton of several genera, nevertheless, avoided to discuss their systematics above the generic level stressing the necessity of studying of the broader material and the impossibility to give appropriate systematic of the group in the current state of knowledge when even the diagnosis of the genus *Peneroplis* could not be appropriately given.

Superfamily Peneroplidea Schultze, 1854

Shell with a simple shell wall, chambers not numerous, not subdivided.

Remarks. Families Fischerinidae, Fischerinellidae and Zoyaellidae are transferred from the suborder Miliolina as having planispiral (or as exclusion – trochospiral) coiling and more multiple and wide chambers per whorl.

Families:Peneroplidae Schultze, 1854 [with Peneroplinae Schultze, 1854 (including *Paraspirolina* Fleury, 1997),Vandenbroeckiinae Mikhalevich, 1988, Renulininae Mikhalevich, 1978, Dendritininae Saidova, 1981], *Fisherinidae Millett, 1899, *Fischerinellidae Saidova, 1981, *Zoyaellidae Saidova, 1981.

Superfamily Meandropsinoidea Henson, 1848 (nom corr. pro Meandropsinidea Saidova, 1981 and Mikhalevich, 1988).

Shell planispiral at the early stage, later may be discoidal, fan-shape, cylindrical or conical; marginal zone of chambers with interseptal pillars and septula partially subdividing the chambers; aperture multiple or rarely single, simple.

Remarks. Contrary to Reichel (1984) the family Rhapydionidae is transferred from Alveolinina based on the structure of the initial part of its shells, which are planispiral, with several wide chambers per whorl. Conical shells have nevertheless clearly expressed features belonging to Miliolata (Soritida). With the questionable inclusion of the family Orduellidae the diagnoses ought to be widened including sphaerical shells possessing stolons as well.

Families: Meandropsinidae Henson, 1848, *Rhapydionidae Keijer, 1945 (with Rhapydioniae Keijer, 1945 (obviously including the genera *Pseudotaberina* Banner et Highton, 1989, *Rhabdorites* Fleury, 1996, *Neotaberina* and *Neorhipidionina* of Hottinger, 2007), Crateritinae Saidova, 1981), Fusarchaiasidae Saidova, 1981, Hottingerinidae Mikhalevich, 1988, ?Orduellidae Sirel, 1998 (nom. corr. pro Orduellinidae Sirel, 1998) (with the type genus *Orduella* Sirel, 1998).

Superfamily Soritidoidea Ehrenberg, 1839

Shell with the chamber lumen subdivided into chamberlets communicating through the stolons, which connect the chamberlets of the preceding and the subsequent rows rather than the chamberlets of the same row.

Families: Soritidae Ehrenberg, 1839 [with Soritinae Ehrenberg, 1839 (icluding *Neotaberina* Hottinger, 2007 and ?*Neorhipidionina* Hottinger, 2007) , Opertorbitolitinae Loeblich et Tappan, 1986], Archaiasidae Cushman, 1927 (possibly with the genera

Praearchaias Sirel, 1996 and *Penarchaias* Hottinger, 2007), Praerhapydionidae Hamaoui et Fourcade, 1973 (obviously including *Praemurgella* Luperto-Sinni, Martin-Chivelet & Gimenez, 2000), Cycledomiidae Mikhalevich, 1988.

INCERTAE SEDIS:

Family: Milioliporidae Bronnimann et Zaninetti, 1971 (with Milioliporinae Bronnimann et Zaninetti, 1971, Galeanellinae Zaninetti, Altiner, Dager et Ducret, 1982, Altinerininae Zaninetti, Ciarapica, Decrouez et Miconnet, 1984, Orthotrinacriinae Zaninetti, Senowbari-Daryan, Ciarapica et Cirilli, 1985, Pseudocucurbitinae Zaninetti, Altiner, Dager et Ducret, 1982).

XI.4.4. CLASS NODOSARIATA Mikhalevich, 1992

Shell unilocular and multi-chambered (pseudo-multi-chambered as an exception), predominant shell types - uniserial, high-trochospiral (in a special polymorphinoid spire, absent in any other taxa), planispiral (nearly exclusively involute), typical trochospiral type is fully absent; chambers often subcircular, triangular, V-shaped, palmate, inner additional skeletal chamber septa only in one fossil taxa (Colaniellidae), outer shell sculpture well developed in secreted forms, but outer additional skeletal plates absent, integrative systems (such as stolons, tunnels, canal systems) not developed (except in one genus – *Delosina* having a special canal system and integrated entosolenian tubes in pleurostomellids); wall monofontinal (= monolamellar), agglutinated in the lower representatives (Hormosinana), may be with the calcareous cement and microgranular, in higher representatives (Nodosariana) - hyaline-radial, with regular pseudopores having cylindrical pseudopore canaliculi but without additional seive-plates (basal seive-plate presenting), growth interrupted, process of forming new chambers unstudied; aperture terminal even in the coiled forms, simple (circular or slit-like or of other configuration) or radial, may be with entosolenian tube (non homologous to the apertural structures of Rotaliata) or with ampula (= apertural chamberlete), slit-like aperture may be asymmetrical, hooked, the last four apertural types (radial, with an entosolenian tube, with ampula, hooked aperture) are unique and are found only in this class, additional apertures absent (except for one case (in *Delosina),* integrated entosolenian tubes also rare (in pleurostomellids*)*; data on nuclear apparatus and gamete structure absent; mostly free living, benthic forms, rarely parasitic .

Ordovician - Holocene.

Remarks. The continuity of the paleozoic and mesozoic nodosariats, their genetically close relationship, gradual disappearance of the microgranular layer of the wall in this line in the process of geological history was searched and documented by Reitlinger, 1965, Basov V.A., 1982, 1990, Kuznetzova & Basov, V.A., 1974, 1982, Grigyalis, 1978 a, Rauser-Chernousova, 1990, Mikhalevich, 1992, 2000, Karavaeva & Nestell, 2007 and in the works of Vachard and co-authors (Vachard & Krainer, 2001, Vachard et al., 2010) who marked that Nodosariata is the unique group with a hyaline wall represented in the Palaeozoic. Hohenegger (1997) and Groves (Groves, 2000, Groves & Boardman, 1999, Groves & Wahlman, 1997, Groves et al., 2003, 2004, 2005) also considered Paleozoic and Mesosoic Lagenida as monophyletic group.

Nodosariats possess a special type of shell symmetry in which the most general and profound morphological shell structure having taxonomic significance is expressed. The most ancient Paleozoic and later nodosariat forms (as well as their agglutinated and calcareous forms) share the common types of shell symmetry and morphology, although the variability and richness of these types is wider in the later Mesozoic and Cenozoic representatives.

The authorship of the class was reinstated in Karavaeva & Nestell, 2007.

Subclass HORMOSINANA Mikhalevich, 1992

Shells with one of Nodosariate types of chamber arrangement and aperural structure; wall monofontinal, agglutinated or microgranular, may be complicated from the inside, pseudopores unstudied.

Ordovician - Holocene.

Remarks. For the exarticulation of the genera belonging to this subclass from the former Textulariacea the additional study of the apertural character at the earlier studies of shell growth is necessary, especially in the families Mayncinidae and Charentiidae.

The system of this subclass is given here up to subfamily level, in Mikhalevich, 2004 - up to the genera level.

Order *SACCAMMINIDA Lankester, 1885

Shell free, unilocular, of more or less regular form – sphaerical, subsphaerical, oval, slightly elongated, with multiple or more often single aperture which may be on a neck or with an entosolenian tube.

Upper Vendian – Holocene.

Remarks. The order is transferred here from the subclass Astrorhizana (see Mikhalevich *et al.*, 2000) and is regarded as the ancestral group for the lagenids.

Families: Saccamminidae Brady, 1884 [with Saccammininae Brady, 1884 (including genus *Lavella* Nestell et Tolmacheva, 2004 and probably the genus *Niveus* (Altin et al., 2009), Ovammininae Mikhalevich, 2003, Caudammininae Mikhalevich, 2003, Pilulininae Brady, 1884], Thuramminidae A.D. Miklukho-Maklay, 1963].

Order HORMOSINIDA Mikhalevich, 1980

Shell multi-chambered or psedo-multi-chambered, represents rectilinear or somewhat curved row of chambers enlarging in size; wall often ferruginous, may be pierced by openings (not of the canalicular character) or alveolar; aperture terminal, rounded, slit-like and asymmetrically slit-like or cribrate, in one genus (*Nodosinum*) with inner radial ribs, in forms with a ramifying end of the shell – two apertures.

Ordovician – Holocene.

Remarks. The uniserial Dusenburyinidae have an inner tooth similar to that of the Stilostomellidae which after investigation of the shell wall of stilostomellid representatives (see above – Part **XI**.2. 3, Mikhalevich, 2011) ought to be placed in the class Nodosariata. Hence according to the new data its agglutinated isomorphic forms (dusenburyinids) contrary to our previous supposition (Mikhalevich, 1998 b) will stay in the subclass Hormosinana.

Superfamily Hormosinelloidea Rauser-Chernousova et Reytlinger, 1986, stat. elevated

Shell with pseudochambers and simple wall.

Family: Hormosinellidae Rauser-Chernousova et Reytlinger, 1986.

Superfamily Hormosinoidea Haeckel, 1894
Shell with true chambers and a simple wall.

Families: Hormosinidae Haeckel, 1894 [(with Hormosininae Haeckel, 1894 (including *Reophanus* – as having true chambers and *Posadia* Guisberti et Coccioni, 2003), Nodosininae Saidova, 1981 (without *Cribratinoides* lacking inner ribs), Polychasmininae Kaminski, 2004)], Ginesinidae Mikhalevich, 2003, (stat. elevated)., Aschemocellidae Vyalov, 1966 (with the new type genus *Arthrodendron* Ulrich, 1904 according to Kaminski et al., 2008), Reophacidae Cushman, 1910 ((with Reophacinae Cushman, 1910 (without *Adelungia*), Bireophacinae Mikhalevich, 2003)), *Oxinoxisidae Vyalov, 1968, *Glaucoamminidae Saidova, 1981 (transferred from the former Textulariacea by Mikhalevich (1999)), Dusenburyinidae Loeblich et Tappan, 1984, Cuneatidae Loeblich et Tappan, 1984, *Earlandinitidae Loeblich et Tappan, 1984, Kunklerinidae Rauser et Reytlinger, 1986 (including *Leptohalysis* from the Reophacinae), Ammolagenidae Kaminski et al., 2009.

Superfamily Cribratinoidea Loeblich & Tappan, 1984
Shell with complicated wall structure – may have canaliculi (as reported for *Thomasinella* (Loeblich&Tappan, 1987*))* or may be alveolar or labyrintic.
Remarks. Type genus of the family Thomasinellidae needs to be restudied

Families: Cribratinidae Loeblich et Tappan, 1984 (*Pseudotriplasia* is transferred from Pavonitininae as having a uniserial shell), Thomasinellidae Loeblich et Tappan, 1984.

Order AMMOMARGINULINIDA Mikhalevich, 2002
Shell multi-chambered, often compressed, may be triangular or quadrangular in transverse section, planispiral, at least initially, later usually elongated, uncoiled into uniserial or rarely biserial part (Orectostominidae), chambers usually narrow, often arched, sutures oblique, more often curved; wall agglutinated; aperture terminal, tending to displacement to the peripheral end, rounded or more often ovate or slit-like, may be cribrate.
Jurassic – Holocene.
Remarks. The forms transferred from the Lituolida differ from planispiral uncoiled lituolid forms in their shell and chamber pattern (usually flat, with elongated oblique or arcuate chambers, sometimes with limbate sutures) and in the pattern and position of the aperture (often elongated and displaced to the peripheral margin), analogous in their features and appearance to some higher secreted Nodosariata (some Vaginulinidae).

Superfamily *Ammomarginulinoidea Podobina, 1978
Shell with a simple wall.
Remarks. Genus *Sepetibaella* is transferred to Bykoviellinae from the Jadammininae as having a characteristic aperture near the peripheral angle.

Families: Ammomarginulinidae Podobina, 1978 ((with Ammomarginulininae Podobina, 1978 (*?Adelungia* (transferred from the Reophacidae looking as having arisen from a coiled forms), Flabellammininae Podobina, 1978 and ?Ammoastutinae Loeblich & Tappan, 1984)),

Bykoviellidae Loeblich & Tappan, 1984 [(reinstated from Vialovinae in which it was placed in synonymy (Loeblich&Tappan, 1987) and given stat. nov.) (with Bykoviellinae Loeblich & Tappan, 1984, ?Ponceammininae Seiglie, 1991 in Seiglie et al., 1991 and Agardhellinae Mikhalevich, 2003,], Orectostominidae Mikhalevich, 2003 (type genus *Orectostomina* Seiglie, 1965 is transferred from the Spiroplectamminidae), Lacroixinidae Saidova, 1981 [with Lacroixininae Saidova, 1981 (reinstated here from the sinonymy of Pseudobolivinidae in Loeblich &Tappan, 1987) and Monotaleinae Brönnimann, Whittaker et Zaninetti, 1992].

Superfamily *Flabellamminopsidoidea Mikhalevich, 2003

Shell with a complicated alveolar wall.

Remarks. Subfamily Amijiellinae transferred from the Hauraniidae Septfontaine, 1988, diagnosis changed, (with genus *Coprolithina,* from Pernerinidae); family Spiropsammiidae transferred from the Pavonitinacea.

Families: Flabellamminopsidae Mikhalevich, 2002 (with Flabellamminopsinae Mikhalevich, 2003 and ?Amijiellinae Septfontaine, 1988), *Spiropsammiidae Loeblich et Tappan, 1961 (from the Pavonitinacea).

Order *NOURIIDA Mikhalevich, 1980

Shell high trochospiral with the chambers in polymorphinoid arrangement or initial part may be triserial or biserial later changing into uniserial, chambers elongated or subshpaerical; aperture terminal, simple, rounded or slitlike.

Upper Triassic – Holocene.

Remarks. The aperture of the initial chambers needs to be specified in many genera of the order. The work of Kaminski et al., 1995, in which the shell and apertural organization of the group of shells which were included later into this taxon was studied in details helps much to establish the new taxonomical composition of the order.

Superfamily Nourioidea Chapman & Parr, 1936 (nom. corr. pro Nouriidoidea Mikhalevich, 2004)

Shell with the simple wall and chamber lumen.

Remarks. Family Nouriidae transferred from the former Textulariacea, the majority of the genera and families are transferred from Verneuilinidae, Valvulinidae, Prolixoplectidae. Family Conotrochamminidae is transferred from Verneuilinidae after EM study of the shell wall and shell and apertural structure at the different stages of its ontogenesis beginning from the initial stages (see above – Part XI, 2, 3). Some subfamilies are placed here questionably, as for instance Spiroplectinatinae whose young individuals obviously also have terminal aperture (judging from the photographs in Tyszka, 2001).

Families: *Nouriidae Chapman et Parr, 1936), Reophacellidae Mikhalevich et Kaminski 2003 in Mikhalevich, 2003 [with Reophacellinae Mikhalevich et Kaminski 2003 in Mikhalevich, 2003 (the character of the inner structure of *Falsogaudryinella* and *Uvigerinammina* Kaminski et al., 1995 entering this subfamily was shown by Kaminski et al., 1995)], Pseudoreophaxinae Mikhalevich et Kaminski, 2003 in Mikhalevich, 2003 and ?Cylindroclavulininae Mikhalevich, 2003 (character of the inner "tooth" needs further study)], Barbourinellidae Saidova, 1981 (with ?Spiroplectinatinae Cushman, 1928 and Barbourinellinae Saidova, 1981), Conotrochamminidae Saidova, 1981.

Superfamily *Liebuselloidea Saidova, 1981, stat. n.

Shell with a complicated wall or subdivided chamber lumen.

Remarks. Genera *Kaeveria* (transferred from the Pernerinidae) and *Alveovalvulinella* from the Textulariellidae are transferred to Liebusellidae questionably, but genera *Jarvisella, Matanzia, Remesella* and *Ruakituria* are excluded from it as having a basal aperture at least initially.

Families: Liebusellidae Saidova, 1981, Marieitidae Loeblich & Tappan, 1986.

Order *PSEUDOPALMULIDA Mikhalevich, 1992

Shell biserial, at least initially, may possess characteristic V-shaped chambers (Pseudopalmulidae); wall microgranular or agglutinated; aperture terminal.

Devonian.

Remarks. Family Semitextulariidae stands apart from the orther representatives of the order and needs further study, its position is questionable. That's why in spite of its more complex inner structure it is not separated here into a separate taxon. Agglutinated material in the shell wall was noted.

Families: Pseudopalmulidae E.V.Bykova, 1959, ? Semitextulariidae Pokorny, 1956.

Incertae Sedis: family Agglutisolenidae Mikhalevich 2003 (type genus *Agglutisolena* Senowbari-Daryan, 1984, transferred from the Pernerinidae, unclear genus with quite unusual structure).

Remarks. If the shell is uniserial throughout, the family belongs to the Hormosinida, otherwise it belongs to the Nouriida.

Subclass NODOSARIANA Mikhalevich, 1992

Shells with one of the Nodosariate types of shell and chamber arrangement and apertural structure; wall monofontinal (monolamellar), cacareous, hyaline-radial, with pseudopores having very narrow canaliculi (up to 0.33 mμ in diameter) with the pseudopore sieve-plate at the base, outer sculpture of the wall well developed (ribs, spines, tuberculi).

Devonian - Holocene.

Remarks. Shell types and especially apertural structures much more developed than in Hormosinana, in one case (*Delosina*) canal system of a special type is developed. The microgranular layer in the shell wall of some genera is most likely the result of recrystallization. In the most widely adapted systems (Loeblich, Tappan, 1987 and some others) many of the forms of the similar shell structure which obviously are closely related are nevertheless placed in the remote taxa based on the peculiarities of their shell wall (for instance, Robuloididae with the one-layered wall are separated from Lenticulinidae with the "double" wall). But this feature does not work well in the taxonomy of nodosariats. Thus the forms with the both types of the wall were included into the family Ichthyolariidae since these both types occur even in the species belonging to one genus. The layers of growth so strongly developed in nodosariats may obscure the character and the origin of the wall layers (which sometimes are called "lamina"). A true bilamellar wall and POM (Primary Organic Membrane) have not yet been discovered and hence we could not consider nodosariate wall as being of bilamellar (bifontinal) type. The monolamellar character of the stilostomellid's wall was mentioned above (Part **XI**. 2. 3).

In spite of some new excellent and detailed last decades works on the ancient Paleozoic (Bartenstein, 1994, Altiner, 1999, Groves, 1999, Groves et al., 2005, 2007, Pronina, 1989, Pronina & Vuks, 1995, Hohenegger, 1997, Pronina & Nestell, 1997, 2001, Nestell & Tolmacheva, 2004, Nestell & Nestell, 2006, Nestell et al., 2009, 2011, Karavaeva & Nestell, 2007, Pinard & Mamet, 1998, Vachard, 1994, Vachard & Krainer, 2001a, Vachard et al., 2001, 2002, Vachard & Cozar, 2010, Vachard et al., 2010, Songzhu et al., 2007 and others) and later, extinct or living nodosarian taxa (Hilterman & Koch, 1957, Ohmert, 1969, Groenlund & Hansen, 1976, Hohenegger, J., 1974, 1978, 1981, Kohl, 1985, Revets, 1989, Clark, 1991, 1995, Grygyalis, 1977, 1995, Herrero, 1995, Herrero &Haynes, 1997, Hayward, 2002, Neagu & Cirnaru, 2001 a, Hayward & Kawagata, 2005, Hayward et al., 2006, 2007, 2010, Popescu, 1983, Popescu &Crihan, 2004 a and others) the taxonomy of subclass still needs the thorough revision proper to the new data and present-day level of investigation just in the way as it was recently done for several nodosarians groups in the monograph of Hayward et al., 2012.

Order LAGENIDA Delage et Herouard, 1896

Shell unilocular; shell wall may be pierced by thinnest tubuli or may be double-layered (not bilamellar); aperture multiple, or more often single, circular (simple or radial, may be on the elongated neck, with a rim or collar) or fissure-like (may be asymmetrical, with a hook), both the circular and fissure-like aperures may have entosolenian tube, circular aperture sometimes with an ampulla (apertural chamberlet).

Upper Silurian - Jurassic – Holocene.

Remarks. Silurian *Eolagena* have outer microgranular layer of the shell wall which, most likely, is recrystallized. The multiple aperture of *Cribrolagena* is considered as primarily multiple.

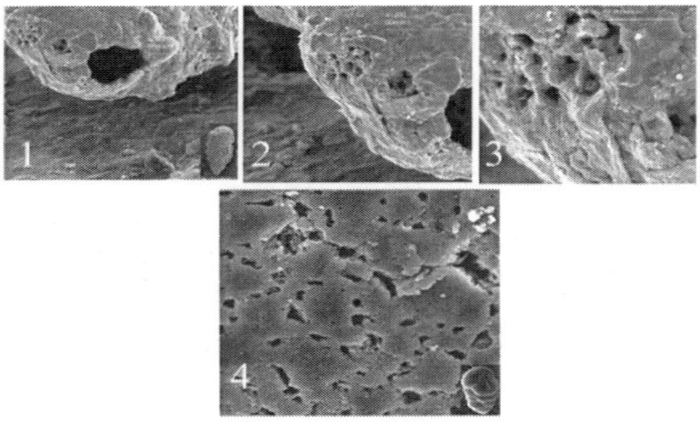

Text Figure 15. Outer surface of the wall of some agglutinated and microgranular species.
1, 2, 3 - wall of *Norvanganina (Textilina) pseudorugosa (Lacroix) fistulosa* Mikhalevich, 1973, "SRT – R- 9086" , 1965 y., station 118, West African Coast, 70 – 72 m, note superficial layer of flat sand particles, and area at the left side with the destroyed upper layer uncovering the pseudopore openings (1 – x 700, 2 x 2400, 3 – 3400), the bigger dark irregulary-oval opening – additional aperture on the peripheral margin of the last chamber, 4 – wall surface of *Karreriella bradyi* (Cushman), 1911, "Ob", station 358 (x 48 000).

The study of this unilocular group has received increased attention in the last decades (Poignant, 1984, Jones, 1984, Knight, 1986, Patterson, 1988, 1990, Patterson & Richardson, 1988, Clark, 1991, Malumian et al., 1991, Boltovskoy &Watanabe, 1993, Popescu & Crihan, 2004 b, Revets, 2005, Karavaeva &Nestell, 2007 and others) revealing in EM variable delicate structures of their aperture and their wall. The discovery by Popescu &Crihan, 2004 b of the smaller hardly visible opening at the side opposite to the aperture gives in principle the new understanding of the group since such forms represent in physiological sense a tube rather than an ampoule. Jones's (1984) opinion of the polyphyletic origin of the group deserves full attention (excluding Eouvigerinid line). In spite of this intensive studies and description of a number of new genera the thorough EM study of the wider representatives of the group is still necessary which inevitably would result in a general taxonomic revision of the group.

Families: Lagenidae Reuss, 1862 [with Lageninae Reuss, 1862 (including *Globulospinella* Patterson, 1988), Cribrolageninae Mikhalevich, 1993], Ellipsolagenidae A.Silvestri, 1923 (with Ellipsolageninae A. Silvestri, 1923, Oolininae Loeblich et Tappan, 1961, Parafissurininae R.W.Jones, 1984), Sipholagenidae Patterson and Richardson, 1987 (including *Nanosilvanella* Patterson, 1990), Asipholagenidae Reddy et Rajshekhar, 2007 (according to the original diagnosis having as in Sipholagenidae a double wall but in distinction of the latter lacking an entosolenian tube).

Order NODOSARIIDA Calkins, 1926

Shell multi-chambered or pseudo-multi-chambered, with chambers in uniserial arrangement. Silurian – Holocene.

Suborder SYZRANIINA Mikhalevich, 1992

Shells consisting of proloculus and a second undivided tubular chamber or pseudo-multi-chambered (pseudochambers are formed by the constrictions of the wall), of irregular form or straight, wall microgarnular entirely or with a radial layer.

Silurian – Devonian - Carboniferous – Permian, Holocene.

Superfamily Caligelloidea Reytlinger, 1959

Shell of irregular form, consisting of proloculus and a second undivided tubular chamber or pseudo-multi-chambered (pseudochambers are formed by the constrictions of the wall), may be attached.

Family: Caligellidae Reytlinger, 1959 (including *Ademassa* Vachard in Vachard et al., 1993).

Superfamily Syzranioidea Vachard, 1981

Shell free, of straight regular form, consisting of proloculus and a second undivided tubular chamber or pseudo-multi-chambered (pseudochambers are formed by the constrictions of the wall), in the transverse section circular or subquadrate with somewhat depressed lateral sides; in some ancient genera the inner microgranular layer of the shell wall is present; aperture circular, simple.

Silurian – Devonian - Carboniferous – Permian, Holocene.

Families: Syzraniidae Vachard, 1981 [including the recent genus *Grigelis* Mikhalevich, 1981 and *Amphoratheca* Mamet et Pinard, 1992], Earlandiidae Cummings, 1955 (including *Lobatiquinella* Vachard, 1994 and *Reitlingerlandia* Vachard, 1994 and *Elevenella* Vachard, 1994), Eogeinitzinidae Vachard, 1994, Paratikhinellidae Loeblich et Tappan, 1984 (with Paratikhinellinae Loeblich et Tappan, 1984, Saccorhininae Mikhalevich, 1995, Juferevellinae Vdovenko et al., 1993).

Superfamily Eonodosarioidea Rauser, 1993 in Vdovenko et al., 1993

Shell may have secondary radially going septae.

Family: Eonodosariidae Rauser, 1993 in Vdovenko et al., 1993.

Suborder NODOSARIINA Calkins, 1926

Shell multi-chambered, sometimes with embracing chambers, in the transverse section circular, oval, flattened, triangular or subquadrate, attached forms may be branching or spreading (Ptychocladiidae); shell wall in the recent forms radial only, in the ancient forms may have microgranular layer; aperture circular (simple, radial or with one peg-like or T-shaped tooth sometimes surrounded by delicate denticles, may have a prominent slanted tooth projecting over the aperture or trematophore), fissure-like or multiple, attached branching forms with no distinct aperture.

Upper Devonian - Triassic – Holocene.

Remarks. The apertural characters are given here a broader exposition on the basis of the new ultrastructural data of Hayward et al. (Hayward et al., 2012, a monograph kindly sent by the authors). Their observations on the reach material (over 20 000 specimens) revealed a picture of variable apertural structures of different nodosariids which, being extremely delicate, may not be preserved in the usual case A new, detailed systematics of this group coming from this monograph and including new taxa widens our knowledge of nodosariins.Practically the whole large taxonomic group is restudied at the new electronic level. The authors had also shown that many of the genera considered previously as living are in fact extinct in spite of their shells being preserved in the superficial sediment (Hayward et al., 2010) (see above – Part XI, 2, 3).

The representatives that have a biserial or triserial shell, even if this occurs only in the initial shell part, are here excluded from the suborder. While in Textulariana (class Rotaliata) their biserial shells originate from the planispiral ones, in Nodosariana they may derive from both planispiral (Palmulidae) and high-spiral forms (Plectofrondiculariidae, Partisaniidae), thus they fall into their respective orders in accord with their origin. Forms with v-shaped (palmate) chambers occur in the both of these lineages, as well as among the uniserial forms.

Superfamily Ptychocladioidea Elias, 1950

Shell attached, branching or spreading, with no distinct aperture.

Family: Ptychocladiidae Elias, 1950.

Superfamily Nodosarioidea Ehrenberg, 1838

Shell free, of regular straight form or slightly curved, in some cases chambers may be embracing, with undivided chamber lumen; aperture simple circular or slit-like opening or radiate.

Remarks. Family Rimulinidae of R.W. Jones is here reinstated from the synonymy of Lingulininae, the latter subfamily (with the Spirolingulininae as its junior synonym) is transferred to the order Vaginulinida at the base of the presence of early planispirally coiled part of the shell, at least in one of generations (see below).

Families: Nodosariidae Ehrenberg, 1838 (including *Dalongella* Songzhu, Quinglai et Weihong, 2007, *Rectoglandulina* Loeblich et Tappan, 1955 reinstated by Karavaeva & G. Nestell, 2007 who followed K. Miklukho-Maklay, 1969 – cited after Karavaeva & G. Nestell, 2007, ?*Cribrella* Gawor-biedowa, 1987 – assigned by the author of the genus to this family, but inner structure not investigated, comparison with externally similar *Cribropleurostomella* not given), ?Dentalinidae Schwager, 1877, Rimulinidae R.W. Jones, 1984 (stat. elevated, reinstated from the synonymy of Lingulininae); Frondiculariidae Reuss, 1860 [belonging of the genus *Acostinella* of Sarkar et Gupta, 2008 to Reussellidae Cushman, 1933 where it was initially placed by the authors of the new genus is doubtful, its radial aperture looks more like nodosariat radial aperture, the fully uniserial hyaline shell with a small pores is also characteristic of the Nodosariana representatives, the genus is closer in its structure to nodosariid *Tristix* and together with the genus *Tribrachia* and the genus *Acostina* (also uniserial entirely and earlier assigned to Reussellidae) it probably ought to be united in a separate subfamily Tristixinae subfam. nov. differing from the Frondiculariinae in its shell form, which is triangular (or quadrangular) rather than strongly flattened, with chambers triangular or quadrangular and with sutures only slightly arched at the center (rather than chambers being extremely narrow and of prominent V-shaped as being characteristic of Frondiculariinae), aperture may be simple (*Tribrachia*), radiate (*Tristix*) or crybrate, deriving from the radiate one (*Acostina*). *Acostinella* probably is synonymic with *Tristix* but a collation of the material is necessary for the final conclusion]. , *Nodosinellidae Rhumber, 1895, Geinitzinidae Bozorgnia, 1973 (including *Nodosinelloides* Mamet et Pinard, 1992, *Vervilleina* Groves in Groves & Boardman, 1999 and the genus *Omoloniella* Karavaeva et Pronina, 2007, *Vervilleina* assigned by the author of the genus to this family as primitively septate form, Vachard & Krainer (2001) placed it in into Syzraniidae – the difference of opinion is connected with the evaluation of the character of the septation), Pachyphloiidae Loeblich et Tappan, 1984, Ichthyolariidae Loeblich et Tappan, 1986 (with *Dagysina* Yadrenkin, 2008, the genus close to *Frondiculinita*), Daucinoididae Mikhalevich, 1993, Protonodosariidae Mamet et Pinard, 1992 (including *Polarisella* Mamet et Pinard, 1992), Glandulonodosariidae Silvestri, 1901 (with *Neugeborina* Popescu in Cicha and others, 1998), Chrysalogoniidae Mikhalevich, 1993 (with 4 new genera in Hayward et al., 2012: *Anastomosa* Hayward, 2012, *Cribroconica* Hayward and Kawagata, 2012, *Lotostomoides* Hayward and Kawagata, 2012, *Scallopostoma* Hayward and Kawagata, 2012).

Superfamily Stilostomelloidea Finlay, 1947

The superfamily differs from the suborder Nodosariina in its more complex apertural structure: aperture with a tooth, of peg-like or T-shaped form sometimes surrounded by delicate denticles, may have a prominent slanted tooth projecting over the aperture or trematophore.

Remarks. Extremely delicate apertural structures are easily broken or dissolved making the process of identification difficult (Hayward et al., 2012).

Family: Stilostomellidae Finlay, 1947 (with the new genera *Carchariostomoides* Hayward and Kawagata, 2012, *Caveastomella* Hayward and Kawagata, 2012, *Toddostomella* Hayward, 2012, *Toddostomella* Hayward, 2012).

Superfamily *Colanielloidea Fursenko, 1959

Shell free, of regular straight form, with the chamber lumen subdivided by the vertical radial inner septa.

Family: Colaniellidae Fursenko, 1959 (with Colaniellinae Fursenko, 1959, Multiseptidinae Rauser, 1993 in Vdovenko et al., 1993).

Order *PALAEOTEXTULARIIDA Hohenegger et Piller, 1975

Shell multi-chambered, straight, biserial, at least in the early part, later may be uniserial; wall micrогranular; aperture simple or often multiple.

Lower Carboniferous – Upper Permian.

Remarks. Rauser-Chernousova et al., 1996 consider Koskinobigenerininae Loeblich et Tappan, 1984 to be a synonym of Palaeotextulariidae, Hance et al., 2011 translated Koskinobigereninae Loeblich et Tappan, 1984 as Koskinotextulariidae. The group Palaeotextulariida unites variable forms and in the strict sense if it were named Paleonodosariida it would have fitted in with its subject and volume better. The belonging of this group to nodosariids is also confirmed with the fact of the presence in its representatives (e.g. *Palaeobigenerina*, Palaeotextulariidae) of the hyaline wall inside the chambers and also in the chamber connections (Piller, 1990).

Superfamily Palaeotextularioidea Galloway, 1933

Chamber lumen not subdivided.

Families: Palaeotextulariidae Galloway 1933 (including *Consobrinella* Mamet et Pinard, 1992).

Superfamily Louisettitioidea Loeblich et Tappan, 1984

Chamber lumen subdivided, secondary partitions perpendicular to the septum produce peripheral chamberletes.

Family: Louisettitidae Loeblich et Tappan, 1984.

Order *BISERIAMMINIDA Mikhalevich, 1981

Shell multi-chambered, biserial, two rows of chambers enrolled in involute spire with the alternating chambers of the last coil visible from the outside; wall microgranular, aperture single.

Lower Carboniferous – Upper Permian.

Remarks. The microgranular wall of the representatives of this order is regarded to-day as recrystallized while their aperture, albeit described as "interiomarginal," may be considered as "interiomarginal" only formally since it is wider and does obviously represent a terminal aperture changing its position in coiled forms; therefore this group is here replaced into the group of ancient nodosarians, *cf.* its previous position in Mikhalevich, 2000. Their valvular projections in the apertural area resemble more the inner tooth (chomata) of endothyrids than the rotaliats inner plates. The evolutionary line of *Globivalvulina* development in Upper Permian was shown in Altiner, 1999. Cozar & Sommerville, 2012 included into biseriamminids (regarded by them as having superfamily level) two families only: agglutinated Biseriamminidae and Globivalvulinidae with secreted wall. They placed genera *Admiranda, Dzhamansorina* and *Ulanbella* described by Marfenkova, 1991 in the synonymy of *Koktjubina* Marfenkova, 1991.

Although the genus *Sengoerina* was originally placed in Dagmaritidae, it is positioned here in Globivalvulinidae following Nestell & Nestell, 2006. This assignment is based on the presence of a well developed coiled part and of the valvular projection clearly seen in Nestell & Nestell, 2006 (Plate 6, Figures 3, 5, 6). The interpretations of the position of Dagmaritinae differ: Rauser-Chernousova et al., 1996, placed it in Palaeotextulariidae while Altiner (1999), into Biseiammnidae and Hance et al., 2011, into Globivalvulinidae. The taxonomy of the orders Palaeotextulariida and Biseriamminida needs further taxonomic elaboration. Their shell morphology, however, as well as the fact that a radial layer may be identified in their shell clearly distinguishes them from Fusulinoida, in which they were previously placed.

Families: Biseriamminidae N. Chernysheva, 1941 (including the genus *Parabiseriella* published by Cozar et Somerville, 2012); Globivalvulinidae Reitlinger, 1950 [with Globivalvulininae Reitlinger, 1950, including *Koktjubina* Marfenkova, 1991, *Charliella* Altiner et Ozkan-Altiner, 2001), *Labioglobivalvulina* Gaillot et Vachard, 2007, *Retroseptellina* Gaillot et Vachard, 2007, *Septoglobivalvulina* Lin, Li et Sun, 1990, *Verispira* Palmieri, 1988), Paraglobivalvulininae Gaillot et Vachard, 2007, Dagmaritinae Bozorgnia, 1973 (including *Sengoerina* Altiner, 1999, *Danielita* Altiner et Ozkan-Altiner, 2010) and Paradagmaritinae Gaillot et Vachard, 2007 (all cited after Hance et al., 2011)].

Order POLYMORPHINIDA Mikhalevich, 1980

Shell multi-chambered, high-spiral (of a special polymorphinoid type) or derived from it (may be biserial, with the final uniserial part or even entirely secondarily uniserial); chambers as a rule elongated; wall hyaline-radial, rarely microgranular wall described; aperture circular (simple or radial) or fissure-like, in both cases may have an entosolenian tube, canal system absent. Permian – Holocene.

Remarks. In Russian literature this polymorphinoid type of trochoid spire is called "spiral-screw" type to distinguish it from the high conical forms of the other types, and I believe it would well fit in with the conventional English-language terminology too.

Suborder POLYMORPHININA Mikhalevich, 1980

Shell with a circular, radial or fissure-like aperture, with or without entosolenian tube, may have an additional apertural chamberlet.

Permian, Triassic – Jurassic, Holocene.

Remarks. The position of the family Ramulinidae is questionable. Most of its representatives obviously represent the final stages of the other nodosariat's forms in which irregular or branching fragile terminal parts may be formed. Since we do not know the structure of their initial parts (it would be the most desirable to see them living in culture) their taxonomic position cannot be properly defined. Loeblich & Tappan's (1987) supposition on their affinity to polymorphinids could be one of possible assumptions. The *Discoramulina* with its initial chamber was studied in detail by Harmelin and Venec-Peyre (Harmelin & Venec-Peyre, 1992), their data give basis to assume another position for this genus, closer to lagenids. Its subcircular proloculus (although attached) with hexagonal pattern of ornamentation having the several openings from which the tubular parts begin to grow resembles some lagenids with hexagonal wall pattern and could to some extent be compared also with thuramminins (agglutinated analogues of some lagenids) with their multiple openings and tubular outgrowths, shorter and simpler, not subdivided into chambers but nevertheless going from the subsphaerical chamber in different directions. The presence of the double wall is also known in lagenids (Asipholagenine). The superficial resemblance of this genus with cyclostomatous Bryozoa was also regarded (Harmelin & Venec-Peyre, 1992). Of course, all these suppositions will stay only a speculation without studying their life cycle or without use of molecular methods.

Families: Polymorphinidae d'Orbigny, 1839 [with Polymorphininae d'Orbigny, 1839 (including the new genus of Neagu et Cirnaru, 2001 – *Scythimorphina*), Falsoguttulininae Loeblich et Tappan, 1986 , Webbinellinae Rhumbler, 1904, Edithaellinae Fuchs, 1967], Glandulinidae Reuss, 1860 [(with Glandulininae Reuss, 1860 (including genus *Bombulina* Mikhalevich, 1983 published in Russian, the publication is hardly available in the West), Entolingulininae Saidova, 1981)], Plectofrondiculariidae Cushman, 1927 (with *Plectolingulina* Hayward, 2012 in Hayward et al., 2012), Partisaniidae Loeblich et Tappan, 1984, ?Ramulinidae Brady, 1884.

Suborder *PLEUROSTOMELLINA Mikhalevich, 1993

Shell having the asymmetrical fissure-like aperture with projecting hood on its terminal end and with its lower side obstructed by one or two teeth, entosolenian tubes of the successive chambers joined into one integrative system, its disposition in the uniserial shells reflect the biserial character of their ancestral forms.

Carboniferous – Holocene.

Remarks. Two genera lacking the entosolenian tubes are here excluded from the pleurostomellids. In *Ellipsolingulina* entosolenian tube is short, probably in consequence of the dissolution of the inner chamber walls in the adult specimens. Shell wall of pleurostomellids is studied insufficiently, earlier they were included into the previous Rotaliacea in spite of the fact that in some cases their wall was described as microgranular.

Families: Pleurostomellidae Reuss, 1860 [with Pleurostomellinae Reuss, 1860 (including *Czarkowyella* Gawor-Biedowa, 1987, *Delphinoidella* of Popescu, 1992, *Neopleurostomella* Srinivasan et Rai, 1992, *Obesopleurostomella* Hayward, 2012, *Ossaggittia* Hayward and Kawagata, 2012), Wheelerellinae Petters, 1954, Cribropleurostomellinae Owen, 1971, *Pazdroellinae Gawor-Biedowa, 1987 (with *Pazdroella*, *Triaperturina* and *Quadryaperturina* of Gawor-Biedowa, 1987, the latter genus with unclear structure of the initial part)];

Ellipsoidinidae Silvestry, 1923, [a family placed by Loeblich and Tappan (Loeblich & Tappan, 1987) into the synonymy of Pleurostomellidae, reinstated by Hayward *et al* who included the new genus *Lateroihiatus* (Hayward, 2012)].

Order DELOSINIDA Revets, 1989, stat. elevated

Shell high-spiral (=spiral-screw), three-serial; chambers inflated, elongated, increasing rapidly in size as added; wall monolamellar; apertural opening absent except for a trematophore-like opaque irregularly outlined area situated at the apex of the shell and extending down the nearest suture, pierced with very fine pore-like openings, with cup-shaped formation beneath this area isolated from the rest of the chamber, additional apertures going in series along the sutures; canal system represented by subsutural tubes connected with the additional sutural apertures and trematophore-like structure.

Holocene.

Remarks. In the only genus of the momotypic family (earlier attributed to Rotalicea) there were discovered canals and a monolamellar (monofontinal) shell wall (Revetes, 1989). Shell structure and chamber arrangement is similar to polymorphinids. Trematophore-like structure sometimes may be absent (Revets, 1989).

Family: Delosinidae Parr, 1950.

Order VAGINULINIDA Mikhalevich, 1993

Shell multi-chambered, planispiral, at least at the early stage or in one of the generations (in the asexual one), in the microsphaeric forms the final part of the shell may be enrolled, sometimes the megalosphaeric shell of the sexual generation may be entirely uniserial (Spirolingulinidae, Palmulidae), like in *Lingulina*; coils involute (with the exception of *Turkmenella);* wall hyaline radial; aperture single (radial or fissure-like) or multiple.

Upper Devonian, Permian, Triassic, Jurassic, Cretaceous, Holocene.

Remarks. Type species of the genus *Lingulina* - *Lingulina carinata* d'Orbigny, 1826, was primarily described as having the uniserial shell arrangement in both generations. In one of the generations of this species (in microsphaeric one) initial planispirally coiled part was discovered (Mikhalevich, 1983, p. 167, Figures 260, 261) and hence the species ought to be excluded from Nodosariidae whose representatives are uniserial entirely in both generations. *Spirolingulinina* have a coiled initial part in both generations and thus *Lingulina* that differ from the above by having no coiling in their macrosphaeric form may be preserved as a valid genus. But Loeblich & Tappan's subfamily Spirolingulininae (1986) as described by these authors later ought to be considered as a junior synonym of their subfamily Lingulininae (1961). At this, the remaining genera of the former Lingulininae, whith their uniserial entirely but compressed shell (in distinction from Nodosariina), require a different name to be applied, and a reinstatement of the previous subfamily Rimulininae introduced by R.W. Jones (Jones, 1984) becomes necessary. Rimulininae lacking the early coiled part in any of their generations ought to be preserved within Nodosariids.

The family Nanicellidae (stat. elevated) is placed here according to its shell structure. This position is in full agreement with the opinion of Vachard et al., 2010, transferring *Nanicella* from fusulinoids into Nodosariata. With the placement of this genus among vaginulinids the geologic range of the order becomes significantly wider. Apertural structure of this ancient genus is still unclear. The gap in the occurrence of such planispirally coiled nodosriate forms since Upper Devonian to Permian is striking and needs further attention and explanation. Some additional findings, however, may cover this gap in the future as the geological record is not yet absolutely complete.

Families: Vaginulinidae Reuss, 1860, Lenticulinidae Chapman, Parr, Collins, 1934, Marginulinidae Wedekind, 1937, Robuloididae Reiss, 1963, Palmulidae Saidova, 1981, Lingulinidae Loeblich et Tappan, 1961, stat. elevated (with Spirolingulininae Loeblich et Tappan, 1986 as its junior synonym), Nanicellidae Fursenko, 1959.

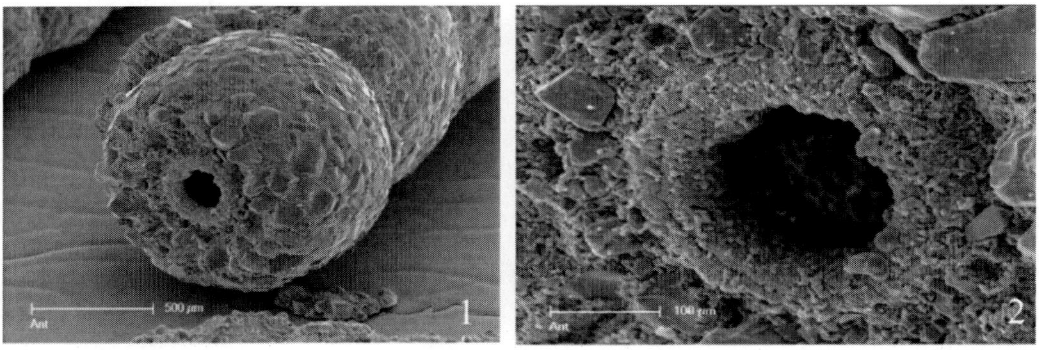

Text Figure 16. Character of the outer shell wall of the chambers and of the apertural area in *Pseudonodosinella margaritaria* (Rhumbler) *margaritaria* Saidova, ("Polarstern", 1996, Station 24), note the finer material around the aperture (1 – x 240, 2 – x 1200).

XI.4.5. CLASS ROTALIATA Mikhalevich, 1980

(= class Globothalamea Pawlowski et al., 2012)

Shell multi-chambered only, all regular types of chamber arrangement well presented (except of the glomerate type which is practically absent), mostly trochospiral, planispiral and, derived from both of them, with biserial, triserial or similar later part (rarely secondarily uniserial), in advanced forms often cyclical; inner complications of chamber lumen are strongly developed: from thickend layers of the wall of different structure (labyrintic, spongy, alveolar - in the subclass Textulariana), pillars, septa and septula (of the first and second order) in higher representatives of the Textulariana and Rotaliana, and also of curved plates, valves, tongues, grooves mostly in higher Rotaliata; in higher representatives chamber number could increase to some hundreds and thousands, and in this case they are often differentiated in their structure (e.g. – embryonic chambers, arcuate median chambers, lateral chambers of varied form in *Orbitoides*, chambers without chamberlets, then chambers with chamberlets and secondary septa in *Heterostegina*); additional skeletal plates above the sutures and umbilicus are often, especially in higher Rotaliata; integrative systems developed strongly (stolons in Textulariana, stolons and very complex canal systems in Rotaliana); wall in Textulariana monofontinal (monolamellar), agglutinated and microgranular, with pseudopores (irregular or as canaliculi), with various complications of the inner wall, which could be cancellate, alveolar, spongy, in Rotaliana – bifontinal (= bilamellar), with true regular pores, having cylindrical pore canaliculi with some transverse sieve-plates besides the basal one, growth interrupted, the contours of every new chamber forming entirely at once, organic and calcareous secretion are divided in time, alternating; aperture usually at the base of the apertural face both in coiled and in some biserial and triserial forms (in the latter could became secondarily terminal), may also be areal and sieve-like, rotaliate aperture may have inner tooth (of another structure and origin than in Miliolata), higher Rotaliata may have various inner tooth-plates and open subcylindrical or closed cylindrical plates, which can be interconnected and can form a single system, additional apertures well developed in both subclasses (sutural, umbilical, peripheral); some groups of Rotaliana gave rise to planctonic forms with morphological adaptations to the pelagic mode of life (subclass Globigerinana); agamonts of the higher Rotaliata multinucleate, heterokaryotic, gametes biflagellate (*Glabratella* has several somatic nuclei and three-flagellate gametes). Mostly free living, benthic forms, rarely attached, a big group (Globigerinana) - planctonic, some parasitic species are known.

Carboniferous - Holocene.

Remarks. Calcareous shells of low-magnesium calcite or, more rarely, aragonite.

Subclass TEXTULARIANA Mikhalevich, 1980

Shell with one of the Rotaliata's types of chamber arrangement and of apertural structure; wall monofontinal, agglutinated or microgranular, may be complicated from the inside (may be cancellate, alveolar, labyrinthic, may have thickenings of the wall, pillars, septa), in many forms pseudopores and canaliculi (sometimes branching) are marked, sculpture of the outer shell surface not developed, but additional skeletal plates in umbilical area could be present; in advanced forms with the great number of chambers stolons as integrative systems are present; inner apertural structures are similar to that of Rotaliana but less developed.

Carboniferous - Holocene.

Remarks. After the investigation of the wall ultrastructure of calcareous stilostomellids (Mikhalevich, 2011) which are isomorphic to dusenburyinds the latter group was excluded from Textulariana and returned to subclass Hormosinana.

The system of this subclass is given here up to subfamily level, in Mikhalevich, 2004 - up to the genera level.

Order NAUTILOCULINIDA Mikhalevich, 2003

Shell planispiral, in one case biserially planispiral, aperture basal (interiomarginal), at least initially, supplementary apertures may present.

Upper Triassic – Holocene.

Remarks. Nautiloculinids were separated from the lituolids (Mikhalevich, 2003, 2004 a, 2005).This order differs from the order Lituolida in the character of its aperture which is really basal (not terminal) at least at the early stages of ontogenetic development or rarely may be interio-areal,; from the order Haplophragmiida Nautiloculinida differs in its planispiral coiling. In some nautiloculinid genera the initial position of the aperture needs to be elucidated.

Superfamily Nautiloculinoidea Loeblich et Tappan, 1985

Shell with a simple wall and chamber interior.

Families: Nautiloculinidae Loeblich et Tappan, 1985, Haplophragmoididae Maync, 1952, ((with Haplophragmoidinae Maync, 1952, (lacking supplementary apertures) (without *Ammosiphonina, Buzasina*)), Debarininae Kaminski, 2004 and Trematophragmoidinae Mikhalevich, 2003).

Superfamily Alveolophragmioidea Stschedrina, 1936

Shell with a complicated wall, more usually alveolar.

Families: Alveolophragmiidae Saidova, 1981 [with Alveolophragmiinae Saidova, 1981 (including *Reticulophragmoides* Gradstein & Kaminski, 1989, without *Popovia* and *Sabellovoluta* having a terminal aperture from the very beginning), Hemicyclammininae Banner, 1966 (without *Flabellamminopsis*)], Discamminidae Mikhalevich, 1980 (without Ammoscalariidae Mikhalevich, 1982 with partially tectinous wall and without *Glaphyrammina*), Cyclamminidae Marie, 1941.

Order SPIROPLECTAMMINIDA Mikhalevich, 1992

Shell elongated, may be widened at the upper part, consists of planispirally coiled initial part and following uncoiled straight biserial part, sometimes changing into a uniserial one; aperture basal, interiomarginal, at least initially, but often in the adult also, single or multiple.

Carboniferous – Holocene.

Remarks. The planispiral part constitutes a significant part of the shell. From the former Spiroplectamminacea only forms with a basal aperture (even if only possessing it initially) are left in this order. Many of the genera need to be restudied concerning the position of their aperture in the initial stage.

Superfamily Spiroplectamminoidea Cushman, 1927

Shell with a simple wall and chamber lumen not subdivided.

Family: Spiroplectamminidae Cushman, 1927 [(with Spiroplectammininae Cushman, 1927 (without *Orectostomina* and questionably including *Cuneospirella* Cherchi, Schroeder et Ruberti, 2009, (the genus needs further study) and *Plectoeratidus* Kaminski, Setoyama et Tyszka, 2009) placed in this subfamily by the authors of the genus), Vulvulininae Saidova, 1981, ?Morulaplectinae Saidova, 1981 (coiling unclear – streptospiral?)].

Superfamily Spirotextularioidea Saidova, 1975

Shell with inner septa or alveolar wall.

Remarks. Family Ecougellidae is transferred here from Loftusiacea as its straight part is biserial.

Families: Spirotextulariidae Saidova, 1975 [with Spirotextulariinae Saidova, 1975, Duquepsammininae Seiglie et Baker, 1987 (*Duquepsammina* Seiglie et Baker, 1987 is very close to *Septigerina*)], Novalesiidae Loeblich et Tappan, 1984, ?Ecougellidae Loeblich et Tappan, 1985, ?Pavonitinidae, Loeblich et Tappan, 1961 (without *Pseudotriplasia*).

Order TEXTULARIIDA Delage et Herouard, 1896

Shell elongated, biserial, more often flattened, may be uniserial at the end, the initial part often having the additional chamber resulting in pseudotriserial appearance; aperture basal, at least initially, later may be terminal, single or multiple, supplementary peripheral apertures may present.

L. Jurassic – Holocene.

Remarks. The forms lacking canaliculi are included here as separate families. Many of the genera need to be restudied concerning the position of their aperture in the initial stage.

Superfamily Textularioidea Ehrenberg, 1938

Shell with a simple wall and undivided chamber lumen.

Families: Pseudobolivinidae Wiesner, 1931 (without *Lacroixina* and *Parvigenerina* with the terminal aperture from the initial stages), Textulariopsidae Loeblich et Tappan, 1982 (without *Plectinella*), Kaminskiidae Neagu, 1999, Textulariidae Ehrenberg, 1938 [with Textulariinae Ehrenberg, 1938, Norvanganinae Mikhalevich, 2003 – shell as in *Textularia*, but with supplementary peripheral apertures at the ends of peripheral outgrowths (type genus *Norvanganina* Mikhalevich, 1981, genus reinstated from the synonymy of the *Textularia* in Loeblich & Tappan, 1987 by Mikhalevich, 2012 on the basis of electron-microscopic reinvestigation of the type species), ?Siphotextulariinae Loeblich et Tappan, 1985 and Planctostomatinae Loeblich et Tappan, 1984], Bigenerinidae Saidova, 1981 [with Bigenerininae Saidova, 1981, subfam. reinstated from the synonymy of Textulariinae in Loeblich & Tappan, 1987 and Cribrobigenerininae Mikhalevich, 2003 (type genus *Cribrobigenerina* Andersen, 1961, from Planctostomatinae), Siphobigenerininae Loeblich et Tappan, 1986 (transferred from the Valvulinidae)], Textularioididae Loeblich et Tappan, 1984 (attached forms).

Superfamily Septotextularioidea Loeblich et Tappan, 1985

Shell with a subdivided chamber lumen.

Families: Septotextulariidae Saidova, 1975, Tawitawiidae Loeblich et Tappan, 1961.

Order VERNEUILINIDA Mikhalevich et Kaminski, 2003 in Mikhalevich, 2003

Shell high trochospiral throughout or only in the initial part, later part may have an increased or decreased number of chambers per whorl or may become uniserial or cyclical; aperture basal at least in the initial part, later may become terminal, single or multiple, some genera with inner apertural structures or supplementary peripheral apertures.

Carboniferous – Holocene.

Remarks. Wall noncanaliculate.

Forms with a terminal aperture from the very beginning of growth and some structural features of the Nodosariata type are transferred into the subclass Hormosinana. In the type species of the family Plectoverneuilinellidae Cetean & Kaminski, 2011 microsphaeric generation initially biserial, becoming trochospiral with 4 chambers per whorl, only in later ontogenetic development, only macrosphaeric generation has proloculus followed by high trochospiral chamber arrangement (Cetean & Kaminski, 2011).

Superfamily Verneuilinoidea Cushman, 1911

Shell with a simple wall and undivided chamber lumen.

Remarks. Wall noncanaliculate.

Families: Verneuilinidae Cushman, 1911 [(without Spiroplectinatinae and Barbourinellinae), with Verneuilinoidinae Suleymanov, 1973 (*Belorussiella* is transferred from the Spiroplectinatinae which are characterized by the terminal aperture), without *Duotaxis* transferred to the Duotaxidae Mikhalevich, 2004, without *Eggerellina, Vialovella* – transferred to the Globotextulariidae, and without *Falsogaudryinella, Paleogaudryina, Paragaudryina, Pseudoreophax, Reophacella, Talimuella, Uvigerinammina* transferred to the subclass Hormosinana) and Verneuilininae Cushman, 1911 (including *Heterostomella* transferred from Barbourinellinae), (without *Latentoverneuilina*)], Prolixoplectidae Loeblich et Tappan, 1985 [with Prolixoplectinae Loeblich et Tappan, 1985 (without *Karrerulina,* including *Neaguammina* Kaminski, Holbourn et Geroch, 1997, *Eobigenerina* Cetean, Setoyama, Kaminski, Neagu, Bubík, Filipescu et Tyszka, 2008, *Rectogerochammina* Kaminski, Cetean et Neagu, 2010 and *Rectoprotomarssonella* Kaminski, Bubík et Cetean, 2008 and some other genera published after 1987 (see Mikhalevich, 2004), Gerochellinae Kaminski, 2004 and Caroniinae Brönnimann, Whittaker et Zaninetti, 1992], Piallinidae Rettori et Zaninetti, 1993, Globotextulariidae Cushman, 1927 (with *Globotextularia, Cribroturretoides, Gravellina, Rhumblerella, Tetrataxiella, Verneuilinulla* and *Eggerellina* and *Vialovella* from Verneuilinidae and *Hagenowella* and *Sabulina* from Ataxophragmiinae), Tritaxidae Plotnikova, 1979, Plectoverneuilinellidae Cetean et Kaminski, 2011.

Superfamily Crenaverneuilinoidea Mikhalevich, 2004

Shell as in Verneuilinoidea but with a subdivided chamber lumen.

Remarks. Wall noncanaliculate.

Type genus *Crenaverneuilina* of the family Crenaverneuilinidae and genus *Hagenowina* (both are transferred from the Pernerinidae), genera *Jarvisella, Matanzia, Remesella, Ruakituria* - from the Liebusellidae, *Alveovalvulina* and *Hagenowinoides* - from the Textulariellidae, - all the genera transferred on the basis of their morphology. The family differs from the Pernerinidae in the character of the coiling (not Ataxophragmium-like) and chamber direction not inclined to the axis of coiling, from the Textulariellidae - in its less regular conical form, inflated chambers and depressed sutures.

Family: Crenaverneuilinidae Mikhalevich, 2003

Superfamily Valvulinoidea Berthelin, 1880

Shell with a simple wall and undivided chamber lumen.

Remarks. Wall canaliculate.

Only forms with an external valvular tooth (flap) are left in the family Valvulinidae. Genus *Colomita* is transferred into the subfamily Colominellinae from Septotextulariinae by Popescu, 2000. In the subfamily Vacuovalvulininae type genus *Vacuovalvulina* Hofker, 1966 is transferred from the Chrysalidinidae, which differs from Vacuovalvulininae in the presence of pillars, ?*Pityusina* transferred from the Ataxophragmiinae, although noncanaliculate but morphologically close to *Vacuovalvulina,* genus *Redmondoides* Banner, Simmons et Whittaker, 1991 - from the Paravalvulininae Banner, Simmons et Whittaker, 1991. Family

Clavulinidae Balakhmatova, 1973 is reinstated from the Valvulinidae in Loeblich & Tappan, 1987, it differs from latter in having an internal tooth and from the other families of the superfamily in the presence of an interconnected system of the inner teeth of the successive chambers running from the aperture to the foramen. Families Eggerellidae and Valvulinidae of the order Eggerellida of Neagu, 2003 are placed here in this superfamily (the third family Valvulamminidae included by Neagu (2003) in this order is placed in the present work in the superfamily Trochamminoidea).

Families: Valvulinidae Berthelin, 1880 (with Valvulininae Berthelin, 1880, Cribrobulimininae Mikhalevich, 2003 and Vacuovalvulininae Mikhalevich, 2003), Eggerellidae Cushman, 1937 [with Dorothiinae Balakhmatova, 1972, Minouxinae Loeblich et Tappan, 1986, Eggerellinae Cushman, 1937,*Biedowiinae nom. n. pro Varsoviellinae Gawor-Biedowa, 1987 with type genus *Biedowia* nom. n. pro *Varsoviella* Gawor-Biedowa, 1987 (named in memory of Gawor-Biedova, the author of the foraminiferal genus *Varsoviella* and her significant contribution to the Foraminiferal taxonomy) (non *Varsoviella* Gieysztor et Wiszniewski, 1947 (Plathyhelmintes) and Colominellinae Popescu, 1998 (in Chicha et al., 1998, p. 71), with the genus *Cubanina* transferred from Ataxophragmiids by Kaminski &Cetean, 2011 on the basis of its canaliculated shell wall], Pseudogaudryinidae Loeblich et Tappan, 1985 (with Pseudogaudryininae Loeblich et Tappan, 1985 and Siphoniferoidinae Loeblich et Tappan, 1985), Clavulinidae Balakhmatova, 1973 (with Clavulininae Balakhmatova, 1973, Goesellinae Mikhalevich, 2003).

Superfamily ?Chrysalidininoidea d"Orbigny, 1839

Shell with the chamber lumen subdivided by partitions or pillars.

Remarks. Wall canaliculate.

Some of the genera show features resembling ataxophragmiids, the question whether it is a result of evolutionary convergence or is based on their relationships needs further study.

Subfamily Chrysalidininae Neagu, 1968 with *Praechrysalidina* from the Ataxophragmiinae, without *Parurgonia* included by Septfontaine (1988) into Parurgoninidae Septfontaine, 1988 and probably includes *?Accordiella* from the Pfenderinidae. Subfamily Chrysalidininae regarded here without the genus *Pseudomarssonella* Redmond, 1965 which was transferred to Paleopfenderinidae by Tasli & Altiner (2010) on the basis of its shell structure. These authors showed that *Pseudomarssonella*, together with *Redmondoides* and *Riyadhella* is phylogenetically and taxonomically connected with pfenderenids.

The new family Olgiidae (type genus *Olgia* Mikhalevich, 2011 with the type species *Clavulina pacifica* Cushman, 1924) was separated from the family Clavulinidae as having chambers with the short septa going from the chamber floor and partially subdividing the chamber lumen, and additional sutural apertures, its apertural apparatus is similar to that of *Clavulinidae* (tooth plates of the adjacent chambers interconnected) (Mikhalevich, 2011).

Families: Chrysalidinidae Neagu, 1968 [with Chrysalidininae Neagu, 1968, ?Pseudodictyopsellinae Septfontaine & De Matos, 1998 with type genus *Pseudodictyopsella* of Septfontaine & De Matos, 1998, Tritaxilininae Loeblich et Tappan, 1986, Paravalvulininae Banner, Simmons & Whittaker, 1991 (without *Pseudomarssonella* and *Riyadhoides*], Olgiidae Mikhalevich, 2011.

Incertae Sedis (to the whole order): *Siphovalvulina* Septfontaine, 1988.

Order TROCHAMMINIDA Saidova, 1981

Shell low trochospiral, often temporarily attached, usually with convex spiral and concave or flat umbilical side, only few genera with convex umbilical and flat spiral side, tendency to uncoil purely expressed, umbilical ends of the chambers often of a special configuration (lobed, curved, elongated); aperture basal, at least initially, at the base of the apertural face more often at the umbilical side, more rarely at the peripheral side, later may be areal, single or multiple, supplementary apertures (umbilical, sutural) well developed, or umbilicus may be opened; additional skeletal plates occur.

Upper Carboniferous – Holocene.

Remarks. Wall noncanaliculate. The group needs to be revised, the apertural position at the early stages of individual development should to be investigated, especially in the forms possessing areal aperture.

Superfamily Trochamminoidea Schwager, 1877

Shell with a simple wall and chamber lumen not subdivided.

Remarks. Genus L*epidodeuterammina* is excluded from the subfamily Polystomammininae, and is transferred into Rotaliamminidae. In Jadammininae the type genus *Jadammina* Bartenstein, 1938 has been placed in the synonymy of *Entzia* Daday, 1884 based on the revision of the type species by Filipescu & Kaminski (2011), *Jadammina* Bartenstein et Brand, 1938 is a junior synonym of *Entzia* Daday, 1883 as was shown by the studies at the morphological and molecular level (Holzmann et al., 2012) but according to the rules of Zoological Record the family name should not be changed.

Families: Trochamminidae Schwager, 1877 [with the subfamilies Trochammininae Schwager, 1877, Trochamminellinae Bronnimann, Zaninetti et Whittaker, 1983 (subfamily reinstated by Bronnimann & Whittaker, 1987, who also placed *Atlantiella* in synonymy of *Pseudotrochammina*), Toretammininae Brönnimann, 1986, Zavadovskininae Brönnimann et Whittaker, 1988, Arenoparrellinae Saidova, 1981, Jadammininae Saidova, 1981, Polystomammininae Bronnimann et Beurlen, 1977, Ammogloborotaloidinae Kaminski et Contreras, 2011 (with the type genus *Ammogloborotaloides* Kaminski et Contreras, 2011 in Kaminski & Contreras, 2011)], Vialovidae Suleymanov, 1983, Rotaliamminidae Saidova, 1981, Adercotrymidae Brönnimann et Whittaker, 1988, Ammosphaeroidinidae Cushman, 1927, Valvulamminidae Loeblich et Tappan, 1986 (including *Histerammina* Neagu, 2000).

Superfamily Remaneicioidea Loeblich et Tappan, 1964

Shell with the chamber lumen subdivided by septula or by the infoldings of the wall.

Remarks. Family Dictyopsellidae is transferred here from the former Ataxophragmiacea, genus *Abuhammadina* Abd-Elsahfy et Ibrahim, 1990 placed by the authors into the Choffatellinae, but having a distinctly trochoid shell is transferred here to Dictyopsellidae.

Families: Remaneicidae Loeblich et Tappan, 1964, Carterinidae (with Carterininae Loeblich et Tappan, 1955 and Zaninettinae Bronnimann et Whittaker, 1983), Asterotrochamminidae Bronnimann, Zaninetti et Whittaker, 1983, Dictyopsellidae Bronnimann, Zaninetti et Whittaker, 1983.

Subclass ROTALIANA Mikhalevich, 1980

Shells with one of the Rotaliata's types of chamber arrangement and apertural structure; wall lime secreted (calcareous), low magnesial (Bowser et Pawlowski, 2006), bifontinal (=bilamellar), with true pores and several pore sieve plates in cylindrical pore canaliculi, outer sculpture of the shell wall well developed, many of the genera with the additional skeletal plates in the umbilical and apertural areas and above the sutures; inner structures of high complexity: besides additional chamber septa there are also valves, grooves, tongues, tooth-plates in the apertural area, sometimes connected into a single system, more complex integrative systems of the higher representatives - stolons, interlocular spaces and well developed canal systems present in advanced forms.

Triassic - Holocene.

Remarks. System of many of the taxa entering the subclass and of the subclass itself needs to be revised based on the study of the outer and inner shell structure, apertural and umbilical systems which are not studied yet in many of its genera, or some other morphological features. Even the ultrastructure of the shell wall needs to be studied for many of their representatives as for the majority of them the presence of a bilamellar shell wall has not been shown conclusively but was extrapolated (thus it was implied that the shell wall of stilostommelids placed among rotalians was bilamellar). Orders Robertinida, Planorbulinida, Glabratellida, superfamilies Gavellinelloidea, Siphoninidoidea need to be revised most of all, some genera also may be found out polyphyletic. Methods of molecular studies would also be useful, although some phylogenetic analysis of Rotaliids gives conflicting results (Ertan et al., 2004, Schweizer et al., 2008). Some lines such as Nummulitids or Bolivinids were shown to be well supported in the molecular studies results (Bowser et al., 2006) although from the morphological point of view these taxa also were not questionable. Thorough comparative morphological analysis of the taxa entering this subclass is also desirable. All the main characters of the shell and apertural shell structures which reflect the plan of the structure of the organism should be considered in such analysis, not the apertural features only as is sometimes understood (Schweizer et al., 2008) as this feature represents only one of the elements of the whole structure, although one of a high functional importance. Such revision has been carried out most recently for their separate groups by Haynes, 1981, Whittaker, 1988, Hayward & Brasier, 1980, Hayward et al., 1997, Barbieri, 1997, Hottinger, 2000, 2001, 2006, Hottinger et al., 1991, Levy et al., 1979, 1986, Hansen & Revets, 1992, Zakrevskaya, 1993, Revets, 1990, 1993 a, b, c, 1994, 1996 a, b, 2001a, b (not seen), 2002 a, b, c, 2005, Rajeshwara Rao & Periakali, 2001, Aguilar et al., 2002, Neagu & Cirnaru, 2002, Tyszka, 2006, Boukhary, 1994, 1996, Boukhary & Scheibner, 2009, Boudagher-Fadel, 2010, Boudagher-Fadel et al., 2000, Hayward, 1990, Hayward et al., 2012, Hohenegger et al., 2000 and by some other authors, but as such work has not been completed it is not yet possible to give the detailed new system of this subclass. Revets (1990, 1993 a, b, c, 1994, 1996 a, b, 2001a, b , 2002 a, b, c, 2005) in his revision devoted a high attention to the apertural structures. He wrote in his article on the bolivinitids revision (Revets, 1996 a): "A rationalisation of the classification is proposed based on overall morphology complemented by toothplate type". In Boix et al., 2009 article four new genera described (*Pyrenerotalia, Iberorotalia, Orbitokathina* and *Calcarinella*) are assigned to rotaliids formally without defining their exact taxonomic position obviously because the accurate comparison with nearest taxa is nearly impossible without the detailed morphological studies of all the closely related forms.

The system given below outlines the main groups (mostly following Loeblich & Tappan, 1987, 1992 classification with some changes, it also differs from the latter in the understanding of the rank of the included taxa), and considers as far as possible the new data available at present literature. In the discussion on the taxonomic value of the tooth plates (Swchweizer et al., 2008, p. 243) it ought to be taken into account that ability of the Rotaliata representatives to form complex tooth plates and other complex inner apertural structures is the taxonomic feature of class (and subclass) rank, such structures are absent in any other foraminiferal class (Mikhalevich, 1981, 1992, 2000, Mikhalevich & Debenay, 2001). Using of this characteristic as diagnostic indications for the lower rank taxa entering Rotaliana (for instance for Buliminids (see Loeblich & Tappan, 1987, and also Haynes, 1981, p. 215, who also mentioned "internal toothplate" in the diagnosis of Buliminida) needs to consider the more minute details of the tooth plates structure, along with the other shell characters, as well as the fact that the toothplates may be formed only in the most developed representatives of the phyletic line, not in all of them.

The major changes made here in the taxonomy of this subclass concern the exclusion of the forms belonging to the subclasses Nodosariana and Spirillinana. The stilostomellids are here excluded from the former Rotaliacea and placed into subclass Nodosariana at the base of its shell structure and shell wall ultrastructure (see above – Part XI, 2, 3) (Pleurostomellids were transferred into Nodosariana earlier (Mikhalevich, 1993), patellinids and chapmaninids (together with their related forms are also excluded from Rotaliacea and transferred into the subclass Spirillinana (see above – Part XI, 4, 2) (Mikhalevich, 1981, 1992, 1999) on the basis of their shell structure. The former inclusion of both these groups (nodosarians and spirillinans) into Rotaliana (the previous Rotaliacea, then - Rotaliina) resulted in the polyphyletic nature of the former rotalians. The possible heterogenous character of some orders or families entering Rotaliana would have another, lower level of polyphyly within the subclass Rotaliana (though it may be that some separate genera belonging to Spirillinata are still awaiting a correct definition). The separation of the rotaliat and spirillinat forms represents in some cases significant difficulties because of existence of convergent forms and some similar tendencies of development in both classes such as trochoid multi-chambered shells or emergence of canal system, even the aperture in some advanced multi-chambered trochoid spirillinat shells may have sometimes the basal position at the umbilical side with only slight differences of that of rotaliat umbilical aperture.

As in the super-multi-chambered forms the embryonal chambers are developed, only super-multi-chambered character of the shell is mentioned in the diagnoses given below.

The new investigations had showed in some cases the wider geologic range of the taxa than that ones known previously (e.g. - extension of the fossil record of the Rosalinidae into the Lower Cretaceous in Revets, 2002). The phylogeny of many of the phyletic branches within Rotaliana is still obscure.

Superorder ?ROBERTINOIDA Mikhalevich, 1980

Shell trochospiral and high trochospiral or nearly planispiral, chambers may be subdivided with partitions of a variable position and direction in different taxa, which may partially subdivide the chamber lumen or form fully isolated chamberlets, the latter more often are formed in the umbilical area; wall mostly aragonite, obviously monolamellar; aperture on the septal surface of the chamber, usually at the base of it, often loop-shaped, may be areal and also may be at the peripheral edges of the chambers, additional sutural apertures may be present in some genera; the primary basic aperture of the preceding chambers is overgrown with the shell matter fully or partially, in the first case the intercameral foramen is later formed as a result of resorbtion and is not homologous to the previous aperture, in the second case foramen is homologous to only a part of the previous aperture.

Middle Triassic – Holocene.

Remarks. The dextral coiling and aragonite shell wall are marked for the majority of the genera. This group with very complex inner structure actually needs further morphological and taxonomical study. Along to the outer similarity of the shell of the representatives of this superorder with the shells of the other representatives of the subclass Rotaliana they have profound distinctions in their inner structure. In spite of the rarely mentioned lamellar character of the wall (*Ceratobulimina, Alliatinella*) the bilamellarity of their shell wall has not yet been demonstrated. All this (along with the aragonite character of their shell wall) makes the position of this superorder among Rotaliana extremely questionable. Starting from the comparative analysis of the morphology of all the foraminiferal large taxa the only other hypothetical position of this group seems to be possible in the class Spirillinata whose calcareous representatives often have aragonite shell wall, trochoid or planispiral shell, in some of their advanced multi-chambered genera the aperture takes the umbilical position. This challenge supposition needs to assume that some features of shell similarity arose in Robertinoida and in some other Rotaliana superorders in parallel. I dare to suppose that some other genera included presently into subclass Rotaliana and having a shell structure unusual for this subclass may belong to Spirillinana. Some similarities between Spirillinana and Rotaliana prompt careful study and may possibly find their close roots in the far past. The border between robertinoids and discorboids also is often unclear. Thus Haynes (1981) wrote that Triassic and Jurassic forms attributed earlier to *Discorbis* should be transferred to Robertinida (=Robertinoida in this study). Further studies of the inner shell structures and ultrustructure of the shell wall of such forms are necessary. The phylogenetic relationship within the Robertinoida also needs further elucidation (Hart, 1984).

Order ROBERTINIDA Mikhalevich, 1980
(= Robertinida Haynes, 1981)
As for the superorder.
Low Cretaceous – Holocene.

Suborder DUOSTOMININA subord. nov.

Shell planispiral to trochospiral and high trochospiral; wall nonlamellar and may be with some foreign particles; aperture single or double, at the base of apertural face.

Middle Triassic – Jurassic.

Families: Duostominidae Brotzen, 1963 (including *Cassianopapillaria* Di Bari et Rettori, 1998), Asymmetrinidae Brotzen, 1963, Oberhauserellidae Fuchs, 1970.

Suborder CERATOBULIMININA Mikhalevich, 1980

Shell low trochospiral, later stage may be uncoiled; aperture going along the base of the septal surface with loop outgoing from it upwards (Ceratobuliminidae), arched opening (Reinholdellidae) either at the peripheral margin or close to it (Epistominidae, Epistominoididae), aperture of the preceding chambers is closed, intercameral foramen (deuteroforamen) is produced by resorbtion above the internal partition subdividing the chambers; partitions may be short and limiting only a part of the chamber or longer and fully closing the separate spaces which could be seen from the outside as a double row of chamberlets or a star-like structure around umbilicus.

Jurassic - Holocene.

Remarks. In the wall of *Ceratobulimina* organic lining is marked going between lamellae, but typical Primary Organic Membranne of the bilamellar wall was not described. The classification within this suborder mostly follows Grigyalis, 1978 b.

Families: Ceratobuliminidae Cushman, 1927, Epistominidae Wedekind, 1937 (with Epistomininae Wedekind, 1937 (obviously including *Carlsiella* Vespermann, 1998, article not available) and Garantellinae Grigyalis, 1977), Reinholdellidae Seiglie et Bermudez, 1965, Epistominoididae Saidova, 1981.

Suborder ASTERIGERININA d'Orbigny, 1839

Shell trochospiral entirely (in the forms with a few chambers) or at the early stage only (in the forms with a large number of chambers and chamberlets); chamber lumen subdivided by secondary septae which may or may not reach the opposite side wall, in the latter case chamberlets are not formed, when septae attach the opposite wall the additional chamberlets are formed; in the multi-chambered representatives stolons are present; canal system absent.

Lower Cretaceous – Holocene.

Remarks. Suborder is placed in robertinoids following Fursenko & Rauser-Chernousova, 1959 and Subbotina et al., 1981. Suborder has much in common with Ceratobuliminina, but has some distinct features: aperture with the addition of the new chambers is closed only partially, and its unclosed part serves as intercameral foramen, in multi-chambered forms stolons are present.

Superfamily Asterigerinoidea d'Orbigny, 1839

Shell trochospiral, biconvex or flattened, usually with one of the sides being involute, more rarely both sides are involute (*Palmerinella*, *Amphistegina*), chamber number is not relatively large, umbilicus in almost all the genera is closed, umbilical ends of the chambers usually simple, without lobes or incurvings, chamber lumen subdivided by secondary septae which may not reach the opposite side wall and form chamberlets (Eponidellinae), or attaching the shell wall in the umbilical area may form secondary chamberlets, these often form star-like structures on the umbilical side ((Alfredinidae, Asterigerinatidae, Asterigerinidae, Amphisteginidae - the latter family possessing interseptal pillars also); stolons and canals absent; aperture at the base of the septal surface of the last chamber, equatorial to extra-umbilical-umbilical, may often end in the fold of septa near the peripheral margin (*Nuttalides*), may go up off the base of this surface and at that be divided longitudinally by the vertical subdivision (*Palmerinella* - Epistomariidae), or may be areal or sieve like, in some representatives (Eponidellinae of Epistomariidae) aperture becomes partially closed with the addition of the new chambers, its remaining part serving as intercameral foramen, additional apertures may be sutural (*Epistomaroides*, *Monspeliensina*), or their position may be connected with the additional inner septae paralleling the peripheral margin ((*Epistomaria*, *Pseudoeponides*).

Remarks. Without Lepidocyclinidae. The ultrastructural studies of amphisteginids by Hottinger (1993) had shown that "The lumen of the main chamber communicates with the lumen of the next instar's stellar chamberlet through the supplementary areal apertures'. He also showed internal pore-pits structures covering the lateral wall surfaces of the stellar and the main chamber cavities including their retral lobes going in parallel to the peripheral margin which (according to Leutenegger, 1984) house diatom symbionts giving the shell its olive-greenish colour.

Families: Asterigerinidae d'Orbigny, 1839, Epistomariidae Hofker, 1954 (with Epistomariinae Hofker, 1954, Eponidellinae Sieglie et Bermudez, 1965, Nuttalidinae Saidova, 1981, Palmerinellinae Loeblich et Tappan, 1984), Asterigerinatidae Reiss, 1963, Amphisteginidae Reiss, 1963 (including *Sanctus* Boukhary, Abd-Elshafi et Mattar, 2011), Alfredinidae S.N.Singh et Kalia, 1972, Actinosiphonidae Adams, 1987.

Superfamily Clypeorbioidea Sigal, 1952

Shell asymmetrical, usually trochoid only at the early stage or in one of generations, later coils may be arranged in one plane, may be embracing, forming the super-multi-chambered shell flattened or sometimes nearly spherical, chambers following the large embryonal chambers are small, multiple; system of stolons may be present.

Remarks. Family Clypeorbidae is transferred from Lepidorbitoididae (Orbitoidina) as being asymmetrical and having early trochoid part.

Families:*Clypeorbidae Sigal, 1952, Boreloididae Reiss, 1963.

Suborder CONORBOIDINA Mikhalevich, 1992

Shell low to high trochospiral (at least initially or in one of the generations), at the later stages with a reduced number of chambers per whorl up to two and one; aperture in trochospiral forms at the umbilical side, extending along the base of the septal margin, in elongated forms – terminal, a prominent tooth plate of hemicylindrycal or pillar-like form projects inward from the aperture as a columella-like structure changing its orientation from chamber to chamber even in uniserial representatives thus reflecting the ancestral coiled state.

Low Cretaceous – Upper Cretaceous.

Family: Conorboididae Thalmann, 1952.

Suborder ROBERTININA Mikhalevich, 1980

Shell trochoid and strongly flattened, with transition to planispiral (Alliatinidae) or high trochoid asymmetrical (Robertinidae); chambers subdivided with the double transverse partition formed by infolding of the wall, and resulting in formation of small additional chamberlets composing the distinct series; aperture going along the base of the septal surface with loop outgoing from it upwards, sometimes aperture areal, septal intercameral foramen is homologous to only a part of the previous aperture, additional apertures sometimes present.

Paleocene – Holocene.

Families: Robertinidae Reuss, 1850, Alliatinidae McGowran, 1966 ((with Alliatininae McGowran, 1966 and Cushmanellinae subfam. nov., type genus *Cushmanella* Palmer et

Bermudez, 1936, (including the genus *Sidebottomina* Seiglie, 1964) – shell nearly planispiral, oval in outline or elongated, biserial (presumably secondarily biserial as the initial part of the shell is slightly asymmetrically curved), with umbilical or peripheral additional chamberlets, main aperture areal or an arch at the base of the terminal face, additional sutural apertures present in both genera of the subfamily distinguishing the new subfamily from the other representatives of the family).

Superorder DISCORBOIDA Ehrenberg, 1838

Shell trochospiral, shell wall calcitic, foramen is homologous to the previous aperture.

Remarks. It seems that this superorder is one which needs taxonomic revision and reclassification even more than all the other foraminiferal groups. As was shown in a set of works (Levy et al., 1979, 1986, Whittaker, 1988, Hansen & Revets, 1992, Revets, 1996, 2001, 2002 a, b, c) even the borders between such widely adapted genera as *Gavelinopsis, Conorbina, Rosalina, Discorbis* are unclear and some genera were transferred by Revets (2002 a, b, c) from discorbids to rosalinids although with significant hesitation (*Conorbina, Albertinopsis*), *Gavelinopsis* was reclassified from Rosalinidae to Ammoniinae (Revets, 2002 a, b, c). Thorough study of the apertural, umbilical and inner shell structures at the recent level of investigation (such as exact position of the aperture, periapertural lip, umbilical plug, umbilical folia, coverplates) was made for a few genera only. The situation is complicated by the fact of the different structure of coverplates whose lamellarity differs in different taxa (Hansen & Revets, 1992). This gave Revets (2002 c)a reason to) to assume the parallel rise of such coverplates in the evolutionary development of the group. Future changes in the generic constitution of the subordinate taxa and reinvestigation of type species of the genera (among which *Rosalina* would be a first priority) will allow the construction of a proper classification. Below is given only an attempt to outline the groups on the base of the Loeblich & Tappan (1987) classification as one covering the widest number of taxa considering where possible later studies on the inner shell structures (or their absence) of the rosalinids, discorbids and rotaliids in the abovementioned works and basing the separation of the orders on the shell structure, the degree of complication of the inner chamber lumen and apertural characters. Many of the groups are still heterogenous. Nevertheless the evolutionary development from simpler forms with undivided chamber lumen to rise of partial partitions, aperture divided into two parts by a valve or foraminal plates, than formation of additional chamberlets, stolons and canals could be researched in some cases more clearly. Hansen & Revets, 1992 in their detailed completed work offer new terminology for such terms as alar projection (folium = "umbilical projection of the lower chamber wall"), paries proximus ("chamber wall coming down at the foraminal side of the chamber at an umbilical position"), chamberlet – "umbilical part of the chamber that is sealed off from the main lumen" (p. 178).They also introduced term "gutter-like structure". I nevertheless cannot accept the uniting of *Discorbis* with *Rotalia* in one family and with *Ammonia* in one subfamily. Although they represent a single wide phyletic line but are of different levels of development of their inner shell structures (from simple without the inner structures (*Rosalina*) to some inner structures (*Discorbis*) up to the higher degree of development of such structures – the presence of canal system having its own walls fully isolating the canal cytoplasm from the rest of the cytoplasm inside the shell). Such developmental stages clearly expressed in the morphological features ought to be reflected in taxonomic constructions. Folia, umbilical coverplates, systems of inner plates could be regarded as integrative systems compared with the apertural integrative systems of plates, for instance, in Buliminids, playing the same role and differentiating the cytoplasmic flows only partially in contrast to canal system. The member taxa of the higher level taxa would be possible to understand after the study of these inner and umbilical structures in the wider representatives. The systems of discorboids as well as that of the whole Rotaliana offered below take only some steps in this direction as the necessary information concerning the whole volume of the member forms is not as yet available.

Order ROSALINIDA Delage et Herouard, 1896

Shell low trochospiral, usually with the convex spiral side and the umbilical side only slightly convex, flat or depressed, in rare cases the shell may be biconcave (Bronnimanniidae) or close to subsphaerical – with the embracing whorls or strongly embracing last chamber (Sphaeroidinoidea), the spire may be enrolled in some genera; umbilicus may be opened or closed, umbilical ends of the chambers may have a widened valve (flap, folium) (some Bagginidae), often resulting in sinuate sutures (Rosalinidae, Discorbinellidae); inner chamber lumen simple, undivided; aperture may be at the base of the septal face ((interiomarginal (on the umbilical side or going from the peripheral margin to the umbilicus), equatorial, it also may be going from the base of the septal face upwards as a high fissure aligned in parallel to the peripheral margin and lying in some depression of the septal surface)), may be on the septal face (areal, sieve-like), sometimes may be terminal (as a rule secondarily terminal), additional apertures more often at the umbilical side (at the umbilical edge of the chambers or in the sutures), more rarely on the spiral side.

Lower Cretaceous – Holocene.

Remarks. Order includes forms lacking internal structure.

Superfamily Rosalinoidea Reiss, 1963

Shell low trochospiral, usually with convex spiral side and the umbilical side only slightly convex, flat or depressed, rarely shell biconvex (in Bronnimanniidae, in *Rectoeponides* with the enrolled final part), spiral side evolute, umbilical side involute, umbilicus may be opened or closed, umbilical end of the chambers may have a widened valve (flap, folium) (some Bagginidae), often resulting in sinuate sutures (Rosalinidae); aperture at the base of the apertural face - interiomarginal (on the umbilical side it may reach the peripheral margin, may band valve-like umbilical end of the chamber and enter the suture between the last and penultimate chambers), rarely may be sieve-like (some Eponididae, in the representatives of this family aperture may also be covered by a perforated plate), additional apertures more often at the umbilical side (in the umbilical area or sutural), more rarely on the spiral side (Heleninidae).

Remarks. Family Ungulatellidae is transferred to the Hergotelloidea (Spirillinana), Pegidiidae and Bueningiidae are removed to the Sphaeroidinoidea (Rosalinida) according to their shell structure. Family Conorbinidae Reiss, 1963 is placed into synonymy of Rosalinidae according to Revets, 2002 b. Family Neoeponididae Hauser et &Keller Grunig., 1993 was proposed by the authors of the family, although the formal taxonomic name was not given but the detailed description and differentiation of the genus *Neoeponides* from the *Eponides* was thoroughly given along with the statement that it deserves to be separated in the new family (Hauser et al., 1993). According to Hansen & Revets, 1992 the superfamily ought to be named Eponidoidea, as the family Eponididae was described earlier. But as the name

Rosalinoidea (= Rosalinacea) is widely in use, as the composition of the superfamily is not yet finally settled and the other member families (as, for example Bagginidae) were described even earlier, the change of the name is here left for the future so as not to have too many nomenclature changes.

The composition of Cancrisidae is here given according to Revets, 1996 (with Gyroidinoidinae Saidova, 1981 and Valvulineriidae Brotzen, 1942 as its synonyms).

Families: Bagginidae Cushman, 1927 (with Bagginiinae Cushman, 1927, Serovaininae Sliter, 1968), Cancrisidae Chapman, Parr et Collins, 1934 (with Gyroidinoidinae Saidova, 1981 and Valvulineriidae Brotzen, 1942 placed into synonymy of Cancrisidae by Haynes 1981 and Revets, 1996), Eponididae Hofker, 1951 (with Eponidinae Hofker, 1951, Sestronophorinae Saidova, 1981, Rectoeponidinae Saidova, 1981), Neoeponididae Hauser & Keller Grunig, 1993, ?Heleninidae Loeblich et Tappan, 1987, Mississippinidae Saidova, 1981 (with Mississippininae Saidova, 1981, Stomatorbininae Saidova, 1981), Rosalinidae Reiss, 1963 [with Conorbinidae Reiss, 1963 as synonym according to Revets, 1996, *Conorbina* Brotzen, 1936, including new genus *Albertinopsis* of Revets, 2002 a, which he assigned to this family, although tentatively, without *Gavelinopsis* Hofker, 1951 (according Revets, 2002 c), with *Hyrrokkin* Cedhagen, 1994 and without *Tretomphalus* ((which either should not receive systematic recognition being only a life stage of *Rosalina* (according Hansen & Revets, 1992) or may be a junior synonym of *Cymbaloporetta* or *Tretomphaloides*, both transferred here, although tentatively, to Cymbaloporidae], Bronnimanniidae Loeblich et Tappan, 1984.

Superfamily Discorbinelloidea Sigal, 1952

Superfamily differs from Rosalinoidea in its apertural structure which is equatorial, areal or going from the base of the apertural face up as a high fissure aligned in parallel to the peripheral margin and lying in some depression (niche) of the septal surface, additional openings beneath the posterior umbilical margin may be present (Discorbinellidae).

Remarks. Subfamily Torresininae of the family Discorbinellidae is transferred here to the order Discorbida as having subdivided chambers. Belonging of the genera *Carlfranklinoides, Discorbitina* and *Colonomilesia* to the family Discorbinellidae is doubtful. Family Parrelloididae Hofker, 1956 is placed by Revets, 1996 into the synonymy of Anomalinidae Cushman, 1927 and is transferred here from the Discorbinelloidea following Revets, 1996.

Families: Discorbinellidae Sigal, 1952, Pseudoparrellidae Voloshinova, 1952 (with Pseudoparrellinae Voloshinova, 1952, Concavellinae Saidova, 1981, ?Stetsoniinae Saidova, 1981), Planulinoididae Saidova, 1981, Svratkinidae Bugrova, 1989.

Superfamily Gavelinelloidea Hofker, 1956, stat. elevated.

Shell trochoid, usually with flattened or only slightly convex spiral side and rather high umbilical side, rarely uncoiling at the later stage (Karreriidae); chambers not subdivided; aperture at the base of the apertural face – equatorial interiomarginal, may extend to the umbilicus usually boarded by a lip or flap, (or may go up the septal face), may be areal and multiple, rarely may have additional (umbilical) apertures.

Remarks. With Gavelinellidae, Heterolepidae, Karreriidae, Trichohyalidae, Oridorsalidae, from the former Chilostomellacea (with now Heterolepidae as synonym of Anomalinidae and Oridorsalidae as synonym of Anomalinidae (part) and Alabaminidae according to Revets, 1996). Family Karreriidae have much in common with Cibicididae and possibly ought to be transferred to Planorbulinina where the family Cibicididae is placed according to Loeblich &Tappan, 1992. However, the planorbulinid group is obviously polyphyletic and is one which mostly needs to be revised. Part of the genera attributed earlier to the Anomalinidae and Cancrisidae are included by Revets, 1996 (p. 72) into Gavelinellidae after thorough morphological revision.

Loeblich & Tappan (1987) after Hansen & Roegl, 1980 a (sited after Revets, 1994) put *Anomalina* as a synonym of *Epistomaroides* and Anomalinidae into the synonymy of Alfrediidae (although the latter has additional chambers in umbilical region while anomalinids shell structure is simple).

With the genus *Anomalina* preserved (and it was not considered invalid by ICZN - see Revets, 1994, 1996) and after Revets's designation of the new type species *A. ariminensis* instead of the lost *A. punctulata* for this genus the family Anomalinidae Cushman, 1927 and superfamily Anomalinacea ought also to be reinstated according Revets (1994, 1996). If after Revets's request the ICZG would set aside *A. punctulata* and reinstate *A. ariminensis* as a new type species for the genus of the superfamily Anomalinacea (Anomalinoidea – according to the ICZN rules for animal groups) should be preserved and obviously Gavelinelloidea will be its later synonym. The genus *Coronatoplanulina* Ujiie, 1990 with additional apertures on the both sides of the shell obviously may be included into Oridorsalidae, although in the future revision of this superfamily it may change its position.

But because the relations between the closely related and other planorbulinid, gavelinellid and discorbinellid genera are uncertain at present and quantity of information concerning their different taxa is inadequate, because many of the discorboid taxa are heterogenous and the whole big group needs critical revision most of the classificatory changes are left here for the future. Revet's authoritative revision based on thorough SEM studies and considering the geological ranges covers only part of the entering taxa, and even he needs to put some genera as Incertae Sedis (Revets, 1996). Among these Incertae Sedis group he placed *Gyroidina*, the majority of the species attributed earlier to this genus he considers as belonging to *Hansenisca* Loeblich et Tappan, 1987.

Families: Gavelinellidae Hofker, 1956 [(with Gavelinellinae Hofker, 1956(without *Gyroidina* according to Revets, 1996 and with *Pilleussella* J. Nikitina in Bugrova, 2001, *Asianella* Bugrova, 1990 and *Primanomalina* Bugrova, 1990), Brotzenellinae Bugrova, 1990, Pseudogavelinellinae Bugrova, 1990, without Gyroidinoidinae Saidova, 1981 placed by Haynes 1981 and Revets, 1996 into the Cancrisidae)], ?Anomalinidae Cushman, 1927 [(with Parrelloididae Hofker, 1956, Heterolepidae Gonzales-Donoso, 1969 and Oridorsalidae Loeblich et Tappan, 1984 as synonyms of Anomalinidae (after Revets, 1996, at that he obviously means only part of the Oridorsalidae as the genus *Oridorsalis* was included by Revets (1996) into Alabaminidae)], Lingulogavelinellidae Scheibnerova, 1972 (fam. reinstated by Bugrova, 1990) (with Lingulogavelinellinae Scheibnerova, 1972, Orithostellinae Bugrova, 1990),?Karreriidae Saidova, 1981 (including *Viveja* Bugrova, 1990), Trichohyalidae Saidova, 1981.

Superfamily Sphaeroidinoidea Cushman, 1927

Shell trochoid, with strongly inflated chambers, may have flattened umbilical side or nearly subsphaerical, with the embracing last whorls or embracing last chamber; aperture as an opening in center of the umbilical area (Bueningiidae), as a fissure above the base of the septal face, may have a flap-like projection (resembling a tooth), rarely aperture as a series of tubes piercing the umbilical area (Pegidiidae), additional sutural apertures on the spiral side may be present.

Remarks. This heterogenous superfamily includes peculiar forms whose phylogenetic relations with the rest of the rosalinids are not clear. In Pegidiidae the earlier chambers are resorbed during shell growth. In Tremachoridae (Miocene) the bilamellar wall was not discovered (Revets, 1991). These facts confirm the supposition that some of these forms may belong to the subclass Spirillinana.

Families: Sphaeroidinidae Cushman, 1927 (including *Eusphaeroidinina* Ujiie, 1990), ?Pegidiidae Heron-Allen et Earland, 1928, Tremachoridae Lipps et Lipps, 1967, Bueningiidae Saidova, 1981.

Superfamily ? Siphoninidoidea Cushman, 1927

Shell trochospiral entirely, with closed umbilicus, or trochoid only at the early stage, later may become nearly planispiral, streptospiral, erected (biserial or uniserial); aperture may be terminal (with hyaline lip) even at the earlier stages.

Remarks. The terminal position of the aperture beginning from the very early chambers in the type genus of the family is a rare exception among Rotaliana, ultrastructure of its shell wall proving its bilamellar character was not demonstrated. All these facts give a basis for doubts as to their belonging to Rotaliana and the supposition of their close relation to Spirillinana. The further study of the shell wall ultrastructure is necessary.

Family: Siphoninidae Cushman, 1927 (with Siphonininae Cushman, 1927, Siphonidinae Saidova, 1981, Siphoninoidinae Loeblich et Tappan, 1984).

Order DISCORBIDA Ehrenberg, 1838

Shell having internal structures.

Eocene – Holocene.

Superfamily Discorbinoidea Ehrenberg, 1838

Shell similar in its structure to that of Rosalinidae, but with the chamber lumen subdivided, in the simpler case by the inward projections of the shell wall on the spiral side (Torresinidae), in more advanced genera (Discorbidae): an internal imperforate proximal wall (paries proximus) extends within the chambers close to the umbilical area and adheres to the septum peripherally, expanding as a lobe toward the chamber base where it may or may not attach to the chamber floor, separating the foramen into two parts, thus in the umbilical area beneath the widened valves of the chambers funnel-like widening spaces appear, which form the continual integrative structure communicating with the umbilical area but being separated from the remaining part of the chamber, a secondary opening at the posterior margin of the umbilical valve (flap) leading into the chamberlet and remaining open as new chambers are added.

Remarks. The complex inner structure of discorbids results in a profound differentiation of the cytoplasmic flows representing a high level of the shell organization. Torresinids with more primitive chamber subdivision than the rest of the discorbinoids probably deserve a separate taxonomic position, they were transferred from the Discorbinellidae whose representatives have a simple, undivided chamber lumen.

Families: Discorbidae Ehrenberg, 1838 (including the genus *Acarotrochus* described by Kawagata, Yamasaki et Jordan in 2005), Torresinidae Loeblich et Tappan, 1984.

Superfamily Alabaminoidea Hofker, 1951

Shell similar to that of the representatives of Gavellinelloidea but with more complex inner structure: aperture with murus reflectus in Globorotalitidae and Osangulariidae, with infolding of the apertural face in Alabaminidae, with integrative system of inner apertural structures in Coleitidae, or chamber lumen with internal partitions (Quadrimorphinidae).

Remarks. All the families and the superfamily itself are transferred from the former Chilostomellacea as having trochospiral shell.

Families: Globorotalitidae Loeblich et Tappan, 1984 (became extinct in Cretaceous), Osangulariidae Loeblich et Tappan, 1964 (Revets, 1996 placed this family into synonymy of Alabaminidae), Alabaminidae Hofker, 1951 (with Oridorsalidae Loeblich et Tappan, 1984 placed by Revets, 1996 as synonym of this family (with the type genus *Oridorsalis*) and partially of the family Anomalinidae – see above), Coleitidae Loeblich et Tappan, 1984, ?Quadrimorphinidae Saidova, 1981.

Order GLABRATELLIDA Mikhalevich, 1994

Shell trochospiral, often high trochospiral (spiral-screw), more rarely flattened and only slightly convex, umbilical side flattened, depressed or slightly convex, umbilicus closed or widely opened , may have small denticles (Rotaliellidae); wall of umbilical side having a sculpture looking like radial striae, costae, rows of pustules or nodes disposed along the cytoplasmic flows and obviously facilitating joining of the two parent gamont shells into syzygy during the plastogamic process; inner chamber lumen without subdivision or paries proximus; aperture at the base of septal face of the last chamber, more often in the form of arch or loop on the umbilical side, going to the umbilicus, more rarely fissure like and disposed closer to the peripheral margin, sometimes umbilical entirely, additional apertures not developed, in Heronallenidae aperture in the umbilical area is covered by the plates which are formed later; sexual reproduction plastogamic (gamontogamic), gametes (if known) - three flagellate.

Lower Eocene – Holocene.

Remarks. Glabratellida is the only taxon not only among rotalians but among all the other foraminifera having three flagellate gametes (although not all the representatives of the order were studied from this point of view). The order is unevenly studied and possibly also includes some heterogenous forms. The position of the order itself may raise questions.

Families: Glabratellidae Loeblich et Tappan, 1964, Rotaliellidae Loeblich et Tappan, 1964 (family transferred to the glabratellids by Pawlowski et Zaninetti from the Discorbacea in Loeblich & Tappan, 1987, with the genus *Rossyatella* of Pawlowski et Zaninetti, 1993),

?Buliminoididae Seiglie, 1970 (without *Elongobula* Finlay, 1939 according to Revets, 1993), ?Heronallenidae Loeblich et Tappan, 1986.

Order PLANORBULINIDA Mikhalevich, 1992

Shell more often attached, trochospiral entirely in the early stage only, usually with the spiral side flattened and evolute, and umbilical side involute and convex, later part may be rectilinear, cyclical, conical, branching or form irregular mass consisting of chambers multiple in number and often losing regular form; inner chamber lumen not subdivided; wall usually with coarse pores; aperture in the entirely trochospiral forms at the base of the septal surface – equatorial or extending to the umbilical or spiral sides, sometimes onto both of them, on the spiral side it may be stretched along the inner ends of several of the last chambers, often boarded by a lip, additional umbilical and sutural apertures may present on the umbilical side, in the uncoiled forms aperture subterminal, in the forms forming irregular mass of chambers aperture multiple, looking like circular openings resembling pore openings ("apertural pores") on the surface of the chamber wall or multiple arch openings either on one of the sides or on the peripheral margin in the sutures between the chambers.

Paleocene – Holocene.

Remarks. Status of Planorbulinacea in the understanding of Loeblich & Tappan, 1987 is elevated here to the order rank, with some changes in its composition (without families Bissaciidae, Cymbaloporidae), nevertheless it still represents rather heterogenous group and needs to be revised. Representatives of Planorbulinida have developed strongly specialized shells adapted to sessile mode of life, some of them are also known as commensals and parasites (see above – Part V, 1).

Suborder PLANORBULININA Mikhalevich, 1992

Shell free living or attached, trochospiral entirely or at the early stage only, initial trochoid part usually with flattened and evolute spiral side and involute convex umbilical side, more rarely partially evolute on the both sides (Planulinidae), final part may be rectilinear (biserial or uniserial), cyclical, may form irregular conglomerated mass where chambers are disposed in one layer or form a weakly elevated cone; aperture single – in trochospiral forms at the base of the septal surface, equatorial or going onto one of the sides or onto both of them, on the spiral side it may go under the inner ends of several of the last chambers, often boarded by a lip, additional apertures on the umbilical side may also present, in the rectilinear forms aperture subterminal, in the cyclical ones and in those spreading as irregular chamber mass - multiple (circular apertural "pores" on the chamber surface or arch openings in the sutures).

Upper Eocene – Holocene.

Remarks. Cymbaloporidae with Fabianiinae and Halkyardiinae are transferred from the former Planorbulinacea into subclass Spirillinana on the base of their peculiar shell structure (see above – Part XI, 4, 2).

Families: Planorbulinidae Schwager, 1877 (with Planorbulininae Schwager, 1877, Caribeanellinae Saidova, 1981), Cibicididae Cushman, 1927, Victoriellidae Chapman et Crespin, 1930 (with Victoriellinae Chapman et Crespin, 1930, Carpenteriinae Saidova, 1981, Rupertininae Loeblich et Tappan, 1961), Planulinidae Bermudez, 1952, Stichocibicididae

Saidova, 1981, Annulocibicididae Saidova, 1981, Stensioeinidae Bugrova, 1990 (nom. correct. pro Stensioinidae Bugrova, 1990).

Suborder ACERVULININA Schultze, 1854

Shell attached, thick, nearly discoidal, with the convex upper surface and only early portion irregularly trochospiral, later of irregular form (mound, irregular mass, spreading in one or more layers, branching and growing up from the attachment), with a huge number of small chambers circular or polygonal in form; wall of bright rose, vinous, orange colour even in the fossil forms (possibly owing to the decay of endosymbiont's chlorophyll pigments), surface of the circular convex chambers may resemble raspberry, sometimes chambers may be flattened and even depressed from the outside and covered with a perforate plate; in the arborescent forms (Homotrematidae) inside the shell imperforate hollow cylindrical double-walled tubes extend upwards from the shell base, these tubes may branch and be encircled by several layers of small chambers; apertures – multiple circular openings in the chamber wall or in the wall of depressed plates closing the entrance to the chambers, apertures are significantly larger than the neighboring pores (earlier they were often described as "large" pores).

Paleocene – Holocene.

Remarks. Young three-chambered form of the *Miniacina* megalosphaeric shell is free living, then settles, and from its third chamber the tubular outgrowth begin to branch with the circular chambers growing on around them; microsphaeric shell settles at the 5 to 9 chambers stage having a distinct trochoid spire. Walls of the chambers may be partially resorbed as layers are added, producing large and irregular passages. Adult shells of Homotrematids reach about 7 mm in height. They differ so markedly from all the other Foraminifera that they were for a long time regarded as corrals. The sessile mode of life resulted in the strongly variable modes of the disposition of the later chambers and to the transformation of the aperture in this group.

Families: Acervulinidae Schultze, 1854 (including *Aberisphaera* Wan, 1991), Menoellidae Cherchi et Schroeder, 2005 (obviously Menaellidae nom corr. pro Menoellidae Cherchi et Schroeder, 2005 as type genus is *Menaella* Cherchi et Schroeder, 2005), Homotrematidae Cushman, 1927.

Order ROTALIIDA Lankester, 1885 (nom. corr. Calkins, 1909 (after Loeblich et Tappan, 1992)

Shell trochospiral, at least at the earlier stage, later may become planispiral, sometimes chambers cyclically arranged, with usual chamber number or super-multi-chambered (many small chambers in numerous whorls), elongated umbilical ends of the chambers with valves; umbilicus closed or may be filled with secondary shell matter, forming vertical pillars (sometimes in several tiers) with fissures between them - primitive canals, besides the umbilical cavities and cavities nearby to it, and besides fissures the interseptal fissures may also be present, majority of the genera possess well developed canal system – spire canal and septal canals connecting the spire canal and branching into the smallest ones; aperture at the base of the spiral face (equatorial or at the umbilical side, may be fissure-like or as a row of circular openings), is often closed and only the canal system functioning, additional apertures may be umbilical or sutural.

Upper Cretaceous – Holocene.

Remarks. Rotaliids are the most complex and highly developed representatives among the calcareous Foraminifera, representing the highest level of organization of the group. Arising in Cretaceous, they in a short period occupied a wide ecologic niche. Families are separated on the basis of the peculiarities of their canal system. Many of the highly organized families are extinct at present, in the Recent oceans only the shells of notorotaliids and calcarinids are known in the shallow tropical seas, and ammoniid shells (especially of the shallow water genus *Ammonia* which is widespread in the warm and moderate waters of the near-continental seas of all the oceans). Families Notorotaliidae, Faujasinidae, Miogypsinidae are transferred here from the former Nonionacea as having trochospiral shell (at least initially) and canal system. Boudagher-Fadel & Price (2010) also placed miogypsinids (with some new genera described in this article and in the earlier works) into the order Rotaliida. Family Chapmaninidae is transferred from the order to the subclass Spirillinana at the base of its inner and outer shell structure (see above – Part XI, 4, 2).

Families: Rotaliidae Ehrenberg, 1839 [with Rotaliinae Ehrenberg, 1839 (including the ?*Rotaliconus* Hottinger, 2007 and *Urnummulites* of Boukhary & Scheibner, 2009), Sakesariinae Bugrova, 1990], Calcarinidae Schwager, 1876 (including *Meghalayana* Matsumaru et Sarma, 2008), Miogypsinidae Vaughan, 1928 (including *Americogypsina* Boudagher-Fadel et Price, 2010, *Paleomiogypsina* Matsumaru, 1996, *Boninella* Matsumaru, 1996, *Miogypsinoidella* Boudagher-Fadel et al., 2000 and *Tania* Matsumaru, 1990), Pseudorbitoididae M.G. Rutten, 1935 [with Pseudorbitoidinae M.G. Rutten, 1935, Vaughanininae MacGillavry, 1963, Pseudorbitellinae Hanzawa, 1962, Pararotaliinae Reiss, 1963 (including *Praepararotalia* Liu, Olsson et Huber, 1998 and *Antarcticella* Loeblich et Tappan, 1988, the authors suggest the benthic habitat of both genera and their close relationship], Faujasinidae Bermudez, 1952, Notorotaliidae Hornibrook, 1961, Cuvillierinidae Loeblich et Tappan, 1964, Ammoniidae Saidova, 1981 (with *Gavelinopsis* Hofker, 1951 according to Revets, 2002 c).

Order ? CHILOSTOMELLIDA Haeckel, 1894

Shell trochospiral entirely or at the early stage only, rounded, streamlined in form, slightly constrained along the axis of coiling or perpendicularly to it, with strongly embracing whorls, chambers few per whorl (three to four, sometimes only two in the last whorl), chambers inflated, rapidly enlarged as added, the last one may be strongly embracing, chamber lumen not subdivided; aperture at the base of the septal face, narrow, fissure-like, with the lip overhanging like a cap peak.

Upper Cretaceous – Holocene.

Remarks. Only families of typical chilostomellid structure are included in this order, the rest are transferred into the superfamily Gavellinelloidea. The genera *Chilostomella* and *Chilostomellus* of the family Chilostomellidae having shell form unusual to rotaliats and resembling in some features some representatives of Hergotelloidea need additional study.

Families: Chilostomellidae Brady, 1881, Pallaimorphinidae Loeblich et Tappan, 1987, stat. nov., Chilostominidae Finger et Gaponoff, 1986, Lublinidae Gawor-Biedowa, 1987.

Superorder NONIONOIDA Saidova, 1981

Shell planispiral, shell wall calcitic, foramen is homologous to the previous aperture.

Order NONIONIDA Saidova, 1981

Shell planispiral, bilaterally-symmetrical or slightly asymmetrical, involute or half involute, flattened or nearly globular and with relatively few chambers; umbilicus opened or filled with skeletal material, may be covered by flap, may have additional skeletal plates covering sutures and umbilical area with additional cavities under them, umbilical area may also be pierced by fissures serving as the passages for the cytoplasm (*Protelphidium*), but lacking the true canal system; aperture at the base of the septal face, equatorial or extending onto the umbilical side, rarely multiple, peripheral or terminal in erected part, in some genera additional sutural or umbilical apertures may be present.

Upper Cretaceous – Holocene.

Suborder NONIONINA Saidova, 1981

Shell without additional skeletal plates.

Superfamily Nonionoidea Schultze, 1854

Shell with the aperture at the base of the septal surface, without additional fissure-like apertures.

Remarks. Subfamily Nonioninae without the genus *Haynesina* in which chamber walls bend inward posteriorly and fuse to the preceding septal face, forming intercameral spaces (lacunae) along the septa. Such an inner structure, representing additional inner spaces of the secondary level is not mentioned in the diagnosis of the subfamily and is absent in the rest Nonioninae genera. This feature makes *Haynesina* closer to elphidiids although this secondary structure is simpler than elphidiids canal system.

Families: Nonionidae Schultze, 1854 (with Nonioninae Schultze, 1854, Pulleniinae Schwager, 1877 (including *Pulleniella* Ujiie, 1990), Melonisinae Voloshinova, 1958), Nonionellidae Voloshinova, 1958, ?Spirotectinidae Saidova, 1981.

Superfamily Almaenoidea Mjatliuk, 1959

Shell with additional fissure-like apertures along the peripheral edges of the chambers and boarded by a lip, additional aperture of the last chamber may be elevated on a neck-like extension, equatorial aperture tends to be displaced to one of the sides.

Remarks. Peripheral apertures of the earlier chambers may be closed by additional shell material. Genus *Saraswati* Kalia, 1982 is transferred into Heleninidae (Rosalinida) at the base of the structure of its trochoid shell and character and position of its additional apertures (sigmoidal and parallel to the periphery), belonging of the genus *Ganella* lacking additional apertures to almenids is doubtful.

Family: Almaenidae Mjatliuk, 1959 (with Almaeninae Mjatliuk, 1959, Anomalinellinae Cushman, 1927).

Suborder ASTRONONIONINA Saidova, 1981

Shell with additional skeletal plates covering the umbilical ends of the sutures and umbilical area and forming cavities opening by the additional apertures in the sutures and in the umbilicus. Meddle Eocene – Holocene.

Remarks. Additional plates may be narrow and tubular in the earlier chambers and in the genus *Pacinonion*. Family Bisacciidae is transferred from the former Planorbulinacea as having planispiral shell and covering plates.

Families: Astrononionidae Saidova, 1981, *Bisacciidae Loeblich et Tappan, 1987.

Order ELPHIDIIDA Saidova, 1981

Shell planispiral, symmetrical, involute, in one genus with the enrolled final part, chamber number relatively not large; umbilicus closed, canal system well developed consisting of the two symmetrical spiral canals, and branches outgoing from these and aligning under the sutures, the latter opening by one or two rows of small septal canaliculi, vertical umbilical canals opening in the umbilical area may be present as well; aperture at the base of the septal face single or multiple, may be areal, small circular openings at the base of the septal surface or above it may be closed and then opened before the formation of the new chamber.

Paleocene – Holocene.

Remarks. New family Haynesinidae with the type genus *Haynesina* Banner et Culver, 1978 (transferred from Nonioninae) is separated here: shell planispiral, at least at the adult, involute or partially evolute (Protelphidium), moderately compressed, with broadly rounded periphery, biumbilicate, may have asymmetrical sides, chambers gradually increasing in size, not numerous (usually 8 to 10), sutures gently curved, deeply incised towards the umbilicus especially in the later chambers, umbilicus slightly depressed, pustules and prominent tubercules which may become pillarlike strongly developed in the umbilical area along the sutures, may also be present on the apertural face; aperture at the base of apertural face, a low narrow arc, may be obscured by pustules, additional integrative apertural systems developed to a different degree (intercameral lacunae along the septa formed by the bend of the chamber walls going inward posteriorly and fusing to the preceding septal face to form intercameral spaces (in *Haynesina*), intercommunicating umbilical cavities between septal flaps and supporting pillars in *Protelphidium*). Paleocene - Holocene.

Such inner structures, represented by additional inner spaces or umbilical cavities represent the formations of the secondary level of organization of apertural system and are not mentioned in the diagnosis of the Nonionidae being absent in the rest of the nonionid genera. These features (differing in each of the aforementioned genera) make *Haynesina* and *Protelphidium* closer to elphidiids although their secondary integrative apertural structures are simpler than elphidiid canal system. Absence of septal or spiral canals, retral processes or fossettes clearly distinguishes the new family from the family Elphidiidae, as well as absence of any canals – from Cribroelphidiidae.

Families: Elphidiidae Galloway, 1933 (with the genus *Bangiana* Drobne, Ogorelec et Riccamboni, 2007); Haynesinidae fam. nov., Cribroelphidiidae Voloshinova, 1958.

Order ORBITOIDIDA Copeland , 1956

Shell discoidal or lenticular, super-multi-chambered, chambers subdivided into chamberlets, system of stolons present, without canal system.

Upper Cretaceous – Upper Eocene.

Remarks. Shells with prominent dimorphism in the gamont and agamont generations. In geologically older forms only openings between chambers present, in later advanced forms the stolons are developed. Embryonic stage with large embryonic chambers as it usually developed in the super-multi-chambered shells is present. Without Clypeorbinae Sigal, 1952 as its shells have initial trochoid stage.

Families: Orbitoididae Schwager, 1876 [including *Novosetia* Özdikmen, 2009 pro *Setia* Ferrandez & Canadell, 2002 (non Adams & Adams, 1852) and obviously *Sirelella* Meric et Inan, 1998 and *Neosivasella* Meric et Coruh, 1998 (the latter genus is doubtful being separated on the basis of its geologic range only], Lepidorbitoididae Vaughan, 1933 [with Lepidorbitoididae Vaughan, 1933(including *Cideina* Sirel, 1992) and Neumannitinae Rahaghi, 1992), Lepidocyclinidae Scheffen, 1932, Linderinidae Loeblich et Tappan, 1984, Droogerinellidae Popescu et Brotea, 1995 (type genus *Droogerinella* Popescu et Brotea, 1995), Orbitosiphonidae Matsumaru & Jauhri 2003. (according to the authors of the family including *Lakadongia* Matsumaru & Jauhri 2003 and *Orbitosiphon*).

Order NUMMULITIDA Carpenter, Parker et Jones, 1862

Shell planispiral, evolute or involute, lenticular, discoidal. subglobular, with very large number of chambers, which may be subdivided into chamberlets, equatorial and final chambers may be differentiated, the latter may be annular, septal flap may be present (as in Rotaliida); canal system of high complexity: spiral marginal cord, system of inner spiral canals and canals going beneath the sutures in geologically earlier representatives, in the later and more advanced genera canals modified or replaced by interseptal canals, stolon system may also be present.

Paleocene – Holocene.

Remarks. Order Nummulitida is the most complex and perfect foraminiferal group. The majority of its families became extinct towards the Eocene but in the family Nummulitidae there are a lot of recent representatives: large beautiful shells of *Assilina*, *Cycloclypeus*, and especially of *Heterocyclina*, *Heterostegina*, *Operculina* and *Nummulites* are abundant in the shallow tropical seas. In the work of Hottinger et al., (2001 b) the specificity of the canal systems at the generic level of the three genera of Pellatispiridae (including the genus *Vacuolispira* reinstated by them from the genus *Pellatispira*) has been studied up to the finest details, showing the presence in these genera of not only the canal system but also the system of interlamellar cavities.

A similar study was made for some other nummulitids (Hansen & Reiss, 1971, Hottinger, 1977, Muller-Merz, 1980, Billman et al., 1980, Hottinger et al., 1993, Hansen, 1999 and Hottinger, 2000. However the inner structure of the majority of nummulitids necessitates such a study at a new technical level. Although the geometry of the canal system reflects the shell geometry, nevertheless "…it is not clear whether the spiral or radial elements are produced the same way, by the same shell elements" (Hottinger, 2001).

Families: Nummulitidae de Blainville, 1827 (including *Decrouezina* Boukhary, 1994 and *Bassiounina* Boukhary, Kulbrok et Kuss, 1998, *Planoperculina* Banner et Hodgkinson, 1991), Heterosteginidae Galloway, 1933 (including *Planostegina* Banner et Hodgkinson, 1991 and *Tansinhokella* Banner et Hodgkinson, 1991), Discocyclinidae Galloway, 1928, Pellatispiridae Hanzawa, 1937, Asterocyclinidae Bronnimann, 1951.

Superorder BULIMINOIDA Saidova, 1981

Shell elongated, early part may be planispiral, later biserial to uniserial or triserial, trochospiral, with high spire, in the three latter cases usually with transition to biserial and uniserial, may consist of two rows of chambers enrolled in involute spire (cassidulinids) also tending to be enrolled; chamber lumen usually (with the exception of one family) not subdivided, primitive canal system is present in the two genera; shell wall calcitic; aperture loop-shaped, going from the base of the apertural face or terminal, single, more rarely multiple, in evolutionary advanced forms with tooth-plates of variable form and complexity (plates, tongues, grooves, hemicylindrical siphons, apertural plates sometimes may be curved, folded, attached to the wall of the previous chamber, may have free outgrowths which may be dentate, may be elevated above the apertural edge), tooth-plates of the successive chambers may be interconnected forming a single inner apertural integrative system, foramen is homologous to previous aperture, additional sutural apertures also rarely occur.

Middle Jurassic – Holocene.

Remarks. Biserial and uniserial forms originated from both the planispiral forms (order Bolivinitida) and from the trochospiral ones (order Buliminida) just as happen in the subclass Textulariana. Among the first order many of the shells are homeomorphic to the shells of spiroplectamminids and textulariids, among the second order – to the shells of vereneuilinids, among Cassidulinida – to plectorecurvoidids. It is possible to see many parallel features in the process of the growing complexity of the tooth plates of Bolivinitida and Buliminida. Tooth plate is usually formed by the inner layer of the shell wall. Among the genera whose tooth plate was studied the genus *Stainforthia* represents an exception as its tooth plate is formed both from the inner and outer layers of the shell wall. This fact gave Revets (1996) a basis for regarding this genus as Incertae Sedis. The fact of the presence of well developed tooth-plates and integrative apertural systems in which the tooth-plates of the adjacent chambers are connected with each other is here regarded as a taxonomic feature of a high rank, but the peculiar features of the tooth-plate structure are characteristic of the lower rank taxa of family or sometimes generic level.

Stilostomellids are here excluded from the former Rotaliacea and placed into subclass Nodosariana (see above – Part XI, 2, 3).

Order BOLIVINITIDA Saidova, 1981

Shell usually flattened, biserial entirely or may have initial planispiral part; chambers more often wide, flattened, later chambers often tending to uniserial arrangement, may become fan-shaped; chamber lumen may have additional subdivisions; canal system not developed; aperture at the base of the apertural face, elevated above it or terminal, more often loop-shaped or fissured, more rarely circular, in the advanced genera may have complex tooth plate, may be multiple.

Cretaceous – Holocene.

Suborder SPIROBOLIVINA Mikhalevich, 1998

Shell with initial planispiral part, later part straight, biserial.

Remarks. Suborder differs from the suborder Bolivinitina in having well expressed initial spiral part. Shells in transverse section flattened or subcylindrical, chamber lumen without inner partitions, aperture loop-shaped, in the majority of the known genera having tooth plate.

Family Spirobolivinidae Saidova, 1981 put previously into the synonymy of Lacosteinidae and subfamily Sigmavirgulininae put into the synonymy of Fursenkoinidae are reinstated here as the first of them differs from Lacosteinidae by its cylindrical shell shape and inflated last chamber, in the flattened character of its shell and chambers more resembling *Bolivina*, and the second one differs from fursenkoinids not only in its flattened chambers and in the form of its chambers and the aperture but also in the presence of the planispiral part.

Families: Lacosteinidae Sigal, 1952, Spirobolivinidae Saidova, 1981 (with Spirobolivininae Saidova, 1981 and Sigmavirgulininae Saidova, 1981).

Suborder BOLIVINITINA Saidova, 1981

Shell biserial entirely or with transition to uniserial arrangement of chambers in the final part, with or without inner tooth-plate.

Remarks. Superfamilies of this suborder differ from each other mainly in the structure of the aperture and of its inner structures, and also in the presence or absence of the secondary subdivision of the chambers.

Superfamily Loxostomatoidea Loeblich et Tappan, 1962

Shell biserial entirely, final chambers tending to uniserial arrangement; chamber lumen lacking the inner subdivisions; aperture simple, without the tooth-plate.

Families: Loxostomatidae Loeblich et Tappan, 1962, Bolivinellidae Hayward, 1980, Tortoplectellidae Loeblich et Tappan, 1985.

Superfamily Bolivinitoidea Cushman, 1927

Shell biserial entirely, two last chambers may have tendency to uniserial arrangement, chamber lumen not subdivided; aperture with well developed tooth apparatus: loop-shaped aperture is encircled by the fully or partially elevated rim with its lower part going inside forming tooth-plate which consists of the two parts: one of them is attached and goes inside to the foramen of the previous chamber, the other one is free, usually elevating above the aperture but sometimes not visible from the outside.

Remarks. The free part of the tooth plate may not be developed in some cases.

In the family Bolivinidae genera *Coryphostoma* and *Cassidella* are transferred from Fursenkoinidae. Superfamily Bolivinitoidea Cushman, 1927 has priority compared with the Bolivinoidea Glaessner, 1937. All the families entering this superfamily are united as having biserial shell and inner tooth-plate. Genus *Brizalina* is considered to be a junior synonym of *Bolivina* having only slight morphologic differences from the latter genus, the molecular data (Ertan et al., 2004) did not supported the validity of *Brizalina*.

Families: Bolivinitidae Cushman, 1927 (with subfamilies Bolivinitinae Cushman, 1927, Sagrininae Revets, 1996 including the genus *Alectinella*, Revets, 1996), Bolivinidae Glaessner, 1937, Bolivinoididae Loeblich et Tappan, 1984.

Superfamily Parabrizalininoidea Revets, 1996, stat. elevated.

Shell biserial entirely or tending to uniseriality, flattened, chambers not secondarily subdivided; aperture usually loop-shaped, horizontal, with the tooth plate in the form of hemicylindrical siphon, siphons of the successive chambers form a single system.

Family: Parabrizalinidae Revets, 1996, stat. elevated.

Superfamily Eouvigerinoidea Cushman, 1927

Shell biserial, tending to turn into uniserial or uniserial at the final part, thick but with flattened lateral sides or cylindrical; chambers not secondarily subdivided; aperture loop-shaped or terminal, may be at the end of the elevated neck, tooth plate in the form of hemicylindrical groove (siphon), siphons of the successive chambers form a single system.

Remarks. Subfamily Tubulogenerininae is excluded here from the family Siphogenerinoididae as having the trochospiral (triserial) initial part, the extinct family Lacosteinidae (Loeblich, Tappan, 1988) which previously was included in this superfamily is here transferred into the suborder Spirobolivinina as its type genus has planispiral initial part.

Families: Eouvigerinidae Cushman, 1927, Siphogenerinoididae Saidova, 1981.

Superfamily Pavonininoidea Eimer et Fickert, 1899, nom. correct.

Shell biserial, later with uniserial part with arched flaring chambers which may result in flabelliform shell; chamber lumen without inner subdivisions; aperture multiple: a row of openings along the apertural face.

Remarks. Family Pavoninidae is here transferred from the former Buliminacea because it has flattened biserial shell and its aperture is without inner tooth or tooth plate.

Family: Pavoninidae Eimer et Fickert, 1899

Superfamily Millettioidea Saidova, 1981

Shell elongated, biserial entirely or only in its initial part with the later part uniserial, circular or oval in the transverse section; chamber lumen subdivided with radial partitions or with partitions going in two directions (vertical and horizontal) forming numerous chamberlets which could be seen through the shell wall as honeycomb cellules; aperture loop-shaped or terminal and circular in outline encircled by a collar and narrow rim, internal tooth plate may be of the complex structure like in bolivinids (*Afrobolivina*) or grooved, may be absent in the initial chambers..

Remarks. Revets (1992) considers the monotypic Millettiidae as *incertae sedis* within the Rotaliina on the basis of its peculiar structure (monolamellar toothplates formed by inner lining are continuous with the septula, the inner lateral side of chambers covered by a weblike coating also continuous with the septula).

Family: Millettiidae Saidova, 1981

Order CASSIDULINIDA d'Orbigny, 1839

Shell subsphaerical, lenticulate, oval or flattened, biserial alternating chambers enrolled into plane spire which is more often flattened, in one case – into trochoid spire, the final part of the shell may be enrolled going in the plane of coiling or perpendicularly to this plane (Ehrenberginidae); aperture oval, loop-shaped or like a fissure, which is often curved and going in parallel to the base of the septal surface or at the angle to it, may have a second additional opening above the first one, in many genera it may be covered from the back of the wall by the apertural plate, on the front side of the wall a small tooth looking like a tongue or denticulate may be seen, in one genus aperture is terminal.

Paleocene – Holocene.

Families: Cassidulinidae d'Orbigny, 1839, Ehrenberginidae Cushman, 1927, Cassidulinitidae Saidova, 1981, Orthoplectidae Loeblich et Tappan, 1984 ((with *Elongobula* Finlay, 1939 according to Revets, 1993), the rank of the subfamily elevated by Revets &Whittaker, 1991), Cassidinidae Boukhary et Scheibner, 2011.

Order BULIMINIDA Saidova, 1981

Shell trochoid with high spire (spiral-screw)or triserial, often tending to form uniserial part, usually circular, oval or triangular in transverse section; chamber more often inflated, subsphaerical or elongated; chambers not subdivided; in a few species of the two genera the primitive canal system present; aperture more often loop-shaped, going from the base of the septal surface, may be terminal at the end of the elongated neck, may be multiple (sieve-like), inner apertural structures from rather simple up to very complex tooth plates, hemicylindrical tubes, both types of inner apertural structures may form a single integrative system, in several genera additional apertures present.

Middle Jurassic – Holocene.

Suborder TURRILININA Saidova, 1981

Shell trochoid, with high spire (spiral-screw)or triserial, circular, oval or triangular in transverse section; chambers rapidly increasing in size, usually strongly inflated, subsphaerical or elongated; aperture loop-shaped at the base of the septal face without a tooth or with a small and weakly developed tooth-plate, sometimes multiple, additional sutural apertures may present. Middle Jurassic – Holocene.

Families: Turrilinidae Cushman, 1927 [with Turrilininae Cushman, 1927 (including *Floresina* Revets, 1990 and *Parkiella* Seiglie, Haman, Barker, 1993) and Sporobuliminidae Mikhalevich, 1998 – with multiple aperture, including genera *Sporobulimina* and *Sporobuliminella*], Tosaiidae Saidova, 1981, Virgulinellidae Loeblich et Tappan, 1984, ?Praeplanctoniidae Georgescu, 2009 (nom corr. pro Praeplanctonidae) (type genus *Praeplanctonia* Georgescu, 2009 with type species *Praeplanctonia globifera* Georgescu 2009).

Suborder BULIMININA Saidova, 1981

Shell like in Turrilinina, in some genera with strongly embracing last chambers; aperture as in Turrilinina (terminal and subcircular in outline disposed at the end of elongated neck, or loop-shaped, may be multiple) but tooth plate always present having a complex structure: its grooved shank (which may be curved, widened, folded or with some other peculiarities of its structure)is attached by its upper part to one side of the edge of the aperture and by its lower part to the inner wall of preceding chamber near the foramen, it may be smooth to dentate, flaring to enrolled and almost tubular, its free border projects out of the aperture forming variable plates (an elevated rim, a flaring lip or serrate cristae).

Paleocene – Holocene.

Remarks. Family Stainforthiidae is transferred from the former Turrilinacea (here suborder Turrilinina) as having well developed tooth plate, family Siphogenerinoididae is excluded from this suborder as subfamily Siphogenerinoidinae and its type genus *Siphogenerinoides* have biserial early stage and is placed into Eouvigerinoidea (suborder Bolivinitina). Family Fursenkoinidae although described as having early biserial part nevertheless in its tooth plate structure and some shell features is closer to buliminins and is placed here with some hesitation though Revets (1996) placed it as subfamily into Bolivinitidae. The SEM data of Revets (1996) showed a special bilamellar character of the *Stainforthia* tooth-plate but such studies have not yet touched on all the number of the related genera.

Families: Buliminidae Jones, 1875, Uvigerinidae Haeckel, 1894 (with Uvigerininae Haeckel, 1894, Angulogerininae Galloway, 1933), Reussellidae Cushman, 1933 (without *Acostina* and *Acostinella* of Sarkar et Gupta, 2008 initially placed by the authors of the new genus into this family - see Part XI. 4. 4)), ?Biedafranciszkinidae Gawor-Biedowa, 1992, Hofker, 1951, Fursenkoinidae Loeblich et Tappan, 1961 (without *Coryphostoma* and *Cassidella,* and belonging of *Coryphostomella* Gawor-Biedova, 1987 placed by the author of the genus in this family is doubtful, its aperture at the earlier stages is unknown, its final aperture more resembles that of some representatives of Nodosariana), Stainforthiidae Reiss, 1963, Trimosinidae Saidova, 1981, Pappinidae Haunold, 1990.

Suborder TUBULOGENERININA Mikhalevich, 1998

Shell with triserial early part which later may become biserial and uniserial, in the transverse section subcircular or oval; aperture usually terminal, often at the end of a short neck, tooth plate as hemicylindrycal groove (siphon), siphons of successive chambers joined in a single integrative system.

Upper Cretaceous – Holocene.

Remarks. Some genera (*Amplectoproductina, Unicosiphonia*) which are uniserial throughout are nevertheless preserved in this suborder because their triserial ancestry is reflected internally by their hemicylindrycal toothplates changing orientation by 120° in successive chambers as in the triserial stage of the rest of the genera of Tubulogenerinina.

Family: Tubulogenerinidae Saidova, 1981 (obviously including the genus *Rectobulimina* from the Turrilinidae).

Suborder CAUCASININA Mikhalevich, 1998

Shell with early stage low trochospiral and widely rounded, consisting of several coils, later high trochospiral, enrolled, rectilinear, triserial or biserial, in transverse section circular; aperture loop-shaped, going from the base of the apertural face, tooth plate usually flap-like.

Remarks. In some genera tooth plate is not known.

Family: Caucasinidae N.K. Bykova, 1959 (without the genus *Tremachora*) (with Caucasininae N.K. Bykova, 1959, Bagatellinae N.K. Bykova, 1959).

Subclass GLOBIGERINANA Mikhalevich, 1980

Shells multi-chambered, trochospiral and planispiral, more rarely derived from them multiserial to uniserial (Heterohelicida), with a small number of chambers per whorl, with the progressive enlarging of the sizes of umbilical area up to nearly uncoiled spire and loose conjunction of the chambers, encircled and supported by the wide cytoplasmic layer or by the vesicular capsula (in advanced specialized forms (*Hastigerina*); chambers strongly flattened in flat forms, in the rest - subsphaerical or inflated and elongated; additional sutural and umbilical skeletal plates well developed; wall calcareous, bifontinal, may have considerable thickening of the outer wall layers during the gametogenesis (crust), may be smooth or hispid and alveolar from the surface, may have long radial needles, immobile or setting loosely in the holes, having the opportunity to turn like in joints, pores regular, usually large, pore canaliculi funnel shaped, with one middle sieve-plate, growth of the new chambers as in Rotaliana; apertures situated at the base of the chamber, usually large, occupying significant part of the shell, additional apertures very often; pseudopodia radial, non anastomosing; agamonts multinucleate, heterokaryotic, in *Globigerina bulloides* and *Globigerinoides ruber* with the polymerization of the somatic nuclei also, unique organelles - cryptosomes and microvillies - are present; gametes biflagellate, their number many times more than in the benthic species; free living pelagic forms.

Jurassic - Holocene.

Remarks. The system of subclass is given here after Loeblich & Tappan, 1992 differing from their previous in the understanding of the rank of taxa and in some small changes. The taxa of the subclass obviously originate from the different Rotaliana groups (and thus the subclass represents a rather heterogenous group) although having their ancestors in the same single subclass Rotaliana. Many of the orders and superfamilies need the consequential revision, their diagnoses are not quite distinctiveб sometimes uniting the representatives of the restricted geological periods and being based mostly on the stratigraphic ranges rather than on morphological features.

The origin of the representatives of the different orders (phylogenetic lineages) of Globigerinana from the different ancestor roots of the subclass Rotaliana is absolutely evident judging from their shell and apertural morphology. This is also confirmed by the molecular rDNA data (Darling et al., 1996, 1997). Their ancestors belonging to the different orders (buliminids, turrilinins, cassidulinids, rosalinids) nevertheless enter the single subclass Rotaliana. Similarly their descendants representing the specialized top parts of the appropriate lineages could be united in a specialized subclass whose representatives acquire some special adaptive features representing the highest level of development of their organization. Pleurostomellids shells differ from all the Rotaliana representatives in their apertural structure (Mikhalevich, 1993) and belonging to Nodosariana subclass have with some Rotaliana representatives (e.g. Fursenkoinids) only convergent resemblance, and contrary to Georgescu (2009 a) I think they could not be regarded as the predecessors of the heterohelicids.

The number of publications on the globigerinans seems 'vastly to exceed the publications on all the other taxonomic groups (Barash, 1970, Darling et al., 1999, 2000, Hart, 1999, Hart et al., 2002, El-Nakhal, 2002, Gonzalez-Donoso et al., 2007, Desmares et al., 2008, Lipson-Benitah, 2008, Georgescu, 2009 a, b, c, 2010, a, b, Premoli Silva et al., 2009, Shuaibi, 2011 and many others). A set of publications devoted to their evolution is also significant (Hart, 1980, Kennet & Srinivasan, 1983, Patterson & Fowler, 1996, Patterson et al., 2004, Price & Hart, 1999 and many others). But an exaggerated attention given to this group which is so useful in stratigraphy and paleoclimatology leads sometimes to the exaggerated understanding of the taxonomic features of the specific level as having generic significance and to the separation of the new genera outnumbering in number the new genera in the other foraminiferal groups.

Order HETEROHELICIDA Mikhalevich, 1992

Shell spirally enrolled at the earlier stage (trochoid or planispiral), later biserial and triserial (or may be biserial and triserial from the very beginning, thus the spiral part may be absent in at least one of the generations), more often with the tendency to proliferation of the chamber number up to the multiple at the later stage, more rarely with the opposite tendency of decreasing chamber number up to the uniserial later stage; chambers strongly inflated, nearly subsphaerical; shell wall costose, more rarely spinose; aperture – an arch opening at the base of apertural face of the final chamber or terminal in uniserial stage, in *Laterostomella* with the inner tooth plate, additional sutural openings may also present.

Lower Cretaceous – Holocene.

Superfamily Heterohelicidoidea Cushman, 1927

Shell spirally enrolled at the earlier stage (trochoid or planispiral), later biserial and triserial (or may be biserial and triserial from the very beginning, thus the spiral part may be absent at least in one of the generations), more often with the tendency of proliferation of the chamber number up to the multiple at the later stage, more rarely with the opposite tendency of decreasing chamber number up to the uniserial later stage; chambers strongly inflated, nearly subsphaerical; shell wall costose, more rarely spinose; aperture – an arch opening at the base of the apertural face of the final chamber or terminal in uniserial stage, in *Laterostomella* with the inner tooth plate, additional sutural openings may also present.

Families: Heterochelicidae Cushman, 1927 (with Heterohelicinae Cushman, 1927, Gublerininae Alyulla, 1977, Pseudogumbelininae Alyulla, 1977), Guembelitriidae Montanaro Gallitelli, 1957, Chiloguembelinidae Reiss, 1963.

Order GLOBOROTALIIDA Mikhalevich, 1980

Shell trochospiral, with usually flattened or only slightly convex spiral side, with chambers flattened on the spiral side, convex or even angular on the umbilical side, or chambers may be strongly inflated, peripheral margin more often narrow, acute, may be killed, double-killed, carinate, may be with imperforate carinal band, in forms with inflated chambers peripheral margin is broadly rounded and usually lobulate; umbilicus opened or closed; wall of variable character but lacking needles; aperture at the base of apertural face extraumbilical-umbilical bordered by a lip, may be umbilical with tegilla of successive chambers covering umbilical area, additional sutural, intralaminal and infralaminal apertures may occur.

Lower Cretaceous - to Upper Cretaceous (Rotaliporoidea, Globorotalioidea).

Lower Paleocene – Holocene (Globorotalioidea).

Superfamily Rotaliporoidea Sigal, 1958

Shell trochoid, low, with usually flattened or slightly inflated spiral side; umbilicus usually opened; aperture extraumbilical-umbilical boarded by a lip, additional sutural apertures at the umbilical side may present.

Families: Rotaliporidae Sigal, 1958 (with Rotaliporinae Sigal, 1958, Rotundininae Bellier et Salaj, 1977, Helvetoglobotruncaninae Lamolda, 1976, Ticinellinae Longoria, 1974), Hedbergellidae Loeblich et Tappan, 1961, Praehedbergellidae Banner et Desai, 1988 (including two genera of Banner et Desai, 1988:*Lilliputianella* and *Blefuscuiana*).

Superfamily Globotruncanoidea Brotzen, 1942

Shell trochoid, low, with usually flattened from the spiral side or more rarely with inflated to globular chambers, may have peripheral imperforate carinal band; aperture umbilical, tegilla of successive chambers cover the umbilical area, additional intralaminal and infralaminal apertures side may be present.

Families: Globotruncanidae Brotzen, 1942 (with Globotruncaninae Brotzen, 1942, Globotruncanellinae Maslakova, 1964, Abathomphalinae Pessagno, 1967, Archaeoglobogerininae Salaj, 1987), Rugoglobigerinidae Subbotina, 1959.

Superfamily Globorotalioidea Cushman, 1927

Shell trochoid, flattened, conical, with keel or evenly inflated and with broadly rounded peripheral margin, umbilicus usually closed; wall with thick outer calcite layers, surface smooth, or may be pitted or pustulose; aperture arch or fissure like opening extraumbilical-umbilical, may be bordered by a lip, additional apertures only weakly developed, rarely additional sutural and infralaminar apertures may occur, tegilla absent.

Families: Globorotaliidae Cushman, 1927, Pulleniatinidae Bronnimann et Bermudez, 1953, Truncorotaloididae Loeblich et Tappan, 1961, Candeinidae Cushman, 1927 (including *Canderotalia* Saito, 1952), Catapsydracidae Bolli, Loeblich et Tappan, 1957 [with Catapsydracinae Bolli, Loeblich et Tappan, 1957 (probably without *Protentella*), Soldaniellinae Bronnimann, Whittaker et Parisi, 1987 (transferred from Globigerinidae by Loeblich & Tappan, 1992, as nonspinose)].

Order HANTKENINIDA Mikhalevich, 1980

Shell planispiral, involute, biconvex, umbilicus may be closed or opened, chambers slightly to strongly inflated, may be elongated in radial direction, may have tubulospines arising from the peripheral margin; aperture at the base of apertural surface, equatorial or rarely very slightly asymmetrical and shortly moving on the umbilical side, may be bordered by a lip, may become cribrate, relict or areal, additional apertures may be present.

Paleocene – Oligocene.

Superfamily Hantkenininoidea Cushman, 1927

Shell planispiral, involute, biconvex, biumbilicate, chambers slightly to strongly inflated, may be elongated in radial direction, may have tubulospines arising from the peripheral margin; aperture equatorial arched or as a high loop, may be boarded by a lip or have apertural flanges at its both sides, may become cribrate.

Remarks. Family Cassigerinellidae is removed from this superfamily as having principally different chamber arrangement.

Families: Hantkeninidae Cushman, 1927, Globanomalinidae Loeblich et Tappan, 1984.

Superfamily Planomalinoidea Bolli, Loeblich et Tappan, 1957

Shell planispiral, chambers inflated or flattened, may be radially elongated, umbilicus closed; aperture equatorial to very slightly asymmetrical, with bordering lip, apertures of earlier chambers may partly remain opened as relict additional apertures.

Remarks. Family Schackoinidae primary placed by Loeblich & Tappan (1987) in this superfamily was later moved by them to the former Rotaliporacea (= Rotalioidea). As the trochoid character of its shell is not distinctly expressed and looks somewhat irregular at the very early stage or is not expressed at all the position of this family is kept here as belonging to Planomalinoidea.

Families: Planomalinidae Bolli, Loeblich et Tappan, 1957, Globigerinelloididae Longoria, 1974 (with Globigerinelloidinae Longoria, 1974,? Eohastigerinellinae Loeblich et Tappan, 1984), Schackoinidae Pokorny, 1958.

Order CASSIGERINELLIDA, ord. nov.

Shell biserial and planispirally enrolled, flattened or subglobular, with chambers from slightly to strongly inflated becoming nearly subsphaerical; umbilicus close; aperture at the base of the apertural face, as arched opening or loop-shaped elongated opening extending up from the base of apertural surface in parallel to the peripheral margin.

Upper Eocene – Lower Miocene.

Remarks. Biserially enrolled chambers clearly distinguishes this order from all the other orders of the subclass Globigerinana. Genus *Dipsidripella* Brotea, 1995 assigned to the family Cassigerinellidae (Pearson et al., 2006) obviously belongs to another family as it is fully trochospiral lacking the biserial arrangement of chambers characteristic for cassigerinellids.

Superfamily Cassigerinellidoidea Bolli, Loeblich et Tappan, 1957

Shell biserial and planispirally enrolled, flattened or subglobular, with chambers from slightly to strongly inflated becoming nearly subsphaerical; umbilicus close; aperture at the base of the apertural face, as arched opening or loop-shaped elongated opening going up from the base of apertural surface in parallel to the peripheral margin.

Remarks. Superfamily removed from the order Hantkeninida whose representatives have planispiral but not biserial shell. This specialized line adapted to pelagic mode of life, supposedly have its predecessors in cassidulinids.

Family: Cassigerinellidae Bolli, Loeblich et Tappan, 1957.

Order GLOBIGERINIDA Carpenter, Parker et Jones, 1862

Shell moderately to high trochoid, in some genera trochoid only at the early stage, later chambers with tendency to planispiral arrangement or the last chamber envelopes all the preceding ones; chambers globular or radially elongated, sometimes only weakly bound with each other; umbilicus narrow or widely opened, yawning or may be covered by additional plate; wall cancellate, in globigerinids may be with a thick calcareous cancellate layer (crust) and large pores, may be covered with long and thin monocrystalline needles; additional skeletal plates may cover aperture and sutures; umbilicus may be narrow or wide yawning, disposed in the umbilical area where the apertures of the preceding chambers are also opened, in the shells with the enveloping last chamber apertures may be multiple, additional sutural and umbilical apertures are widely developed.

Middle Eocene – Holocene.

Superfamily Favuselloidea Longoria, 1974

Shell trochospiral entirely, moderately to high trochospiral, chambers strongly inflated, nearly subsphaerical; wall cancellate with fields between costellae or ridges abundantly perforated and rare pores on the elevated reticulum (Favusellidae) or may only appear cancellate (with imperforate small tubercules and pores grouped in local areas as in Favusellidae (Globuligerinidae). umbilicus narrow to wide and deep; aperture umbilical to slightly extraumbilical or high umbilical arch, in both cases boarded by imperforate lip.

Remarks. Additional apertures, needles or crust not developed. Wall may be aragonitic (Grigelis & Gorbatchick, 1980).

Families: Favusellidae Longoria, 1974 (including *Ascoliella* Banner et Desai, 1988), Globuligerinidae Loeblich et Tappan, 1984.

Superfamily Globigerinoidea Carpenter, Parker et Jones, 1862

Shell trochoid, sometimes only at the early stage, later chambers with tendency to planispiral arrangement or the last chamber envelopes all the preceding ones; chambers globular or radially elongated, sometimes only weakly bound with each other and being empty (without the cytoplasm) could be easily broken (*Boliella*); umbilicus usually widely opened, yawning or may be covered by additional plate; wall may be with a thick calcareous cancellate layer (crust) and large pores in each cell, may be covered with long and thin monocrystalline needles set loosely in the holes of elevated papilla (Globigerinidae) and jointed movable or in some genera needles are fastened and immobile, in transverse section needles may be circular, trihedral, triradial, may be disposed all over the shell surface or at the peripheral ends of the chambers only, adult stage may develop a thick outer calcite layer; additional skeletal plates are often covering aperture and sutures; aperture usually huge, yawning, disposed in the umbilical area where the apertures of the preceding chambers are also opened, in the shells with the enveloping last chamber apertures may be multiple (multitudinous circular openings all over the surface of the enveloping chamber which are significantly larger than pore openings), additional sutural and umbilical apertures are widely developed.

Remarks. In the subfamily Orbulininae genus *Biorbulina* is regarded as phenotypic variation of *Orbulina* (see Caron et al., 1987, Spero, 1988). The latter author explains the formation of the double sphere as a result of the greater energy reserve of the individual rather than its growth in sub-optimal conditions. The genus *Globigerinatella* is regarded here as belonging more likely to the cymbaloporids.

Families: Globigerinidae Carpenter, Parker et Jones, 1862 (with Globigerininae Carpenter, Parker et Jones, 1862, Porticulasphaerinae Banner, 1982), Marginotruncanidae Pessagno, 1967 (with *Pessagnoites* AlShuaibi, 2011), Orbulinidae Schultze, 1854 (without the genus *Globigerinatella* which is transferred here to Cymbaloporids), Hastigerinidae Bolli, Loeblich et Tappan, 1957, Globigerinitidae Bermudez, 1961 1961 [with the subfamilies Globigerinitinae Bermudez, 1961 (including *Tenuitellinata* Li, 1987) and *Tenuitellinae Banner, 1982 transferred by Li, 1987 from the Candeinidae and including *Tenuitellita* Li, 1987 and *Praetenuitella* Li, 1987)].

Chapter XII

SUMMARY

This book deals with testate Foraminifera, the most abundant shelled organisms in the modern oceans and whose fossil shells are documented in the geological record since the Cambrian (i.e., for more than 500 million years). A new synthesis, the Foraminiferal Macrosystem Concept, examines the taxonomy and evolution of this exceptionally useful lineage of protists. The synthesis is based upon shell morphologies and characteristics of the soft cell body, as well as upon the more limited information on life cycles and biology.

In this synthesis, the Foraminifera are regarded as a phylum. The peculiarities of the nuclear apparatus, their exceptional diversity and specialization, as well as the originality of morpho-physiological trends in evolution that have created multi-chambered shells and the canal system - an integrative system of the organism level – are all unique among unicellular Eukaryotes and provide many analogies to evolutionary trends in multicellular organisms. Along with the range of peculiarities of the structure of their granuloreticulopodia, their unusual life cycles (with antithetic change of generations with the intermediate reduction), this range of complex characteristics gives strong support to elevate the rank of Foraminifera to the phylum level (Margulis, 1974, Margulis et al., 1990, Mikhalevich, 1980, 1981, 1992, 2000, Page, 1987).

Consistent with the elevation to phylum rank, class- to subfamily-level taxonomy is proposed for the Foraminifera. Five classes (Astrorhizata Saidova, 1981, Spirillinata Mikhalevich, 1992, Miliolata Saidova, 1981, Nodosariata Mikhalevich, 1992, Rotaliata Mikhalevich, 1980) and 11 subclasses (Astrorhizana, Saidova, 1981, Lagynana Mikhalevich, 1980, Spirillinana Mikhalevich, 1992, Ammodiscana Mikhalevich, 1980, Miliolana Saidova, 1981, Miliamminana Mikhalevich, 1980, Nodosariana Mikhalevich, 1992, Hormosinana Mikhalevich, 1992, Rotaliana Mikhalevich, 1980, Textulariana Mikhalevich, 1980 and Globigerinana Mikhalevich, 1980) are defined. These include 73 orders, 27 suborders, 98 superfamilies, 499 families and 368 subfamilies, among which 2 orders (Cymbaloporida, Cassigerinellida), suborder Duostominina, two families (Cymbaloporettidae, Haynesinidae), and two subfamilies (Cushmanellinae and Tristixinae) are new.

Historically, agglutinated lineages were classified as distinct from calcareous lineages. The definition of the Textulariacea (later progressively elevated to an order - Loeblich & Tappan, 1992) was based only on the common feature of the agglutinated character of the shell wall. However, detailed analysis of the multitude of foraminiferal morphotypes has revealed that morphological trends are similar for the agglutinated and calcareous forms within the separate classes. The differences between the agglutinated subclasses of the different classes proposed, in fact, are greater than between the agglutinated and calcareous subclasses within a single class. Thus, the taxon based mainly on the only common feature (agglutinated character of the shell wall) was considered to be polyphyletic (Mikhalevich, 1980, 1981, 1992, 2000, 2004).

The basic hypothesis presented in this book is that the different kinds of the calcareous wall structures evolved independently in different phyletic lines and at different times in geological history (Mikhalevich, 1981, 1992, 2000) (e.g., the Spirillinata's calcareous wall emerged in the Upper Devonian, Nodosariata's earlier in the Devonian, Rotaliata's in the Triassic). Thus, the former Textulariacea here are distributed among the five different classes according to their morphologies, as the rise of agglutinated shell wall apparently occurred independently and in parallel in several phyletic lineages (Mikhalevich, 1980, 1981, 1992, 2000, 2004). The close affinity of agglutinated subclasses with their analogous calcareous subclasses within each class is demonstrated based on morphologies, similar tendencies in ontogenetic and phylogenetic development, as well as the timing of appearances and diversifications in geological history.

The former fusulinoids are also shown to be a heterogenous group. Paleozoic representatives of Nodosariata and Spirillinata were moved from the fusulinoids to their corresponding classes, consistent with the hypothesis of evolutionary continuity of the ancient nodosariate and spirillinate lineages to the recent forms (Mikhalevich, 1992, 1999, 2000, 2009). The primary fusulinoid group here is placed in the class Miliolata, based on the hypothesis that some characteristics of typical fusulinids represent a divergent Miliolata lineage with a strong and peculiar direction in the development of complex inner structures in their advanced forms (Mikhalevich, 2004, 2006, 2009).

A basic premise for this group also is that its classification should be constructed mainly on their morphologic features, rather than on stratigraphic ranges or the number and character of the layers of their shell wall, which can change during the ontogenesis of the shell, as happens in recent miliolids. The lack of communication between specialists studying foraminifera of restricted periods in geological history and those studying the recent forms has often resulted in the interpretation of artificial borders within phyletic lineages

Some taxa with hyaline calcitic shell walls are transferred from Rotaliata to Spirillinata and Nodosariata, corresponding to shell morphologies and also considering the characteristics of shell-wall ultrastructure. Additional changes in the composition of the orders and families are also presented. Taxa whose position is still doubtful are marked with an interrogation mark to attract the attention of future researchers. The new system is here presented to the subfamily level; the class Astrorhizata, four agglutinated subclasses (Ammodiscana, Miliamminana, Hormosinana and Textulariana) and milioloid group of superorders of subclass Miliolana were earlier defined to the generic level (Mikhalevich, 1985, 1988, 2004).

The taxonomic significance of different morphological features is discussed in detail. Contrary to the majority of the classifications adapted previously, the whole building plan structure of the organism in connection with its functions which is reflected in the basic morphological features of the shell is considered as having preferential significance compared with the ultrastructure of the shell wall. These shell features include the number of chambers (one, two or many), the mode of coiling, shell and chamber form, the structure and position of the main and additional apertures, subdivision of the chamber lumen, and the presence or absence of integrative systems, as well as the complexes of all these features. The shell wall is important, but is of subordinate value and may serve as the diagnostic feature of taxa of lower rank. Thus, agglutinated/microgranular shell wall characterizes lower subclasses of each class, and special types of calcareous microstructure – the more advanced subclasses of the same phyletic line. Diagnoses of the classes are based mostly on their shell morphology. Among the taxonomically significant features enumerated above, the plan of the structure of the shell (which is considered as the disposition of its parts in connection with their function) reflecting the plan of the organism (including form and number of chambers, and mode of their arrangement in connection with the apertural position and structure) is regarded here as the basic feature (but not the aperture only as was understood by some authors, the latter, although being very important functionally, nevertheless represent only one of the elements of the whole structural plan of the organism).

Thus, when assessing taxonomic rank based on shell structure and shell-wall ultrastructure, priority is given here to the former. Forms with different kinds of agglutinated (agglutinated/microgranular) wall are considered to be most likely ancestral stages (the preceding steps) in the evolutionary development of various phyletic lineages (classes), and they acquire special types of calcareous shell microstructure only in their most advanced forms. This supposition is supported by the biostratigraphic record: usually in each phylogenetic line under consideration, the agglutinated forms were found earlier than their calcareous secreted analogues. In some cases, however, forms with agglutinated /microgranular walls may be sister groups to their analogues with special ultrastructural types of their calcareous wall.

Taxa of equal taxonomic rank are separated using the categories of the features for that level (for instance, superfamilies are separated based on the complication of their chamber lumen and inner chamber wall). Paleozoic and later forms are viewed in a unified system with the recent forms. The calcareous analogues of the agglutinated forms, viewed previously as a result of evolutionary convergence, here are regarded as related and are united within each class according to their specific organization. These changes (as well as the analysis of the cases of convergence and parallelism within the different taxa) clarified a significant number of parallelisms of the previous systems and known polyphyletic groups.

The above mentioned analyses were focused primarily upon defining the classes within the Foraminifera. Further elucidation of the phylogenetic development within the classes and subclasses, and revision of the subordinate groups, generally are not discussed in detail and should be regarded among the next goals for future taxonomic studies.

The goal of taxonomy as a science is to describe the diversity of life forms within the hierarchy of taxa that are related genetically. A taxonomic structure should reflect the main features of the evolutionary process giving the general picture of the development of living matter in the space-time continuum. That is why it is important to stress the significance of the hierarchical composition of classification proposed. Ranking of the taxa is not a formal technique only. The natural system of organisms reflects in its ideal variant the phylogeny of the group. Indeed, life is a hierarchical structure. The hierarchical relations of the taxa of the lower taxonomic level (from species up to the order) in the traditional foraminiferal classifications were usually adequately ranked. But the hierarchy at the higher taxonomic level (superorder, class, phylum) has mostly been neglected. Thus, regarding the bulk of the Foraminifera as belonging to superfamilies or orders of equal rank distorts their actual relationships.

Moreover, classifying the whole foraminiferal group as an order or a class instead of a phylum limits consideration of phylogenetic relationships of the different taxa within the Foraminifera, as well as the position of Foraminifera among the other Eukaryotes.

The accuracy of the classification depends of the right hyerachy of the diagnostic features. Thus the change of the understanding of hierarchy of the features used in foraminiferal taxonomy (shell structure/ shell wall versus shell wall/shell structure) fundamentally changes the structure of the foraminiferal classification.

The new classification allows a distinctive diagnosis of each of the classes proposed, as well as their evolutionary progressions. The diagnoses of the high rank taxa are based on features of their skeleton structure which are of evolutionary significance. For instance complication of the shell structure by subdivision of the chamber lumen, such as formation of the complex apertural and canal systems, promotes differentiation and compartmentalization of the cytoplasm. Moreover, the integrative systems provide for more efficient interaction of the different parts of the organism and of the whole shell with the environment. Additional apertures increase communication with the surroundings. Taking into account their great functional and evolutionary significance, their presence should have high taxonomic value, the specific character and position of these apertures may serve as a feature of some lower rank taxa. Cytological data were considered in the diagnoses of the high rank taxa where possible. Some of these skeletal as well as some cytoplasmic features were not used previously in the diagnoses of the highest foraminiferal taxa.

The foraminiferal macrosystem is further examined in the context of evolution. Evolutionary regularities in the development of living organisms have previously been studied primarily at the multicellular level. The regularities of such trends in the evolution of unicellular eukaryotes are much less commonly considered. The Phylum Foraminifera provides the broadest possibilities for studies of evolutionary trends, given its wide distribution, diversity, and numerous examples of independent development of cytoplasmic and skeletal complexity, the later achieving the level striking for unicellular organisms. The whole process of foraminiferal evolution is documented in the geological record, which is the most detailed among the fossil groups and provides an objectively controlled historical background.

The present study represents the results of the comparative morphological analysis of the Foraminifera, revealing the amazingly vast array of evolutionary trends possible at the unicellular level. Emphasis is placed on the primary trends – polymerizations (i.e., duplication of features, the most basic being multi-chambered shells), differentiation and integration – that represent the mainstream of evolutionary development that permit structures to achieve higher levels of organization. The phenomenon of aromorphosis (term of Severtzov, 1925, 1939 = key evolutionary change), permitting substantive change in the structural plan of the organism and its basic properties, is partly touched on. (These trends have a general character and can be applied to the social level of life of organisms as well.)

Comparative morphological analysis of representatives of each class provides examples of polymerization, with the subsequent differentiation of different structures (number of chambers, number of volutions, number of apertures, number of nuclei, etc.), rise of different types of integrative systems providing the process of integration, as well as aromorphoses in each phyletic line. This analysis also permits evaluation of the significance of different features in taxonomic representations.

These analyses reveal that many of the trends occurred in different foraminiferal classes independently and in parallel, and not simultaneously. Examples include the rise of multi-chambered shells, agglutinated and then fully secreted calcareous shell wall, compartmentalization of chambers, increase of shell complexity, appearance of inner apertural structures, of integrative systems of different type, and complexity. The results are similarity of trends within the constraints of the foraminiferal bauplan and, at the same time, resulting in specific morphological features in each class (Mikhalevich, 1981, 1992, 2000, 2005). Thus, agglutinated/microgranular shell walls emerged in different classes and were replaced by fully secreted calcareous shell walls in the classes Spirillinata, Miliolata, Nodosariata and Rotaliata, also in parallel and at different geologic periods, although the ultrastructure of the calcareous wall in each of these classes has its own peculiar features.

Apertural modifications also emerged in several classes in parallel, bearing its specificity in each of them. Similarly, inner apertural structures in the different classes also have special characteristics that can be used as informative taxonomic features. Even such complex structures as canal systems arose in parallel in the subclasses Spirillinana and Rotaliana [in the Paleozoic in Lasiodiscids (Spirillinana) and the upper Cretaceous in the Rotaliana]. As can be seen throughout the book, including from the comparison of the variability and complexity of genera within two subclasses in each class, the agglutinated shell wall has less potential for structural modifications than a wholly secreted calcareous wall. Under the calcareously secreted wall the considerable number of structural prohibitions evidently proved to have been eliminated.

A table summarizing the Foraminiferal Macrosystem Concept (Plate **1a**) and the rise of complexity in its separate branches (Plates **36, 38, 39**) can be compared to the periodic table of elements, where the increase of the level of organization goes along each morphological row (class), as well as in the smaller lines within them. And although the increasing level of organisation in the foraminiferal rows are not expressed as strictly as in chemistry, it, nevertheless, is regular and, to great extent, even predictable.

Through their evolutionary histories, the different classes have achieved degrees of complexity. The most primitive shell morphology is that of the class Astrorhizata, having unilocular shells with agglutinated or soft organic walls. The appropriate system of this class, in which the diversity of its representatives is not yet fully known, remains to be defined using combined cytological, molecular and morphological methods of study.

Representatives of the other classes achieved multi-chambered organization, with super-multi-chambered shells in the most advanced of them – Miliolata (fusulinoids group, alveolinids) and Rotaliata (nummulitids). The most complex apertural structures and integrative systems of the shell are achieved in the representatives of the class Rotaliata. The bilamellar (=bifontinal – in Russian language literature) character of their calcareous shell wall gives them the possibility of creating more complex shell constructions with less weight. These advanced features allowed members of the Rotaliata to occupy a vast, new ecospace – the pelagic realm.

The qualitative originality of evolution is most easily discerned in the more primitive subclasses Lagynana and Astrorhizana, with their nearly exclusively unilocular (or rarely pseudo-multi-chambered or pseudocolonial) shells often of irregular form, simple aperture and tectinous or agglutinated, rarely microgranular wall. Their agamonts remain for quite a long time mononuclear, they lack nuclear dimorphism, and their cytoplasm is not differentiated into endo- and ectoplasm.

Homokaryosity is also characteristic of the other ancient and peculiar foraminiferal branch, the Spirillinata, possessing pseudo-two-chambered shell with a long tubular second chamber (or derived from such forms), with a simple aperture, but possessing well-ordered radial ultrastructure of the wall acting as a single crystal in their advanced forms. Their ancient forms combine a set of plesiomorphic features the (presence of the tubular chamber in many of them at least at the earlier stage, pseudo-chamberedness, primitive septa, simple aperture, glomerate coiling) with some advanced apomorphic structures (special trochospiral shell types, multi-chamberedness in their advanced groups, the presence of tunnels and canals, additional apertures).

In miliolats and nodosariats, the shell types are rather diverse with the predominance of a specific coiling, chamber form and apertural structures in each, with very purely developed integrative systems. The shell wall in representatives of their calcareous subclasses is monolamellar, with peculiar packing of crystals in each of them. The nuclear apparatus of miliolats acquires nuclear dimorphism, data on Nodosariata are not available.

Whatever evolutionary trends considered – nuclear apparatus, shell, apertural and integrative structures, wall ultrastructure – all of them have the most development, diversity and complexity in Rotaliana group. The rise of the bilamellar calcite wall allowed their shells to be strong with minimum thickness and weight. This adaptation gave them the possibility of aromorphic changes, resulting in sharp rise of their evolutionary rates, increase in the number of taxa both of higher and lower rank, adaptive radiation, and the advent of the pelagic forms. The origin of the latter from the different branches of benthic rotaliats (close to some discorbids, nonionids, bolivinids) could be supported by their morphological resemblance and by the presence of a benthic stage in their life cycle. Such a close relationship was conclusively and graphically demonstrated in the work of Darling et al. (2009) documenting the genetic and morphologic identity of the *Bolivina/ Streptochilus* species living usually as benthic forms but also surviving and growing as plankton. Obviously such a double lifestyle is not so rare a phenomenon among the Rotaliana and possibly may also occur in some other foraminiferal taxa, providing the wider distribution of those species. The more specialized planktonic forms, living fully as pelagic forms and acquiring special adaptive features changing significantly their shell morphology, provides a basis for separating them as an independent taxon (subclass Globigerinana) within the class Rotaliata.

The origin of Miliolata is, according to Arnold (1978) from ancestral forms similar to *Periptigma* (Lagynana). He was one of the first to suggest the independent emergence of Miliolata and Rotaliata from soft-shelled Lagynana (Arnold, 1978, 1979, 1984). Other foraminiferal classes also likely originated from ancient Lagynana representatives, and these events may have occurred prior to the Cambrian (Mikhalevich – 1980, 1999, 2000).

The primitive prototypes of the morphologic forms of the higher testate subclasses could be found among the naked or soft-walled unilocular predecessors. The roots of Spirillinata and Nodosariata obviously could also be discovered among the Recent Lagynana or their ancient forerunners. The supposition of traditional taxonomists that ancestral forms of the later appearing skeletal foraminiferal groups (representing separate phyletic lineages – classes) existed among the unilocular soft-walled forms representing distinct phyletic lines, has been supported by molecular analysis (Habura et al., 2004).

The transition from one organizational level to a new one occurs when possibilities of the previous structures are exhausted. It happens as a result of aromorphoses and usually goes through a nonequilibrium stage (Prigogene & Stengers, 1986). Thus, the eukaryotic

Rotaliana's cells of higher complexity represent the highest level of organization at the unicellular level, and the mainstream of evolution of life went later the way of multicellularity just as the previous Procaryota level was followed by the eukaryotic way. The same mechanisms could be applied and searched for within different organizational levels of varying complexity.

This book presents a new approach to foraminiferal systematics and evolution. The implications require profound revisions of many taxa, especially of agglutinated groups, of the class Spirillinata, and of several major groups within the subclass Rotaliana. The unifying trends and patterns revealed in this study provide a roadmap for such revision.

LIST OF REFERENCES USED AND RECOMMENDED[1]

Achnazarov, E. B. (2002). *Kontury evolucii.*[*The contours of evolution*]. Nedra, St. Petersburg. 438 (In Russian).

Adl, S. M., Simpson, A. G. B., Lane, C. E., Lukes, J., Bass, D., Bowser, S. S., Brown, M. W., Burki, F., Dunthorn, M., Hampl, V., Heiss, A., Hoppenrath, M., Lara, E., Le Gall, L., Lynn, D. H., McManus, Y., Mitchell, E. A. D., Mozley-Stanridge, S. E., Parfrey, L. W., Pawlowski, J., Rueckert, S., Shadwick, L., Schoch, C. L., Smirnow, A. & Spiegel, F. W. (2012). The Revised Classification of Eukaryotes. *J. Eukaryot. Microbiol.*, *59*(5), 429 – 493. DOI: 10.1111/j.1550-7408.2012.00644.x.

Afanasieva, M. S. (1982). Structurnye elementy stenki rakovin buliminid [Structural elements of the buliminid's shell wall]. *Paleontologicheskij jurnal*, *4*, 14 – 20.

Aguilar, M., Bernaus, J. M., Caus, E. & Hottinger, L. (2002). *Lepitorbitoides minima* Douville from Mexico, a foraminiferal index fossil for Campanian. *Journ. Foram.Res.*, *32* (2), 126 - 134.

Akimoto, K., Matsui, C., Shimokava, A. & Furukawa, K. (2002). *Atlas of Holocene benthic foraminifers of Shimbara Bay, Kyushu, Southwest Japan*. The Kagoshima University Museum Monographs, N *2*, 1 – 112.

Aldredge, A. L. & Jones, B. M. (1973). *Hastigerina pelagica*: Foraminiferal habitat for planktonic dinoflagellates. *Marine Biology*, *22* (2), 131 - 135.

Alexander, S. P. & Delaca, T. E. (1987). Feeding adaptations of the foraminiferan *Cibicides refulgens* living epizoically in the antarctic scallop *Adamussium colbecki*. *Biol. Bull.*, *173*. (1), 136 - 159.

Allen, R. D. (1964). Cytoplasmic streaming and locomotion in marine foraminifera. In: Allen R.D., Kamiya N. *Primitive Motile Systems in Cell Biology. Acad.* Press. New York, 407 - 433.

Allen, K., Roberts, S. & Murray, J. W. (2000). Analysis of organic components in the test wall of agglutinated Foraminifera by Fourier Transform Infrared and Pyrolisis Gas Chromatography Mass Spectrometry. Eds. Hart M.B., Kaminski M.A., Smart C.W. *Proc. Fifth Intern. Workshop on Agglutinated Foraminifera.* Grzybowski Foundation Spec. Publ., *7*, 1 – 13.

---

[1] The present list includes mostly literature on the cytology and biology of Foraminifera as well as the works on taxonomy published after 1988 year becaus the full bibliography of he preceeding references on the foraminiferal systematic and morphology is represented in the Loeblich & Tappan (1987) monograph.

Allison, N., Austin, W., Paterson, D. & Austin, H. (2010). Culture studies of the benthic foraminifera *Elphidium williamsoni*: Evaluating pH, Δ[CO32−] and inter-individual effects on test Mg/Ca. *Chemical Geology*, *274* (1 – 2), 87 - 93.

AlShuaibi, A. (2011). *Pessagnoites*, a new foraminifer genus from the Cretaceous of the Austin Chalk, northeast Texas. *Micropaleontology*, *57*, 87 – 91.

Altenbach, A. V., Heeger, T., Linke, P., Spindler, M. & Thies, A. (1993). *Miliolinella subrotunda* (Montagu), a miliolid foraminifer building large detritic tubes for a temporary epibenthic lifestyle. *Marine Micropaleontol.*, *20*, 293 – 301.

Altenbach, A. V., Struck, U. (2001). On the coherence of organic carbon flux and benthic foraminiferal biomass. *Journ. Foram. Res.*, *31*, 79 – 85.

Altin, D. Z., Habura, A., Goldstein, S. T. (2009). A new allogromiid foraminifer *Niveus flexilis* nov. gen., nov. sp., from Coastal Georgia, USA: fine structure et gametogenesis. *Journ. Foram.Res.*, *39* (2), 73 - 86.

Altiner, D. (1988). Pseudovidalinidae n. fam. and *Angelina* n. gen. from the upper Permian of south and southeast Turkey. *Rev de Paleobiol., Sp. vol. 2, (Benthos '86),* 25 – 36.

Altiner, D. (1999). *Sengoerina argandi*, n. gen., n. sp., and its position in the evolution of Late Permian biseriamminid foraminifers. *Micropaleontology*, *45* (2), 215 – 220.

Altiner, D. & Özkan-Altiner, S. (2001). *Charliella rossae* n.gen., n.sp., from the Tethyan realm: remarks on the evolution of Late Permian biseriamminids. *Journ. Foram. Res.*,31 (4), *309 - 314.*

Altiner, D. & Özkan-Altiner, S. (2010). *Danielita gailloti n. gen., n. sp.,* within the evolutionary framework of Middle and Late Permian Dagmaritins. *Turkish Journ. of Earth Sciences, 19*, 497 – 512.

Alve, E. (1991). Benthic Foraminifera in sediment cores reflecting heavy metal pollution in S0Rfiord, Western Norway. *Journ. Foram.Res.*, *21* (1), 1 - 19.

Alve, E. (1995). Benthic foraminiferal responses to estuarine pollution: a review. *Journ. Foram. Res.*, *25* (3), 190 - 204.

Alve, E. (1999). Colonization of new habitats by benthic foraminifera: a review. *Earth-Sci. Rev.*, *46*, 167 – 185.

Alve, E. & Olsgard, F. (1999). Benthic foraminiferal colonization in experiments with copper-contaminated sediments. *Journ. Foram. Res.*, *29* (3), 186 - 195.

Alve, E. & Murray, J. W. (2001). Temporal variability in vertical distributions of live (stained) intertidal foraminifera, Southern England. *Journ. Foram.Res.*, *31*, 12 – 24

Alve, E. & Goldstein, S. T. (2002). Resting stage in benthic foraminiferal propagules: a key feature for dispersal? Evidence from two shallow-water species. *J.Micropalaeontol.*, *21*, 95 – 96.

Alve, E. & Goldstein, S. T. (2003). Propagule transport as a key method of dispersal in benthic foraminifera. *Limnol. Oceanogr. 48*, 2163 – 2170.

Alve, E. & Goldstein, S. T. (2010). Dispersal, survival and delayed growth of benthic foraminiferal propagules. *Journ.of Sea Research*, *63* (1), 36 - 51

Andersen, H. V. (1984). The wall ultrastructure of Superfamily Miliolacea and its lineage in the Gulf Coastal Plain. *Tulane studies Geol. and Paleontol.*, *18*, 1 – 19.

Anderson, O. R. (1984). Cellular specialization and reproduction in planktonic Foraminifera and Radiolaria. In: Marine Plankton Life Cycle Strategies. *Boca Raton. Florida*, 35 - 66.

Anderson, O. R. (1988). *Comparative Protozoology - ecology, physiology, life history.* Springer Verlag. New York.

Anderson, O. R. & Be, A. W. H. (1976 a). A cytochemical fine structure of phagotrophy in a planktonic foraminifer, *Hastigerina pelagica* (d'Orbigny). *Biol. Bull.*, *151* (3), 437 - 449.

Anderson, O. R. & Be, A. W. H. (1976 b). The ultrastructure of a planktonic foraminifer , *Globigerinoides sacculifer* (Brady), and its symbiotic dinoflagellates. *Journ. Foram. Res.*, 6 (1), 1 - 21.

Anderson, O. R. & Be, A. W. H. (1978). Recent advances in foraminiferal fine structure Researches. In: Hedley R.H., Adams C.G. (Eds). *Foraminifera.*, Vol. *3*, 121 - 203.

Anderson, O. R. & Tuntivate-Choy, S. (1984). Cytochemical evidence for peroxisomes in planktonic Foraminifera. *Journ. Foram. Res.*, *3*, 203 - 206.

Angell, R. W. (1967). The text structure and composition of the Foraminifer *Rosalina floridana*. *Journ. Protozool.*, *14*, 299 - 307.

Angell, R. W. (1967). The process of chamber formation in the Foraminifer *Rosalina floridana*. *Journ. Protozool.*, *14* (4), 566 - 574.

Angell, R. W. (1971). Observations on gametogenesis in the foraminifer *Myxotheca*. *Journ.Foram. Res.*, *1* (1), 39 - 42.

Angell, R. W. (1979). Calcification during chamber development in *Rosalina floridana*. *Journ. Foram. Res.*, *9* (4), 341 - 353.

Angell, R. W. (1980). Test morphogenesis (chamber formation) in the foraminifer *Spiroloculina hyalina* Schultze. *Journ.Foram. Res.*, *10* (2), 89 - 101.

Angell, R. W. (1990). Observations on reproduction and juvenile test building in the foraminifer *Trochammina inflata*. *Journ. Foram. Res.*, *20*, (3), 246 - 247.

Angel, D. L., S. Verghese, S., Lee, J. J., Saleh, A. M., Zuber, D., Lindell, D. & Symons, A. (2000). Impact of a net cage fish farm on the distribution of benthic foraminifera in the Northern Gulf of Eilat (Aqaba, Red Sea). *Journ. Foram. Res.*, *30* (1), 54 – 65.

Archambault-Guezou, T. & Lefevre, R. (1981). Some applications of ion microscopy to biomineralization involved in sedimentary rocks. *Scanning Electron Microscopy.*, *1*, 503 - 514.

Armynot du Chatelet, E. & Debenay, J. -P. (2010). Anthropogenic impact on the western French coast as revealed by foraminifera: a review. *Rev. Micropal.*, *53*, 129 – 137.

Arnaud -Vanneau, A. (1988). Grands Foraminifères du Crétacé. *"Benthos' 1986". Rev. Paléobiol,.* Vol. Spécial, *2*, Part 1, 1 – 351.

Arnaud -Vanneau, A., Boisseau, T. & Darsac, C. (1988). Le genre *Trocholina* Paalzow 1922 et ses principales espèces au Crétacé. *"Benthos' 1986". Rev. Paléobiol.* Vol. Spécial *2*, Part. 1, 353 – 377.

Arnaud-Vanneau, A. & Prestat, B. (1984). *Thomasinella* and Co. *International Symposium on Benthic Foraminifera (2nd : 1983 : Pau, France) & Oertli, H. J & Société nationale Elf-Aquitaine (Production) &Total Compagnie Francaisedes Petroles & Esso Rep.(1984). Benthos '83: 2nd International Symposium on Benthic Foraminifer , Pau (France), 11-15/4/1983. Elf Aquitaine, Esso REP, Total CFP, Pau and Bordeaux, France*, 19 – 26.

Arnold, A. J., Ecrivan, F. & Parker, W. C. (1985). Predation and avoidance responces in the Foraminifera of the Galapagos Hydrothermal Mounds. *Journ. Foram. Res.*, *15* (1), 38 – 42.

Arnold, Z. M. (1954). *Discorinopsis aguayi* (Bermudez) and *Discorinopsis vadescens* Cushman and Bronnimann: a study of variation in cultures of living foraminifera. *Contr. Cushm. Found. Foram. Res.*, *5*, 4 – 13.

Arnold, Z. M. (1954). A note on foraminiferan sieve-*plate*s. *Contr. Cushm. Found.Foram. Res.*, *5* (2), 77.

Arnold, Z. M. (1954). Variation and isomorphism in *Allogromia laticollaris*: A clue to foraminiferal evolution . *Contrib. Cushman Found. Foraminiferal Res.*, *5*, 78 - 87.

Arnold, Z. M. (1955). An unusual feature of Miliolid reproduction. *Contr. Cushm.Found. Foram. Res.*, *6*, 94 – 96.

Arnold, Z. M. (1964). Biological observations on the foraminifer *Spiroloculina hyalina* Schulze. *Univ. Calif. Publ. Zool.*, V. *72*, 93.

Arnold, Z. M. (1967). Biological observations on the foraminifer *Calcituba polymorpha Roboz. Arch. Protistenkunde*, *110*, 280 – 304.

Arnold, Z.M. (1978 a). An Allogromiid ancestor of the Miliolidean Foraminifera. *Journ. Foram. Res.*, *8* (2), 83 – 97.

Arnold, Z. M. (1978 b). Biological evidence for the origin of polythalamy in Foraminifera. *Journ. Foram. Res.*, *8* (2), 147 – 166.

Arnold, Z. M. (1979). Biological clues to the origin of miliolidean foraminifera. *Journ. Foram. Res.*, 9 (4), 302 – 321.

Arnold, Z.M. (1982. *Psammophaga simplora* n. gen., n. sp., a polygenomic Californian saccamminid. *Journ. Foram. Res.*, *12* (1), 72 - 79.

Arnold, Z. M. (1984). The gamontic cariology of the saccamminid foraminifer *Psammophaga simplora* Arnold. *Journ. Foram. Res.*, *14* (3), 171 - 187.

Arnold, Z. J., d'Escrivan, F. & Parker, W. C. (1985). Predation and avoidance responses in the Foraminifera of the Galapagos hydrothermal mounds. *Journ. Foram. Res.*, *15* (1), 38 - 42.

Ashraf M. Samir (2000). The response of benthic Foraminifera and ostracodes to various pollution sources: a study from two lagoons in Egypt. *Journ. Foram. Res.*, *30*, 83 – 98.

Austin, H. A., Austin, W. E. N. & Paterson, D. M. (2005). Extracellular cracking and content removal of the benthic diatom *Pleurosigma angulatum* (Quekett) by the benthic foraminifera *Haynesina germanica* (Ehrenberg), *Marine Micropaleontology*, *57*, 68–73.

Azbel, A. Ja. (1986). Novye foraminifery Oksforda Russkoj platformy. [New Foraminifera of the Oxford of the Russian Platform]. *Paleontologicheskij Journal*, *1*, 27 – 32. (In Russian).

Banner, F. T. (1971). A new genus of the Planorbulinidae an endoparasite of another foraminifer. *Rev. Esp. Micropal.*, *3* (2), 113 - 128.

Banner, F. T., Sheehan, R. & Williams, E. (1973). The organic skeletons of Rotaline Foraminifera: a review. *Journ. Foram. Res.*,*3* (1), 30 - 42.

Banner, F. T. & Pereira, C. P. G. (1981). Some biserial and triserial agglutinated smaller foraminifera: their wall structure and its significance. *Journ. Foram. Res.*,*11* (2), 85 - 117.

Banner, F.T. & Desai, D. (1988). A review and revision of the Jurassic-Early Cretaceous Globigerinina, with a special reference to the Aptian assemblages of Speeton (North Yorkshire, England). *Journ. Micropal., 7*, pt. 2, 324 p.

Banner, F.T. & Highton, J. (1989). On *Pseudotaberina malabarica* (Carter) (Foraminiferida). *Journal of Micropaleontology*, 8, (1), 113 - 129.

Banner, F.T. & Hodgkinson, R.L. (1991). A revision of the foraminiferal subfamily Heterosteginidae. *Rev. Espagnola Micropaleontol., 23*, 101 - 140.

Banner, F. T., Simmons, M. D. & Whittaker, J. E. (1991). The Mesozoic Chrysalidinidae (Foraminifera, Textulariacea) of the Middle East: the Redmond (Aramco) taxa and their relatives. *Bull. British Mus. Nat. Hist. (Geology)*, *47*, 101 – 152.

Banner, F. T. & Whittaker, J. E. (1991). Redmond's "new lituolid foraminifera" from the Mesozoic of Saudi Arabia. *Micropaleontology*, *37*, 41 – 59.

Banner, F.T.; Whittaker, J.E.; Boudagher-Fadel, M.K. and Samuel, A. (1997). *Socotraina*, a new hauraniid genus from the upper Lias of the Middle East (Foraminifera, Textulariina). *Rev. de Micropaleontol.*, 115 – 124.

Barash, M. S. (1970). *Planktonnye foraminifery v osadkach severnoy chasti Atlantiki. [Planktic Foraminifera in the deposits of the North part of Atlantic*]. Moscow. Nauka. 103 (In Russian).

Barker, R. W. (1960). Taxonomic notes on the species figured by H.B.Brady in his Repport on the Foraminifera dredged by H.M.S. "Challenger" during the years 1873 - 1876. *Soc. Econ. Paleontol. and Mineral. Spec. Publ.*, *9*. 1 - 238.

Barmawidiaja, D. M., Jorissen, F. J., Puskaric, S. & Van der Zwaan, G. J. (1992). Microhabitat selection by benthic foraminifera in the Northern Adriatic Sea. *Soc. Econ. Paleontol. and Mineral*, *22* (4), 297 - 317.

Barras, C., Geslin, E., Duplessy, J. -C. & Jorissen, F. J. (2009). Reproduction and growth of the deep-sea benthic foraminifer *Bulimina marginata* under different laboratory conditions. *Journ. Foram. Res.*,*39* (3), 155 - 166.

Barbieri, R. (1997). Taxonomy of the Neogeneee foraminiferal genus *Spiroloxostoma* (Buliminacea) and description of a new species from the upper Miocene-Pliocene of Morocco. *Rev. Paleobiol.*, *Geneve*, *16* (2), 425 – 432.

Bartenstein, H. (1994). Die Foraminiferen-Art *Falsopalmula tenuistriata* (Franke 1936) und ihre Unterarten im Toarcium Mitteleuropa. *Senckenbergiana lethaea*, *74* (1/2), 33 – 42.

Basov, V. A. (1982). Sostojanie systematiki nodosariid i nekotorye zadachi ee dalnejshej razrabotki. [The state of the nodosariid systematic and some problems of its further elaboration]. *In: Otchet Komissii po mikropaleontologii za 1981 god. [Report of the commission on Micropaleontology for 1981 year. Moscow*, 12 – 15. (In Russian).

Basov, V. A. (1990). Klassifikaciya i phylogenez Lagenid. [Classification and phylogeny of Lagenids]. *Thez. Of XI All-Union micropaleontol. Conference*, Moscow: 21. (In Russian)

Bassi, D., Hottinger, L. & Iryu, Y. (2009). Reassessment of "Boueina" pcifica Ishijima (1978 (Orbitolininae, Foraminiferida), formely considered a green halimedacean algae. *Journ. Foram. Res.*, *39* (2), 120 - 127.

Bassoullet, J.P. (1998). A new larger foraminifer *Bostia irregularis* n.gen., n.sp. (Lituolacea, Hauraniidae) from the Middle Jurassic of Dordogne (France). *Rev. de Micropaleontol.*, *41* (3), 181 – 195.

Baumfalk, Y. A., Fortuin, A. R. & Mok, R. P. (1982). *Talpinela cunicular*ia n. gen., n. sp., a possible foraminiferal parasite of Late Cretaceous *Orbitoides. Journ Foram. Res.*, *12*, 185 – 196.

Be, A. W. H. (1980). Gametogenic calcification in a spinose planktonic foraminifer *Globigerinoides sacculifer* (Brady). *Marine Micropaleontol.*, *5* (3), 283 - 310.

Be, A. W. H. (1982). Biology of planktonic foraminifera. In Foraminifera: *Notes for a short Course. Studies in Geol., Univ. Of Tenessee, Khoxville*, *6*, 51 – 92.

Be, A. W. H. & Hemleben C. (1970). Calcification in a living planktonic foraminifera, *Globigerinoides sacculifer* (Brady). *Neues Jahrb. Geol. Paleontol. Abh.*, *134*, 221 – 234.

Be, A. W. H., Anderson O. R. & Roger O. (1976). Gametogenesis in planktonic foraminifera. *Science, 192* (4242), 890 - 892.

Be, A. W. H., Hemleben, C., Anderson, O. R., Spindler, M., Hacunda, J. & Tuntivate-Choy, S. (1977). Laboratory and field observations of living planktonic Foraminifera. *Micropaleontology, 23* (2), 155 - 179.

Be, A. W. H., Hemleben, C., Anderson, O. R. & Spindler, M. (1979). Chamber formation in planktonic Foraminifera. *Micropaleontology, 25* (3), 294 - 307.

Be, A. W. H., Hemleben, C., Anderson, O. R. & Spindler, M. (1980). Pore structure in planktonic Foraminifera. *Journ. Foram. Res., 10* (2), 117 - 128.

Be, A. W. H., Spero, H. T. & Anderson, O. R. (1982). Effects of symbiont elimination and reinfection on the life processes of the planktonic foraminifer Globigerinoides sacculifer. *Marine Biol., 70* (1), 73 – 86.

Beklemishev, V. N. (1964). Osnovy sravnitelnoj anatomii bespozvonochnych, 3rd edn (Fundamentals of the comparative analysis of Invertebrata). LA Zenkevich (ed) Akademiya Nauk SSSR, otdelenie biologicheskich nauk. Nauka, Moscow . (In Russian).

Beljaeva, N. V. & Saidova, Ch. M. (1967). Foraminiferovye pojasa mirovogo okeana. [Foraminiferal belts of the Worls Ocean]. *Priroda, 1*, 117 – 118.(In Russian).

Bell, R. N. & Winchester-Seeto, T. M. (1999). Linings of agglutinated Foraminifera from the Devonian: taxonomic and biostratigraphic implications. *Journ. Micripaleont., 18*, 37 – 43.

Bellemo, S. (1974a). Studies on the ultrastructures of calcareous Foraminifera with hyaline perforate walls. *Paleontol. inst Univ.Uppsala, 153*, 14 P.

Bellemo, S. (1974b). Ultrastructure in the foraminifer *Cibicides floridanus* (Cushman). *Paleontol. inst Univ.Uppsala, 154*, 11.

Bellemo, S. (1976). Wall ultrastructure in the foraminifer *Cibicides floridanus* (Cushman). *Micropaleontology, 22*, 352–362.

Bellier, J. P., Matheieu, R. & Granier, B. (2010). *Short Treatise on Foraminiferology (Essential on modern and fossil Foraminifera).* Carnets de Geologie – Notebooks on Geology, *2*, 104.

Bender, H. (1992). Chamber formation and biomineralization in *Textularia candeiana* d'Orbigny (Sarcodina: Textulariina). *Journ. Foram.Res., 22* (3), 229 - 242.

Bender, H. (1995). Test structure and classification in agglutinated foraminifera. In: Kaminsky M.A., Geroch S., Gasinski M.A. (Eds).*Proc. Fourth Intern.Workshop on Agglutin. Foram. Krakow, Poland.* Sept. 12 – 19. 1993. *Grzybowski Found. Sp. Publ., 3*, 27-70.

Bender, H. & Hemleben, C. (1988). Calcitic cement secreted by agglutinated foraminifers grown in laboratory culture. *Journ. Foram. Res., 18* (1), 42 - 46.

Berg, L. S. (1922). *Nomogenez ili evoluciya na osnove zakonomernostej (Nomogenez or the evolution based on regularities*). St Petersburg. (In Russian).

Berg, L. S. (1977). *Trudy po teorii evolucii, 1922-1930. [Studies in evolutionary theory*]. Nauka, Leningrad. (In Russian).

Berger, W. H., Buzas, M. A., Douglas, R. G. & Ross, C. A. (Eds.) (1979). Foraminiferal Ecology and Paleoecology. *SEPM Short Course, 6*, 105 – 155.

Berger, W. H., Bonneau, M. C. & Parker, F. L. (1982). Foraminifera on the deep-sea floor: lysocline and dissolution rate. *Oceanol. Acta, 5* (2), 249 – 258.

Bermudez, P. & Seiglie, A. (1963). Estudio systematico de los Foraminiferos del golfo del Carisao. *Bol. Inst. Oceanogr. Univ. Oriente*, *2* (2), 1 – 267.

Bernhard, J. M. (1996). Microaerophilic and facultative anaerobic benthic Foraminifera: A review of experimental and ultrastructural evidence. *Rev. Micropal.*, *15*, 261 – 275.

Bernhard, J. M. (2003). Potential symbionts in bathyal foraminifera. *Science*, *299*, 861

Bernhard, J. M. & Bowser, S. S. (1992). Bacterial biofilms as a trophic resource for certain benthic foraminifera. *Mar. Ecol. Prog. Ser.*, *83*, 263 – 272.

Bernhard, J. M. & Alve, E. (1996). Survival, ATP pool, and ultrastructural characterization of benthic fora,minifera from Drammensfjord (Norway), response to anoxia. *Marine Micropal.*, *28*, 5 – 17.

Bernhard, J. M., Sen Gupta, B. K. & Borne, P. F. (1997). Benthic foraminiferal proxy to estimate dysoxic bottom-water oxygen concentrations; Santa Barbara Basin, U.S. Pacific continental margin. *Journ. Foram.Res.*, *27*, 301 – 310.

Bernhard, J. M. & Bowser, S. S. (1999). Benthic foraminifera of dysoxic sediments: chloroplast sequestration and functional morphology. *Earth-Science Reviews*, *46*, 149 – 165.

Bernhard, J. M., Buck, K. R., Farmer, M. A. & Bowser, S. S. (2000). The Santa Barbara Basin is a symbiosis oasis. *Nature*, *403*, 77 — 80.

Bernhard, J. M., Habura, A. & Bowser, S. S. (2006). An endosymbiont-bearing allogromiid from Santa- Barbara Basin: implications for the early diversification of foraminifera. *Journ. Geophysial Res.*, *111*, G03002, doi: 10.1029/2005JG000158.

Bernhard, J. M., Edgcomb, V. P., Casciotti, K. L., McIlvin, M. R. & Beaudoin, D. J. (2012 a). Denitrification likely catalyzed by endobionts in an allogromiid foraminifer. *ISME Journ.*, *6*, 951 – 960.

Bernhard, J. M., Casciotti, K. L., McIlvin, M. R., Beaudoin, D. J., Visscher, P. T. & Edgcomb, V. P. (2012 b). Potential importance of physiologically diverse benthic foraminifera in sedimentary nitrate storage and respiration. *Journ. Geophys. Res.*, *117*, G03002, doi:10.1029/2012JG001949.

Berthold W. -U. (1971). Untersuchungen uber die sexuelle Differerierung der Foraminifere *Patellina corrugata* Williamson mit einen Beitrag zum Entwicklungsgang und Shalenbau. *Arch. Protistenk.*, *113* (3), 147 – 184.

Berthold, W. U. (1976). Biomineralisation bei milioliden Foramini feren und die MatrizenHypothese. *Naturwissenschaften*, *63*, 196.

Berthold, W. U. (1976). Ultrastructure and function of wall perforations in *Patellina corrugata* Williamson, Foraminifera. *Arch. Protistenk.*, *6* (1), 22 - 29.

Berthold, W. -U. (1977). Synaptinemale komplexe bei *Patellina corrugata* Williamson (Protozoa, Foraminifera).*Cytobiology*, *14*, 253 – 258.

Berthold, W. -U. (1978). Ultrastructure analysis of the endoplasmic algae in *Amphistegina lessonii* d'Orbigny, Foraminifera (Protozoa) and their systematic position (in German with English summary). *Arch. Protistenkund.*, *120*, 16 - 62.

Billman, H., Hottinger, L. & Oesterle, H. (1980). Neogeneee Recent Rotaliid Foraminifera from the Indopacific Ocean; their Canal System, their Classification and their stratigraphic use. *Schweizerishe Palaeontol*. V (*101*), 71 –113.

Bian, N. & Martin, P. A. (2010). Investigating the fidelity of Mg/Ca and other elemental data from reductively cleaned planktonic foraminifera. *Paleoceanography*, *25*, PA2215, 1 – 21.

Bilyard, G. R. (1974). The feeding habits and ecology of Dentalina entale stimpsoni Henderson. *Velliger*, *17*, 126 – 138.

Blau, J. (1987 a). Neue Foraminiferen aus dem Lias der Lienzer Dolomiten. Teil I: Die Foraminiferenfauna einer rotten Spaltenfullung in Oberhatkalken. *Jahrb. Geol. Bundesanstalt*, *129,* 495 – 523.

Blau, J. (1987 b). Neue Foraminiferen aus dem Lias der Lienzer Dolomiten. Teil II (Schluss):Foraminiferen (Involutinina, Spirillinina) aus der Lavanter Breccie (Lienzer Dolomiten) und der Nordlichen Kalkalpen. *Jahrb. Geol. Bundesanstalt, 130,* 5 – 23.

Blau, J. & Grun, B. (1996). Sedimentologische Beobachtungen im Ro-Grau-Schnoll-Bruch (Hettangium/Sinemurium) von Adnet (Osterreich). *Giessener Geol.en Schriften, 56 (Festschrift KNOBLICH,* 95 – 106.

Blau, J. & Grun, B. (1997). Neue Involutininen (Foraminifera) aus dem marmoreal-Hartgrund (Hettangium/Sinemurium, Lias) von Adnet (Osterreich). *Neu. Jahrb. Geol.Palaonto. Abhandl., 204,* 247 – 262.

Blau, J. & Wernli, R. 1999. New Spirillinidae (Foraminifera) from the Dogger of Lokut (Transdanubian Central Range, Hungary). *Rev. Paleobiologie*, *Geneve*, 18, 535 – 546.

Boersma, A. (1990). Late Oligocene to Late Pliocene bentic foraminifers from depth traverses in the Central Indian Ocean. In: Duncan, R.A., Backman, J., Peterson, L.C. et al., *Proc. OceanDrilling Program*, *Sci. Results*, Washington, D.C., U.S Government Printing Office, *115*, 315 – 379.

Bogdanovich, A. K. (1952). *Miliolidy I peneroplidy. [Miliolids and peneroplids*]. Iskopaemye foraminifery SSSR. Trudy Vses. Neftjan. NII geologo-razvedk [Proc. All-Union Oil Inst (VNIGRI)] , nov. ser., *64*, 1- 338.

Boix, C., Villalonga, R., Caus, E. & Hottinger, L. (2009). Late Cretaceous rotaliids (Foraminiferida) from the Western Tethys. *Neu. Jahrb. Geol. Palaont. Abh., 253* (2 - 3), 197 – 227.

Boltovskoy, E. & Wright, R. (1976). Recent Foraminifera. Dr.W.Junk b.v.- publishers.Hague. 515.

Boltovskoy, E., Giussani, G., Watanabe, S., & Wright, R. (1980). *Atlas of benthic shelf foraminifera of the Southern Atlantic*. Dr. W. Junk Publishers, Hague-Boston-London. 147.

Boltovskoy, E. & Watanabe, S. (1993). Cenozoic monothalamous foraminifers from DSDP Site 525 (Southern Australia). *Micropaleontology*, *39* (1), 1 – 27.

Bosak, T., Lahr, D.J.G., Pruss, S.B., Macdonald, F.A., Gooday, A.J., Dalton, L. & Matys, E.D. (2012). Possible early foraminiferans in post-Sturian (716 – 635 Ma) cap carbonates. *Geological Soc. of America, GEOLOGY*, 40 (1), 67 – 70. doi:10.1130/G32535.1

Boudagher-Fadel, M.K. (2008). Evolution and geological significance of larger benthic foraminifera. P.B. Wignall (ed) *Developments in paleonthology and Stratigraphy*. Elsevier. 559.

Boudagher-Fadel, M.K., Noad J.J., Lord, A.R. (2000). Larger foraminifera from Late Oligocene-Earliest Miocene reefal limestones of North East Borneo. *Rev. Espanol. Micropaleontol., 32* (3), 341 – 361.

Boudagher-Fadel, M. K. & Price, G. D. (2010). American Miogypsinidae: an analysis of their phylogeny and biostratigraphy. *Micropaleontology*, *56* (6), 567 – 586.

Boudreau, R. E. A., Pattersen, R. T., Dalby, A. P. & McKillop, W. B. (2001). Non-marine occurrence of the foraminifer *Cribroelphidium gunteri* in Northern Lake Winnipegosis, Manitoba, Canada. *Journ Foram. Res.*, *31* (2), 108 – 120.

Boukhary, M. (1994): *Decrouezina* n.gen: descendant of *Ranikothalia*in Nummulitacea. *Micropaleontology.*, *40* (1), 95 – 99.

Boukhary, M., Kulbrok, F. & Kuss, J. (1998): New nummulitids from lower Eocene limestones of Egypt (Monastery of St. Paul, Eastern Desert). *Micropaleontology 44*, 99 – 108.

Boukhary, M. & Scheibner, C. (2009). On the origin of *Nummulites*: *Urnummulites schaubi* n.gen. n. sp., from the late Paleocene of Egypt. *Micropaleontology, 55*, 413 - 420. http://www.google.ru/search?q=Boukhari,+M.,+Scheibner,+C.,+2009.+On+the+origin+of+Nummulites:+Urnummulites+schaubi+n.+gen.+n.+sp.,+from+the+late+Paleocen+of+Egypt.+Micropaleontology,+55+%284%29:+413+%E2%80%93+420.&newwindow=1&client=firefox&hs=BKm&rls=org.mozilla%3Aru%3Aofficial&tbm=bks&hl=ru&sa=X&as_q=&spell=1&ei=u-lJUtP-JoKbtQbS_YGoCQ&ved=0CCIQBSgA

Boukhary, M., Abd-Elshafi. E. & Mattar, Y. (2011). *Sanctus sinaicus* n. gen., n. sp. (Foraminiferida, Amphisteginidae) from Late Eocene of Sinai, Egypt. Micropaleontology, *57* (6), 537 – 542.

Bowser, S. S., Israel, A., Samuel, M., Russel, M. G. & Reider, C. L. (1984). Surface transport properties of reticulopodia: do intracellular and extracellular motility share a common mechanism? *Cell. Biol. Intern. Rep.*, *8* (12), 1051 - 1063.

Bowser, S. S. & Rieder C. L. (1985). Evidence that cell surface motility in *Allogrmia* is mediated by cytoplasmic microtubules. *Can. Journ. Biochem. Cell Biol.*, *63*, 608 – 620.

Bowser, S. S., McGee-Russell, S. M. & Rieder, C. L. (1985). Digestion of pray in Foraminifera is not anomalous: A correlation of light microscopic, cytochemical and HVEM technics to study phagotrophy in two allogromiids. *Tiss. Cell.*, *17*, 823 – 839.

Bowser, S. S., Travis, J. L. & Rieder, C. L. (1988. Microtubules associated with actin-containing filaments at duscrete sites along the ventral surface of *Allogromia* reticulopods. *Journ Cell Sci.*, *89*, 297 – 307.

Bowser, S. S., Alexander, S. P., Stockon, W. L. & Delaca, T. E. (1992). Extracellular matrix augments mechanical properties of pseudopodia in the carnivorous foraminiferan *Astramminarara*: role in prey capture: *Journal of Protozool.*, *39*, 724–732.

Bowser, S. S. & Bernhard, J. M. (1993). Structure, bioadhesive distribution , and elastic properties of the agglutinated test of *Astrammina rara*, Protozoa, (Foraminifera). *Journ. Eikaryotic Microbiol.*, *40*, 121 – 131.

Bowser, S. S., Gooday, A. J., Alexander, S. P. & Bernhart, J. M. (1995). Larger agglutinated foraminifera of McMurdo Sound, Antarctica: are *Astrammina rara* and *Notodendrodes antarcticus* Allogromiids? *Marine Micropaleontology*, *26*, 75 – 88.

Bowser, S. S. & Travis, J. L. (2000). Methods for structural studies of reticulopodia, the vital foraminiferal "soft part". *Micropaleontology*, *46* (suppl. 1), 47 - 56

Bowser, S. S., Bernhard, J. M., Habura, A. & Gooday, A. J. (2002). Structure, taxonomy and ecology of *Astrammina triangularis* (Earland), an allogromiid-like agglutinated foraminifer from Explorers Cove, Antarctica. *Journ.Foramin. Res.*, *32*, 364 – 374.

Bowser, S. S. & Travis, J. L. (2002). Reticulopodia: Structural and behavioral basis for the suprageneric placement of granuloreticulosan protists. *Journ. Foramin. Res.*, *32*, 440 – 447.

Bowser, S., Habura, A. & Pawlowski, J. (2006). *Molecular evolution of Foraminifera*. In: Genomics and Evolution of Microbial Eukaryotes. Katz L.S., Bhattacharya, D. (eds). Oxford Biology. Chapter 5, 78 – 93.

Brady, H. B. (1884). *Report on the Foraminifera dredged by H.M.S. Challenger*, during the years 1873 – 1876, in Report on the Scientific Results of the Voyage of the H.M.S. Challenger during the years 1873 – 1876. Zoology.

Brand, T. E. & Lipps, J. H. (1982). Foraminifera in the trophic structure of shallow-water communities. *Journ. Foram. Res.*, *12* (2), 96 - 105.

Brenckle, P. L. (1997). What is *Urbanella*? In: Ross, J.R.P., and Brenckle, P.L., (eds). Late Paleozoic for aminifers, their biostratigraphy, evolution and paleoecology, and the mid-Carboniferous boundary. *Cushm. Found.Foram.Res.*, *Sp.Publ.*, *36*, 11 - 15.

Brenckle, P. L. (2004). Late Visean (Mississippian) calcareous microfossils from the Tarim Basin of Western China. *Journ. Foram.Res.*, *34* (2), 144 - 164.

Brenckle, P. L. (2005). A compendium of Upper Devonian-Carboniferous type foraminifers from the former Soviet Union. *Cushm. Found.Foram.Res.*, *spec. publ.*, *38*, 1–153.

Brenckle, P. L. & Hance, L. (2005). New and revised Tournaisian (early Mississippian) foraminiferal taxa from Belgium. *Rev. Ital.Paleontol. e Stratigrafia*, *11* (2) (197 - 214.

Bresler, V. & Yanko, V. (1995). Chemical ecology: a new approach to the study of living benthic epiphytic foraminifera. *Journ. Foram.Res.*, *25* (3), 267 - 280.

Brodsky, A. L. (1928). Foraminifera (Polythalamia) v kolodcach pustyni Kara-Kum.[Foraminifera (Polythalamia) in the wells of the Kara-Kum desert]. *Proc. State Middle-Asatic Univ.*, *ser. 8 a*, *Zoology* (5), 1 – 16. (In Russian).

Brokeland, W., Gudmundsson, G. & Svavarsson, J. (2010). Diet of species of deep-sea isopods (Crustacea: Malacostraca: Peracarida) in the South Atlantic and the Southern Ocean. *Mar. Biol.*, *157*, 177 – 187. DOI 10.1007/s00227-009-1308-9

Bronnimann, P., Zaninetti, L. & Whittaker, J. E. (1983). On the classification of the Trochamminacea (Foraminiferida). *Journ.Foram.Res.*, *13*, 202 – 218.

Bronnimann, P. & Whittaker, J. E. (1987). Remarks on the subfamily Trochamminellinae Bronnimann, Zaninetti and Whittaker (Protista: Foraminiferida). *Revist. Espan. Paleontol.*, *2*, 87 – 93.

Bronnimann, P., Whittaker, J. E. & Parisi, G. (1987). *Soldaniella toscanae* n. gen. n. sp., a new planktonic foraminifer (Globigerinacea) from the Pliocene of Southern Tuscany, central Italy. *Rev.Paleobiologie*, *Geneve*, *6*: 169 – 175.

Bronnimann, P., Whittaker, J. E. & Zaninetti, L. (1992). Brackish water foraminifera from mangrove sediments of Southwestern Viti Levu, Fiji Islands, Southwest Pacific.*Revue de Paleobiologie*, *11* (1), 13 – 65.

Brotea, D. (1995). A new planktonic foraminifer in upper Eocene deposits from north Transylvania, Romanian . *Journ. Paleontol.*, *76*, 31 – 33.

Brummer, G. -J., Hemleben, C. & Spindler, M. (1987). Ontogeny of extant spinose planktonic foraminifera (Globigerinidae), A concept examplified by *Globigerinoides sacculifer* (Brady) and *G. ruber* (d'Orbigny). *Marine Micropaleontol.*, *12* (4), 357 – 381.

Brynov, O. P. (1984). Kristallicheskie edinicy v strukture stenki rakoviny chetvertichnych foraminifer semejstva Cassidulinidae i Islandiellidae. [Crystal units in the shell wall structure of the Tertiary Foraminifera of the families Cassidulinidae and Islandiellidae]. *News of the Moscw Univ.*, *Ser. 4*, *Geology*, *1*, 87 – 90. (In Russian).

Bubík, M. (1997). What is *Rothina silesica*? A taxonomic revision of the genus *Rothina* (Foraminiferida). *Annales Societatis Geologorum Poloniae*, *67*, 175-182.

Bubík, M. (2000). New observations on the type specimens of Recurvoidinae (Foraminiferida) described by Hanzlikova (1966, 1972 and 1973). *In:* Hart, M.B., Kaminski, M.A., & Smart, C.W. (eds), Proceedings of the Fifth International Workshop on Agglutinated Foraminifera. Grzybowski Foundation Special Publication, 7, 59-70.

Buchanan, J. B. & Hedley, R. H. (1960). A contribution to the biology of *Astrorhiza limicola* (Foraminifera). *Journ Marin. Biol. Assn, U.K.*, *39*, 549 – 560.

Bugrova, E. M. (1989). Melkie bentosnye foraminifery zonalnych kompleksov Eocena juga USSR. [Small benthic foraminifera of the Eocene zonal complexes of the South USSR]. *Ezhegodnik VPO*, *32*, 13 – 29. (In Russian) .

Bugrova, E. M. (1990). Novoe v systeme rotaloidnych foraminifer. [News in the system of the rotaliid Foraminifera]. In: *Problems of the recent micropaleonthology.* Leniingrad. Nauka, 22 – 36. (In Russian).

Bugrova, E. M. (2001). Novye taksony bentosnych foraminifer Paleogena SNG. [New taxa of benthic Foraminifera from the Peleogene of the SNG]. *Paleontologicheskij zhurnal*, *6*, 3 – 9. (In Russian)

Bugrova, E. M. & Bugrova, I. Ju. (2011). *Mikropaleontologiya. Foraminifery I infusorii mesozoya I kaynozoya.[Micropaleontology. Foraminifera and infusoria of the Mesozoic and Cenozoi]c*. St. Petersburgian State University, St. Petersburg, 1 – 103. (In Russian).

Burki, F., Shalchian-Tabrizi, K., Minge, M., Skjaeveland, A., Nikolaev, S. I., Jakobsen, K. S. & Pawlowski, J. (2007). Phylogenomics reshuffles the eukaryotic supergroups. *PLoS ONE*, *2*, e790. doi:10.1371/journal.pone.0000790

Burki, F., Kudryavtsev, A., Matz, M. W., Aglyamova, G. V., Bulman, S., Keeling, P. J. & Pawlowski, J. (2010). Evolution of Rhizaria: new insights from phylogenomic analysis of uncultivated protists. *BMC Evol. Biol.*, *10*, 377.

Calvez, H. (1988). *Pseudonummoloculina aurigerica* n. gen., n. sp. et *Dobrogelina* ? *angulata* n. sp., deux Foraminifères nouveaux de l'Albien calcaire des Pyrénées franco-espagnoles. *"Benthos' 1986". Rev. Paléobiol,*. Vol. Spécial *2*, Part. 1, 391-399.

Canerot, J. (1984). Fluctuations marines et evolution biologique: exemple du Neocomien des Iberides Orientales (Espagne). *Benthos, '83, 2nd Intern. Symp. Benthic Foram. (Pau, April 1983)*, 131 – 139.

Caron, D. A., Faber, W. W., Jr. & Be, A. W. H. (1987). Growth of the spinouse planktonic foraminifer *Orbulina universa* in laboratory culture and the effect of temperature on life process. *Journ. Marine Biol. Ass. U.K.*, *67*, 343 – 358.

Carpenter, W. B. (1861). On the systematic arrangement of the Rhizopoda. *Natural History Rev.*, *1*, 456 – 472.

Carpenter, W. B., Parker, W. K. & Jones, T. R. (1862). *Introduction to the study of Foraminifera*. Roy.Soc.London. 319.

Cartwright, N. G., Gooday, A. J. & Jones, A. R. (1989). The morphology, internal organasation, and taxonomic position of *Rhizammina algaeformis* Brady, a large agglutinated deep-sea foraminifer. *Journ. Foram. Res.*, *19* (2), 115 - 125.

Cavalier-Smith, T. (1993). Kingdom Protozoa and its 18 phyla. *Microbiol. Rev.*, *57*, 953 – 994.

Cavalier-Smith, T. (2002). The phagotrophic origin of eukaryotes and phylogenetic classification of Protozoa. *Int. Journ. Syst. Evol. Microbiology*, *52*, 297 – 354.

Cavalier-Smith, T. (2003). Protist phylogeny and the high-level classification of Protozoa. *European Journal of Protistology*, *34* (4), 338–348

Cavalier-Smith, T. & Chao, E. E. (2003). Phylogeny and classification of the phylum Cercozoa (Protozoa). *Protist*, *154*, 341 – 358.

Cedhagen, T. (1994). Taxonomy and biology of *Hyrrokkin sarcophaga* gen. and sp. n., a parasitic foraminiferan (Rosalinidae). *Sarsia*, *79*, 65 – 82.

Cedhagen, T. & Frimanson, H. (2002). Temperature dependence of pseudopodial organelle transport in seven species of foraminifera and its functional consequences. *Journ Foram. Res.*, *32*, 434 – 439. doi:10.2113/0320434

Cedhagen, T., Goldstein, S. & Gooday, A. (2002). A theme issue on the biology and biodiversity of "Allogromiid" foraminifera. *Journ. Foram. Res.*, *32* (4), 331 – 333.

Cedhagen, T. & Pawlowski, J. (2002). *Toxisarcon synsuicidica* n. gen., n.sp., a large monothalamous foraminiferan from the west coast of Sweden. *Journ. Foram. Res., 32* (4), 351 – 357.

Cesana, D. (1972). Ultrastructure des gametes chez un Foraminifere *Iridia lucida* Le Calvez. *C.R. Acad. Sci. Paris. Ser. D.*, *274*, 1044 - 1047.

Cesana, D. (1975). Les stades initaux de la gamogonie chez *Iridia lucida* Le Calvez (Foraminifera. Lagenidae). *C.R. Acad. Sci. Paris. Ser. D.*, *281* (4), 263 - 266.

Cesana, D. (1978). La mitose gamogonique chez *Iridia lucida* (Foraminifera, Lagynidae). *Ann. Sci. Natur. Ser.* 12. , *20* (3), 287 - 320.

Cesana, D. (1981). Real structure and new interpretation of Foraminiferal test. *C.R. Acad. Sci. Paris.*, *292*. Ser. 3 (2) (197 - 200.

Cetean, C.G., Setoyama, E., Kaminski, M.A., Neagu, T., Bubík, M., Filipescu, S. & Tyszka, J. (2008). *Eobigenerina* n.gen., a cosmopolitan deep-water agglutinated foraminifer, and remarks on species formerly assigned to the genera *Pseodobolivina* and *Bigenerina*. *In: Filipescu, S. & Kaminski, M.A. (eds), Eighth International Workshop on Agglutinated Foraminifera, Abstract volume. Grzybowski Foundation Special Publication*, *14*, 6 - 7.

Cetean, C. G. & Kaminski, M. A. (2011). New deep-water agglutinated foraminifera from the Upper Oligocene of offshore Angola. *Micropaleontology*, *57* (3), 255 - 262.

Cetean, C. G., Setoyama, E., Balc, R., Tyszka, Ja. & Kaminski, M. A. (2011). Foraminifera and calcareous nannofossils from the lower Jurassic Rosso Ammonitico Umbro-Marchigiano unit, Marche, Italy. *In: Bak, M., Kaminski, M.A., Waskowska, A. (eds). Integrating Microfossil Rrecords from the Oceans and Epicontinental Seas. Grzybowski Foundation Spec. Publ.*, *17*, 82 – 83.

Chai, J. & Lee, J. J. (2000). Recognition, establishment and maintenance of diatom symbiosis in foraminifera. *Micropaleontology*, *46*, suppl. 1, 182 – 195.

Chaikovskij, Ju.V. (2010). *Diatropica, evolucija i systematika (k jubileju Mejena). Sbornik statej s prilozheniem polnogo spiska trudov S.V. Mejena, sostavlennogo I.A. Ignat'evym. [Diatropica, Evolution and Systematic (to the Mejen's jubilee). Collective book with the attachment of the full list of S.V.Mejen's publications prepared by I.A. Ignatiev.* Tovaristschestvo nauchnych izdanij KMK.Moscow: 407 p. (In Russian).

Chave, K. E. (1954). Aspects of the biogeochemistry of magnesium. 1. Calcareous marine organisms. *Journ. Geol.*, *62*, 266 – 283.

Cherchi, A. (1983). Evolution et valeur stratigraphique du genre Coskinolinella Delmas & Deloffre dans l'Aptien-Albien du SW de L'Europe. *Benthos'83, 2nd Int. Symp. Benthic Foraminifera (Pau, April 1983)*, 153 – 158.

Cherchi, A. & Schroeder, R. (1999). *Montseciella,* a new orbitolinid genus Foraminiferida) from the Uppermost Hauterivian - Early Barremian of SW Europe. *Treb. Mus. Geol. Barcelona*, 8, 5 – 23.

Cherchi, A. & Schroeder, R. (2005). Revision of *Keramosphaerinopsis haydeni* (H. Douville), larger foraminifer (Miliolacea) from the Paleocene of southern Tibet (Tethys Himalaya). *Boll. Soc. Paleontol. Italiana*, *44* (2), 175 – 183.

Cherchi, A. & Schroeder, R. (2009). Revision of the orbitolinid foraminiferal genus *Praeorbitolinoides* Matsumaru 2005 from the Aptian of Hokkaido, Japan. *Micropaleontology*, *55* (4), 421 – 424.

Cherchi, A., Schroeder, R. & Ruberti, D. (2009). *Cuneospirella samnitica* gen. n., n. sp. (Foraminiferida) from the Santonian of the Matese Mountains (Molise, central Italy). *Rivista Italiana di Paleontologia e Stratigrafia, 115* (1), 59 - 65.

Christiansen, O. (1971). Notes on the biology of Foraminifera. In: *Vie et milieu. Suppl*., *22* (2), 465 – 478.

Chusid, T. A. (1988). Rol foraminifer v bentose okeana. [The role of Foraminifera in the benthos of the ocean]. In: Morskaya stratigraphiya I micropaleontologiya [Sea stratigraphy and micropaleontology]. Krasheninnikov V.A., Basov I.A.(eds). *Russian Academy of Sciences*, *Comission on the problems of the World Ocean. Moscow. Nauka*: 191 – 114. (In Russian).

Clark, F. E. (1991). Notes on the benthic foraminiferal genus *Heteromorphina* R.W. Jones, and *Heteromorphina amplivestibulata* sp. nov. *Journ. Micropaleontology*, *9* (2), 221 – 225.

Clark, F. E. (1993). Morphological details of the genus *Nubeculina* Cushman (1924. *Journ. Foram. Res*., *23* (3), 201 - 204.

Clark, F. E. (1995). New species of unilocular calcareous foraminifera from the Holocene of the Southwest Pacific Ocean. *Journ. micropaleontol*., *14* (1), 1 – 5.

Colom, G. (1974). Foraminiferos Ibericos. Introduccion al estudio de las especies bentonicas recientes. *Investigac. Pesquera. Barcelona*, *38* (1), 245.

Coleman, A. R. (1980). Test structure and function of the agglutinated foraminifera *Clavulina*. *Journ. Foram. Res*., *10* (2), 143 - 152.

Collen, J. D. (1998). *Metarotaliella tuvaluensis* sp.nov. from Funafuti Atoll, western Pacific Ocean: relationship to miliolid foraminifera. *Journ. Foram. Res*., *28* (1), 66 – 75.

Collen, J. D. & Newell, P. (1999). *Fissurina* as an ectoparasite. *Journ. Micropalaeontology*, 18:110. doi:10.1144/jm.18.2.110.

Cornelius, N., Gooday, A. J. (2004). 'Live' (stained) deep-sea benthic foraminifera in the western Weddell Sea: trends in abundance, diversity and taxonomic composition in relation to water depth. *Deep-Sea Research*, II *51*, 1571 – 1602.

Cozar, P. (2000). Endothyracea y Tournayellacea (foraminiferos) del Mississippiense del area del Guadiato (Cordoba, Espana). *Rev. Espanol. Micropaleontol*., *32* (3), 319 – 341.

Cozar, P., Merino-Tome, O. & Villa, E. (2007). New data (smaller foraminifera and correlation with depositional sequences) on the Las Llacerias section (Late Pennsylvanian, Cantabrian Mountains, NW Spain). *Facies*, *53*, 427 – 450.

Cozar, P. & Somerville, J. D. (2006). Significance of the Bradyinidae and *Parajanischewskina* n. gen.for biostratigraphic correlations of the Late Visean (mississippean) in Western paleothethyan basins. *Journ. Foram. Res*., *36*, 262 – 272.

Cozar, P. & Somerville, J. D. (2012). Mississippian Biseriamminacea and their evolutionary development. *Journ Foram. Res.*, *42* (3), 216 – 233.

Cozar, P. & Vachard, D. (2001). Dainellinae subfam. nov. (Foraminiferida du Carbonifere inferieur), revisio et nouveaux taxons. *Geobios*, *34* (5), 505 – 526.

Culver, S. J. (1993). *Foraminifera.* In: Lipps J.H. Fossil Prokaryotes and Protists. Blockwell. Boston: 203 - 247.

Culver, S. J. & Buzas, M. A. (1995). The effects of anthropogenic habitat disturbance, habitat destruction, and global warming on shallow marine benthic foraminifera. *Journ. Foram. Res.*, *25* (3), 204 – 212.

Culver, S. J. & Buzas, M. A. (2003(. Predation on and by Foraminifera. In Predator-Prey Interactions in the Fossil Record. *Topics in Geobiology*, *20*, 7 - 32.

Cushman, J. A. (1948). *Foraminifera. Their classification and Economic use.* 4 th ed. Cambridge. Mass. Harvard Univ. Press. 605.

Cuvier, G. (1801). Lecons d'anatomie comparee. Vol. *1*, Paris.

Dabbous, A. & Scott, D. B. (2012). Short-term monitoring of Halifax)Nova Scotia, Canada) pollution remediation using benthic foraminifera as proxies. *Journ. Foram. Res.*, *42*, 187 - 205.

Dahlgren, L. (1964). On nuclear cytology and reproduction in the monothalamous foraminifer *Ovammina opaca* Dahlgren. *Zool. Bidrag. Uppsala*, *36* (3), 315 - 334.

Dahlgren, L. (1967 a). On the ultrastructure of the gamontic nucleus and the adjacent cytoplasm of the monothalamous foraminifer *Ovammina opaca* Dahlgren. *Zool. Bidrag. Uppsala*, *37* (2), 77 - 112.

Dahlgren, L. (1967 b). On the nuclear distribution of RNA and DNA and on the ultrastructure of nuclei and adjancent cytoplasm of the Foraminifers *Hipocrepinella alba* Heron-Allen and Earland and *Globobulimina turgida* (Bailey). *Zool. Bidrag. Uppsala*, *37* (2), 113 - 138.

Daniels, R. A. & Lipps, J. H. (1978). Predation on Foraminifera by Antarctic fish. *Journ. Foram. Res.*, *8* (2), 110 - 113.

Darling, K. F., Kroon, D., Wade, C. M. & Leigh Brown, A. J. (1996). Molecular evolution of planktic foraminifera. *Journ. Foram. Res.*, *26*, 324 - 330.

Darling, K. F., Wade, C. M., Kroon, D. & Leigh Brown, A. J. (1997). Planktic foraminiferal molecular evolution and their polyphyletic origin from benthic taxa. *Marine Micropaleontol.*, *30*, 251 – 266.

Darling, K. F., Wade, C. M., Kroon, D., Leigh Brown, A. J. & Bijma, J. (1999. The diversity and distribution of modern planktic foraminiferal small subunit ribosomal RNA genotypes and their potential as tracers of present and past ocean circulation. *Paleoceanog.*, *14*, 3 – 12.

Darling, K. F., Wade, C. M., Stuart, I. A., Kroon, D., Dingle, R. & Leigh Brown, A. J. (2000). Molecular evidence for genetic mixing of Arctic and Antarctic subpolar populations of planktonic foraminifers. *Nature* , *405*, 43 – 47.

Darling, K. F., Thomas, E., Kasemann, S. A., Seears, H., Smarb, C. W. & Wade, C. M. (2009). Surviving mass extinction by bridging the benthic/planktic divide. *Proc. Natl. Acad. Sci.USA*, *106*, 12629 – 12633.

Darling, K. & Leckie, R. M. (2010). Bridging the Benthic/Planktic divide. In: *Forams 2010 Abstracts volume with program. Baylaydruk, Bonn*, 74.

Darwin, C. (1859). The origin of species by means of natural selection, or the preservation of favoured races in the struggle of life. John Murray, London

Davydov, V. I., Belasky, P. & Karavaeva, N. (1996). Permian fusulinids from the Koryak terrane, Northeastern Russia, and their paleobiogeographic affinity. *Journ. For a. Res., 26* (3), 213 - 243.

Davydov, V., Krainer, B., Krainer, K. & Erlangen, I. (1999). Fusulinid assemblages and facies of the Bombasa Formation and basal Meladis Formation (Moscovian-Kasomovian) in the central Carnic Alps (Austria/Italy). *Facies, 40*, 150 – 183.

Debenay, J. -P. (1990). Recent foraminiferal assemblages and their distribution related to environmental stress in the paralic environments of Western Africa. *Journ. For a. Res., 20* (3), 267 - 282.

Debenay, J. -P. (2000). Foraminifers in Paralic Tropical environments. *In* : Biology of Foraminifera 2nd edition J. Lee (ed). *Micropaleontology, 46* (suppl. 1), 153 – 160.

Debenay, J. -P. (2009). *Foraminifera, chapter thirteen*. In: Environmental assessment of estuarine ecosystems: a case study (Amiard C. & Rainbow P.S., eds). Taylor and Francis, Boca Raton, Florida, 255 - 280.

Debenay, J. -P., Pawlowski, J. & Decrouez, D. (1996 a). *Les Foraminiferes actuels*. Collect. 3 Cycl. et Recherche, 352.

Debenay, J. -P., Guilou, J. J. & Lesourd, M. (1996 b). Colloidal calcite in foraminiferal tests: crystallization and texture of the test. *J. Foram Res, 26* (4), 277 - 288.

Debenay, J. -P., Redois, F. (1997). Recent Foraminifera of the northern continental shelf, off Senegal. *Revue de micropaleontologie, 40*(1), 15 - 38.

Debenay, J-P., Duleba, W., Bonetti, C., De Melo, S. H., Souza, E. & Eichler, B. B. (2001). *Pararotalia gananeiansis* n.sp.: indicator of marine influence and water circulation in Brazilian coastal and paralic environments. *Journ. Foram.Res., 31*, 152 – 163.

Debenay, J. -P. & Payri, C. (2010). Epiphytic foraminiferal assemblages on macroalgae in reefal environments of New Caledonia. *Journ. Foram. Res., 40* (1), 36 - 60.

Debenay, J-P., Guillou, J-J., Geslin., E., Lesourd, M. & Redois, F. (1998). Processus de cristallisation de plaquettes rhombohédriques à la surface d'un test porcelané de foraminifère actuel. *Géobios, 31*(3), 295-302.

Debenay, J. -P., Guillou, J. J., Geslin, E. & Lesourd, M. (2000). Crystallization of calcite in foraminiferal tests. *Micropaleontology, 46* (Suppl. 1), 87 – 94.

Debenay, J. -P. & Guillou, J. J. (2002). Ecological transitions Indicated by Foraminiferal Assemblages in Paralic Environments. *Estuaries, 25* (6A), 1107 – 1120.

Debenay, J. -P., Carbonel, P., Morzadec-Kererfourn, M. -T., Cazaubon, A., Denefle, M. & Lezine, A-M. (2003). Multi-bioindicator study of a small estuary in Vendée (France), *Estuarine, Coastal and Shelf Sciences, 58*, 843 - 860.

Debenay, J. -P., Jouanneau, J-M., Sylvestre, F., Weber, O. & Guiral, D. (2007). Biological Origin of Rhythmites in Muddy Sediments of French Guiana. *Journ.Coastal Res., 23*, 1431 - 1442.

Debenay, J. -P., Sigura, A. & Justine, J-L. (2011). Foraminifera in the diet of coral reef fish from the lagoon of New Caledonia: predation, digestion, dispersion. *Rev. Micropaléontol., 54*, 87 – 103.

Delaca, T. E., Karl, D. M. & Lipps, J. H. (1981). Direct use of dissolved organic carbon by agglutinated benthic foraminifera: *Nature, 289*, 287–289.

Delaca, T. E. (1982). Use of dissolved amino acids by the foraminifer *Notodendrodes antarctikos*. *Amer. Zoologist*. *22* (3), 683 - 690.

Delaca, T. E. (1986). The morphology and ecology of *Astrammina rara*. *Journ. Foram.Res.*, *16*, 216–224.

Delaca, T. E., Karl, D. M. & Lipps, J. H. (1981). Direct use of dissolved organic carbon by agglutinated benthic Foraminifera. *Nature*, *289*, 287 - 289.

Delaca, T. E., Bernhard, J. M., Reilly, A. A. & Bowser, S. S. (2002). *Notodendrodes hyalinosphaira*(sp. nov.), Structure and autecology of an allogromiid-like agglutinated foraminifer. *Journ. Foram.Res.*, *32* (2), 177 – 187.

Desai, D. & Banner, F. T. (1987). The evolution of Early Cretaceous Dorothiinae (Foraminiferida). *Journ. Micropaleontol.*, *6*(2), 3-27.

Desmares, D., Grosheny, D. & Beaudoin, B. (2008). Ontogeny and phylogeny of Upper Cenomanian rotaliporids (Foraminifera). *Marine Micropaleontol.*, *69*, 91 – 105.

Dettmering, C., Rџttger, R. http://www.sciencedirect.com/science/article/pii/ S093247399 8800047 Hohenegger, J. & Rolf Schmaljohann, R. (1998). The trimorphic life cycle in foraminifera: Observations from cultures allow new evaluation. *European Journ. Protistol.*, *34* (4), 363-368.

Deutsch, S. & Lipps, J. H. (1976). Test structure of the foraminifer *Carterina*. *Journ. Paleontol.*, *50*, 312 – 317.

De Vargas, C., Norris, R., Zaninetti, L., Gibb, S. W. & Pawlowski, J. (1999). Molecular evidence of cryptic speciation in planktonic foraminifers and their relation to oceanic provinces. *Proc. Natl. Acad. Sci. USA*, *96*, 2864 – 2868.

De Vargas, C., Renaud, S., Hilbrecht, H. & Pawlowski, J. (2001). Pleistocene adaptive radiation in *Globorotaliatruncatulinoides*: genetic, morphologic, and environmental evidence. *Paleobiology*, *27*, 104 – 125.

Di Bari, D. (1998). *Ornatoconus francorussoi* gen. et sp.nov. (Involutinacea, Foraminiferida) from the Upper Triassic (Carnian) of the northeastern Dolomites, Italy. *Rev. Paleobiol.*, *17* (1), 23 – 33.

Di Bari, D.& Rettori, R. (1998). *Cassianopapillaria*, new name for the foraminiferal genus *Papillaria* Di Bari & Rettori, 1996. *Revue de Paléobiologie* 17(2), 565.

Dietz-Elbraechter, G. (1971). Untersuchungen uber die zooxanthella der Foraminifere *Heterostegina depressa* d'Orbigny 1826. *"Meteor" Forschergeben (c) b*, 41 - 47.

Dissard, D., Nehrke, G., Reichart, G. J. & Bijma, J. (2010). The impact of salinity on the Mg/Ca and Sr/Ca ratio in the benthic foraminifera *Ammonia tepida*: Results from culture experiments. *Geochimica et Cosmochimica Acta*, *74* (3), 928-940.

Dixon, M. & Haig, D. W. (2004). Foraminifera and their habitats within a cool-water carbonate succession following glaciation, Early Permian (Sakmarian), Western Australia. *Journ. Foram. Res.*, *34* (4), 308 – 324.

Dobson, M. & Haynes J. (1973). Association of foraminifera with hydroids on the deep shelf. *Micropaleontology.*, *19* (1), 78 - 90.

Dogiel, V. A. (1929). Polymerization als ein Prinzip der progressiven Entwicklung bei Protozoa. *Biol. Zbl.*, *49* (8), 451 – 459.

Dogiel, V. A. (1954). *Oligomerizacija gomologichnych organov kak odin iz glavnych putej evolutcii zhivotnych [Oligomerization of the homologous organs as one of the main ways of the animal evolution]*. A.A. Strelkov (ed.) Izdatelstvo Leningradskogo Universiteta, Leningrad. (In Russian).

Dreyer, F. (1898). *Eine studie zur biologischen Morphologie und zur speciesfrage*. Leipzig. W. Engelmann:119.

Drobne, K., Ogorelec, B., Riccamboni, R., 2007. *Bangiana hanseni* n. gen. n. sp.(Foraminifera) an index fossils species of Danian age (Lower Palaeocene) from the Adriatic Carbonate Platform (SW Slovenia, NE Italy, Herzegovina). Razprave IV. Razreda SAZU *48* (1), 5 – 71.

Douglas, R. G. & Woodruff, F. (1981). Deep-Sea benthonic foraminiferen. In Emiliani (Ed). *The Sea*, *7*. Wiley Interscience. N. Y., 1233 - 1327.

Dujardin, F. (1835). Observations nouvelles sur les prйtendus Cйphalopodes microscopiques. *Annales Sci. Nat.*, *2* (3), 312 – 314.

Dupuy, C., Rossignol, L., Geslin, E. & Pascal, P. Y. (2010). Predation of mudflat meio-macrofaunal metazoans by a calcareous foraminifer, *Ammonia tepida* (Cushman, 1926). *Journal of Foraminiferal Research*, *40* (4), 305-312.

Eiland, M. & Gudmudsson, G. (2004). Taxonomy of some recent Nodosariinae (Foraminifera) from the North Atlantic, with notes on wall lamination. *Micropaleontology*, *50* (2) (195 – 210.

Ellison, R. L., Broome, R. & Ogilvie, R. (1986). Foraminiferal response to trace metal contamination in the Potapsco River and Baltimore Harbour, Maryland. *Marine Pollution Bull.*, *17* (9), 419 - 423.

Ellis, B. F. & Messina, A. R. (1940 and follow). *Cataloge of Foraminifera*. Amer. Mus. Natur. Hist., *30* volumes and supplements.

El-Nakhal, H. (1990). *Septloculina*, a new genus Hauerininae (Foraminifera). *Micropaleontology, 36* (1), 88 – 95.

El-Nakhal, H. (2002). Classification of the meridionally costellate Cretaceous planktic foraminifera. *Journ. Micropaleontol.*, *21*, 1 – 8.

Ernst, S. R., Bours, R., Duijnstee, I. A. P. & Van der Zwaan, G. J. (2005). Experimental effects of an organic matter pulse and oxygen depletion on a benthic foraminiferal shelf community. *Journ. Foram. Res.*, *35*, 177 – 197.

Ertan, K. T., Hemleben, V. & Hemleben, C. (2004). Molecular evolution of some selected benthic foraminifera as inferred from sequences of the small subunit ribosomal DNA. *Marine Micropaleontology*, *53*, 367–388.

Evans, K. M., Bud, C., Scweitzer, M., Austin, W. & Darling, K. (2012). Investigating Elphidium diversity in European seas using Scanning Electron Microscopy and DNA sequencing. *The Micropaleontol Soc. Foraminifera and Nannofossil Groups Joint Meeting 2012*, *Interdisciplinary advances in Foraminiferal and nannofossil Res.*, *Pollock Halls*, *Univ. Edinburg*, 17 - 18.

Faber ,W. W. Jr. (1991). Distribution and substrate preferance of *Peneroplis planatus* and *P. arietinus* from the Halophila Meadow near Wad: Taba, Eilat, Israel. *Journ. Foram. Res.*, *21* (3), 218 - 221.

Faber, W.W. Jr. & Lee, J. J. (1991a). Feeding and growth of *Peneroplis planatus* (Fichtel and Moll). Montfort. *Symbiosis.*, *10*, 63 - 82.

Faber, W.W. Jr. & Lee, J. J. (1991b). Histochemical evidence for digestion in *Heterostegina depressa* and *Operculina ammonoides* (Foraminifera). *Endocytobiol. Cell. Res.*, *8*, 53 - 59.

Faber, W.W. Jr. & Lee, J. J. (1993). Contributions to the Life Cycle of *Peneroplis planatus* (Fichtel and Moll) Montfort. *Journ. Eukaryotic Microbiol.*, *40*, 718–721.

Fahrni, J. F., Pawlowski, J. (1995). Identification of actins in Foraminifera: Phylogenetic Perspectives. *Europ. Journ. Protistol.* Vol. *31.*, 161 – 166.

Fahrni, J. F., Pawlowski, J., Richardson, S., Debenay, J. -P. & Zaninetti, L. (1997). Actin Suggests *Miliammina fusca* (Brady) is related to porcellaneous rather than to agglutinated foraminifera. *Micropaleontology*. Vol. *43.*, N 2. P. 211 – 214.

Febvre-Chevalier, C. (1971). Constitution ultrastructurale de *Globigerina bulloides* d'Orbigny 1826 (Rhizopoda - Foraminifera). *Protistologica*, *7* (3), 311 - 324.

Fedonkin, M. A. (2009). Problema proischozhdenija zhizni i paleontologiya[Problem of the origin of the life and paleontology]. In: Paleontology and perfection of the stratigraphic base of the geologic mapping. *Materials LV session of Paleontol. Soc., Rus. Acad. Sci, Paleontol. Soc., VSEGEI (All-Russian Geol. Inst. in hon. A.P.Karpinski), 6 – 9 Apr. 2009*, St. Petersburg: 155 – 157. (In Russian).

Fernandez-Canadell, C. (2002). New Paleocene orbitoidal foraminifera from the Punjab Salt Range, Pakistan. *Journ. Foram. Res., 32* (1),11 – 21.

Fernandez-Canadell, C. & Serra-Kiel., J. (1992). Morphostructure and paleobiology of *Discocyclina* Gumbel, 1870. *Journ. Foram. Res.*, *22* (2), 147-165.

Filipescu, S. & Kaminski, M. A. (2011). Re-discovering *Entzia*, an agglutinated foraminifer from the Transylvanian salt marshes. *In: Kaminski, M.A., Filipescu, S. (eds), Proceedings of the Eighth International Workshop on Agglutinated Foraminifera.* Grzybowski Foundation Special Publications, *16*, 29 - 35.

Fillon, R. H. (1974). Late Cenozoic foraminiferal paleoecology of the Ross Sea, Antarctica. *Micropaleontology*, *20* (2), 129 – 151.

Finger, K. L. (1990). *Atlas of California Neogeneee foraminifera.* Cushm.Found. Foram. Res., Sp. Publ., *28*, 271.

Flakowski, J., Bolivar, I., Fahrni, J. & Pawlowski, J. (2005). Actin phylogeny of Foraminifera. *Journ. Foram. Res.*, *35*, 93–102.

Fleury, J.-J. (1996). Convergence morphologique Alveolinacea-Soritacea : Les genres *Rhapydionina* Stache et *Rhabdorites* n. gen. *Rev. Micropaléontologie*, *Paris, 39*,1, 41 – 51.

Fleury, J.J. 1997. Eocene soritids with ribs and/or faint sub-epidermal partitions: the genera *Spirolina, Praerhapydionina* and *Paraspirolina* n. gen. in the middle to upper Eocene e "Facies a imperfores" of Greece. *Revue de Micropaleontologie*, 297 - 311.

Fourcade, E., Mouty, M. and Teheran, K. (1997). Jurassique et du Crétacé du Moyen-Orient *Levantinella* nov. gen. and revision of the genus *Mangashtia* Henson, large foraminifers from the Jurassic and Cretaceous of theMiddle-East. *Geobios, 30*, 179 – 192

Foeyn, B. (1936). Foraminiferen studien. 1. Der Lebenzyklus von *Discorbina vilardeboana* d'Orbigny. *Bergens Mus. Arbok. Naturvidens kapeling rekke*, *2*, 1 - 22.

Fugagnoli, A. (1964). First record of *Evertycyclammina* Redmond, (E. praevirguliana n. sp., Foraminifera) from the Early Jurassic of the venetian Prealps (Calcari Grigi, Trentio platform, North Italy). *Journ. Form. Res.*, *30* (2), 126 – 135.

Fugagnoli, A., Giannetti, A., Rettori, R. (2003). A new foraminiferal genus (Miliolana) from the Early Jurassic of the Southern Alps (Calcari Grigi formation, Northeastern Italy). *Rev. Espanol. Micropaleontol., 35* (1), 43 – 50.

Fujita, K. & Hallock, P. (1999). A comparison of phytal substrate preferences of *Archaias angulatus* and *Sorites orbicularis* in mixed macroalgal-seagrass beds in Florida Bay. *Journ. Foram. Res.*, *29* (2), 143-151

Fursenko, A. V. (1958). Osnovnye etapy razvitiya faun foraminifer v geologicheskom proshlom [Fundamental state of development of foraminiferal faunas in the geological past]. *Trudy Instituta Geologicheskich nauk, Akademia Nauk Beloruskoi SSR*, Minsk, *1*, 10 – 29. (In Russian).

Fursenko, A. V. (1978). *Vvedenie v izuchenie foraminifer.* [*Introduction to the foraminiferal study*]. Novosibirsk. "Nauka". Proc. Inst. Geol. Geophys. SO USSR Acad. Sci., *391*, 242 p. (In Russian).

Gaillot, J. & Vachard, D. (2007). The Khuff formation (Middle East) and time-equivalens in Turkey and South China: biostratigraphy from Capitanian to Changhsinian times (Permian), new foraminiferal taxa, and paleogegraphical implications. *Col. Paleont.*, *57*, *Madrid*: 37 – 223.

Gauthier-Lievre, L. (1935). Sur une des singularites de l'oued Rhir: des Foraminiferes Thalassoides vivants dans les eaux sahariennes. *Bull. Soc. H. N. Afrique du Nord*, XXVI, 142 – 147.

Gausher, C. & Sprechmann, P. (1999). Upper Vendian skeletal fauna of the Arroyo del Soldado Group, Uruguay. *Beringeria, 23*:55 – 91.

Gawor-Biedowa, E. (1984). Foraminiferal zonation of the Upper- Cretaceous deposits in Poland (except for the Carpathians and Sudeten). *Rev. Paleobiol., Benthos, 1986, 2nd Intern. Symp. Benthic Foram. (Pau, April 1983)*, 213 – 223.

Gawor-Biedowa, E. (1992). Campanian and Maastrichtian foraminifera from the Lublin upland, eastern Poland. *Pal. Polonica, Warsaw, Poland, 52*, 3 – 187.

Georgescu, M. D. (2009 a). On the origins of superfamily Heterohelicacea Cushman (1927 and the polyphyletic nature of planktic foraminifera. *Rev. Espanol. Micropaleontol., 41* (1 – 2), 107 – 144.

Georgescu, M. D. (2009 b). Upper Albian-lower Turonian non-schackoinid planktic foraminifera with elongate chambers: morphology reevaluation, taxonomy and evolutionary classification. *Rev. Esp. Micropaeontology, 41* (3), 255 - 293.

Georgescu, M. D. (2009 c). On the origins of Superfamily Heterohelicacea Cushman, 1927 and the polyphyletic nature of planktic foraminifera. *Rev. Esp. Micropal., 41* (1 – 2), 107 - 144.

Georgescu, M. D. (2010 a). Evolutionary classification of the Upper Cretaceous (Turonian – lower Campanian) planktic foraminifera with incipient meridional ornamentation. *Journ. Micropaleontol., 29* (2), 149 – 161.

Georgescu, M. D. (2010 b). Origin, taxonomic revision and evolutionary classification of the late Coniacian–early Campanian (Late Cretaceous) planktic foraminifera with multichamber growth in the adult stage. *Rev. Espanol. Micropaleontol., 42* (1), 59 – 118.

Geroch, S. & Nowak, W. (1988). Proposal of zonation for the Late Tithonian – Late Eocene, based upon arenaceous foraminifera from the outer Carpathians, Poland. *Rev. de Paleobiol., 2nd Int. Symp. Benthic Foraminifera (Pau, April 1983), Benthos '1986, Spec. v. 2, Mus. Hist. Natur., Geneve-Svisse*: 225 – 239.

Geslin, E., Debenay, J. -P., Duleba, W. & Bonetti, C. (2002). Morphological abnormalities of foraminiferal tests in Brasilian environments: comparison between polluted and non-polluted areas. *Mar. Micropal., 45*, 151 – 168.

Gibson, T.G. (1989). Miocene evolution of *Tubulogenerina* in the Indo-Pacific and African areas. *Journ. Foram. Res., 19* (2), 126 - 143.

Gibson, J. A. E., Quilty, P. G., Swadling, K. M., Newman, L. & Paterson, K. S. (2008). *Paratrochammina minutissima* n. sp. in brackish, marine-derived lakes of the Vestfold Hills, East Antactica. *Journ. Foram. Res.*, *38* (4), 292 - 297.

Goldstein, S. T. (1988). On the life cycle of *Saccammina alba* Hedley 1962. *Journ. Foram. Res.*, *18* (4), 313 - 327.

Goldstein, S. T. (1997). Gametogenesis and the antiquity of the reproductive pattern in the Foraminiferida. *Journ. Foram. Res.*, *27* (4), 319 – 328.

Goldstein, S. T. (1999). *Foraminifera: a biological overview*. In: Sen Gupta, B.K. (ed), Modern Foraminifera. Kluwer Academic Publishers, New York, 37 – 55.

Goldstein, S. & Barker, W. W. (1988). Test ultrastructure and taphonomy of the monothalamous agglutinated foraminifer *Cribrorobulina*, n. gen., alba (Heron-Allen and Earland). *Journ. Foram. Res.* , *18* (2), 130 - 136.

Goldstein, S. T. & Barker W. W. (1990). Gametogenesis in the monothalamous agglutinated foraminifer *Cribrothalammina alba. Journ. Protozool.*, *37* (1), 20 - 27.

Goldstein, S. T. & Harben E. B. (1993). Taphofacies implications of infaunal foraminiferal assemblages in a Georgia salt marsh, Sapelo Island. *Micropaleontology,. 39* (1), 53 - 62.

Goldstein, S. & Moodley, L. (1993). Gametogenesis and life cycle of the foraminifer *Ammonia beccarii* (Linne) forma *tepida* (Cushman). *Journ. Foram. Res., 23* (4), 213 - 221.

Goldstein, S. T. & Corliss, B. H. (1994). Deposit feeding in selected deep sea and shallow-water benthic foraminifera. *Deep-Sea Res.*, *41*, 229 – 241.

Goldstein, S. T., Watkins, G. T. & Kuhn, R. M. (1995). Microhabitats of salt marsh foraminifera: St. Catherines Island, Georgia, USA. *Mar. Micropal.*, *26*, 17 – 29.

Goldstein, S. T. & Bernhard, J. M. (Editors) (1997). Theme Issue: Biology of Foraminiferida: Applications in Paleoceanography, Paleobiology, and Environmental Sciences: *Journ. Foram. Res.*, *27*(4), 254 - 328.

Goldstein, S. T., Habura, A., Richardson, E. A. & Bowser, S. S. (2006). A new Allogromiid foraminifer from coastal Georgia, USA: the case for an unusual form of chloroplast equestration. *Microscopy and Microanalysis*, *12*, suppl. 2, 266 – 267.

Goldstein, S. T. & Richardson, E. (2002). Comparison of test and cell body ultrastructure in three modern Allogromiid foraminifera: application of high pressure freezing and freeze substitution. *Journ. Foram. Res.*, *32* (4), 375 – 383.

Goldstein, S. T., Habura, A., Richardson, E. A. & Bowser, S. S. (2010). *Xiphophaga minuta*, and *X. allominuta*, nov. gen, nov. spp., new monothalamid foraminifera from coastal Georgia (USA), Cryptic species, gametogenesis, and an unusual form of chloroplast sequestration. . *Journ. Foram. Res.*,*40* (1), 3-15

Golubev, S. N. C. H. (1981). *Realnye kristally v skeletach kokkolitophorid. [Real crystals in the Coccolitophorids skeleton*]. Moscow. "Nauka": 161 (In Russian).

Golubowski, M. D. (1994). Classicheskaya I sovremennaja genetika: evoluciya vzgljadov na nasledstvennuju izmenchivost [Classical and modern genetics: evolution of views of hereditary variation]. *Tr .Spb Obstv. Estestvoispyt.*, *90*, 37—48. (In Russian).

Gonzalez-Donoso, J. M., Linares, D. & Robaszynski, F. (2007). The rotaliporids, a polyphyletic group of Albian-Cenomanian planktic foraminifera: emendation of genera. *Journ. Foram. Res.*, *37*, 175 – 186.

Gooday, A. J. (1986). Meiofaunal foraminiferans from the bathyal Porcupine Seabight (north-east Atlantic), size structure, standing stock, taxonomic composition, species diversity and vertical distribution in the sediment. *Deep-Sea Research*, *33* (10), 1345 – 1373.

Gooday, A. J. (2002). Organic-walled allogromiids:aspects of their occurrence, diversity and ecology in marine habitats. *Journ. Foram. Res.*, *32*, 384 — 399.

Gooday, A. J. (2012). Benthic foraminiferal squatters: species that inhabit the tests of other foraminifera. Alegret, L., Ortiz, S., Kaminski, M.A. (eds), Ninth Intern. Workshop Aggl. Foram., *Grzybowski Foundation Sp. Publ.*, *18*, 33 – 34..

Gooday, A. J., Levin, L. A., Linke, P. & Heeger, T. (1992). *The role of benthic foraminifera in deep-sea food webs and carbon cycling.* In: Rowe, G. T. and Pariente, V., eds, Deep-Sea Food Chains and the Global Carbon Cycle. Kluwer Academic Publishers, the Netherlands.: 63 – 91.

Gooday, A. J., Bernhard, J. M. S. S. & Bowser, S. S. (1995). The taxonomy and ecology of *Crithionina delacai* sp. nov., and abundant large agglutinated foraminifer from Explorers Cove, Antarctica. *Journ.Foram. Res.*, *25*, 290 – 298.

Gooday, A. J., Bowser, S. S. & Bernhard, J. M. (1996). Benthic foraminiferal assemblages in Explorers Cove, Antarctica: A shallow-water site with deep-sea characteristics. *Prog. Oceanography*, *37*, 117–166.

Gooday, A. J., Shires, R., Alick, R. & Jones, A. R. (1997). Large, deep-sea agglutinated Foraminifera; two differing kinds of organization and their possible ecological significance. *Journ. Foram. Res.*, *27*, 278 – 291.

Gooday, A. J. & Smart, C. W. (2000). Wall structure and test morphology in three deep-sea agglutinated foraminifera, *Rhabdammina parabyssorum* Stschedrina 1952, *R. abyssorum* Sars 1869 and *Astrorhiza granulosa* (Brady 1879) (Foraminiferida, Textulariina). *In:* Hart, M.B., Kaminski, M.A., & Smart, C.W. (eds) Proceedings of the Fifth International Workshop on Agglutinated Foraminifera. *Grzybowski Foundation Special Publication*, *7*, 105- 115.

Gooday, A. J. & Pawlowski, J. (2004). *Conqueria laevis* gen. & sp. nov., a new soft-walled, monothalamous foraminiferan from the deep Weddell Sea. *Journal of the Marine Biological Association of the United Kingdom*, *84*, 919-924.

Gooday, A. J. & Bowser, S. S. (2005). The second species of *Gromia* (Protista) from the Deep Sea:its natural history and association with the Pakistan margin oxygen minimum zone. *Protist*, *156*, 113 — 126.

Gooday, A. J., Bowser, S. S., Cornelius, N., Hald, M., Korsun, S. & Pawlowski, J. (2005). Monothalamous foraminiferans and gromiids (Protista) from western Svalbard: a preliminary survey. *Marine Biol. Res.,1,* 290 – 312.

Gooday, A. J., Cedhagen, T., Kamenskaya, O. E. & Cornelius, N. (2007 a). The biodiversity and biogeography of komokiaceans and other enigmatic foraminiferan-like protists in the deep Southern Ocean. *Deep-Sea Res*, II *54*, 1691 – 1719.

Gooday, A. J., Kamenskaya, O. E. & Cedhagen, T. (2007 b). New and littleknown Komokiacea (Foraminifera) from the bathyal and abyssal Weddell Sea and adjacent areas. *Zool J Linn Soc-Lond*, *151*, 219 – 251.

Gooday, A. J., Nomaki, H. & Kitazato, H. (2008). Modern deep-sea benthic foraminifera: a brief review of their morphology-based biodiversity and trophic diversity. *Geol Soc Lond., Spec Publ*, *303*, 97 – 119.

Gooday, A. J., Todo, Y., Uematsu, K. & Kitazato, H. (2008). New organic-walled Foraminifera (Protista) from the ocean's deepest point, the Challenger Deep (western Pacific Ocean) *Zoological Journal of the Linnean Society*, *153*, 399–423.

Gooday, A. J, Aranda de Silva, A., Koho, K. A., Lecroq, B. & Pearce, R. B. (2010 a). The 'Mica sandwich'; a remarkable new genus of Foraminifera (Protista, Rhizaria) from the Nazarй Canyon (Portuguese margin, NE Atlantic). *Micropaleontology*, *56* (3-4), 345 – 357.

Gooday, A. J., Anikeeva, O. V. & Pawlowski, J. (2010 b). New genera and species of monothalamous Foraminifera from Balaclava and Kazach'ya Bays (Crimean Peninsula, Black Sea). *Marine Biodiversity*, DOI 10.1007/s12526-010-0075-7, 14.

Gooday, A.J., Bett, B.J., Jones, D.,O.,B., Kitazato, H. (2012). The influence of productivity on abyssal foraminiferal biodiversity. *Marine Biodiversity*. 42: 415 – 431.

Gooday, A.J., Jorissen, F.J. (2012). Benthic foraminiferal biogeography: controls on global d istribution patterns in deep-water settings. *Annual Review Marine Science*. *4,* 237 – 62.

Gradstein, F. & Kaminski, M. (1989). Taxonomy and biostratigraphy of new and emended species of Cenozoic deep-water agglutinated foraminifera from the Labrador and North Sea. *Micropaleontology*, *35*, 72 – 92.

Grell, K. G. (1954). Der Generationswechsel der polythalamen Foraminifere *Rotaliella heterocariotica. Arch. Protistenkund.*, *100* (2), 268 - 286.

Grell, K. G. (1956). Der Generationwechsel der polythalamen Foraminifere *Rotaliella heterocaryotica. Arch. Protistenkund.*, *100*, 211 - 235.

Grell, K. G. (1973). *Protozoology*. Berlin – Heidelberg – New York, N.Y. Springer – Verlag, 554.

Grell, K. G. (1979). Cytogenetic systems and evolution in Foraminifera. *Journ. Foram. Res.*, *9* (1), 1 – 14.

Grell ,K. G. (1988). The life-cycle of the monothalamous foraminifer *Heterotheca lobata*, n. gen., n. sp. *Journ. Foram. Res.*, *18* (1), 54 - 74.

Grigyalis, A. A. (1977). Klassifikaciya I istoricheskoe razvitie nadsemejstva Nodosariacea (Foraminifera). [Classification and historical development of the superfamily Nodosariacea (Foraminifera)]. *Paleontol. Journ.*, *1*, 5 – 13. (In Russian)

Grigyalis, A. A. (1978 a). Higher foraminiferal taxa. *Paleontological Journal*, *121*, 1 – 9.

Grigyalis, A. A. (1978 b). Morphology and classification of Ceratobuliminacea. *Journ. Foram.Res.*, *8* (4), 275 – 285.

Grigyalis, A. A. (1995). *Foraminifery jurskich otlozhenij Jugo-Zapadnoj Pribaltiki. [Foraminifera of the Jurassic deposits of the South-West ajancent to the Baltic sea*]. Vilnius: 240 (In Russian).

Grigyalis, A. A. & Gorbatchick, T. N. (1980). Morphology and taxonomy of Jurassic and Early Cretaceous representatives of the superfamily Globigerinacea (Favusellidae). *Journ. Foram. Res.*, *10* (3), 180 - 190.

Groenlund, H. & Hansen, H. J. (1976). Scanning electrone microscopy of some recent and fossil nodosariid foraminifera. *Bull. Geol. Soc. Denmark*, *25*, 121 – 134.

Gross, O. (2001). Foraminifera. *In: Costello, M.J. et al. (Ed.) (2001). European register of marine species: a check-list of the marine species in Europe and a bibliography of guides to their identification. Collection Patrimoines Naturels*, *50*, 60-75.

Groussin, M., Pawlowski, J. & Yang, Z. (2011). Bayesian relaxed clock estimation of divergence times in foraminifera. *Molecular Phylogenetic and Evolution*, *61*, 157 – 166.

Groves., J. R. (1999). Suborder Lagenina and other smaller foraminifera from the uppermost Pennsylvanian-lower Permian rocks of Kansas and Oklahoma. *Micropaleontology*, *46*, 285 – 326.

Groves., J. R. (2000). Suborder Lagenina and other smaller foraminifers from uppermost Pannsylvania-Lower Permian rocks of Kansas and Oklahoma. *Micropaleontology*, *46* (4), 285 – 326.

Groves, J. R. & Wahlman, G. P. (1997). Biostratigraphy and evolution of the Late Carboniferpus and early Permian smaller foraminifers from Barebtz Sea (offshore Arctic Norway). *Journ. Paleontol.*, *71*, 758 – 779.

Groves, J. R. & Boardman, D. R., II (1999). Calcareous smaller foraminifers from the Lower Permian Council Grove Group near Hooser, Kansas. *Journ Foram. Res.*, *29* (3), 243 – 262.

Groves, J. R., Altiner, D. & Rettori, R. (2003). Origin and early evolutionary radiation of the order Lagenida (Foraminifera). *Journ. Paleontol.*, *77* (5), 831 – 843.

Groves, J. R., Rettori, R. & Altiner, D. (2004). Wall structure in selected Paleozoic Lagenide Foraminifera. *Journ. Paleontol.*, *78* (2), 245 – 256.

Groves, J. R., Altiner, D. & Rettori, R. (2005). Extinction, survival, and recovery of lagenide foraminifera in the Permiaan-Triassic boundary interval, Central TauRides, Turkey. *Journ. Paleontol.*, *79* (Suppl. 4), 37 pp.

Groves, J. R., Kulagina, E. I. & Vella, E. (2007). Diachronous appearance of the Pennsylvanian fusulinid *Profusulinella* in Eurasia and North America. *Journ. Paleontol.*, *81* (2), 7 – 17.

Groves, J. R., Rettori, R., Payne, J. L., Boyce, M. D. & Altiner, D. (2007). End-Permian mass-extinction of Lagenidae foraminifers in the Southern Alps (northern Italy). *Journ. Paleontol.*, *81*, 415 – 434.

Groves, J. R. & Lee, A. (2008). Accelerated rates of foraminiferal origination and extinction during the late Paleozoic ice age. *Cushm. Found. Foram. Res.*, *38* (1), 74 – 84.

Gubenko, T. A. (1988). Microstructura stenki rakoviny nekotorych permskich i triasovych miliolat [Wall ultrastructure of the shell of some Permian and Triassic Miliolata]. *In:* Mikhalevich, V.I. (Ed.), *In*: Sistematika, ekologia i stratigrafiya miliolyat (Foraminifery). *Trudy Zoologicheskogo Instituta AN SSSR*, *184*, 44- 49 + 1 tabl. (In Russian).

Gubenko, T. A. (1989). Tipy stroenija rakovin rannekamennougolnych foraminifer. [Types of the shell structure in the foraminifera of the Lower Carboniferous]. Doct. Thes.., Vladivostok: 1 - 20. (In Russian).

Gudmundsson, G. (1994). Phylogeny, ontogeny and systematics of Rcent Soritacea Ehrenberg, 1839 (Foraminiferida). *Micropaleontology*, *40* (2), 101 – 155.

Gudmundsson, G. Von Schmalensee, M. & Svavarsson, J. (2000). Are foraminifers (Protozoa) important food for small isopods (Crustacea) in the deep sea? *Deep-sea research. Part 1. Oceanographic research papers*, *47* (11), 2093 - 2109.

Guilbault, J-P. & Patterson, R. T. (1998). *Rudoloculina hooperi*, a new miliolid with an agglutinated outer surface from the NorthEastern Pacific ocean. *Journ. Foram.Res.*, *26* (4), 306 – 311.

Guisberti, L. & Coccioni, R. (2003). *Posadia feroniensis* nov. gen., n. sp. (Lituolida, Hormosinidae) from the Bathonian of Sardinia, Italy. *Journ. Foram.Res.*, 33 (3), 211 – 218.

Gupta, A. K. (1993). Txonomy and bathymetric distribution of Holocene deep-sea benthic foraminifera in the Indian Ocean and the Red Sea. *Micropaleontology*, *40* (4), 351- 367.

Gussone, N. & Filipsson, H. L. (2010). Calcium isotope ratios in calcitic tests of benthic foraminifers. *Earth and Planetary Science Letters*, *290* (1-2), 108-117.

Gustafsson, M. & Nordberg, K. (2001). Living (stained) benthic foraminiferal response to primary production and hydrography in the deepest part of the Gullmar Fjord, Swedish West Coast, with comparison to Hoeglund's 1927 material. *Journ. Foram. Res.*, *31* (1), 2 – 11.

Habura, A., Pawlowski, J., Haynes, S. D. & Bowser, S. S. (2004). Unexpected foraminiferal diversity revealed by small subunit rDNA analysis of Antarctic sediment. *Journ. Eukaryotic Microbiology*, *51* (2), 173 – 179.

Habura, A., Wegener, L., Travis, J. L. & Bowser, S. S. (2005). Structural and functional implications of an unusual foraminiferal Я-tubulin. *Molecular Biology and Evolution*, *22*, 2000-2009.

Habura, A., Goldstein, S., Broderick, S. & Bowser, S. S. (2008). A bush not a tree: the extraordinary diversity of cold-water basal foraminiferans extends to warm-water environments. *Limnology and Oceanology.*, *9*, 79 – 87.

Habura, A., Hou, Y., Reilly, A. A. & Bowser, S. S. (2011). Highthroughput sequencing of *Astrammina rara*: sampling the giant genome of a giant foraminiferan protist. *BMC Genomics*, *12*, 169.

Haig, D.W. & McCartain, E. (2010). Triassic organic-cemented siliceous agglutinated foraminifera from Timor Leste: conservative development in shallow marine environments. *Journ. Foram. Res.*, *40* (4), 366- 392.

Hallock, P. (1979). Trends in test shape with depth in large, symbiont-bearing foraminifera. *Journ.Eukaryotic Microbiology*, *9* (1), 61 - 69.

Hallock, P. (1981a). Light dependence in *Amphistegina*. *Journ.Eukaryotic Microbiology*, *11* (1), 40 - 46.

Hallock, P. (1981b). Algal symbiosis: a mathematical analysis. *Marine Biol.*, *62* (4), 249 - 255.

Hallock, P. (1982). Evolution and extinction in larger Foraminifera. *Third North Am. Paleont. Conv., Proc.*, *1*, 221 - 225.

Hallock, P. (1984). Distribution of selected species of living algal symbiont-bearing Foraminifera on two Pacific coral reefs. *Journ. Foram. Res.*, *14* (4), 250 - 261.

Hallock, P. (1985). Why are larger foraminifera large?*Paleobiology*, *11*, 195 – 208.

Hallock, P. (1987). Fluctuations in the trophic resource continuum: a factor in global diversity cycles? *Paleoceanography*, *2*, 457 – 471.

Hallock, P. (2000). Symbiont-bearing foraminifera: harbingers of global change? *Micropaleontology*, *46*, suppl. I, 95 – 104.

Hallock, P. (2000). *Larger foraminifera as indicators of coral reef vitality*. In: Environmental Micropaleontology: The application of microfossils to Environmental Geology, Martin, R.E. (ed), Kluwer Academic/Plenum Publishers, 121 – 150.

Hallock, P. & Hansen (1979). Depth adaptation in *Amphistegina*: change in lamellar thickness. *Bull. Geol. Soc. Denmark*, *27*, 99 – 104.

Hallock P., Cottey, T. L., Forward, L. B. & Halas, J. (1986). Population biology and sediment production of *Archaias angulatus* (Foraminiferida) in Largo Sound, Florida. *Journ. Foram. Res.*, *16* (1), 1 – 8.

Hallock, P., Premoli Silva, I. & Boersma, A. (1991). Similarities between planktonic and larger foraminiferal evolutionary trends through Palaeogene paleoceanographic changes. *Palaeogeography, Palaeoclimatology, Palaeoecology*, *83*, 49 – 64.

Hallock, P. & Peebles, M. W. (1993). Foraminifera with chlorophyte endosymbionts: Habitat of six species in the Florida Keys. *Marine micropaleontology*, *20*, 277 – 292.

Hallock, P. & Talge H. K. (1994). A predatory foraminifer *Floresina amphiphaga* n. sp., from the Florida Keys. *Journ. Foram. Res.*, *24* (3), 210 - 213.

Hallock, P., Talge H. K., Cockey E. M. & Muller R. D. (1995). A new disease in reef-dwelling foraminifera: implications for coastal sedimentation. *Journ. Foram. Res.*, *25* (3), 280 - 286.

Hance, L., Hou, H. & Vachard, D. (2011). *Upper famenian to Visean Foraminifers and some carbonate Microproblematica from South China – Hunan, Guangxi and Guizhou*. Hance, L., Hou, H., Vachard, D., (eds). Beijing:Geological Publishing House, China: 359.

Hamaoui, M. (1984). On the dynamics of some major architectural trends in the Alveolinidae Foraminifera. *International Symposium on Benthic Foraminifera (2nd : 1983 : Pau, France) & Oertli, H. J & Société nationale Elf-Aquitaine (Production) & Total compagnie française des pétroles & Esso REP (1984). Benthos '83 : 2nd International Symposium on Benthic Foraminifer, Pau (France), 11-15/4/1983. Elf Aquitaine,Esso REP, Total CFP, Pau and Bordeaux, France.*

Hamaoui, M. (1984). A selection of biostratigraphic markers from the Upper Paleozoic of Tunisia. *International Symposium on Benthic Foraminifera (2nd : 1983 : Pau, France) & Oertli, H. J & Société nationale Elf-Aquitaine (Production) & Total compagnie française des pétroles & Esso REP (1984). Benthos '83 : 2nd International Symposium on Benthic Foraminifer , Pau (France), 11-15/4/1983. Elf Aquitaine,Esso REP, Total CFP, Pau and Bordeaux, France*, 281 - 287.

Hansen, H. J. (1972). Two species of Foraminifera of the genus *Turrilina* with different wall structure. *Lethaia.*, *5*, 39 - 45.

Hansen, H. J. (1975). On feeding and supposed buoyancy mechanism in four recent Globigerinid Foraminifera from the Gulf of Elat, Israel. *Rev. Espan. Micropal.*, *2*, 325 - 341.

Hansen, H. J. (1979). Test structure and evolution in the Foraminifera. *Lethaia*, *12* (2), 173 - 182.

Hansen, H. J. (1999). *Shell construction in modern calcareous foraminifera*. In: Sen Gupta, B.K.., ed., Modern Foraminifera. Dordrecht: Kluwer: 57 70.

Hansen, H. J. & Hanzlikova, E. (1974). Ultrastructure of some silicieous Foraminiferida. *Rev. Esp. Micropal.*, *6* (3), 447 - 466.

Hansen, H. J., Reiss, Z. & Schneiderman N. (1969). Ultrastructure of bilamellar walls in Foraminiferida. *Rev. Esp. Micropal.*, 1 (3), 293 - 316.

Hansen, H. J. & Reiss, Z. (1971 a). Electron microscopy of Rotaliacean wall structures. *Bull. Geol. Soc. Denmark*, *20*, 329 - 346.

Hansen, H. J. & Reiss, Z. (1971 b). Scanning electrone microscopy of wall structures in some benthic and planktonic Foraminiferida. *Rev. Esp. Micropal.*, *4* (2), 169 - 181.

Hansen, H. J. & Reiss, Z. (1972). Scanning electron microscopy of wall structures in some benthonic and planctonic Foraminiferida. *Rev. Esp. Micropal.*, *4* (2), 169 – 181.

Hansen, H. J. & Lykke-Andersen, L. (1976). Wall structure and classification of fossil and recent elphidiid and nonionid Foraminifera. *Fossils and Strata*, *10*, 1 – 37.

Hansen, H. J. & Revets, S. (1992). A revision and reclassification of the Discorbidae, Rosalinidae and Rotaliidae. *Journ. Foram. Res.*, *22* (2), 166 - 181.

Hansen, H. J. & Dalberg, P. (1979). Symbiotic algae in milioline foraminifera: CO@ uptake and shell adaptations. *Bull. Geol. Soc. Danmark*, *28*, 47 – 55.

Harmelin, J-G. & Venec-Peyre, M-T. (1992). Morphology, ecology and biostratigraphy of *Discoramulina bolli* Seiglie, 1964, a cyclostomata-like foraminifer. *Journ. Foram. Res.*, *22* (2), 181 - 186.

Harney, J. N., Hallock, P. & Talge, H. K. Observation on a trimorphic life sycle in *Amphistegina gibbosa* population from the Florida Keys. *Journ. Foram. Res.*, *28* (2), 141 -147.

Hart, M. B. (1980). A water depth model for the evolution of the planktonic foraminiferida. *Nature*, *286*, 252 – 254.

Hart, M. B. (1984). The superfamily Robertinacea in the lower Cretaceous of the UK and Adjacent areas of NW Europe. *Benthos' 83*, *2nd Intern. Symp. Benthic Foraminifera (Pau, April 1983*), 289 – 298.

Hart, M. B. (1999). The evolution and biodiversity of Cretaceous planktic Foraminiferida. *Geobios*, *32*, 247 – 255.

Hart, M. B., Oxford, M. J. & Hudson, W. (2002). The early evolution and paleobiogeography of Mesozoic planktic foraminifera. In: Cramer, J.A., Owen, A.W., Eds., *Paleobiogeography and biodiversity change: the Ordovician and Masozoic-Cenozoic radiations*, *London: The Geological Soc. of London*, Sp. publ. *194*, 115 – 125.

Harward, N. J. B. & Haynes, J. R. (1976). *Chlamys opercularis* (Linnaeus) as a mobile substrate for Foraminifera. *Journ.Foram. Res.*, *6* (1), 30 - 36.

Haugen, J. -E., Sejrup, H. -P. & Vogt, N. B. (1989). Chemotaxonomy of Quaternary benthic Foraminifera using amino acids. *Journ. Foram. Res.*, *19* (1), 38 - 52.

Hauser, E. H. & Keller Grunig, A. (1993). Eponides and some related genera (Cretaceous to Recent), A Taxonomic Revision. *Journ. Foram. Res.*, *23* (4), 238 - 253.

Hay, W. W., Towe, K. M. & Wright, R. (1963). Ultrastructure of some selected foraminiferal tests. *Micropaleontology*, *9* (2), 171 - 196.

Haynes, J. R. (1965). Symbiosis, wall structure and habitat in foraminifera. *Contr. Cushm. Found. Foram. Res.*, *16*, 40 – 43.

Haynes, J. R. (1981). *Foraminifera*. New York. John Wiley and Sons. 433.

Haynes, J. R. (1990). The classification of the Foraminifera - a review of historical and philosophical perspectives. *Paleontology*, *33* (3), 503 - 528.

Hayward, B. W. (1990). *Taxonomy, paleobiogeography and evolutionary history of the Bolivinellidae (Foraminiferida).* N.Zealand Geol. Surv. Paleontol. Bull 63, 132.

Hayward, B. W. (2002). Late Pliocene to Middle Pleistocene extinctions of deep-sea benthic foraminifera ("*Stilostomella*" extinction) in the Southwest Pacific. *Journ. Foram.Res.*, *32* (3), 274 – 307.

Hayward, B. W. & Brasier, R. C. (1980). Taxonomy and distribution of present day *Bolivinella. Journ. Foram. Res.*, *10* (2), 102 - 116.

Hayward, B. W., Hollis, C. J. & Grenfell, H. R. (1997). *Recent Elphidiidae (Foraminiferida) of the South-west Pacific and fossil Elphidiidae of New Zealand.* Institute of Geological and Nuclear Sciences Monograph: 1 – 170.

Hayward, B. W. & Kawagata, S. (2005). Extinct foraminifera figured in Brady's Challenger Report: *Journal of Micropalaeontology*, *24*, 171-175.

Hayward, B. W., Kawagata, S., Grenfell, H. R., Droxler, A. W. & Shearer, M. (2006). Mid-Pleistocene extinction of bathyal benthic foraminifera in the Caribbean Sea. *Micropaleontology*, *52* (3), 245 – 266.

Hayward, B. W., Kawagata, S., Grenfell, H. R., Sabaa, A. T. & O'Neil, T. (2007). The last global extinction in the deep sea drilling the mid-Pleistocene climate transition. *Paleooceanographym*, *22*, PA3103, doi:10.1029/2007PA001424.

Hayward, B. W., Sabaa, A. T., Thomas, E., Kawagata, S., Nomura, R., Schroder-Adams, C., Gupta, A. K. & Johnson, K. (2010). Cenozoic record of elongate, cylindrical, deep-sea benthic foraminifera in the Indian Ocean (ODP Sites 722, 738, 744, 758, and 763). *Journ.Foram.Res.*, *40* (2), 113 – 133.

Hayward, B. W., Cedhagen, T., Kaminski, M. & Gross, O. (2011). World Modern Foraminifera database. Available online at http://www.marinespecies.org/foraminifera.

Hayward, B. W., Kawagata, S., Sabaa, A., Grenfell, H., Van Kerckhoven, L., Johnson, K. & Thomas, E. (2012). *The last global extinction (Mid-Pleistocene) of deep-sea benthic foraminifera (Chrysalogonidae, Ellipsoidinidae, Glandulonodosariidae, Plectofrondiculariidae, Pleurostomellidae, Stilostomellidae), their Late Cretaceous-Cenozoic history and Taxonomy.* Cusman Foundation for Foraminiferal Research Special Publication, *43*, 408.

Hayward, B. W., Cedhagen, T.,Kaminski, M., Gross, O. (2013). Word modern Foraminifera database. Accessed through World Register of Marine Species at http://www.marinespecies.org/aphia.php?p=taxdetails&id=533807 on 2013-07-12

Hedley, R. H. (1962). The significance of an" inner chitinous lining" in Saccamminid organisation with special reference to a new species of *Saccammina* (Foraminifera) from New Zealand. *New Zealand Oceanograph. Inst. Contrib.*, *127. New Zealand Journ. of Sci.*, *3*, 375 - 389.

Hedley, R. H. (1963). Cement and iron in the arenaceous Foraminifera. *Micropaleontology.*, *9* (4), 433 – 441.

Hedley, R. H. (1964). *The biology of Foraminifera.* In: Intern. Rev.Gener. Experim. Zoology. Felts W.J.L., Harrison R.J. (Eds). New-York - London., 1 - 48.

Hedley, R. H., Ogden C. G. & Wakefield J. St. J. (1973). Shell ultrastructure in allogromiid Foraminifera (Protozoa). *Bull. Brit. Mus. Nat. Hist. Zool.*, *24* (9), 467 - 474.

Hedley, R. H., Parry D. M. & Wakefield J. St. J. (1968). Reproductiion in *Boderia turneri* (Foraminifera). *Journ. Nat. Hist.*, *2*, 147 - 151.

Heinz, P., Kitazato, H., Schmiedl, G. & Hemleben, C. (2001). Response of deep-sea benthic Foraminifera from the Mediterranean sea to simulated phytoplankton pulses under laboratory conditions. *Journ. Foram.Res.*, *31*, 210 – 227.

Hemleben, C. (1969). Zur morphogenese planktonicher Foraminiferen. *Zitteliana*, *1*, 91 – 133.

Hemleben, C. (1975). Spine and pustule relationships in some Recent planktonic Foraminifera. *Micropaleontology*, *21* (3), 334 - 341.

Hemleben, C., Be, A. W. H., Anderson, O. R. & Tuntivate, S. (1977). Test morphology, organic layers and chamber formation of planktonic foraminifer *Globorotalia menardii* (d'Orbigny). *Journ. Foram. Res.*, *7* (1), 1 - 25.

Hemleben, C. & Spindler M. (1983). Recent advances in research on living planktonic Foraminifera. In: Meulenkamp J.E. (Ed) Reconstruction of marine palaeoenvironments. *Utrecht Micropal. Bull.*, *30*, 141 - 170.

Hemleben, C., Anderson, O. R., Berthold, W. U. & Spindler, M. (1986). Calcification and chamber formation in Foraminifera – a brief overview. *In:* Leadbeater B.S.C. & Riding R., (eds), Biomineralization in plants and animals. *The Systematics Association Special* Volume *30*, 237 – 249.

Hemleben, C., Kaminski, M. A., Kuhnt, W. & Scott, D. B. (eds) (1989). *Paleoecology, biostratigraphy, paleoceanography and Taxonomy of agglutinated foraminifera.* Series c: Mathematical and physical sciences, *327*, 1 – 1017.

Hemleben ,C., Spindler M. & Anderson O. R. (1989). *Modern Planktonic Foraminifera.* Springer-Verlag. New-York., 363.

Hemleben, C. & Kitazato, H. (1995). Deep-sea foraminifera under long time observation in the laboratory. *Deep Sea Research Part I: Oceanographic Research Papers*, *42* (6), 827 – 832.

Henbest, L. G. (1963). Biology, mineralogy and diagenesis of some typical Late Paleozoic sedentary Foraminifera and algal-foraminiferalmcolonies. *Cushm. Found. Foram. Res., spec. Publ.*, *6*, 1 – 44.

Hengeveld, R. & Fedonkin, M. A. (2004). Causes and consequences of eukaryotization through mutualistic endosymbiosis and compartmentalization. *Acta Biotheor.*, *52* (2), 105-154.

Herrero, C. (1995). *Eoguttulina palomerensis*, a new foraminiferal species from the Plienbachian- Toarcian boundary (Early Jurassic) of SW Europe. *Journ. Micropaleontol.*, *14* (1), 53 – 57.

Herrero, C. & Haynes, J. R. (1997). Revision of the Gault foraminifera from the Hollis and Neaverson Collection (1921). *Journ. Micropaleontol.*, *16*, 73 – 84.

Hickman, C. S. & Lipps, J. H. (1983). Foraminiferivory: selective ingestion of Foraminifera and test alternations produced by the neogastropod *Olivella*. *Journ. Foram. Res.*, *13* (2), 108 – 114.

Hilterman, V. H. & Koch, W. (1957). Revision der Neoflabellinen (Foram.). *Geol. Jb.*, 74: 269 – 304.

Hodgkinsson, R. L. (1992). W.K. Parker's collection of foraminifera in the British Museum (Natural History). *Bull. Brit. Mus. Nat. Hist.*, *(Geol)*, *48* (2), 45 – 78.

Hofker, J. (1927). *The foraminifera of the Siboga expedition. Part. I. Tinoporidae, Rotaliidae, Nummulitidae, Amphisteginidae.* Uitk. Zool. Bot. Oceanogr. Geol. Geb., monogr. IV a: 1 – 78.

Hofker, J. (1930 a). *Notizen uber die Foraminiferen des Golfes von Neapel.* Publ. Staz. Zool. Di Napoli, *10*, 365 – 406.

Hofker, J. (1930 b). *The foraminifera of the Siboga expedition.* Pt. II. Monogr. IVa. Leiden, Brill.: 79 – 170.

Hofker, J. (1930 c). Der generationswechsel von *Rotalia beccarii* var. *flevensis*, nov. var. *Zschr. Zellforsch. Micr. Anat.*, *10*, 756 - 768.

Hofker, J. (1951). The toothplate Foraminifera. *Arch. Neerland. Zool.*, *8*, 353 - 373.

Hofker, J. (1971). Studies of Foraminifera. Pt. III. Systematic problems. *Publ. Naturhist. Genootschap Limburg. Reeks*, XXI (1,2,3), 202 S.

Hofker, J. (1983). Zoological exploration of the continental shelf of Surinam: the foraminifera of the shelf of Surinam and the Guyanas. *Zoologische Verhandelingen , Rijksmuseum van Natuurlijke Historiete Leiden*, *201*, 1 – 75.

Hohenegger, J. (1974). Zur Anwendung stochastischer Methoden in der Taxonomie obernorischer Nodosarien. *Verh. Geol. B.-A.*, *1*, 25 – 69.

Hohenegger, J. (1978). Populationsgenetische Deutung des morphologischen Wandels der triassischen Foraminifere *Nodosaria striatoclavata. Lethaia*, *11*, 199 – 215.

Hohenegger, J. (1981). *Ichtyolaria denticostata* n. sp., eine charakteristische Foraminifere des Unteren Lias Mitteleuropas. *Stutgarter Beitr. Naturkund.*, *Ser. B (Geol. und Paleontol.)*, *74*, 1 – 33.

Hohenegger, J. (1990). *On the way to optimal suprageneric classification of agglutinated Foraminifera*. In: Hemleben C., Kaminski M.A., Kuhnt W., Scott D.B. (Eds). Palaeoecol., Biostratigr., Palaeoceanogr. and Taxonomy Aggl. Foram. Kluwer Acad. Publishers. Netherlands. 77 - 105.

Hohenegger, J. (1993). Determination of Upper Triassic and Lower Jurassic Ichtyolarias using morphogenetic programs. *Micropaleontology*, *39* (3), 233 - 262.

Hohenegger, J. (1994). Distribution of living larger foraminifera NW of Sesoko-Jima, Okinawa Japan. *P.S.Z.N.*, *I. Marine Ecology*, *15*, 291 – 334.

Hohenegger, J. (1995). Depth estimation by proportions of living larger foraminifera. *Marine Micropaleontol.*, *26*, 31 – 47.

Hohenegger, J. (1997). Morphologicas niches as tools for phylogenetic analysis : Permian and Triassic Lagenina as a case study. In: Ross, C.A., Ross, J.R.P., Brenckle, P.L., Eds. Late Paleozoic foraminifera – their biostratigraphy, evolution, and paleoecology and the Mid-Carboniferous boundary. *Cushm. Found. Foram. Res.*, *Sp. Publ.*, *36*, 63 – 70.

Hohenegger, J. (1999). Larger Foraminifera – Microscopical greenhouses indicating shallow-water tropical and subtropical environments in the present and past. *In: Hatta, A., Oki, K. (eds). Foramiifera as indicators of marine environments in the present and past: Kagoshima University Research Center of the South Pacific, Occasional Papers*, *32*, 19 – 45.

Hohenegger, J. (2000). Coenoclines of larger Foraminifera. *Micropaleontology*, *46*, suppl. 1, 127 – 151.

Hohenegger, J. (2004). Depth coenoclines and environmental considerations of Western Pacific larger foraminifera. *Journ. Foram. Res.*, *34* (1), 9 - 33.

Hohenegger, J. (2011). *Large Foraminifera. Greenhouse constructions and gardeners in the oceanic microcosm.* The Kagoshima University Museum Bulletin. No 5.

Hohenegger, J. & Piller W. (1975). Wandstructuren und Grosgliederung der Foraminiferen. *Oster. Acad. Wiss. Sitzungsberichte. Abt 1*, *184*, 1 (5), 67 - 98.

Hohenegger, J. & Piller W. (1975). Diagenetische Veranderungen bei obertriadischen Involutinidae (Foraminifera). *N. Jb.Geol. Palaeont.Mh.*, *1*, 26 – 39.

Hohenegger, J. & Piller W. (1977). Die Stellung der Involutinidae Butschli und Spirillinidae Reuss im System der Foraminiferen. *N. Jb.Geol. Palaeont.Mh.*, *7*, 407 – 418.

Hohenegger, J., Piller W. & Baal C. (1993). Horizontal and vertical spatial microdistribution of Foraminifers in the shallow subtidal Gulf of Trieste, North Adriatic Sea. *Journ. Foram. Res.*, *23* (2), 79 - 102.

Hohenegger, J., Yordanova, E. & Hatta, A. (2000). Remarks on West Pacific Nummulitidae (Foraminifera). *Journ. Foram. Res.*, *30*, 3 - 28, doi:10.2113/0300003

Holbourn, A. E. L. & Kaminski, M. A. (1997). *Lower Cretaceous deep-water benthic foraminifera of the Indian Ocean.* Grzybowski Foundation, Spec. publ. *4*, 1 – 172.

Holzmann, M. & Pawlowski, J. (2000). Taxonomic relationships in the genus *Ammonia* (Foraminifera) based on ribosomal DNA sequences. *Journ. Micropal.*, *19*, 85 – 95.

Holzmann, M. & Pawlowski, J. (2002). Freshwater foraminiferans from Lake Geneva:past and present. *Journ. Foram. Res.*, *34*, 344 – 350.

Holzmann, M., Habura, A., Giles, H., Bowser, S. S. & Pawlowski, J. (2003). Freshwater Foraminiferans revealed by analysis of Environmental DNA Samples. *Journ. Eukaryotic Microbiology*, *50* (2), 135 – 139.

Holzmann, M., Hohenegger, J. & Pawlowski, J. (2003). Molecular data reveal parallel evolution in nummulitid foraminifera. *Journ. Foram. Res.*, *33* (4), 277 – 284.

Holzmann, M., Kaminski, M. A., Filipescu, S. & Pawlowski, J. (2012). A molecular comparison of *Entzia tetrastomella* and *Jadammina macrescens*. Alegret, L., Ortiz, S., Kaminski, M.A. (eds), Ninth Intern. Workshop Aggl. Foram., *Grzybowski Foundation Sp. Publ.*, *18*, 42.

Hornibrook, N. & de, B. (1864). A revision of the Oligocene and Miocene Foraminifera from New Zealand described by Karrer and Stache in the Reports of "Novara" Expedition. *New Zealand Geol. Surv*, *Paleontol. Bull.*, *43*, 1 – 85.

Hottinger, L. (1976). An Early Umbilical Canal System in *Trocholina schouberti* n. sp. from the Lower Cretaceous of North Eastern Morocco. *Eclog. Geol. Helvetiae*, *69* (3), 815 - 820.

Hottinger, L. (1978). Comparative anatomy of elementary shell structures in selected larger Foraminifera. In: Hedley R.H., Adams C.G. (Eds). Foraminifera. London. Pt. *3*, 203 - 206.

Hottinger, L. (1979). Araldit als Helfer in der Mikropalaeontologie. Aspekte. Giba-Geigy AG. Division Kunststoffe und Additive, Nachdruck von Wort und Bild mit Quellenangabe gestattet. *Uber nur mit Genehmigung der Redaktion*, *Basel*, *3*, 1-10

Hottinger, L. (1982). Larger Foraminifera, Giant Cells with a Historical Background. *Naturwissenschaften*, *69*, 361 - 371.

Hottinger, L. (2000). Functional morphlogy of benthic foraminiferal shells, envelopes of cells beyond measure. *Micropaleontology*. vol. *46*, suppl. 1, 57 – 87.

Hottinger, L. (2001). The shell cavity systems in Elphidiid and Pellatispirine bilamellar Foraminifera. *Micropaleontology*, *47* (4), 1 – 4.

Hottinger, L. (2006). The "face" of benthic foraminifera. *Bolletino della societa Paleontologica Italiana*, *45* (1), 75 -89.

Hottinger, L. (2006). *Illustrated glossary of terms used in foraminiferal research*. Carnets de Geologie/Notebooks on Geology, Brest, Memoir 2006/02, (CG 2006_MO2), 126 p.

Hottinger, L. (2007). Revision of the foraminiferal genus *Globoreticulina* Rahaghi, 1978, and of its associated fauna of larger foraminifera from the late Middle Eocene of Iran. Carnets de Géolog. / Notebooks on Geology - Article 2007/06 (CG2007_A06).

Hottinger, L. & Dreher D. (1974). Differentiation of Protoplasm in Nummulitidae (Foraminifera) from Elat, Red Sea. *Marine Biol.*, *25*, 41 - 61.

Hottinger, L. & Drobne, K. (1980). *Early Tertiary conical imperforate foraminifera*. Slovenska AKad. Znanosti in Umetnosti, Acad. Scientiarium et Artium Slovenica, Razred za prirodoslovne vede classis IV: Histor.Naturalis, Razprave Disserationes, XXII/3, ed. Ak. I. Racovec, 90 p, 22 tables.

Hottinger, L. & Leutenegger, S. (1980). The structure of calcarinid foraminifera. *Schweizerische Palaentologische Abhandl.*, *101*, 115 – 151.

Hottinger, L., Drobne, K. & Caus, E. (1989). Late Cretaceous, larger, complex miliolids (Foraminifera)endemic in the Pyrenean faunal province. *Facies*, *21*, 99 – 133.

Hottinger, L., Halicz, E. & Reiss, Z. (1991). Architecture of *Eponides* and *Poroeponides* (Foraminifera) reexamined. *Micropaleontology*, *37* (1), 60 – 75.

Hottinger, L., Halicz, E. & Reiss, Z. (1993). Recent Foraminifera from the Gulf of Aqaba, Red Sea. Akad. dr. Mario Plenicar (ed). *Slovenske Akademije Znanosti in Umetnosti - S.A.Z.U.*, *Ljubljana*, *Dela Opera*, *33*, 179.

Hottinger, L., Reiss, Z. & Langer, M. (2001 a). Spiral canals of some Elphidiidae. *Micropaleontology*, *47* (Suppl. 2), 5 – 34.

Hottinger, L., Romero, J. & Caus, E. (2001 b). Architecture and revision of the Pellatispirines, planispiral canaliferous foraminifera from the Late Eocene Tethys. *Micropaleontology*, *47* (Suppl. 2), 55 – 77.

Hyams-Kaphzan, O. & Lee, J. J. (2008). Cytological examination and location of symbionts in "living sands" – *Baculogipsina*. *Journ. Foram. Res.*, *38* (4), 298 – 304.

Isakova, T. N. & Filimonova, T. V. (2003). Characteristics of the Test Wall of Several Fusulinoida (Foraminifers) Studied with a Scanning Electron Microscope (SEM).. *International Congress on Carboniferous and Permian Stratigraphy*, *Univ. Utrecht*, *Netherlands*, *August 10–16*, (Univ. Utrecht, Utrecht, 2003), 248–249.

Ishitani, Y., Ishikawa, S. A., Inagaki, Y., Tsuchiya, M., Takahashi, K. & Takishita, K. (2011). Multigene phylogenetic analyses including diverse radiolarian species support the "Retaria" hypothesis: the sister relationship of Radiolaria and Foraminifera. *Mar. Micropaleontol.*, *81*, 32 – 42.

Ismail, A. A., Hussein-Kamel, Y. F., Boukhary, M. & Ghandour, A. A. (2009). Late Cenomanian-Early Turonian foraminifera from Eastern Desert, Egypt. *Micropaleontology*, *55* (4), 396 – 412.

Jones, R. W. (1984). A revised classification of the unilocular Nodosariida and Buliminida (Foraminifera). *Revista Espanol Micropaleontol.*, 16: 91 – 160.

Jones, R. W. (1994). "*The Challenger Foraminifera.*" Oxford University Press: 149.

Jorgensen, B. B., Erez, J., Revsbech, N. P. & Cohen, Y. (1985). Symbiotic photosynthesis in planktonic Foraminiferan, *Globigerinoides sacculifer* (Brady), studied with microelectrodes. *Limnology and Oceanography*, *30*, 1253 – 1267.

Kamenskaya, O. E. (1987). Xenophyophoria and Komokiacea in trophic chains of the deep-sea benthos. In: *Feeding of marine invertebrates and its significance in formation of communities*. Academy of Sciences of the USSR, P.P. Shirshov Institute of Oceanology, Moscow: 15 – 22.

Kamenskaya, O.E., Bagirov, N.E. & Simdianov, T.G. (2002). *Luffammina atlantica* gen. et sp. nov. (Foraminiferida, Arboramminidae) from the hydrothermal area Rainbow. In: Kuznetsov, A.P., & Zezina, O.N. (eds), *Adaptational aspects of the Evolution of Marine Fauna: Collected Proceedings. VNIRO Publishers, Moscow*, 144 - 152.

Kamenskaya, O. E., Gooday, A. J., Radziejewska, T. & Wawrzyniak-Wydrowska, B. (2012). Large, enigmatic foraminiferan-like protists in the eastern part of the Clarion-Clipperton Fracture Zone (abyssal north-eastern subequatorial Pacific), biodiversity and vertical distribution in the sediment. *Senckenberg, Gesellschaft fur Naturforschung and Springe, Mar Biodiv:* DOI 10.1007/s12526-012-0114-7.

Kamenskaya, J. B. & Saidova, Ch. M. (1998). Pereopisanie vida *Psammina planata* (Saidova, 1970) – ultrastruktura abissalnoj ksenofioforii iz Kurilo-Kamchatskogo zheloba.

[Redescription of the species Psammina planata(Saidova, 1970) – ultrastructure of the abyssal xenophiophorian from the Kurilo-Kamchatskij depression]. In: *Benthos of the high latitude regions*. Moscow. (In Russian).

Kaminski, M. A. (2000). The New and Reinstated Genera of Agglutinated Foraminifera published between 1986 and 1996. In: Hart, M.B., Kaminski, M.A., & Smart, C.W. (eds), *Proceedings of the Fifth International Workshop on Agglutinated Foraminifera. Grzybowski Foundation Special Publication*, *7*, 185-219.

Kaminski, M.A. (2008). The new and reinstated genera of agglutinated foraminifera published between 2000 and 2004. *In: Kaminski, M.A. & Coccioni, R. (eds), Proceedings of the Seventh International Workshop on Agglutinated Foraminifera. Grzybowski Foundation Special Publication, 13*, 47 - 55.

Kaminski, M. A. (2004). The new and reinstated genera of agglutinated Foraminifera published between 1996 and 2000. *In:* Bubík, M. & Kaminski, M.A. (eds), *Proceedings of the Sixth International Workshop on Agglutinated Foraminifera. Grzybowski Foundation Special Publication*, 8.

Kaminski, M. A. (2004). The year 2000 Classification of the Agglutinated Foraminifera. In: Bubik, M. and Kaminski, M. A. (eds),. *Proceedings of the Sixth International Workshop on Agglutinated Foraminifera. Grzybowski Foundation Special Publication*, *8*, 317 – 349.

Kaminski, M. A., Boersma, A., Tyszka, J. & Holbourn, A. E. L. (1995). Response of deep-water agglutinated Foraminifera to dyoxic conditions in the California Borderland Basins. In: Kaminski M.A., Geroch S., Gasinski M.A. (Eds). *Proc. Fourth Intern. Workshop Aggl. Foram. Krakow, Poland, Sept. 12 – 19, 1993. Grzybowski Found. Sp. Publ.*, *3*, 131 - 140.

Kaminski, M. A. & Kuhnt, W. (1995). Tubular agglutinated Foraminifera as indicators of organic carbon flux. In: Kaminski M.A., Geroch S., Gasinski M.A. (Eds). *Proc. Fourth Intern. Workshop Aggl. Foram. Krakow, Poland, Sept. 12 – 19, 1993. Grzybowski Found. Sp. Publ.*, *3*, 141 - 144.

Kaminski, M. A., Neagu, T. & Platon, E. (1995). A revision of *Falsogaudryinella* from the Lower Cretaceous of the North Sea and Romania, and its relationship to *Uvigerinammina. In:* Kaminski M.A., Geroch S., & Gasinski M.A. (eds) Proceedings of the Fourth International Workshop on Agglutinated Foraminifera. *Grzybowski Foundation Special Publication*, *3*, 145- 157.

Kaminski, M. A., Kuhnt, W. & Radley J. D. (1996). Paleocene-Eocene deep water aglutinated foraminifera from the Numidian Flysch (Rif, Northern Morocco), their significance for the palaeoceanography of the Gibraltar gateway. *Journal of Micropaleontology*, *15*, 1 – 19.

Kaminski, M. A., Holbourn, A. E. L. & Geroch, S. (1997). *Neaguammina* n.gen., a new agglutinated foraminiferal genus from the Lower Cretaceous of DSDP Site 263 (Indian Ocean*). Annales Societatis Geologorum Poloniae*, *67*, 231 – 235.

Kaminski, M. A. & Filipescu, S. (2000). *Praesphaerammina*, a new genus of Cenozoic deep-water agglutinated foraminifera from the Carpathian flysch deposits. *Micropaleontology*, *46*, (4), 353-359.

Kaminski, M.A. & Wetzel, A. (2004). A tubular protozoan predator: A burrow selectively filled with tubular agglutinated protozoans (Xenophyophorea, Foraminifera) in the abyssal South China Sea. In; Bubik, M. & Kaminski, M.A., (Eds), *Proceedings of the*

Sixth International Workshop on Agglutinated Foraminifera. Grzybowski Foundation Special Publication, 8, 287 - 293.

Kaminski, M. A. & Gradstein, F. M. (2005). Cenozoic cosmopolitan deep-water agglutinated foraminifera. Grzybowski Foundation Special Publication, *10*, 547.

Kaminski, M. A., Armitage, D. A., Jones, A. P. & Coccioni, R. (2008). Shocked Diamonds in Agglutinated Foraminifera from the Cretaceous/Palaeogene Boundary, Italy – a preliminary report. In: Kaminski M.A., Coccioni, R. (Eds).*Proc.* Sevehtn *Intern. Workshop Aggl.Foram.Urbino, Italy, Octob. 2 – 8, 2005. Grzybowski Found,. Sp. Publ., 13*, 57 - 61.

Kaminski, M.A., Bubik, M. & Cetean, C.G. 2008. *Rectoprotomarssonella* n.gen., a new agglutinated foraminiferal genus from the Upper Cretaceous of the Carpathian flysch. *Micropaleontology*, *53* (6), 517 – 521.

Kaminski, M. A., Uchman, A., Neagu, T. & Cetean, C. (2008). A larger agglutinated foraminifer originally described as a marine plant: the case of *Arthrodendron* Ulrich, 1904 (Foraminifera), its synonyms and homonyms. *Journ. of Micropaleontol, 23*, 103 – 110.

Kaminski, M.A., Setoyama, E. & Cetean, C. (2008). Revised Stratigraphic Ranges and the Phanerozoic Diversity of Agglutinated Foraminiferal Genera. *In: Kaminski, M.A. & Coccioni, R. (eds), Proceedings of the Seventh International Workshop on Agglutinated Foraminifera. Grzybowski Foundation Special Publication, 13*, 70 - 106.

Kaminski, M.A., Cetean, C.G. and members of International Working Group on Foraminiferal Systematics Bubik, M., Filipescu, S., Hohenegger, J., Mikhalevich, V., Nagy, J., Pawlowski, J., Pronina-Nestell, G., Rettori, R., Setoyama, E., Tyszka, J. & Whittaker, J. (2008). A Catalogue of Agglutinated Foraminiferal genera – Towards a revision of Loeblich & Tappan (1987). *In: Filipescu, S. & Kaminski, M. (Eds). Eighth International Workshop on Agglutinated Foraminifera, Abstract Vol., Grzybowski Foundation Special Publication, 14*, 27.

Kaminski, M. A., Silye, L. & Kender, S. (2009). Miocene deep-water agglutinated foraminifera from the Lomonosov Ridge and the opening of the Fram Strait. *Micropaleontology*, *55* (2 3), 117 – 136.

Kaminski, M. A., Henderson, A. S., Cetean, C. G. & Waskowska, A. (2009). The Ammolagenidae n. fam., a new family of agglutinated foraminifera. In: Pisera, A., Bitner, M.A., Halamski, S.T. (eds), *Ninth Paleontological Conference, 10 – 11 October, 2008, Polish Academy of Sciences, Warszawa, Abstracts*: 41 – 43.

Kaminski, M. A., Cetean, C. G. & Tyszka, Ja. (2011). Nomenclature to describe the transition from multiserial to uniserial chamber arrangement in benthic foraminifera. *Journ. Micropaleontology*, *30*, 7 – 10.

Kaminski, M. A. & Cetean, C. G. (2011). The systematic position of the foraminiferal genus *Cubanina* Palmer, 1936 and its relationship to *Colominella* Popescu, 1998. *Acta Paleontol. Romaniae*, *7*, 231 – 234.

Kaminski, M. A. & Contreras, J. (2011). The new foraminiferal genus *Ammogloborotaloides* n. gen. and subfamily Ammogloborotaloidinae n. subfam. from the Neogeneee of Venezuela: an example of isomorphism between agglutinated and planktonic Foraminifera. *Journal of Micropaleontology*, *30*, 11 -16.

Karavaeva, N. I. & Nestell, G. P. (2007). Permian foraminifers of the Omolon Massif, northeastern Siberia, Russia. *Micropaleontology*, *53* (3), 161 – 211.

Karpov, C. A. (1990). *Systema protistov*. [*System of Protists*]. Omsk. Pedagog. Inst., Ser. boil., (1), 192 (In Russian).

Kathal, P. K. (2002). Taxonomy, distribution patterns and ecology of Recent littoral foraminifera of the east coast of India. *N. Jb. Geol. Paleontol. Abh.*, *224* (1), 115 – 160.

Kawagata, S., Hayward, B. W., Grenfell, H. R. & Sabaa, A. T. (2005). Mid-Pleistocene extinction of deep-sea foraminifera in the North Atlantic Gateway (ODP Sites 980 and 982). *Palaeogeography*, *Palaeoclimatology*, *Palaeoecology*, *221*, 267-291.

Kawagata, S., Yamasaki, M. & Jordan, R. (2005). *Acarotrochus* lobatulus, a new genus and species of shallow-water benthic foraminifer from Mecherchar Jellyfish Lake, Palau, NW equatorial Pacific Ocean. *Journ. Foiram. Res.*, *35* (1), 44 – 49.

Kawagata, S., Hayward, B. W. & Gupta, A. K. (2006). Benthic foraminiferal extinctions linked to late Pliocene-Pleistocene deep-sea circulation changes in the northern Indian Ocean (ODP Sites 722 and 758), *Marine Micropaleontology*, *58*, 219-242

Keeling, P. J. (2001). Foraminifera and Cercozoa are related in actin phylogeny: two orphans find a home? *Mol. Biol. Evol.*, *18*, 1551 – 1557.

Kender, S., Kaminski, M. A. & Jones, R. W. (2006). Four new species of deep water agglutinated foraminifera from the Oligocene-Miocene of the Congo Fan (offshore Angola). *Micropaleontology*,. *52*, 465 - 470, doi:10.2113/gsmicropal.52.5.465.

Kender, S., Kaminski, M. A. & Jones, R. W. (2008). Oligocene deep-Water Agglutinated Foraminifera from the Congo Fan. In: Kaminski, M.A., Coccioni, R. & Marsili, A. *Proceedings of the Seventh International Workshop on Agglutinated Foraminifera. Grzybowski Foundation Special Publication*, *13*, 107 – 156.

Kennet, J. P. & Srinivasan, S. (1983). *Neogeneee Planktonic Foraminifera: A Phylogenetic Atlas*. Hutchinson Ross Publishing Company, Stroudsburg, Pennsylvania: 265.

King, K. Jr. (1977). Amino acid survey of recent calcareous and siliceous deepsea microfossils. *Micropaleontology*, *23* (2), 180 - 193.

King, K. Jr. & Hare, P. E. (1972). Amino acid compposition of planktonic Foraminifera: A palaeobiochemical approach to evolution. *Science*, *175* (4029), 1461 - 1463.

Kitazato, H. & Matsushita, S. (1996). Laboratory observations of sexual and asexual reproduction of *Trochammina hadai* Uchio.*Trans.Proc. Paleont. Soc. Jpn*, *182*, 454 – 466.

Klein, C. & Mutterlose, J. (2001). Benthic foraminifera: indicators for a long-term improvement of living conditions in the Late Valanginian of the NW German Basin. *Journ. Micropaleontol.*, *20*, 81 – 95.

Kloos, D. P. (1980). Studies on the foraminifer Sorites orbicularis. *Geol. Mijubouw.*, *59* (4), 375 - 383.

Knight, R. (1986). Apertural characteristics of certain unilocular foraminifera: methods of study, nomenclature and taxonomic significance. *Journ Micropaleontol.*, *5* (2), 37 – 47.

Kobayashi, F. & Altiner, D. (2007). Late Carboniferous and early Permian fusulinoideans in the central Taurides, Turkey: biostratigraphy, faunal composition and comparison. *Journ. Foram. Res.*, *37* (1), 69 – 89. DOI: 10.2113/gsjfr.37.1.69

Kobayashi, F., Ross., C. A. & Ross, J. R. P. (2010). Classification, phylogeny, and paleobiogeography of the new subfamily Gifuellinae and a revision of the family Neoschwagerinidae (superorder Fusulinoida); Guadalupian (Middle Permian). *Jour. Foram. Res.*, *40* (3), 283 – 300.

Kohl, B. (1985). Early Pliocene benthic Foraminifera from the Slina Basin, Southern Mexico. *Bull. Amer. Paleontol.*, *88* (322), 1 - 173.

Kohler-Rink, S. & Kuhl, M. (2000). Microsensor studies of photosynthesis and respiration in larger symbiotic foraminifera. I. The physico-chemical microenvironment of *Marginopora vertebralis*, *Amphistegina lobifera* and *Amphisorus hemprichii*. *Marine Biology*, *137*, 437 - 486.

Koho, K. A., Moodley, L., Pozzato, L., Middelburg, J. J. & Reichart, G-J. (2012). Uptake of particulate and dissolved organic matter by protists (Foraminifera and Gromia), 13 C-tracer study from the Arabian Sea. *The Micropaleontol Soc. Foraminifera and Nannofossil Groups Joint Meeting 2012*, *Interdisciplinary advances in Foraminiferal and nannofossil Res.*, *Pollock Halls*, *Univ. Edinburg*: 24.

Kotlyar, G. V., Zakharov, Yu. D., Kropacheva, G. S., Pronina, G. P., Chedya, I. O. & Burago, V. I. (1989). *Evolucia pozdne-permskoy bioty. [Evolution of the Latest Permian biota].* Midian regional state in the USSR Leningrad. Nauka.184 (In Russian).

Krasheninnikov, V. F. (1956). Mikrostruktura stenki nekotorych kajnozojskoch foraminifer i metodika ee izuchenija v poljarizovannom svete. [Wall microstructure of of some Cenozoic Foraminifera and the methods of its study in the polarized light]. *Voprosy mikropaleontologii.*, *1*, 37 – 48. (In Russian).

Kremer, B. P., Schmaljohann, R. & Roettger, R. (1980). Features and nutritional significance of photosyntheites produced by unicellular algae symbiotic with larger Foraminifera. *Marine Ecol. Progress Ser.*, *2* (3), 225 - 228.

Krueger, R., Roettger, R., Lietz, R. & Hohenegger, J. (1997). Biology and reproductive processes of the larger foraminiferan *Cycloclypeus carpenteri* (Protozoa, Nummulitidae). *Archiv Protistenkunde*, 307-321.

Krylov, M. V. (1996). *Opredelitel paraziticheskich prostejshich cheloveka*, *domashnich zhivotnych I selskochozjajstvennych rastenij. [Key-book of parasitic protists of the human beings*, *domestic animals and agricultural plants*]. Russian Academy of Sciences, Zoological Institut, St. Petersburg: 602 (In Russian).

Krylov, M. V. (2005). [Initial Principles of the Evolution of Living Matter]. *Proc. Zool. Inst. Russ. Acad. Sci.*, *308 (Zool.Sessions. Ann. Rep. 2004)*, 35–40.

Krylov, M. V. & Libenson, M. N. (2002). Continium evolucionnych processov zhivoj I nezhivoj materii [Continium of the evolutionary processes of the organic and inorganic matter]. In: MN Libenson (Chairman of the Programme Commettee) Problemy i perspectivy mezhdisciplinarnych fiindamentalnych issledovanij. *Materials of the second scientific conference of S.Petersburg Association Scientists and Scolars (SPASS)*, *11-12 April 2002. Sankt-Peterburgskiy Sojuz Uchenych*, S. Petersburg: 55. (In Russian).

Krylov, M. V., Dobrovolsky, A. A., Issi, I. V., Mikhalevich, V. I., Podlipaev, S. A., Reshetnyak, V. V., Seravin, L. N., Starobogatov, Ja. I., Sculman, S. S. & Jankovski, A. W. (1980). Novye predstavlenija o systeme odnokletochnych zhivotnych. [New conception of the system of unicellular animals].]. *In:* Principy postroeniya makrosistemy odnokletochnyih zhivotnyh. *Trudy Zoologicheskogo Instituta*, *AN SSSR*, *94*, 42–61 (In Russian).

Kuile, B. (1991). Mechanisms for calcification and carbon cycling in algal symbiont-bearing Foraminifera. In: *Biology of Foraminifera* (Lee, J.J. , Anderson, O.R. (eds). *Academic press*, London, 73 – 89.

Kuile, B. & Erez, J. (1987). Uptake of inorganic carbon and internal carbon cycling in symbiont-bearing benthonic Foraminifera. *Marine Biol. 94* (4), 499 - 509.

Kulagina, E. I. (2012). Evolucija kamennougolnych foraminifer nadsemejstva Archodiscoidea [Evolution of the Carboniferous foraminifers of the superfamily Archodiscoidea]. *Trudy XV Vserossijskogo mikropaleontol Sovestschanija, Gelendzhik, Sept. 2012*, 90 – 92.

Kusakin, O. N. & Drozdov, A. L. (1998). *Filema organicheskogo mira*. [*Phylema of the organic world*]. Part 2. *Prokaryotes, Eukaryotes: Microsporobiontes, Archemonadobiontes, Euglenobiontes, Myxobiontes, Rhodobiontes, Alveolates, Heterokontes*. St. Petersburg, Nauka: 358 (in Russian).

Kuznetzova, K. I. & Basov, V. A. (1974). K sistematike rodovych taksonov nodozariid. [On the systematic of the generic taxa of Nodosariids]. *Voprosy micropaleont., 17*, 84 – 89. (In Russian).

Kuznetzova, K. I. & Basov, V. A. (2001). Svjaz taksonomicheskogo sostava I raznoobrazija jurskich foraminifer s paleogeograficheskimi I facialnymi obstanovkami. [Dependence of the taxonomic composition and diversity of the Jurassic Foraminifera of the paleogeographic and facial conditions]. *In: Biodiversity in the Earth history. Thes. XLVII session of Paleontol Soc., St. Petersburg. Krymgolc, N.G., Modzalevskaya, T.L. (eds). Rus. Acad. Sci.,Paleontol. Soc. RAN, VSEGEI*: 49 – 50. (In Russian).

Laatsch, H. & Schwab, D. (1984). Identification and quantification of carotenoids and carotenoid esters from the foraminiferen *Allogromia laticollaris. Journ. Foram. Res., 14* (1), 36 - 49.

Labas, Ju. A. & Mikhalevich, V. I. (1987). Rakovinoobrazovanie u Textulariata i Rotaliata (Foraminifera). [Formation of the shell wall in Textulariata and Rotaliata (Foraminifera)]. *Thes. Dokladov i Soobstchenij IV VOPR (All-Russian Protozzolog. Soc.) Congress, Leningrad, "Nauka":* 11. (In Russian).

Langer, M. R. (1992). New Recent foraminiferal genera and species from the lagoon at Madang, Papua New Guinea. *Journ.of Micropalaeontol.,11*: 85 - 93.

Langer, M. R. (1992). Biosynthesis of glycosaminoglycans in foraminifera: A review. *Marine Micropaleontol., 19*, 245 – 255.

Langer, M. R. (1993). Epiphytic foraminifera. *Marine Micropaleontol., 20*, 235 – 265.

Langer, M. R. & Gehring, C. A. (1993). Bacteria farming: a possible feeding strategy of some smaller motile Foraminifera. *Journ. Foram. Res., 23* (1), 40 – 46.

Langer, M. R., Lipps J. H. & Piller W. E. (1993). Molecular palaeobiology of protists: amplification and direct sequencing of foraminiferal DNA. *Micropaleontology, 39* (1), 63 - 68.

Langer, M. R., Lipps, J. H. & Guillermo, M. (1995). Predation on foraminifera by the dentaliid deep-sea scaphopod *Fissidentalium megathyris. Deep-Sea-Research, 6*, 849 – 857.

Langer, M. R., Silk, M. T. B. & Lipps, J. H. (1997). Global ocean carbonate and carbon dioxide production: the role of reef Foraminifera. *Journ.Foram.Res., 27* (4), 271–277.

Langer, M. R. & Hottinger, L. (2000). Biogeography of selected "larger" foraminifera. *Micropaleontology, 46*, suppl. 1, 105 – 126.

Langer, M. R., Makled, V. A., Pietsch, S. J. & Weinmann, A. (2009). Asynchronous calcification in juvenile megalospheres: an ontogenetic window into the life cycle and polymorphism of *Peneroplis. Journ. Foram. Res., 39* (1), 8 – 14.

Larin, J. S. (1977). Proischozhdenie i zhiznedejatelnost kletochnoj organizacii v svete idej polymerizacii i oligomerizacii V. A. Dogielja [The origin and vital activity of the cellular organization in the light of the ideas of the polymerization and oligomerization of V. A. Dogiel]. In: Significance of the processes of polymerization and oligomerization in evolution]. In: OA Scarlato (ed) *Published by Zoological Institute of the Academy of Sciences USSR*, Leningrad: 89-93. (In Russian).

Lea, D. W., Martin, P. A., Chan, D. A. & Spero, H. J. (1995). Calcium uptake and calcification rate in the planktonic foraminifer *Orbulina universa*. *Journ. Foram. Res.*, *25* (1), 14 – 23.

Lecroq, B., Lejzerowicz, F., Bachar, D., Christen, R., Esling, P., Baerlocher, L., Osteras, M., Farinelli, L. & Pawlowski, J. (2011). Ultra-deep sequencing of foraminiferal microbarcodes unveils hidden richness of early monothalamous lineages in deep-sea sediments. *PNAS*, *108* (22), 13177 – 13182.

Le Cadre., V. & Debenay, J. -P. (2005). Morphological and cytological responses of *Ammonia* (Foraminifera to copper contamination: implication for the use of foraminifera as bioindicators of pollution. *Environmental pollution*, *143* (2), 304 – 317.

Le Calvez, J. (1935). Les gamètes du quelques Foraminifères. *Académie des Sciences*, *Paris*, *Comptes Rendus*, *201*, 1505 – 1507.

Le Calvez, J. (1938). Recherches sur les Foraminiferes. I. Development et Reproduction. *Arch. Zool. Exper. Gen.*, *80*, 163 - 333.

Le Calvez, J. (1947). Entosolenia marginata, foraminnifere apogamique ectoparasite d'un autre foraminifere *Discorbis vilardeboanus*. *Comt. Rendu Hebdomadaire Seanc. Acad. Sci. Paris*, *224*, 1448 - 1450.

Le Calvez, J. (1950). Recherches sur les Foraminiferes. II. Place de meiose et sexualite. *Arch. Zool. Exp. Gen.*, *87* (4), 211 - 243.

Le Calvez, J. (1953). Ordre des foraminiferes. In: *Traite de Zoologie (ed. P. Grasse)*, *1* (2), 149 – 265.

Le Calvez, Y. (1977). Revision des Foraminiferes de la collection d'Orbigny. II – Foraminiferes de l'ile de Cuba. *Cahiers de Micropaleontol*, *1*, 1 – 127.

Lee, J. J. (1980). Nutrition and Physiology of the Foraminifera. *In: Biochemistry and physiology of Protozoa*, *2 nd Edition*, *3*, 43 - 66.

Lee, J. J. (1990). *Phylum Granuloreticulosa (Foraminifera).* In: Corliss J.O. Handbook of protoctista. Boston. 914.

Lee, J. J. (1990). Fine structure of the Rhodophycean *Porphiridium purpureum in situ* in *Peneroplis pertusus* (Forscal) and *P. acicularis* (Batch) in axenic culture. *Journ. Foram. Res.*, *20*, 162 – 169.

Lee, J. J. (1993). "On a Piece of Chalk" - Up dated. *Journ. Eukaryotic Microbiol.*, *40* (4), 395 - 409.

Lee, J.J. (1998). "Living sands" – larger Foraminifera and their endosymbiotic algae. *Symbiosis*, *25*, 71 – 100.

Lee, J. J. (2006). Symbiotic: Forms of life. *In: Seckhbach, J. (ed). Life as we know it. Springer*, *Dordrecht*, 307 – 324.

Lee, J. J. & Anderson, O. R. (Ed). (1991). Biology of Foraminifera. Acad. Press. London.

Lee, J. J., Anderson, R. O., Karim, B. & Beri, J. (1991 a). Additional insight into the structure and biology of *Abyssotrema pacifica* (Bronnimann, Van Dover and Whittaker) from the East Pacific Rise. *Micropaleontology*, *37* (3), 197 - 134.

Lee, J. J., Sang, K., Kuile, B. Ter., Strauss, E., Lee, P. J. & Faber, Jr., W. W. (1991 b). Nutritional and related experiments on laboratory maintenance of three species of symbiont-bearing , large foraminifera. *Marine biology*, *109*, 417 – 425.

Lee, J. J., Faber., W. W. & Lee, R. E. (1991 c). Granular reticulopodian digestion – a possible preadaptation to benthic foraminiferal symbiosis? *Symbiosis*, *10*, 47 – 61.

Lee, J. J. & Bock, D. (1976). The importance of feeding in two species of Soritid Foraminifera with algal symbionts. *Bull. Marine Sci.*, *26* (4), 530 - 537.

Lee, J. J., Freudental, H. G., Kossoy, V. & Be, A. W. H. (1965). Cytological observations on two planktonic Foraminifera, *Globigerina bulloides* d'Orbigny,1826 and *Globigerinoides ruber* (d'Orbigny, 1839) Cushman,1927. *Journ. Protozool.*, *12* (4), 531 - 542.

Lee, J. J., Freudental, H. D., Muller, W. A., Pierce, S. & Grossman, R. (1963). Growth and physiology of Foraminiifera in the laboratory. Part 3. Initial studies of *Rosalina floridana* (Cushman). *Micropaleontology*, *9* (4), 449 - 466.

Lee, J. J. & Hallock, P. (1987). Algal symbiosis as the driving force in the evolution of larger Foraminifera. In: Lee J.J., Frederik J.F. (Eds). Endocytobiology. III. *Annals New York Acad. Sci.*, *503*, 330 - 347.

Lee, J. J., Hunter, S. H. & Bovee, E. C. (1985 a). An illustrated guide to the Protozoa. Lawrence, Kansas. *Soc. Protozool.*, 16 - 17.

Lee, J. J., Lee, M. J. & Weis, D. S. (1985 b). Possible adaptive value of endosymbionts to their protozoan hosts. *Journ. Protozool.*, *32* (3), 380 - 382.

Lee, J. J., McEnery, M. E. & Garrison, J. R. (1980). Experimental studies of larger Foraminifera and their symbionts from the Gulf of Elat on the Red Sea. *Journ. Foram. Res.*, *10* (1), 31 - 47.

Lee, J. J., McEnery, M. E., Kahn, E. & Shuster, F. (1979). Symbiosis and the evolution of larger Foraminifera. *Micropaleontology*, *25* (2), 118 - 140.

Lee, J. J., McEnery, M. E., Lee, M. J., Reidy, J. J., Garrison, J. R. & Roettger, R. (1980). Algal symbionts in larger Foraminifera. In: *Endocytobiology*, *Endosymbiosis and Cell Biology*, 113 - 124.

Lee, J. J., McEnery, M., Pierce, S., Freudental, H. D. & Muller, W. A. (1966). Tracer experiments in feeding littoral Foraminifera. *Journ. Protozool.*, *13* (4), 659 - 670.

Lee, J. J., McEnery M. E. & Rubin H. (1969). Quantitative Studies on the growth of *Allogromia laticollaris* (Foraminifera). *Journ. Protozool.*, *16* (2), 377 - 395.

Lee, J. J. & Muller W. A. (1973). Trophic dynamics and niches of salt marsh Foraminifera. *Amer. Zool.*, *13*, 215 - 223.

Lee, J. J., Muller W. A., Stone, R. J., McEnery, M. E. & Zucker, W. (1969). Standing crop of Foraminifera in sublittoral epiphytic communities of a Long Island salt marsh. *Marine Biol.*, *4*, 44 - 61.

Lee, J. J., Pierce, S., Tentchoff, M., McLaughlin, J. J. A. (1961). Growth and physiology of foraminifers in the laboratory. Part 1. Collection and maintenance. *Micropaleontology*, *7* (4), 461 – 466.

Lee, J. J. & Pierce, S. (1963). Growth and physiology of Foraminifera in the laboratory . IV. Monoxenic culture of an allogromiid with notes on its morphhology. *Journ. Protozool.*, *10*, 404 - 411.

Lee, J. J., Reimer, Ch. W. & McEnery, M. E. (1980). Diatoms endosymbionts from larger Foraminifera of the Red Sea. *Botanica Marina*, XXIII, 41 - 48.

Lee, J. J., Saks, N. M., Kapiotou, F., Wilen, S. M. & Shilo, M. (1984). Effects of host cell extracts on cultures of endosymbiotic diatoms from larger Foraminifera. *Marine Biol., 82* (2), 113 - 120.

Lee, J. J., Tietjen, J. H., Stone, R. J., Muller, W. A., McEnery, M., Saks, N., Mastropaol,o C. & Kennedy, E. (1971). Effects on environmental stress on the community structure, productivity, energy flow, and mineral cycling in salt marsh epyphytic communities. *Proc. third Nation. Symp.Radioecol., 2*, 816 - 828.

Lee, J. J. & Zucker, W. (1969). Algal symbiosis in the foraminifer *Archaias. Journ. Protozool., 16* (1), 71 – 81.

Lee, J. J., Morales, J., Bacus, S., Diamont, A., Hallock, P., Pawlowski, J. & Thorpe, J. (1997). Progress in characterizing the endosymbiotic dinoflagellates of soritid Foraminifera and related studies on some stages in the life cycle of *Marginopora vertebralis*. *Journ.Foram. Res., 27*, 254 - 263.

Lee, J. J., Fine, M., Levy, O. & Morales, J. (2009). A note on asexual reproduction of a *Marginopora* sp. from a modern deep-water population in the Heron-Wistari channel, Australia. *Journ.Foram. Res.,39* (1), 4 – 7.

Lehmann., G., Roettger, R. & Hohenegger. (2006). Life cycle variation including trimorphism in the foraminifer *Trochamina inflata* from the North European salt masrshes. *Journal of Foraminif. Res., 36* (4), 279 – 290.

Leiter, C. & Altenbach, A. V. (2010). Benthic Foraminifera from the Diatomaceous Mud Belt off Namibia: Characteristic Species for Severe Anoxia. *Palaeontologia Electronica, 13* (2), 19.

Lejzerowicz, F., Pawlowski, J., Fraissinet-Tachet, L. & Marmeisse, R. (2010). Molecular evidence for widespread occurrence of Foraminifera in soils. *Environ. Microbiol., 12* (9), 2518 – 2526. doi: 10.1111/j.1462-2920.2010.02225.x

Lena, H. (1972). Cytological studies in *Allogromia flexibilis* (Wiesner)(Foraminifera). Int. *Rev. Hydrobiol., 57* (4), 637 - 644.

Lena, H. & Freire, F. (1974). Estudios citologicos en *Allogromia laticollaris* Arnold (Protozoa, Foraminifera). *Phys. Seccion A., 33* (86), 1123 - 133.

Lengsfeld, A. M. (1969). Zum Feinbau der Foraminifere *Allogromia laticollaris*. I. Mitt. : Zellen mit ausgestreckten und eingezogenen Rhizopodien. *Helgoland Wiss. Meeresunters., 19* (2), 230 - 261.

Leontiev, K . (1996). *Vostok, Rossiya i slavjanstvo* [*The East, Russia and Slavdom*]. Respublika, Moscow. (In Russian).

Leppig, U., Forke, H. C., Montenary, M. & Fohrer, B. (2005). A three- and two- dimensional documentation of structural elements in schwagerinids (superfamily Fusulinoidea) exemplified by silicified material from the Upper Carboniferous of the Carnic Alps (Austria/Italy), a comparison with verbeekinoideans and alveolinids. *Facies, 46*, 1 – 13.

Leutenegger, S. (1977). Ultrastructures de Foraminiferes perfores et imperfores ainsi que de leurs symbiotes. *Cahiers Micropaleontol., 3*, 1 - 52.

Leutenegger, S. (1983). Specific host- symbiont relationship in larger Foraminifera. *Micropaleontology, 29* (2), 111 - 125.

Leutenegger, S. (1984). Symbiosis in benthic foraminifera: specificity and host adaptations. *Journ. Foraminiferal Res., 14* (1), 16 – 35.

Leutenegger, S. & Hansen, H. J. (1979). Ultrastructural and radiotracer studies of pore function in Foraminifera. *Marine Biol., 54* (1), 11 - 16.

Leven, E. Ya. (1963). O filogenii vysshich fusulinid i podrazdelenii verchne-paleozojskich otlozhenij Tetisa.[About the phylogeny of higher fusulinids and subdivision of the Upper Paleozoic deposits of the Tethys]. *Voprosy Micropaleontologii*, *7*, 57 – 70. (In Russian)

Leven, E. Ya. (1970). O proischozhdenii vysshich fusulinid. [On the Origin of Higher Fusulinids]. Paleontol. Zhurnal, 3: 18–25. (In Russian).

Leven, E. Ya. (1997). Permian stratigraphy and Fusulinida of Afganistan with their paleogeographic and paleotectonic implications. *Geol. Soc. of America*, *Spec. Paper*, *316*, 134.

Leven, E. Ya. & Davydov, V. (2001). Stratigraphy and Fusulinids of the Kasimovian and Upper Gzhelian (Upper Carboniferous) in the Southwestern Darvaz (Pamir). *Riv. Ital. Paleontol. Stratigr.*, *107* (1), 3–46.

Levy, A., Mathieu, R., Poignant, A., Rosset-Moulinier, M. & Rouvillois, A. (1979). Revision de quelques genres de la famile Discorbidae (Foraminiferida), fondee sur l'observation de leur architecture interne. *Rev. Micropaleontologie*, *22*, 66 – 88.

Levy, A., Mathieu, R., Poignant, A. & Rosset-Moulinier, M. (1986). Discorbidae and Rotaliidae: a classification to be revised. *Journ. Foram. Res.*, *16* (1), 63 - 71.

Lewis, K. B. (1970). A key to the recent genera of the Foraminiferida. *New Zealand Dept. Sci. Industr. Res. Bull. 196. New Zealand oceanogr. Inst. Mem.*, *45*, 88.

Li, Q. (1987). Origin, phylogenetic development and systematic taxonomy of the *Tenuitella plexus* (Globigerinitidae, Globigerinina). *Journ. Foram. Res.*, *17* (4),

Libenson, M. N. & Przhibelskii, S. G. (2003). Concepciya sovremennogo estestvoznaniya [Concepts of modern natural science. Part II. Physics]. *International Banking Institute*, St Petersburg, 84 (In Russian).

Linke, P. & Lutze, G. F. (1993). Microhabitat preferences of benthic foraminifera – a static concept or a dynamic adaptation to optimize food acquisition? *Marine Micropaleotol.*, *20*, 215 – 234.

Linke, P., Altenbach, A. V., Graf, G. & Heeger, T. (1995). Response of deep-sea benthic foraminifera to a simulated sedimentation event. *Journ. Foram. Res.*, *25* (1), 75 - 82.

Lipps, J. H. (1966). Wall structure systematics, and phylogeny studies of Cenozoic planktonic Foraminifera. *Journ. Paleontol.*, *35*, 245 - 330.

Lipps, J. (1973). Text structure in foraminifera. *Ann. Rev. Microbil.*, *27*, 471 – 488.

Lipps, J. H. (1975). Feeding strategies and test function in foraminifer. *Proc. "Benthonics 75"*: 26.

Lipps, J. H. (1983). Biotic interactions in benthic Foraminifera. In: Tevesz M.J.S.,McCall P.L.(Eds). Biotic interactions in Recent and fossil benthic communities. *Plenum Publishing Corp.*, 331 – 376.

Lipps, J. H. (1987). Prokaryotes and Protista. In: Lipps J.H. (Ed). *Fossil procaryotes and Protists. Tenessee*, *18*, 1 - 19.

Lipps J. H. & Erskian, M. G. (1969). Plastogamy in Foraminifera *Glabratella ornatissima. Journ. Protozool.*, *16* (3), 422 - 453.

Lipps, J. H. & Valentine, J. W. (1970). The role of Foraminifera in the trophic structure of marine communities. *Lethaia*, *3* (3), 270 – 286.

Lipps, J. H. & Ronan, T. E. Jr. (1974). Predation on Foraminifera by the Polychaete worm, Diopatra. *Journ. Foram. Res.*, *4* (3), 139 - 143.

Lipson-Benitah, S. (2008). Phylogeny of the middle Cretaceous (late Albian- late Cenomanian) planktic foraminiferal genera *Parathalmaninella* nov. gen. and *Thalmaninella*. *Journ. Foram.Res.*, *38*, 183 – 189.

Lister, J. J. (1895).Contributions to the life history of the Foraminifera. *R. Soc.London Phil. Trans. Ser. B.*, *186*, 401 - 453.

Lith, Y., Langezaal, A. M., de Nooijer, L. J. & Vanderzwaan, G. J. (2009). Benthic foraminiferal effect on nitrogen and carbon cycling. *Jour.Foram.Res.*, *39* (2), 97– 111.

Liu, C., Olsson, R.K. & Hubert, B.T. (1998). A benthic paleohabitat for *Praepararotalia* gen.nov and *Antarcticella* Loeblich and Tappan *Journ. Foram.Res.*, *28* (1), 3 – 18.

Loeblich, A. R., III. (1974). Protistan Phylogeny as indicated by the fossil record. *Taxon.*, *23* (2), 277 – 290.

Loeblich, A. R., Jr. & Tappan, H. (1953). Studies of Arctic Foraminifera. *Smithsonian Miscellaneous Collections*, *121* (7), 1-150.

Loeblich, A. R. & Tappan, H. (1964). *Part C. Protista 2. Chiefly "Thecamoebians" and Foraminiferida. In:* Moore, R.C. (Ed.), Treatise on Invertebrate Paleontology. The Geological Society of America and the University of Kansas. Lawrence Kansas. Vol. *1*, 2: 900 p.

Loeblich, A. R., Jr. & Tappan, H. (1985). Some new and redefined genera and families of agglutinated foraminifera. *Jour.Foram.Res.*, *15*, 91 – 104.

Loeblich, A. R., Jr. & Tappan, H. (1987). Foraminiferal genera and their claassification. Department of Earth and space sciences and center for the study of evolution and origin of the life. Univ. California. Los Angeles. Van Nostraand Company. N.Y. Vol. *1*, *2*, 970.

Loeblich, A. R., Jr. & Tappan, H. (1989). Implications of wall composition and structure in agglutinated foraminifera. *Paleont.*, *53* (6), 769 – 777.

Loeblich, A. R. & Tappan, H. (1992). Present status of foraminiferal classification. Studies in Benthic Foraminifera, *BENTHOS '90, Sendai, 1990. Tokai University Press:* 93 – 102.

Loeblich, A. R., Jr. & Tappan, H. (1994). Foraminifera of the Sahul Shelf and Timor sea. *Cushm. Found. For. Res. Spec. Publ.*, *31*, 661.

Longet, D., Burki, F., Flakowski, J., Berney, C., Polet, S., Fahrni, J. & Pawlowski, J. (2004). Multigene evidence for close evolutionary relations between *Gromia* and Foraminifera. *Act.Protozool.*, *43*, 303 – 311.

Lopez, E. (1979). Algal chloroplasts in the protoplasm of three species of benthic Foraminifera: taxonomic affinity, variability, and persistence. *Marine Biol.*, *53*, 201 - 211.

Loubere, P., Meyers, P. & Gary, A. (1995). Benthic foraminiferal microhabitat selection, carbon isotope values, and association with larger animals: a test with Uvigerina peregrine. *Jour.Foram.Res.*, *25* (1), 83– 95.

Luperto Sinni, E., Martín-Chivelet, J. & Giménez, R. (2000). *Praemurgella valenciana* nov. gen. et sp. (Foraminifera) in the Sierra de Utiel Formation (Coniacian - Santonian) of the Prebetic domain (SE Spain). *Geobios 32* (2), 145 - 151.

Lutze, G. F. & Thiel, H. (1989). Epibenthic Foraminifera from elevated microhabitats: *Cibicidoides wuellerstorfi* and *Planulina ariminensis. Journ. Foram. Res.*, Vol. *19*. N 2. 153 - 159.

Lutze, G. F. & Weber, G. (1980). Habitat and asexual reproduction of *Cyclorbiculina compressa* (Orbigny), Soritidae. *Journ. Foram. Res.*, *10* (4), 251 - 260.

Lynn, D. H. & Small, E. B. (1981). Protists kinetids structural conservatism, kinetid structures and ancestral stages. *BioSystems*, *14*, 277 – 385.

Macintyre, I. G. & Reid, R. P. (1998). Recrystallization in living porcelaneous foraminifera (*Archaias angulatis*), Textural changes without mineralogical alteration. *Journal of Sedimentary Research*, *68*, (1), 11-19.

Majewski, W. (2005), Benthic foraminiferal communities: distribution and ecology in Admiralty Bay, King George Island, West Antarctica. *Polish Polar Res.,26*, 159 – 214.

Makled, W. A. & Langer, M. R. (2010). Preferential selection of titanium-bearing minerals in agglutinated Foraminifera: Ilmenite (FeTiO3) in *Textularia hauerii* d'Orbigny from the Bazaruto Archipelago, Mozambique. *Rev.Micropaléontologie*, *53* (3), 163 – 173.

Malumian, N., Nanez, C. & Carames, A. (1991). Unilocular foraminifera of reticular surface from Argentina. *Micropaleontology*, *37* (4), 393 – 406.

Mamet, B. & Pinard, S. (1992). Note sur la taxonomie des petits foraminiferes du Paleozoique superior. *Bul. Societe belge de Geolog.*, *99* (1990), 373 – 398.

Mantzurova, V. N. & Gjrbatchik, T. N. (1982). Novye dannye o stroenii rakovin trocholin. [New data on the shell structure of trocholins]. *Voprosy mikropaleontologii*, *25*, 116 – 129. (In Russian).

Marfenkova, M. M. (2012). Visean and Serpukhovian Koktjubinids (Biseriamminacea, Foraminifera) from Kazakhstan, Paleontological Journal, *43* (2), 251 – 257. DOI:10.1134/S0031030109030022

Margulis, L. (1974). Five-kingdom classification and the origin of evolution in cells. *Evolutionary Biology*, *7*, 45-78.

Margulis, L. & Schwartz, K. (1988). *Five Kingdoms: An illustrated guide to the Phyla of life on Earth.* (second edition), W.H. Freeman & Co., New York, 376.

Margulis, L., Schwartz, K. (1988). *Five Kingdoms: An illustrated guide to the Phyla of life on Earth.* (second edition), W.H. Freeman & Co., New York: 376.

Margulis, L., Corliss, J. O., Melkonian, M. & Chapman, D. J. (1990). *Handbook on Protoctista* Boston. Jones and Bartlett, 914.

Marshalek, D. S. (1969). Observations on Iridia diaphana, a marine foraminifer. *Journ. Protooozool.*, *16* (4), 599 - 612.

Martins, V., Ferreira Da Silva, E., Sequeira, C., Rocha, F. & Duarte, A. C. (2010). Evaluation of the ecological effects of heavy metals on the assemblages of benthic foraminifera of the canals of Aveiro (Portugal). *Estuarine, Coastal and Shelf Science*, *87* (2), 293 – 304.

Maslakova, N. I. (1978). *Globotrunkanidy juga evropejskoj chasti SSSR [The Globotrunkanids of the South of the European part of the USSR*]. Moskow. Nauka. 166 (In Russian).

Maslakova, N. I. (1983). K revisii pozdnemelovykh planktonnykh foraminifer semeystva Marginotruncanidae. [Revision of Late Cretaceous planktonic foraminifers of the family Marginotruncanidae]. *Trudy Instituta Geologii i Geofisiki, Akademiya Nauk SSSR, Sibirskoe otdelenie*, *559*, 23 – 31. (In Russian).

Maslakova, N. I. (1990). Kriterii vydelenija vysshich taksonov foraminifer. [Criteria of separation of the foraminiferal higher taxa]. In: Sistematika i philogenija bespozvonochnych. Kriterii vydelenija vysshich taksonov. [Sytem and phylogeny of invertebrata. Criteria of separation of the higher taxa]. Moscow. „Nauka". *Academy of Sciences of USSR. Moscow Soc. of Nature Investigators*, 22 – 27. (In Russian). (In Russian).

Mateu-Vicens, G., Box, A., Deudero, S. & Rodriguez, B. (2010). Comparative analysis of epiphytic foraminifera in sediments colonized by seagrass *Posidonia oceanica* and invasive macroalgae *Caulerpa* spp. *Journ. Foram. Res.*, *40* (2), 134 - 147.

Matsumaru, K. & Anil K. Jauhri, A.K. (2003). *Lakadongia*, a new Orbitoidal foraminiferal genus from the Thanetian (Paleocene) of Meghalaya, NE India. *Micropaleontology, 49*(3):277 - 292.

Matsushita, S. & Kitazato, H. (1990). Seasonality in the benthic foraminiferal community and the life history of *Trochammina hadai* Uchio in Hamana Lake, Japan. In: Hemleben C. et al.,(Eds).*Pa- laeoecology, Biostratigraphy, Palaeoceanography and Taxonomy of Agglutinated Foraminifera. NATO ASI. Ser. C.*, 327. Kluwer Acad. Publ. 695 - 715.

McEnery, M. E., Lee, J. J. (1976). *Allogromia laticollaris*: a foraminifer with unusual apogamic metagenic life cycle. *Journ. Protozool.*, *23* (1), 94 - 108.

McEnery, M. E. & Lee, J. J. (1981). Cytological and fine structural studies of three species of symbiont bearing larger Foraminifera from the Red Sea. *Micropaleontology*, *27* (1), 71 - 83.

McGann, M. (2002). Historical and modern distributions of benthic foraminifers on the continental shelf of Monterey Bay, California. *Marine Geology*, *181*, 115 - 156.

McGann, M. (2007). Foraminiferida, in Carlton, James T., ed., The Light and Smith Manual (4th ed.), Intertidal invertebrates from central California to Oregon. *University of California Press*, 46 - 69.

McGann, M. (2008). High-resolution foraminiferal, isotopic, and trace element records from Holocene estuarine deposits of San Francisco Bay, California. *Journ.Coastal Research*, *24* (5), 1092 - 1109.

McGann, M. (2009). Review of impacts of contaminated sediment on microfaunal communities in the southern California Bight. *In Lee, H., and Normark, W.R., eds., Earth Science in the Urban Ocean; the Southern California Continental Borderland. Geol. Soc. of America*, *Spec. Paper 454*, 413 - 455.

McGann, M. (2010). Chapter III: 1998 Central San Francisco Bay Foraminiferal Study, in Chin, J.L., Woodrow, D.L., McGann, M., Wong, F.L., Fregoso, T., and Jaffe, B.E., 2010, Estuarine sedimentation, sediment character, and foraminiferal distribution in central San Francisco Bay, California. U.S. *Geological Survey Open-File Report* 2010-1130, p. 35-58 [URL http://pubs.usgs.gov/of/2010/1130/].

McGann, M. Benthic foraminifera identify source area of earthquake- and storm-induced turbidite events, Monterey Canyon, California, In Press, 47.

McGann, M. & Sloan, D. (1996). Recent introduction of the Foraminifer *Trochammina hadai* Uchio in San Francisco Bay, California, USA. *Mar Micropaleontol.*, *28*, 1 – 3.

McNeil, D.H. (1997). New foraminifera from the Upper Cretaceous and Cenozoic of the Beaufort-MacKenzie Basin of Arctic Canada. *Cushman Foundation for Foraminiferal Research Special Publication*, *35*, 95 pp.

Meric, E. & Inan, N. (1998). *Sirelina orduensis* (foraminifera) a new genus and species from the Maastrichtian of northeast Anatolia (Golkoy-Ordu). . *Micropaleontology, 44* (2), 195 – 206.

Meric, E. & Coruh, T. (1998). *Neosivasella sungurlui,* a new genus and species from the upper Paleocene of southeast Turkey. *Micropaleontology, 44* (2), 187 – 194.

Meric, E., Avsar, N., Yokes, M. B., Tugrul, A. B., Bayari, S., Ozyurt, N., Barut, I. F., Balkis, N., Uysal, K. & Kam, E. (2008 a). Morphological abnormalities in benthic foraminifers of the Antalya coast. *Micropaleontology*, *54* (3 – 4), 241 – 276.

Meric, E., Goermus, M., Avsar, N., Yokes, M. B. & Dincer, F. (2008 b). Twin, triplet and quadruplet teratogens in benthic foraminifera from Antalya. *Micropaleontology*, *54* (3 – 4), 293 – 306.

Meyn, H. & Vespermann, J. (1994). Taxonomische Revision von Foraminiferen der Unterkreide SE-Niedersachsens nach Roemer (1839, 1841, 1842), Koch (1851) und Reuss (1863). *Senckenbergiana Lethaea*, *74* (1/2), 49 – 272.

Meisterfeld, R., Holzmann, M. & Pawlowski, J. (2001). Morphological and molecular characterization of a new terrestrial allogromiid species: *Edaphoallogromia australica* gen. et spec. nov. (Foraminifera) from Northern Queensland (Australia). Protist, *152*, 185 – 192.

Mikhalevich, V. I. (1972 a). Vozrastnaja izmenchivost antarkticheskich vidov roda *Trochammina* Parker et Jones I ejo znachenie dlja systematiki podsemejstva Trochammininae (Foraminifera). V kn.: Resultaty biologicheskich issledovanij Sovetskich Antarkticheskich expedicij. [Age variability of Antarctic species of the genus Trochammina and its significance for the systematic of the subfamily Trochammininae (Foraminifera. In: Results of the biological investigations of the Soviet Antarctic Expeditions].*Issledovanie fauny morej*, *11* (19), vyp. 5: 5 – 40. (In Russian).

Mikhalevich, V. I. (1976). New data on the Foraminifera of the Ground waters of the Middle Asia. *Intern. Journ. Speleol.*, *3*, 167 - 175.

Mikhalevich, V. I. (1978). Zoogeographija foraminifer shelfov tropicheskoj Atlantici. [Zoogeography of the foraminifera of the shelves of the tropical Atlantic]. *Proc. Zool. Inst. Russian Acad. Sci.*, *78*, 59 – 96. (In Russian).

Mikhalevich, V. I. (1978). Osobennosti raspredelenija trech podotrjadov foraminifer (Textulariina, Miliolina, Rotaliina). [Peculiarities of the distribution of the the three foraminiferal orders (Textulariina, Miliolina, Rotaliina)]. *Oceanology*, XVIII (5), 909 – 912. (In Russian).

Mikhalevich, V. I. (1979). Verticalnoe raspredelenie i osobennosti ecologii donnych foraminifer shelfov tropicheskoj Atlantiki. [Vertical distribution and peculiar features of the ecology of benthic foraminifera of the shelves of the tropical Atlantic]. *Biology of the Sea*, *3*, 3 – 10. (In Russian)

Mikhalevich, V. I. (1980 a). Sistematika i evolyuciya foraminifer v svete novyikh dannyikh po ih citologii i ul'trastrukture [Systematics and evolution of the Foraminifera in view of the new data on their cytology and ultrastructure]. *Materialy VIII mikropaleontologicheskogo sovestchaniya "Systematika I morphologiya mikroorganismov". 21 – 24 October, Baku: ELM*: 77. (In Russian).

Mikhalevich, V. I. (1980 b). Sistematika i evolyuciya foraminifer v svete novyikh dannyikh po ih citologii i ul'trastrukture [Systematics and evolution of the Foraminifera in view of the new data on their cytology and ultrastructure]. *In:* Principy postroeniya makrosistemy odnokletochnyih zhivotnyh. *Trudy Zoologicheskogo Instituta*, *AN SSSR*, *94*, 42–61 (In Russian).

Mikhalevich, V. I. (1981). Parallelism i konvergencia v evolucii skeletov Foraminifer [The parallelism and convergence in the evolution of the Foraminiferal skeleton]. *Trudy Zoologicheskogo Instituta*, *AN USSR*, *107*, 19–41 + 2 pls. (In Russian).

Mikhalevich, V. I. (1983). *Donnyie Foraminiferyi shel'fov tropicheskoj Atlantiki [The bottom foraminifera of the shelves of the Tropical Atlantic]*. Skarlato, O. A. (ed), Zoological Institute of USSR Academy of Sciences, Leningrad, 247 p. + 26 tables. (In Russian)

Mikhalevich, V. I. (1986). Novye taksony miliolat. [New taxa of Miliolata]. Thes. X Mikropaleontol. Conference, Leningrad, 145 – 146. (In Russian).

Mikhalevich, V. I. (1988). Sistema podklassa Miliolata (Foraminifera) [The systematics of the subclass Miliolata (Foraminifera)]. *In:* V.I. Mikhalevich (Ed.), Sistematika, ekologia i stratigrafiya miliolyat (Foraminifery). *Trudy Zoologicheskogo Instituta, AN SSSR, 184*, 77–110 + 8 pls. (In Russian).

Mikhalevich, V. I. (1990).Ultrastruktura foraminifer. {Foraminiferal ultrastructure]. In: *Problems of the recent micropaleonthology*. Leniingrad. Nauka: 14 - 22. (In Russian).

Mikhalevich, V. I. (1991). Foraminifery (Foraminifera) zaliva Rybij chvost. [Foraminifera (Foraminifera) of the Fish Tail Bay]. *Inform. Bull. Sov. Antarct. Exped., St. Petersburg, Gidrometeoizdat, 116*, 27 – 38. (In Russian).

Mikhalevich, V. I. (1992). Makrosistema foraminifer [The macrosystem of the Foraminifera] Doct. Thes. St.-Pb., 1 – 43. (In Russian).

Mikhalevich, V. I. (1993 a). New higher taxa of the subclass Nodosariata (Foraminifera). *Zoosystematica Rossica, 2*, 5 - 8.

Mikhalevich , V. I. (1993 б). New higher taxa of the class Spirillinata (Foraminifera). *Zoosystematica Rossica, 2*, 9 - 11.

Mikhalevich, V. I. (1994). Novyj rod i vid semejstva Ammoscalariidae (Foraminifera). [New genus and species of the family Ammoscalariidae (Foraminifera)]. *Zool. Journ., 73* (6), 18 – 21. (In Russian)

Mikhalevich, V. I. (1995). A new classification of the class Astrorhizata (Foraminifera). *Zoosystematica Rossica, 3* (2)), 161 - 174.

Mikhalevich V. I. (1997 a). The new conception of the Foraminiferal macrosystem. *Applic. Micropaleontol. Environment Stud.*, 88 – 89.

Mikhalevich, V. I. (1997 b). Main directions of the Foraminiferal evolution. *Evol. Of Life on the Earth*, 44 – 45.

Mikhalevich, V. I. (1998 a). On taxonomic position of the family Dusenburyinidae (Foraminifera). *Zoologicheskij journ., 77* (8), 976 – 978. (In Russian).

Mikhalevich, V. I. (1998 b). Makrosistema foraminifer [The Macrosystem of the Foraminifera*]. Izvestija Russian Academy Sciences. Ser. Biol.*, (*2*), 266-271. (In Russian).

Mikhalevich, V. I. (1999). Sistema i philogenia Foraminifer [The Foraminiferal System and Phylogeny]. Lichachev. S. F. (ed), Suchanova, K. M. & Starobogatov, Ja. I.(rev-s). *.Zoological Institute RAS, Omskij Gosudarstvennyj Pedagogicheskij Universitet. St. Petersburg – Omsk. Biologicheskaja Ser. Vyp. 3*, 1 - 185 + 28 pls. (In Russian, English summary p. 11 - 30).

Mikhalevich, V. I. (2000). Tip Foraminifera d'Orbigny, 1826 – Foraminifery [The phylum Foraminifera d'Orbigny, 1826 – Foraminifers]. *In:* Alimov, A.F. (ed.), *Protisty: Rukovodstvo po Zoologii*, pt. 1. Nauka Publishers, St. Petersburg, 533-623. (In Russian, with English summary 611 - 616)

Mikhalevich V. I. (2002). The evolution of the foraminiferal shell wall. *Proceedings of the International Symposium "Evolutionary ideas in Biology", in memoir. of N.N. Voronzcov*, 46 – 62.

Mikhalevich, V. I. (2003). *Systema chetyreoch podklassov s agglutinirovannoj stenkoj rakoviny (Ammodiscana, Miliamminana, Hormosinana, Textulariana)(Foraminifera). [System of the four subclasses with the agglutinated shell wall (Ammodiscana, Miliamminana, Hormosinana, Textulariana)(Foraminifera)*]. Rossiskaja Akademija Nauk, Izvestija Zoologicheskogo instituta. *7*, 46.

Mikhalevich, V. I. (2004 a). On the heterogeneity of the former Textulariina (Foraminifera). In: Bubik, M. and Kaminski, M. A. (eds),. *Proceedings of the Sixth International Workshop on Agglutinated Foraminifera. Grzybowski Foundation Special Publication, 8*, 317 – 349.

Mikhalevich, V. I. (2004 b). The general aspects of the distribution of Antarctic foraminifera. *Micropaleontology*, *50* (2), 179 – 194, 5 plates.

Mikhalevich, V. I. (2004 c). On the new understanding of the order Lituolida Lankester, 1885 (Foraminifera). *Acta Paleontol. Romaniae*, *4*, 247 – 267.

Mikhalevich, V. I. (2005). Polymerization and oligomerization in foraminiferal evolution. *Stud Geol Polonica*, *124*, 117-141.

Mikhalevich, V. I. (2005). The New System of the Superfamily Quinqueloculinoidea Cushman, 1917 (Foraminifera). *Acta Palaeontol. Roman.*, *5*, 303–310.

Mikhalevich, V. I. (2006). The similarity of Fusulinoida and Milioloida—convergence or parallelism? *Proc Zool Inst Russ Acad Sci.*, *310*, 133-137. (In Russian).

Mikhalevich, V. I. (2008 a). General Evolutionary Regularities of Organic and Social Life. *Evolutionary Biology from Concept to Application*, Part I. Modelization of Evolution. P. Pontarotti (ed.),_Springer-Verlag Berlin Heidelberg: 73 – 94.

Mikhalevich, V. I. (2008 b). Novye podsemejstva podotrjada Miliolina (Foraminifera).[New subfamilies of the suborder Miliolina (Foraminifera). *Zool. Zhurn.*, *87*, (11), 1 – 6. (In Russian).

Mikhalevich, V. I. (2009). The taxonomic position of the superorder Fusulinoida Fursenko in Foraminiferal system. *Paleontol. zhurn.*, *43* (2), 117 – 128. (In Russian).

Mikhalevich, V. I. (2009). Comparison between the test and wall structure of the Miliolata and Fusulinoida (Foraminifera), (including the data on the Antarctic Miliolids). *Micropaleontology*, *55* (1), 1 – 21, pl-s 1 -10.

Mikhalevich,V. I. (2010). Results of the electron microscopic study of the Antarctic species *Conotrochammina antarctica* Saidova, 1975 (Foraminifera). *In: "Synthetic and Integrative Biology: Parts & Systems, Design Theory and Applications." Editors: James T. Gevona. Chapter 5*, 77-91

Mikhalevich, V. I. (2011 a). Wall ultrastructure of some Antarctic foraminiferal Genera. In: Kaminski, M..A., S. Filipescu. (eds), *Proceedings of the Eighth International Workshop on Agglutinated Foraminifera. Grzybowski Foundation Special Publication, 16*, 115 – 129.

Mikhalevich, V. I. (2011 b). The new genus *Olgia* and new family Olgiidae (Textulariana, Foraminifera). *Micropaleontology*, *57* (2), 173 – 182.

Mikhalevich, V. I. (2011 c). Some fundamental regularities of Evolution of Geo-, Bio-, and Noosphere. *Proceedings of Sixth International Conference "Environmental Micropaleontology, Microbiology and Meiobentology, EMMM-2011", September 19 – 22, Avalon Inst. of Applied Sci., Borissiak Paleontological Inst. Rus. Akad. Sci., Moscow:* 187 – 190.

Mikhalevich, V. I. (2011 d). Test wall ultrastructure and taxonomic position of the foraminiferal genus *Strictocostella* Patterson 1987 *Micropaleontology*, *57*, (6), 543-555, plates 1- 4, text-fig. 1.

Mikhalevich, V. I. (2012). Reinvestigation of the two agglutinated genera of the order Textulariida Delage et Herouard, 1896 (Foraminifera). Alegret, L., Ortiz, S., Kaminski, M.A. (eds), Ninth Intern. Workshop Aggl. Foram., *Grzybowski Foundation Sp. Publ.*, *18*, 64 - 65.

Mikhalevich, V. I. (2012). O sistematicheskom polozhenii nekotorych odnorjadnych rakovin foraminifer iz sostava prezhnich Rotaliana (po materialam Sovietskich Antarcticheskich expedicij). [On the systematic position of some uniserial foraminiferal shells of the former Rotaliina (from the materials of the Soviet Antarctic Expeditions]. (On the *Trudy* XV Vserossijskogo mikropaleontol. Sovestschanija, Gelendzhik, Sept. 2012, 109 – 112. (In Russian).

Mikhalevich, V. I. & Voronova, M. N. (1982). Gametogenez u gametogamnych foraminifer. [Gametogenezis in the gametogamic Foraminifera*]. In: Thez. of Materials of the 3d VOPR (All- Russian Protozoological Society) Meeting*, *Vilnius*, 233. (In Russian).

Mikhalevich, V. I. & Voronova, M. N. (1985 Б). Arcticheskie cibicidesy. [Arctic Cibicidesis]. *In: Problemy chetvertichnoj paleoecologii I paleogeographii Barencova I Belogo morej. [Problems of the Quaternary paleoecology and paleogeography of the Barentz and White seas]. Thes. Vsesojuzn. Conference*, *Murmansk*, 94 – 95. (In Russian).

Mikhalevich, V. I., Rodionova M. K. & Sinjakova G. N. (1986). Ultrastructura stenki dvuh agglutinirovannyh rodov foraminifer quinqueloculinovogo tipa stroenija [The wall ultrastructure of the two agglutinated foraminiferal genera with the quinqueloculine type of the shell]. *Trudy Zoologicheskogo Instituta*, *AN USSR. 144*, 66–71. (In Russian).

Mikhalevich, V. I. & Voronova, M. N. (1999). O systematicheskom polozhenii roda *Pelosina* Brady, 1879 (Xenophyophoria, Protista, Inc. Sedis).[On the systematic position of the genus *Pelosina* Brady, 1879 (Xenophyophoria, Protista, Inc. Sedis)]. *Zool. Zhurn.*, *78* (2), 133 – 141. (In Russian).

Mikhalevich, V. I. & Debenay, J. -P. (1998). Parallelism I konvergencija v razvitii ust'ja foraminifer. [Parallelism and convergence in the development of the foraminiferal aperture]. In: *Polevye I eksperimentalnye biologicheskie issledovanija.* [*Field and experimental biological investigations*]. *Ros. Gos. Pedagogic. Univ. in honour of Gertzen*, *Omskij Gos. Pedagogic. Univ.*, *S. Petersburg – Omsk*, *Ser. Biol.*, (4), 3 – 31. (In Russian).

Mikhalevich, V. I., Pronina, G. P. & Nestell, M. (2000). Novyj antarcticheskij vid *Saccammina basispiculata* sp. nov. i ego polozhenie v sisteme foraminifer [The new antarctic species *Saccammina basispiculata* sp. nov. and its taxonomic position (Foraminifera)]. *Intern. Congr. – 2000 "Fundamental problems of the natural sciences and technic"*, *1*, *(1)*, *S-Petersburg. Russia*: 184 – 187. + 1 pl. (In Russian).

Mikhalevich, V. I. & Debenay J. -P. (2001). The main morphological trends in the development of the foraminiferal aperture and their taxonomic significance. *Journal of Micropalaeontology*, *20*, 13– 28.

Mikhalevich, V. I. & Kaminski, M. A. (2008). Revised systematic of Schlumbergerinida (phylum Foraminifera). In: Kaminski, M.A., Coccioni, R. & Marsili, A. *Proceedings of the Seventh International Workshop on Agglutinated Foraminifera. Grzybowski Foundation Special Publication*, *13*, 157 – 166.

Mikhalevich, V. I. & Peryt, D. (2008). Shell wall ultrastructure in some genera of the order Verneuilinida Mikhalevich et Kaminski, 2004 Foraminifera). Part I. In: Kaminski, M.A., Coccioni, R. & Marsili, A. *Proceedings of the Seventh International Workshop on Agglutinated Foraminifera. Grzybowski Foundation Special Publication, 13*, 167 – 182.

Milam, R. W. & Anderson, J. B. (1981). Distribution and ecology of recent benthonic foraminifera of the Adelie-Georg V continental shelf and slope. *Marine Micropaleontology, 6*, 297 – 325.

Misik, M. & Sotak, J. (1998). "Microforaminifers" – a specific fauna of organic-walled Foraminifera from the callovian-oxfordian limestones of the pieniny Klippen Belt (Western Carpathians). *Geologica Carpathica, 49* (2), 109 – 123.

Moodley, L. (1990). "Squatter" behavior in soft-shelled Foraminifera. *Marine Micropal., 16*, 149 – 153.

Morigi,, C., Jorissen, F. J., Gervais, A., Guichard, S. & Borsetti, A. M. (2001). Benthic foraminiferal faunas in surface sediments off NW Africa:relationship with organic flux to the ocean floor. *Journ. Foram. Res., 31*, 350 – 368.

Movellan, A. & Howa, S. H. (2012). First-order estimate of the planktic foraminifer biomass in the modern global oceans. *The Micropaleontol Soc. Foraminifera and Nannofossil Groups Joint Meeting 2012*, Interdisciplinary advances in Foraminiferal and nannofossil Res., Pollock Halls, Univ.Edinburg, 33.

Muller, W. A. (1975). Competition for food and other Niche-Related studies of three species of salt- marsh Foraminifera. *Marine Biol., 31*, 339 - 351.

Muller, W. A. & Lee, J. J. (1969). Apparent indespensability of bacteria in foraminiferan nutrition. *Journ. Protozool., 16* (3), 471 - 478.

Muller-Merz, E. (1980). Strukturanalyse ausgewahlter rotaloider Foraminiferen. *Mem. Suisses de Paleontol., 101*, 5 – 70.

Muller-Merz, E. & Lee, J. J. (1976). Symbiosis in the larger foraminiferan *Sorites marginalis* (with notes on *Archaias* sp.). *Journ. Protozool., 23* (3), 390 - 396.

Mullineaux L. S. (1988). Taxonomic notes on larger agglutinated foraminifers encrusting manganese nodules, including the description of a new genus, *Chondrodapis* (Komokiacea). *Journ. Foram. Res., 18* (1), 46 - 53.

Munier-Chalmas, E. (1880). Sur le dimorphisme des Nummulites. *Bull. Soc. Geol. France. Ser. 3*, 8, 300 - 301.

Murray, J. W. (1973). Wall structure of some agglutinated Foraminiferida. *Paleontology, 16* (4), 777 – 786.

Murray, J. W. (1973). *Distribution and ecology of living benthic Foraminiferids*. New York, Crane Russak, Co, 274.

Murray, J. W. (2000). The enigma of the continued use of total assemblages in ecological studies of benthic Foraminifera. *Journ. Foram.Res., 30* (3), 244 – 245.

Murray, J. W. (2006). *Ecology and Applications of Benthic Foraminifera*. Univ. of Southampton. Myers, E.H., 1935. The life history of *Patellina corrugata* Williamson, a foraminifer. *Bull. Scripps. Inst. Oceanogr. Univ. California. Techn. ser., 3* (15), 355 - 391.

Myers, E. H. (1936). The life-cycle of *Spirillina vivipara* Ehrenberg with notes on morphogenesis, systematics and distribution of the Foraminifera. *Journ. Roy. Microsc. Soc., 56*, 120 - 146.

Myers, E. H. (1940). Observations on the origin and fate of flagellate gametes in Multiple tests of *Discorbis* (Foraminifera). *Journ. Marine Biol. Ass.*, *24* (1), 201 – 226.

Myers, E. H. (1943). Biology, ecology and morphogenesis of a pelagic Foraminifera. *Stanford Univ. Publ. Biol. Sci. Ser. IX.*, *1*, 1 - 30.

Mylnikova, Z. M. & Mylnikov, A. P. (2010). Biologija I morphologia presnovodnogo chistchnogo zhgutikonosca *Colponema* aff. *loxodes* Stein (Colponemida, Alveolata). [Biology and morphology of the fresh-water carnivorous flagellate *Colponema* aff. *loxodes* Stein (Colponemida, Alveolata)]. *Biologija morja [Biology of the Sea*], *1*, 23 – 29.

Nagy, J. & Alve, E. (1987). Temporal changes in Foraminiferal faunas and impact on pollution in Sandebukta, Oslo Fiord. *Marine Micropaleontol.*, *12* (2), 109 - 128.

Nagy, J. & Johansen, H. O. (1991). Delta-influenced foraminiferal assemblages from the Jurassic (Toarcian-Bajocian) of the Northern Sea. *Micropaleontology*, *37* (1), 1 – 40.

Nagy, J. & Basov, V. A. (1998). Revised foraminiferal taxa and biostratigraphy of Bathonian to Ryazanian deposits in Spitsbergen. *Micropaleontology*, *44*, (3), 217 – 255.

Naumov, A. D., Borkin, L. J. & Podlipaev, S. A. (1977). Principy polymerizacii i oligomerizacii: processy i systemy (Principles of polymerization and oligomerization: processes and systems). In: Znachenie processov polimerizacii i oligomerizacii v evolucii (Significance of the processes of polymerization and oligomerization in evolution). O.A. Scarlato (ed) *Published by Zoological Institute of the Academy of Sciences USSR*, Leningrad: 5-8. (In Russian).

Neagu, T. (1968). Study of the Miliolidaceae in the Lower Cretraceous (Barremian) of Southern Dobrogea. In: "The Centenuary Grigore Antipa, 1867 – 1967". *Traveaux du Museum d'Histoire Naturelle "Grigore Antipa"*. Bucharest. *8*, 563 – 572.

Neagu, T. (1970). The genus *Hechtina* (For a., Lower Cretaceous) and its systematical position. *Senckerbengiana Lethaea*, *51* (5/6), 417 – 427.

Neagu, T. (1975). Monographie de la Faune des Foraminiferes des Eocretaces du couloir de Dimbivicioara, de Codlea et des Monts Persani (Coushes de Carhaga). *Mem. Inst. Geol. si Geofiz. Bucaresti*, *25*, 1 – 141.

Neagu, T. (1979). Donnees nouvelles concernant les representants de la famile des Pfenderinidae de l'eocretace de la Dobrogea Meridionale (Roumanie). *Rev. Español. Micropaleontol.*, *11*, (3), 479 – 504.

Neagu, T. (1984). Nouvelles donnees sur la morphologie du test, sur la systematique et la nomenclature des Miliolides agatistegues du Mesozoique. *Rev. Español. Micropaleontol.*, *16*, 75 –90.

Neagu, T. (1985). Berriasian-Valanginian Miliolid fauna in Southern Dobrogea (Romania). *Rev. Español. Micropaleontol.*, *17* (2), 201 – 220.

Neagu, T. (1986). Barremian-lower Aptian Miliolid fauna in Southern Dobrogea (Romania). *Rev. Español. Micropaleontol.*, *18*, (3), 313 – 348.

Neagu, T. (1994). Genera *Haplophragmoides* Cushman, 1910; *Recurvoides* Earland, 1934; *Thalmannammina* Pokorny, 1951; *Plectorecurvoides* Noth, 1952 and *Pokornyammina* N.Gen. from Upper Cretaceous flysch facies, Eastern Carpathians, Romania. *Rev. Espanol. Micropaleontol.*, *26* (1), 5 – 30.

Neagu, T. (1994). Early Cretaceous *Trocholina* group and some related genera from Romania. Part. I. *Rev. Espanol. Micropaleontol.*, *26* (3), 117 – 143.

Neagu, T. (1995). The Cretaceous *Trocholina* group and some related genera from Romania. Part II. *Rev. Espanol. Micropaleontol., 27* (2), 5 – 40.

Neagu, T. (1997). Lower Cretaceous agglutinated foraminifera from the superfamilies Verneuilinacea and Ataxophragmiacea; southern Dobrogea, Romania. *Annales Societatis Geologorum Poloniae, 67*, 307 - 323.

Neagu, T. (1999). *Barbuspira complanata* n.g., n. sp., a new cornuspirid foraminifera from lower Valanginian, Cernavoda – Pod (Southern Dobrogea) – Romania. *Acta Paleontol. Romaniae, 2*, 305 – 307.

Neagu, T. (2000). Lower Cretaceous calcareous agglutinated Foraminifera from Southern Dobrogea, Romania. *In:* Hart, M.B., Kaminski, M.A., & Smart, C.W. (eds), *Proceedings of the Fifth International Workshop on Agglutinated Foraminifera. Grzybowski Foundation Special Publication, 7*, 363 - 386.

Neagu, T. (2003). New opinion for more natural systematic of the agglutinated foraminifera. *Proc. Romanian Acad., Ser. B:chemistry, life sciences and geosciences, 5* (3), 127 – 130.

Neagu, T. & Platon, E. (1994). Genera *Haplophragmoides* Cushman, 1910; *Recurvoides* Earland, 1934; *Thalmannammina* Pokorny, 1951; *Plectorecurvoides* Noth, 1952 and *Pokornyammina* n. Gen. From Upper Cretaceous flysch facies, Eastern Carpathians, Romania. *Revista Española de Micropaleontologia, 26*, (1), 5 – 30.

Neagu, T. & Cirnaru, P. (2001 a). Early Cretaceous benthic calvareous foraminifera from Southern Dobrogea, Romania. Part I. Superfamily Polymorphinacea. *Rev. Roman. Geologie, 45*, 69 – 84.

Neagu, T. & Cirnaru, P. (2001 b). Benthic calcareous foraminifera from the lower Cretaceous deposists – Southern Dobrogea – Romania. II – Spirillinida and Rotaliida (Placentulinidae). *Acta Paleontol. Romaniae, 3*, 283 – 297.

Neagu, T. & Cirnaru, P. (2002). Early Cretaceous benthic calvareous foraminifera from Southern Dobrogea, Romania. Part III. Lagenida, Robertinida, Rotaliida. *Rev. Roman.Geologie, 46*, 95 – 114.

Neagu, T., Popescu, D. P., Crihan, H. M. & Popescu, G. (2011). Upper Eocene agglutinated foraminifera from Buciumeni section (Ialomita Valley, Dambovija District, Romania. In: Kaminski, M..A., S. Filipescu. (eds), *Proceedings of the Eighth International Workshop on Agglutinated Foraminifera. Grzybowski Foundation Special Publication, 16*, 151 – 172.

Nestell, G. P., Sudar, M. N., Jovanovic, D. & Kolar-Jurkovsek, T. (2009). Latest Permian foraminifers from the Vlasic mountain area, northwestern Serbia. *Micropaleontology, 55*, 495 – 513.

Nestell, G. P. & Nestell, M. K. (2006). Middle Permian (Late Guadalupian)Foraminifers from Dark Canyon, Guadalupe Mountains, New Mexico. *Micropaleontology, 52* (1), 1 – 50.

Nestell, G. P. & Tolmacheva, T. Yu. (2004). Early Ordovician foraminifers from the Lava River Section, northwestern Russia. *Micropaleontology, 50* (3), 253 – 280.

Nestell, G. P., Mestre, A. & Heredia, S. (2009). First Ordovician Foraminifera from South America: A Darriwilian (Middle Ordovician) fauna from the San Juan Formation, Argentina. *Micropaleontology, 55* (4), 329 – 344.

Nestell, G. P., Kolar-Jurkovsek, T., Jurkovsek, B. & Aljinovic, D. (2011). Foraminifera from the Permian-Triassic transition in western Sloveniia. *Micropaleontology, 57* (3) 197 – 222.

Newell, R. G. (1970). *Biology of intertidal animals*. London, Logos Press: 555.

Neymayr, M. (1889). *Die Staemme des Thierreiches Wirbellose. Thiere. 1*. Wien. F.Tempsky. S. 153 – 206.

Nielsen, K. S. S., Collen, J. D. & Ferland, M. A. (2002). *Floresina*: a genus of predators, parasites or scavengers? *Journ. Foram. Res.*, *32* (1), 93 – 95.

Nikolaev, S. I., Berney, C., Fahrni, J. F., Bolivar, I., Polet, S., Mylnikov, A. P., Aleshin, V. V., Petrov, N. B. & Pawlowski, J. (2004). The twilight of Heliozoa and rise of Rhizaria, an emerging supergroup of amoeboid eukaryotes. *PNAS, 101* (21), 8066 – 8071.

Nomaki, H., Ogawa, N. O., Ohkouchi, N., Suga, H., Toyofuku, T., Shimanaga, M., Nakatsuka, T. & Kitazato, H. (2008). Benthic foraminifera as trophic links between phytodetritus and benthic metazoans: carbon and nitrogen isotopic evidence. *Marine Ecology Progress Series*, *357*, 153 –164.

Nozawa, F., Kitazato, H. & Gooday, A. J. (2006). Benthic Foraminifera at three abyssal sites in the equatorial Pacific nodule province: abundance, diversity and taxonomic composition. *Deep-Sea Res*, I *53*, 1406 – 1422.

Nyholm, K. G. (1957). Orientation and binding power of recent monothalamous foraminifera in soft sediments. *Micropaleontology*, *3* (1), 75 – 76.

Nyholm, R. -G. (1962). A study of the foraminiferan *Gypsina. Zool. Bidr. Upps*., *33*, 201 – 206.

Nyholm, K. G. & Nyholm, P. G. (1975a). Ultrastructure of monothalamous Foraminifera. *Zoon*. *3* (2), 141 - 154.

Nyholm, K. G. & Nyholm, P. G. (1975b). On the microtubules of some monothalamous Foraminifera, especially *Cylindrogullmia alba*. *Zoon*., *3* (2), 151 - 154.

Ohmert, V. W. (1969). Die Neoflabellinen (Foraminifera) des bayerischen Coniac-Santons. *Mitt.Bayer. Staatssamml. Palaeontol. hist. Geol*., *9*, 3 – 32.

d'Orbigny, A. (1826). Tableau methodique de la classe des Cephalopodes. *Ann. Sci. Nat. Paris*. *Ser*. 1. *7*, 245 - 314.

d'Orbigny, A. (1839). Foraminiferes des Iles Canaries. In: Barker-Webb et Berthelot." *Hist. Nat. Iles Canaries" Zoologie. Paris*., *2* (2), 119 - 146.

Orokos, D. D., Cole, R. W. & Travis, J. L. (2000). Organelles are transported on sliding microtubules in *Reticulomyxa*. *Cell Motility and the Cyoskeleton*, *47*, 296 - 306.

Özdikmen, H. (2009). Substitute names for some unicellular animal taxa (Protozoa). *Munis Entomology & Zoology*, *4* (1): 233 – 256.

Ozgen, N. (2000). *Nurdanella boluensis* n.gen., n.sp., a miliolid (Foraminifera) from the Lutetian of the Bolu Area (northwestern Turkey). *Revue de Paleobiologie, 19* (1), 79 – 85.

Page, F. C. (1987). The classification of "naked" amoebae (Phylum Rhizopoda). *Arch. Protistenk*., *133*, 199 – 217.

Page, F. C. (1988). *A new key to fresh-water and soil gymnamoebae with instruction for culture*. Amblleside: Freshwater Biological Association, 122.

Palmieri, V. (1988). Globivalvulinid foraminifera from the Permian of Queensland . *Alcheringa, Sydney, Australia, 12* (1), 38 - 39.

Papp, A. & Schmid, M. E. (1985). *The Fossil Foraminifera of the Tertiary Basin of Vienna. Revision of the monograph by Alcide d'Orbigny (1846) by Adolf Papp & Manfred E. Scmid.* Abh. Geologisch. Bundesanstalt, Wien, *37*, 311.

Parfrey, L. W., Grant, J. & Katz, L. A. (2012). Ribosomal DNA is differentially amplified across life- cycle stages in the foraminifer Allogromia laticollaris strain CSH. *Journ. Foram. Res.*,. *42*, 151 - 155, doi:10.2113/gsjfr.42.2.151.

Parker, J. H. (2009). *Taxonomy of Foraminifera from Ningaloo reef, Western Australia.* Mem. Assoc. Australasian Paleontol., *36*, 810.

Patterson, R. T. (1988). *Globulospinella*, a new unilocular foraminiferal genus, and designation of a neotype for *Palliolatella avita* Patterson ans Richardson. *Journ. Paleontol., Tulsa, Oklahoma, 62* (4), 529 – 530.

Patterson, R. T. (1990). *Nanosylvanella*, a new genus of the Sipholageninae (Foraminiferida). *Amer. Microsc. Soc., Trans., Lawrence, Kansas, 109* (3), 326.

Patterson, R. T. & Richardson, R. H. (1988). Eight new genera of Unilocular foraminifera. *Trans. Amer. Microscop. Soc.*, *107* (2), 240 – 258.

Patterson, R. T. & Fowler, A. D. (1996). Evidence of self organization in planktic foraminiferal evolution: implications for interconnectedness of paleoecosystems. *Geology*, *24*, 215 – 218.

Patterson, R. T., Fowler, A. D. & Huber, B. T. (2004). Evidence of hierarchial organization in the planktic foraminiferal evolutionary record. *Journ. Foram. Res.*, *34* (2), 85 – 95.

Pawlowski, J. (2000). Introduction to the molecular systematics of foraminifera. *Micropaleontology*, *46* (suppl. 1), 1 – 12.

Pawlowski, J. & Lee J. (1991). Taxonomic notes on some tiny shallow water Foraminifera from the Northern Gulf of Elat (Red Sea). *Micropaleontology.*, *37* (2), 149 - 163.

Pawlowski, J. & Lee J. J. (1992). The life cycle of *Rotaliella elatiana* n. sp. - a tiny macroalgavorous foraminifer from the Gulf of Eilat. *Journ. Protozool.*, *39* (1), 131 - 143.

Pawlowski, J. & Zaninetti, L. (1993). *Rossyatella*, N. Gen. (Foraminiferida) and other tiny Glabratellacea from the Mediterranean Sea. *Journ. Foram. Res.*, *23* (4), 231 – 237.

Pawlowski, J., Bolivar, I., Fahrni, J. & Zaninetti, L. (1994). Taxonomic identification of Foraminifera using ribosomal DNA sequenses. *Micropaleontology.*, *40* (4), 373 - 378.

Pawlowski, J., Boliva,r I., Fahrni, J., Zaninetti, L. & Kitazato, H. (1995 a). Partial LSUrDNA sequences of *Trochammina* sp. (Foraminiferida). In: Kaminski M.A., Geroch S., Gasinski M.A. (Eds).*Proc. Fourth Intern. Workshop Aggl. Foram. Krakow, Poland. Sept. 12 - 19 1993. Grzybowski Found. Sp. Publ. 3*, 227 - 232.

Pawlowski, J., Swiderski, Z. & Lee, J. (1995 b). Observations on the ultrustructure and reproduction of *Trochammina* sp. (Foraminiferida). In: Kaminski M.A., Geroch S., & Gasinski M.A. (eds) Proceedings of the Fourth International Workshop on Agglutinated Foraminifera. Kraków Poland, Sept. 12 – 19, 1993. *Grzybowski Foundation Special Publication*, *3*, 233 – 237.

Pawlowski, J., Bolivar, I., Fahrni, J., Vargas, C., Gouy, M. & Zaninetti, L. (1997). Extreme Differences in Rates of Molecular Evolution of Foraminifera Revealed by Comparison of Ribosomal DNA Sequences and the Fossil Record. *Mol. Biol. Evol.*, *14* (5), 498 – 505.

Pawlowski, J., Bolivar, I., Fahrni, J. F., de Vargas, C. & Bowser, S. S. (1999). Molecular evidence that *Reticulomyxa filosa* is a freshwater naked foraminifer. *Journ. Eukaryotic Microbiol.*, *46* (6), 612 – 617.

Pawlowski, J., Holzmann, M., Fahrni, J. F. & Hallock, P. (2001). Molecular identification of algal endosymbionts in large miliolid foraminifera. *Journ. Eukaryot.Microbiol.*, *48* (3), 362 – 367.

Pawlowski, J., Holzmann, M., Fahrni, J. F., Pochon, X. & Lee, J. J. (2001). Molecular identification of Algal Endosymbionts in Large Miliolid Foraminifera: 2. Dinoflagellates. *Journ. Eukaryot. Microbiol.*, *48* (3), 368 – 373.

Pawlowski, J. & Holzmann, M. (2002). Molecular phylogeny of Foraminifera – a review. *Europ. Journ. Protistol.*, *38*, 1 – 10.

Pawlowski, J., Holzmann, M., Berney, C., Fahrni, J., Gooday, A. J., Cedhagen, T., Habura, A. & Bowser, S. S. (2003). The evolution of early foraminifera. *PNAS*, *100* (20), 1494 – 11498.

Pawlowski, J., Holzmann, M., Fahrni, J. & Richardson, S. L. (2003). Small Subunit Ribosomal DNA Suggests that the Xenophyophorea *Syringammina corbicula* is a foraminiferan. *Journ. Eukaryot. Microbiol.*, *50* (6), 423 – 437.

Pawlowski, J. & Gooday, A. J. (2009). Precambrian biota: Protistan Origin of Trace Fossils? *Current Biology*, *19* (1), R28 – R30.

Pawlowski, J. & Burki, F. (2009). Untangling the Phylogeny of Amoeboid Protists. *Journ. Eukaryot. Microbiol.*, *56* (1), 16 – 25.

Pawlowski, J. & Majewski, W. (2011). Magnetite-bearing foraminifera from Admiralty Bay, West Antarstica, with description of *Psammophaga magnetica*, sp. nov. *Journ. Foram. Res.*, *41* (1), 3 – 13.

Pearson, P. N., Olsson, R. K., Huber, B. T., Hemleben, C. & Berggren, W. A. (2006). *Atlas of Eocene Planktonic Foraminiferas*. Cushm. Found. Spec. Publ.: 1 – 513.

Pecheux, M. J-F. (1995). Ecomorphology of a recent large foraminifer, *Operculina ammonoides*. *Geobios*, *28*, 529 – 566.

Pelissie, T., Peybernes, B. & Rey, J. (1988). Les grands Foraminiferes benthiques du Jurassique Moyen/Superieur du Sud-Ouest de la France (Aquitaine, Causses, Pyrenees). Interet biostratigraphique, paleoecologique et paleobiostratigraphique. *Rev. Paleobiol.*, *Benthos*, *1986*, *2nd Intern. Symp. Benthic Foram. (Pau, April 1983)*, 479 – 489.

Peybernes, B. (1976). *Le Jurassique et le Cretace inferieur des Pyrenees franco-espagnoles, entre la Garonne et la Mediterranee.* These Sciences Toulouse, 1 – 459.

Peybernes, B. (1988). Foraminiferes benthiques nouveax de L'Albian Superieur du massif du Turbon (Pyrenees Espagnoles). Benthos '83, 2nd Intern. Symp. Benthic Foraminifera (Pau, April 1983), 491 – 499.

Pierce, S. & Nathanson, M. E. (1974). Electron microscopical investigations on surface-related structures in *Allogromia* (Protozoa: Foraminiferida). *Trans. Amer. Micr. Soc.*, *93* (2), 170 - 180.

Piller, W. E. (1978). *Involutinacea (Foraminifera) der Trias und Lias*. Beitrage zur palaeontologie Oesterreich, Wien, *5*, 1 – 164.

Piller, W. E. (1990). Wall structures of Palaeotextulariid Foraminifers and discussion of Microgranular test walls. Hemleben et al., ed-s. *Palaeoecology*, *Biostratigraphy*, *Palaeooceanography and Taxonomy of Agglutinated Foraminifera.*, 25 – 35.

Piña-Ochoa, E., Høgslund, S., Geslin, E., Cedhagen, T., Revsbech, N. P., Nielsen, L. P., Schweizer, M., Jorissen, F., Rysgaard, S. & Risgaard-Petersen, N. (2010). Widespread occurrence of nitrate storage and denitrification among Foraminifera and Gromiida. *PNAS*, *107*(3), 1148 – 1153.

Pinard, S. & Mamet, B. I. (1998). *Taxonomie des petits foraminiferes du Carbonifere superieur-Permien inferieur du basin de Sverdrup, Arctique Canadiana*. Paleontographica Canadiana, *15*, 253.

Plewes, C. R., Palmer, T. J. & Haynes, J. R. (1993). A boring foraminiferan from the Upper Jurassic of England and Northern France. *Micropaleontology*, *12* (1), 83 – 89.

Plotnikova, L. F. (1991). O kriterijach systematiki ataxophragmiid. [On the criterions of the taxonomy of ataxophragmiids]. In: *Paleontological and stratigraphic investigations on the Ukranian territory. Ukraina. Kiev*, 74 – 76.

Plotnikova, L. F. (1991). Vendskie foraminifery Podolii. [Vend foraminifera of Podolia]. *Geol. Journ.*, Kiev, *3*, 25 – 42.

Plotnikova, L. F. & Egorova, S. I. (1982). Osobennosti stroenija stenki rakovin Ataksophragmiid. [Peculiarities of the shell wall structure of Ataxophragmiids]. *Paleontol. and Stratigraphy of the Mazozoic of Ukraine*, *Kiev*, *Inst. of Geol. Sci.*, 24 – 25. (In Russian).

Poag, C. W. (1971). Notes on the morphology asnd habit of *Vasiglobulina alabamensis* (Foraminiferida). *Journ. Paleontol.*, *45*, 961 – 962.

Pochon, X., Pawlowski, Ja., Zaninetti, L. & Rowan, R. (2001). High genetic diversity and relative specificity among Symbiodinium-like endosymbiotic dinoflagellates in soritid foraminiferans. *Marine Biology*, *139*, 1069 – 1078.

Pochon, X., LaJeunesse, T. C. & Pawlowski, Ja. (2004). Biogeographic partitioning and host specialization among foraminiferal dinoflagellate symbionts (*Symbiodinium*; Dinophyta). *Marine Biol.*, *146* (1), 17 – 28.

Podlipaev, S. A., Naumov, A. D. & Borkin, L. Ja. (1974). K opredeleniju ponjatij polimerizacii i oligomerizacii [On the definition of the notions of polymerization and oligomerization]. *Zhurn. Obstsch. Biol.*, *35* (1), 100-113. (In Russian).

Poignant, A. (1984). La morphologie externe et interne des Oolininae – quelques aspects du tube entosolenian. [The external and internal morphology of the Oolininae – views of the entosolenian tube]. *Benthos '83; 2nd International Symposium Benthic Foraminifera* (Pau, April 1983), 501 – 509.

Poljanskiy, Ju. (1970). O morphologicheskich zakonomernostjach evolucii prostejshich. [On the morphological regularities in the evolution of protists]. *Zool. Joourn.*, *49* (4), 560 – 569. (In Russian).

Poljanskiy, Ju. & Raikov, I. B. (1977). Polimerizacijai oligomerizacija v evolucii prostejshich. [Polymerization and oligomerization in the evolution of protists]. In: *Znachenie processov polimerizacii i oligomerizacii v evolucii.* [*The role of the precosses of polymerization and oligomerization in evolution*]. Leningrad: 29-32. (In Russian).

Polovodova, I. & Schoenfeld, J. (2008). Foraminiferal test abnormalities in the Western Baltic Sea. *Journ. Foram. Res.*, *38* (4), 318 – 336.

Ponomareva, G. Ju., (2012). O nachodkach Eosigmoilina (Foraminifera) v razrezach srednego Urala [About find of Eosigmoilina (Foraminifera) in the sections of the middle Ural]. *Trudy XV Vserossijskogo mikropaleontol Sovestschanija*, *Gelendzhik*, *Sept. 2012*: 146 – 149.

Popescu, G. (1983). Marine Middle Miocene monothalamous foraminifera from Romania.*Mem. Inst. Geol. si Geofzsica*, Bucharest, *31*, 261 – 280.

Popescu, G. (1992). *Delphinoidella*, a new genus of the Family Pleurostomellidae (Foraminifera). *Rom. Journ. Paleontology*, *75*, 45 - 46.

Popescu, G. (1995). Contribution to the knowledge of the Sarmatian Foraminifera of Romania. *Roman. J. Paleontology*, *76*, 85 - 98.

Popescu, G. & Brotea, D. (1995). *Droogerinella* n. gen., an Orbitoidal foraminifer from the Mesogean Upper Eocene. *Rev. Espanol. Micropaleontol., 27* (1), 5 – 16.

Popescu, G. (2000). Lower and Middle Miocene agglutinated foraminifera from the Carpathian area. *Acta Palaeontologica Romaniae, 2*, 407 – 425.

Popescu, G., Cicha, I. & Rögl, F. (1998). Systematic Notes. *In:* Cicha, I., Rögl, F., Rupp, C., & Cytroka, J. (eds), Oligocene – Miocene foraminifera of the Central Paratethys. *Abhand. senckenbergischen naturforschenden Gesellschaft, 549*, 69-325.

Popescu, G. & Crihan, I-M. (2002). Contributions to the knowledge of the marine Middle Eocene Miliolida from Romania. *Acta Paleontologica Romaniae, 3*, 371 – 397.

Popescu, G. & Crihan, I-M. (2004 a). Contribution to the knowledge of the Miocene Foraminifera from Romania: superfamily Nodosariacea (fam. Nodosariidae and Vaginulinidae). *Acta paleontologica Romaniae, 4*, 385 – 402.

Popescu, G. & Crihan, I-M. (2004 b). Contribution to the knowledgeof the calcareous unicameral Foraminifera from Middle Miocene of Romania.). *Acta paleontologica Romaniae, 4*, 403 – 421.

Pozaryska, K. & Voigt, E. (1978). Bryozoans as substratum fossil fistulose Foraminifera (Fam. Polymorphinidae). *Lethaia, 18*, 155 - 165.

Preece, R. C., Kaminski, M. A. & Dignes, T. W. (1999). Miocene benthic foraminiferal morphogroups in an oxygen minimum zone, offshore Cabinda. In: The oil and gas habitats of the South Atlantic. Cameron, N.R., Bate, R.H., Clure, V.S. (eds). *Geological Society Special Publication, 153*, 267 – 282.

Premoli Silva, I., Caron, M., Leckie, R. M., Rose Petrizzo, M., Soldan, D. & Verga, D. (2009). *Paraticinella* n. gen. and taxonomic revision of *Ticinella bejaouensis* Sigal, 1966. *Journ.Foram. Res.*, (*2*), 128 – 137.

Preobrazhenskaya, T. V. & Tarasova, T. S. (2004). Polovoe razmnozhenie foraminifery *Planoglabratella opercularis* (d'Orbigny, 1839) v prirodnych uslovijach. [Sexual reproduction of the foraminifer *Planoglabratella opercularis* (d'Orbigny, 1839) in nature. *Biologiya morja, 30* (5), 375 – 379. (In Russian).

Price, G. D. & Hart, M. B. (1999). Low-latitude sea-surface temperatures for the mid-Cretaceous and the evolution of planktic foraminifera: Comment and reply. *Geology, 27*, 857.

Prigogine, I. & Stengers, I. (1986). *Porjadok iz chaosa*. [*Order out of chaos*]. Progress, Moscow. (In Russian).

Pronina, G. P. (1988). Pozdnepermskie milioljaty Zakavkaz'ja. {Late Permian Miliolata of the Trasncaucasian region].)]. *In:* V.I. Mikhalevich (Ed.), Sistematika, ekologia i stratigrafiya miliolyat (Foraminifery). *Trudy Zoologicheskogo Instituta, AN SSSR*, 184, 49 – 54. (In Russian).

Pronina, G. P. (1989). [Foraminifera of the *Paratriolites kinli* zone of the Dorashamian of Late Permian of Transcaucasia]. *Ezhegodnik Vsesojuznogo Paleontol. Obstschestva, Leningrad, 32*, 30 – 37. (In Russian).

Pronina, G. P. (1994). Klassifikacija i phylogenija otrjada Hemigordiopsida (Foraminifera). [Classification and phylogeny of the order Hemigordiopsida (Foraminifera)]. *Paleontologicheskiy Zhurnal, 3*, 13 – 24. (In Russian)

Pronina, G. P. (1995). Genus *Sphaerionia* and its stratigraphic significance. *4th Intern. Sympos. Albrechtsberg, (Austria) 8 – 11 Sept. 1994. Suppl. agli Annati dei Musei Civici di Rovereto Sezione Archeologia, Storia e Scienze Naturali, 1*, 105 – 117.

Pronina, G. P. (1999). Novye vidy verchney Permi Zakavkaz'ja.[New species of small foraminifers from the Upper Permian of Transcaucasus].*Voprosy Micropaleontologii, 11*, 4 – 14. (In Russian)

Pronina, G. P. & Vuks, V. Ja. (1995). New data on the Triassic foraminifers of Crimea. *4th Intern. Sympos. Albrechtsberg, (Austria) 8 – 11 Sept. 1994. Suppl. agli Annati dei Musei Civici di Rovereto Sezione Archeologia, Storia e Scienze Naturali, 11*, 215 – 227.

Pronina, G. P. & Nestell, M. N. (1997). Middle and Late Permian foraminifera from exotic limestone blocks of the Alma River basin, Crimea. *In: Ross, C.A., Ross, J.R.P., Brenckle, P.L., (eds) Late Paleozoic Foraminifera: their biostratigraphy, evolution and Paleoecology and the Mid-Carboniferous boundary. Cushman Foundation Foraminiferal Res., SpecialPubl., 36*, 11 – 114.

Pronina - Nestell, G. P. & Nestell, M. N. (2001). Late Changsingian foraminifera of the northwestern Caucasus. *Micropaleontology, 47*, 205 – 234.

Rahaghi, A. (1992). Remarks on the genera *Sirtina, Vanderbeekia*, *Iranites*, and *Neumannites* (Foraminifera) from Upper Cretaceous of Iran with suggestion of a new subfamily Neumannitinae. *Rev. Espan. Micropaleontol., 24* (2), 119 – 129.

Rajeshwara Rao & Periakali, N. P. (2001). *Cocoarota madrasensis* Rajeshwara Rao and Revets n.sp. – a new foraminiferal taxon from the inner shelf of the bay of Bengal, off Madras, India. *Journ. Foram. Res., 31*, 319 – 323.

Raikov, I. B. (1978). *Jadro prostejshich. Morfologija I evolucija*. [*Nucleus of Protists. Morphology and evolution*]. Leningrad. Nauka. 326 (In Russian).

Raikov, I. B., Karajan, B. P. & Mikhalevich, V. I. (1998). Ultrastructure of the gamont Shell and Nucleus in the Polythalamous Foraminifer *Elphidium ponticum. European Journ. of Protistology., 34*, 153 – 161.

Raskin, R. (2001). *Rossijskaja imperiya XIXnachala XX veka kak systema gosudarstvennych uchrezhdenij, sluzhby, soslovij, gosudarstvennogo obrazovaniya i elementov grazhdanskogo obstschestva [Russian empire of XIXbeginning of XX centuries as a system of the state institutions, service, classes, state education and the elements of the civil society*]. Mellen, (USA), Lewiston (UK). 449 (In Russian).

Rathburn, A. E., Levin, L. A., Held, Z. & Lohmann, K. C. (2000). Benthic foraminifera associated with cold methane seeps on the northern California margin: Ecology and stable isotopic composition. *Marine Micropaleontol., 38*, 247 – 266.

Rauser-Chernousova, D. M. (1990). O systematicheskom polozhenii Nodosariidej I o sostave nadotrjada Lagenoida (Foraminifery). [On the systematic position of the Nodosariids and on the compostion of the superorder Lagenoida (Foraminifera)]. *Thes. XI Vsesojuzn. Sovestch., Moscow*: 109. (In Russian)

Rauser-Chernousova, D. M. & Reytlinger, E. A. (1986). Systematica nadrodovych taksonov otrjada Hormosinida (Foraminifera). [Systematic of the suprageneric taxa of the order Hormosinida (Foraminifera)]. *Paleontol. Journ., 20* (4), 11 – 16. (In Russian).

Rauser-Chernousova, D. M. & Fursenko, A. V. (eds) (1959). *Osnovy paleontologii. Obshcaya chast', Prosteyshie. [Principles of Paleontology. General part and Protozoa.*] Moscow, 482 (In Russian).

Rauser-Chernousova, D. M., Bensh, F. R., Vdovenko, M. V., Gibshman, N. B., Leven, E. Ja., Lipina, O. A., Reytlinger, E. A., Solovjeva, M. N. & Chedija, I. O. (1996). *Spravochnik po systematike foraminifer paleozoja (Endotyroidy, Fusulinoidy)* [*Reference-book on the systematics of Paleozoic foraminifera (Endothyrida, Fusulinida*)]. Geologicheskiy institute Rossiyskoy Akademii Nauk. Nauka, Moscow, 207 (In Russian).

Reddy, P. P. & Rajshekhar, C. (2007). New subfamily Asipholageninae from Holocene deposits of the gulf of Kachchh, India. *Journ. Foram. Res.*, *37* (1), 4 – 7.

Reichel, M. (1984). Le crible apertural de *Rhapydionina liburnica* Stache du Maastrichtien de Vremski- Britof, Yougoslavia. *Benthos '83, 2nd Int. Symp. Benthic Foraminifera (Pau, April 1983)*, 525 – 532.

Reiss, Z. (1963). Reclassification of perforate Foraminifera. *Bull.Geol. Surv. Israel.*, *35*, 1 - 111.

Reiss, Z. & Hottinger L. (1984). *The Gulf of Aqaba. Ecological Micropaleontology*. In: Ecological Studies, Analysis and Synthesis.Vol. *50*. Springer-Verlag. Berlin, Heidelberg, New York, Tokyo. 354.

Reitlinger, E. A. (1965). Evolucia foraminifer v epochu pozdnej permi i rannego triasa na territorii Zakavkaz'ja. [Evolution of foraminifera in the latest Permian and earliest Triassic epochs of the territory of the Transcaucasus. *Voprosy mikropaleontologii*, *52*, 45 – 70.

Remizova, S. T. (2004). *Fusulinoida of Timan. Evolution, stratigraphy and paleogegraphy*. Russian Academy of Sciences, Ural Department, Komi Scientific Center, Institute of Geology. Ekaterinburg, 217 (In Russian).

Resig, J. M. & Glenn, C. R. (1997). Foraminifera encrusting phosphoritic hardgrounds of the Peruvian upwelling zone: taxonomy, geochemistry, and distribution. *Journ. Foram. Res.*, *27* (2), 133 – 150.

Rettori, R. (1994). Replacement name *Hoyenella*, gen. n. (Triassic Foraminiferida, Miliolana) for Glomospira sinensis Ho, 1959. *Bull. Soc. Paleontol. Italiana, 33* (3), 341 – 343.

Rettori, R. & Zaninetti, L. (1993). *Paulbronnimannia* n. gen. (Foraminiferida, Ammodiscidae, Paulbronnimanninae n. Subfam., Trias) un nuovo genere per *Agathammina judicariensis* Premoli Silva, 1971. *Rev. Paleobiologie*, *12*, 176 – 177.

Rettori, R., Angiolini, L. & Muttoni, G. (1994). Lower and Middle Triassic Foraminifers from the Eros Limestone (Hydra Island, Greece). *Journ Micropaleontol.*, *13* (2), 25 -46.

Revets, S. A. (1989). Structure and taxonomy of the genus *Delosina* Wiesner, 1931 (Protozoa: Foraminifera). *Bull Br Mus(Nat Hist) Zool Ser.*, *55*, 1-9.

Revets, S. A. (1990). The genus *Floresina*, gen. nov. *Journ. Foram. Res.*, *20* (2), 157 – 161.

Revets, S. A. (1991). The reclassification of *Tremachora* Lipps and Lipps, 1967 in the suborder Delosinina. *Journ. Foram. Res.*, *21* (3), 252 – 255.

Revets, S. A. (1992). The structure and taxonomic position of *Millettia* Schubert, 1911 (Foraminiferida). *Journ. Micropaleontol.*, *11* (1), 37 – 46.

Revets, S. A. (1993 a). The Revision of the genus *Elongobula* Finlay, 1939. *Journ. Foram.Res.*, *23* (4), 254 – 266.

Revets, S. A. (1993 b). The foraminiferal toothplate, A Review.*Micropaleontology*, *12* (2), 155 – 168.

Revets, S. A. (1993 c). The revision of the genus *Buliminella* Cushman & Stainforth, 1947, and *Eobuliminella* gen. nov. *Journ. Foram.Res.*, *23* (2), 141 – 151.

Revets, S. A. (1994). The status of the genus *Anomalina* d'Orbigny, 1826. *Journ. Foram. Res.*, *24* (4), 296 – 304.

Revets, S. A. (1996). The generic Revision of the Bolivinitidae Cushman, 1927. *Cushm. Found. Foram. Res., Spec. Publ., 0* (34), 1 – 66.

Revets, S. A. (1996). The generic Revision of Five Families of Rotaliine Foraminifera. Part. I. The Bolivinitidae. Part II. The Anomalinidae, Alabaminidae, Cancrisidae and Gavelinellidae. *Cushm. Found. For. Res. Sp. Publ.*, *34*, 113.

Revets, S. A. (1996). The generic consept of *Hanzawaia* Asano, 1944 and the new genus *Riminopsis*. *Journ. Foram. Res.*, *26* (1), 3– 8.

Revets, S. A. (2001 a). The genus *Rotorbinella* Bandy, 1944 and its classification. *Journ. Foram.Res.*, *31*, 315 – 318.

Revets, S. A. (2001 b). *The revision of Gavelinella Brotzen, 1942, Berthelina Malapris, 1965 and Lingulogavelinella Malapris, 1965*. Cushman Foundation Spec. Publication, *37*, 1 – 110.

Revets, S. A. (2002 a). The genus *Albertinopsis*, gen. nov. and the relations between Gavelinellidae and Rosalinidae. *Journ. Foram. Res.*, *32* (1), 51 – 56.

Revets, S. A. (2002 b). The systematic position of *Conorbina* Brotzen, 1936. *Journ. Foram. Res.*, *32* (1), 57 – 60.

Revets, S. A. (2002 c). On the rotaliine nature of *Gavelinopsis* Hofker, 1951. *Journ. Foram. Res.*, *32* (1), 61 – 67.

Revets, S. A. (2005). A key to the unilocular hyaline Foraminifera. *Journ. Micropaleontol.*, *24*, 145 – 158.

Revets, S. A. & Whittaker, J. E. (1991). The taxonomic position of *Orthoplecta* Brady, 1884 (Foraminiferida). *Journ Micropaleontol.*, *9* (2), 167 – 172.

Reuss, A. E. (1862). Entwurf einer systematischen Zusammenstellung der Foraminiferen. *Sitzungberichte der Katiserlichen Akademie der Wissenschaften in Wien*, *Mathematisch-Naturwisseschaftliche Classe (1861)*, *44*, (1), 355 – 396.

Rhumbler, L. (1895). Entwurf eines naturlichen systems der Thalamorphoren.Nachrichten von der Koniglichen Gesellschaft der Wissenschaften zu Gottigen, *Mathematisch-Physikalische Klasse 1895*, (1), 51–98

Rhumbler, L. (1911). Die Foraminiferen (Thalamophoren) der Plankton-Expedition. *Ergebnisse. Plankton-Exped. Humbolt-Stiftung.*, *3*, 1 - 476.

Ribes, T., Salvado, H., Romero, J. & Gracia, M. P. (2000). Foraminiferal colonization on artificial seagrass leaves *Journ. Foram. Res.*, *July 2000*, *30*, 192 – 201.

Richard, K., Olsson, J. D., Wright & Miller, Kenneth G. (2001). Paleobiogeography of *Pseudotextularia elegans* during the latest Maastrichtian Global warming event. *Journ. Foram.Res.*, *31*, 275 – 282.

Richardson, S. L. (2001). Endosymbiont change as a key innovation in the adaptive radiation of Soritida (Foraminifera). *Paleobiology*, *27* (2), 262 – 289.

Rigaud, S., Martini, R. & Rettori,R. (2012). Parvalamellinae, a new subfamily for Triassic glomospiroid Involutinidae. *Journ. Foram. Res.*, *42*, 245 - 256.

Risgaard-Petersen, N., Langezaal, A. M., Ingvardsen, S., Schmid, M. C., Jetten, M. S., Op den Camp, H. J., Derksen, J. W., Pina-Ochoa, E., Eriksson, S. P., Nielsen, L. P., Revsbech, N. P., Cedhagen, T. & van der Zwaan, G. J. (2006). Evidence for complete denitrification in a benthic foraminifer. *Nature*, *443* (7107), 93–96.

Rivkin, R. B. & DeLaca, T. E. (1990). Trophic dynamics in Antactic benthic communities. I. In situ injestion of microalgae by Foraminifera and metazoan meiofauna. *Marin. Ecol. Prog. Ser.*, *64*, 129 – 136.

Robert, E. A., Boudreau, R. E. A. ., Patterson, R. T., Andrew, P., Dalby, W. & McKillop, B. (2001). Non-marine occurrence of the foraminifer *Cribroelphidium gunteri* in Northern lake Winnipegosis, Manitoba, Canada. *Journ. Foram. Res.*, *31*, 108 – 119.

Roberts, S. & Murray, J. W. (1995). Characterization of cement mineralogy in agglutinated foraminifera (Protista) by Raman spectroscopy. *Journ. Geol. Soc.*, *152*, 7 – 9.

Roettger, R. (1974). Larger Foraminifera: reproduction and early stages of development in *Heterostegina depressa*. *Marine Biol.*, *26* (1), 5 - 12.

Roettger, R. (1978). Unusual multiple fission in the gamont of the larger Foraminiferan *Heterostegina depressa*. *Journ. Protozool.*, *25* (1), 41 – 44.

Roettger, R. & Hallock, P. (1982). Shape trends in *Heterostegina depressa* (Protozoa, Foraminifera). *Journ.Foram. Res.*, *12* (3), 197 – 204.

Roettger, R. & Spindler, M. (1976). Development of *Heteristegina depressa* individuals (Foraminifera, Nummulitidae) in laboratory cultures. 1st Int. Symp. on Benthonic Foraminifera of Continental margins. Part A. Ecology and Biology. *Maritime Sediments*, Sp. publ. *1*, 81 – 87.

Roettger, R., Spindler, M., Schmaljohann, R., Richwien, M. & Fladung, M. (1984). Functions of the canal system in the rotaliid foraminifer *Heterostegina depressa*. *Nature*, 309/597, 789 – 791.

Roettger, R., Schmaljohann, R. & Zacharias, H. (1989). Endoreplication of zygotic nuclei in the larger foraminifer *Heterostegina depressa* (Nummulitidae). *Europ. Journ. Protistology*, *25*, 60 – 66.

Roettger, R. & Krueger, R. (1990). Observations on the biology of Calcarinidae (Foraminiferida). *Marine Biology*, *106*, 419 – 425.

Roettger, R., Krueger, R. & De Rijk, S. (1990). Trimorphism in Foraminifera (Protozoa), verification of an old hypothesis. *Europ. Journ. Protistol.*, *25*, 60 – 66.

Roettger, R., Dettmering, C, Krueger, R., Schmaljohann, R. & Hohenegger, J. (1998). Gametes in nummulitids (Foraminifera). *Journ. Foram. Res.*, *28* (4), 345 – 348.

Ross, C. A. (1972). Biology and Ecology of *Marginopora vertebralis* (Foraminiferida), Great Barrier Reef. 1. *Protozool.*, *19*, 181-192.

Ross, C. A. & Haman, D. (1989). Suprageneric ranges of Foraminiferids. *Journ. Foram. Res.*, *19* (1), 72 – 84.

Rossignol, L., Dupuy, C., Pascal, P. Y. & Debenay, J.-P. (2007). *Hydrobia ulvae*: a deposit feeder for cleaning living hard-shelled foraminifera: *Journ. Foram.Res.*, *37*, *8*–11.

Rupp, G., Bowser, S. S., Manella, A. & Rieder, C. L. (1986). Naturally occuring Tubulin-containing paracrystals in *Allogromia*: Immunocytochemical identification and functional significance. *Cell Motil. Cytoskel.*, *6*, 363 – 375.

Sabbatini, A., Bonatto, S., Gooday, A. J., Morigi, C., Pamcotti, I., Pucci, F. & Negri, A. (2010). Modern benthic foraminifers at Northern shallow sites of Adriatic Sea and soft-walled, monothalamous taxa: a brief overview. *Micropaleontology*, *56* (3-4), 359 – 376.

Saidova, Kh. M. (1976). *Bentosnye foraminifery Mirovogo okeana. [Benthonic Foraminifera of the Worls Ocean*]. Moscow. Nauka: 153 (In Russian).

Saidova, Kh. M. (1980). Mikrostruktura stenki rakovin vidov nadsemeystva Miliolidea (Foraminifera). [Microstructure of the shell wall of the species of superfamily Miliolidea (Foraminifera). *Materialy VIII mikropaleontologicheskogo sovestchaniya "Systematika I morphologiya mikroorganismov*". 21 – 24 October, Baku: ELM: 92 - 93. (In Russian).

Saidova, Kh. M. (1981). *O sovremennom sostojanii nadvidovych taksonov kajnozojskych bentosnych Foraminifer* [On the present state of the supraspecific taxa of the Cenozoic benthonic Foraminifera]. Rauser-Chernousova, D.M., Isakova, T.N., Reitlinger, E.A. (eds) Moscow: Inst. Okeanologii P.P. Shirshova, Akademiya Nauk SSSR, 1 -73. (In Russian).

Saidova, Kh. M. & Beklemishev, K. V. (1953). O nachozhdenii v morskich osadkach foraminifer, prosverlennych molod'ju gastropod. [On the occurrence in the sea sediments of foraminifers drilled by the young of gastropods]. *Doklady Academii Nauk USSR*, *92* (5), 1061 – 1063. (In Russian).

Salami, M. B. (1976). Biology of *Trochammina* cf. *T. quadriloba* Hoeglund (1947), an agglutinated foraminifer. *Cell Motil. Cytoskel.*, *6*, 142 - 153.

Sarkar, S. & Gupta, A. K. (2008). Five new species of benthic foraminifera from upper Pleistocene sequence in ODP Hole 716A, Maldives Ridge, equatorial Indian Ocean. *Micropaleontology*, *54* (2), 176 – 184.

Schaudinn, F. (1894). Die Fortpflanzung der Foraminiferen und eine neue. *Art der Kernteilung Biologishes Centralbla*t, *14*, 161 – 166.

Schaudinn, F. (1894). Uber die systematische und Fortpflanzung von *Hyalopus* n. gen. *Gromia dujardini* Schultze. *Naturwissenchaftiche Wochenschrift*, *95*, 196.

Schaudinn, F. (1895). Untersuchungen an Foraminiferen. I. *Calcituba polymorpha. Z. Wiss. Zool.*, *59*, 191 - 232.

Schlagintweit, F., Velić, I. (2011). New and poorly known Middle Jurassic larger benthic foraminifera from the Karst Dinarides of Croatia. *Geologia Croatica*, *64* (2), 81 – 99. doi: 10.4154/GC.2011.08

Schmalchgausen, I. I. (1939). *Putii zakonomernosti evolucionnogo processa. Izbrannye trudy. [Ways and regularities of the evolutionary process. Selected works].* AN SSSR, Sekcia chimikotechnologicheskich i bilogicheskich nauk In-t evolucionnoy morphologii i ekologii zhivotnych im. A.N. Severcova, Nauka, Moscow. (In Russian).

Schmalchgausen, I. I. (1946). *Factory evolucii* [*Factors of evolution*]. Izdatelstvo Academii Nauk SSSR, Moscow-Leningrad. (In Russian).

Schmalchgausen, I. I. (1969). *Problemy darvinisma.* [Problems of darvinism]. Leningrad. "Nauka": 492 (In Russian).

Schneider, A. (1878). Beitrage zur Kenntnis der Protozoen. Zeits. f. wiss. Zool., *30* (SupPlate), 446-456.

Schroeder, R. A. & Bada J. L. (1977). Kinetics and mechanism of the epimerization and decomposition of threonine in fossil Foraminifera. *Geochim. et Cosmochim. Acta.*, *41* (8), 1087 - 1095.

Schroeder, R. A., Clavel, B. & Charollais, J. (1990). *Praedictyorbitolina carthusiana* n. gen. n. sp., Orbitolinidae (Foraminiferida) de la limite Hauteriven-Barremien des Alpes occidentales. *Palaeontologische Zeitschrift*, *64*, 193 – 202.

Schultze, M. S. (1854). *Ueber den Organismus der Polythalamien (Foraminiferen) nebst Bermerkungen ueber die Rhizopoden im Allgemeinen.* W.Engelmann. Leipzig. 68 S.

Schultze, M. (1875). Rhizopodenstudien. III. *Arch. mikr. Anat.*, *11*, 94 – 139.

Schwab, D. (1969). Electronen microscopische Untersuchungen an der Foraminifere *Myxotheca arenilega* Schaudinn. *Z. Zellforsch microsk. Anat.*, *96*, 295 - 324.

Schwab, D. (1971). Elektronenmikroskopische Untersuchungen intracellular lebenden Einzellern in Foraminiferen. *Z. Naturforsch.*, *26b* (12), 1341 - 1344.

Schwab, D. (1972). Electron microscopic studies on the foraminifer *Allogromia laticollaris* Arnold. Mitosis in agamonts. *Protoplasma. 75*, 79 - 89.

Schwab, D. (1973). Centrosomal bodies during meiosis in the foraminifer *Myxotheca arenilega* Schaudinn. *Protoplasma. 78* (3), 339 - 341.

Schönfeld, J. (2001). Benthic Foraminifera and pore-water oxygen profiles: a re- assessment of a species boundary conditions at the Western Iberian margin. *Journ. Foram.Res., 31*, 86 – 107.

Schwab, D. (1976). Gametogenesis in *Allogromia laticollaris. Journ.Foram. Res., 6* (4), 251 – 257.

Schwab, D. & Hofer, H. W. (1979). Metabolism in protozoan *Allogromia laticollaris* Arnold. *Z. Mikroskop.-anat.Forschung., 93* (4), 715 - 727.

Schwab, D. & Plapp, R. (1983). Quantitative chemical analysis of the shell of the monothalamous foraminifer *Allogromia laticollaris* Arnold. *Journ. Foram. Res., 13* (1), 69 - 72.

Schwab, D. & Schlobach, H. (1973). Carotene wax pigments in the foraminiferan *Allogromia laticollaris* Arnold. *Journ. Protozool., 2* (2), 339 - 341.

Schwab, D. & Schlobach, H. (1979). Pigments in monothalamous Foraminifera. *Journ. Foram. Res., 9* (2), 141 - 146.

Schwab, D. & Schwab-Stey, H. (1981). Hexagonal structures at the plasma membrane of the foraminifer *Allogromia laticollaris* Arnold. *Journ. Foram. Res., 11* (3), 212 - 216.

Schwab-Stey, H. & Schwab, D. (1979). The transformation of Cytoplasmic Microtubules into Helices and Paracrystals by Halothane in the Foraminifer *Allogromia laticollaris* Arnold. *Z. mikroskop.- anat. Forschung.* Leipzig. *93* (4), 751 - 762.

Schwager, C. (1877). Quadro del proposto sistema de classificazione dei foraminiferi con guscio. *R. Comitato Geol. Italia, Bull. 8* (1 – 2), 18 - 27.

Schweizer, M., Pawlowski, J., Kouwenhoven, T., Guiard, J. & van der Zwaan, B. (2008). Molecular phylogeny of Rotaliida (Foraminifera) based on complite small subunit rDNA sequences.*Marine Micropaleontol., 66*, 233 – 246.

Schweizer, M., Pawlowski, J., Kouwenhoven, T. & Van Der Zwaan, b. (2009). Molecular phylogeny of common cibicids and related Rotaliida (Foraminifera) based on small subunit rDNA sequences. *Journ. Foram. Res., 39* (4), 300 – 315.

Schweizer, M., Bird, C., Evans, K., Austin, W. & Darling, K. (2012). Phylogeography of selected rotaliids (benthic Foraminifers) from the North East Atlantic. *The Micropaleontol Soc. Foraminifera and Nannofossil Groups Joint Meeting 2012*, Interdisciplinary advances in Foraminiferal and nannofossil Res., Pollock Halls, Univ. Edinburg, 39.

Scott, G. (2002). Upper Albian benthic foraminifera new in West Texas. *Journ. Foram. Res., 32* (1), 43 – 50.

Scott, G., Thompson L., Hitchin, R. & Scourse, J. (1998). Observations on selected salt-marsh and shallow-marine species of agglutinated foraminifera:grain size and mineralogical selectivity. *Journ. Foram. Res., 28* (4), 261 – 267.

Seiglie, G. A. (1971). A preliminary note on the relationships between Foraminifers and pollution in two Puerto Rican Bays. *Carib. Journ. Sci., 11* (2), 93 - 98.

Seiglie, G. A. (1982). A note on the homeomorphism of three soritid genera (Foraminiferida). *Micropaleontol., 28* (4), 431 – 434.

Seiglie, G. A., Baker, M. B. & Haman, D. (1991). *Ponceammina vancouveringi* n.gen. n.sp. (Ponceamminidae, Foraminiferida) from the Early Miocene of Puerto Rico. *Tulane Studies in Geology and Paleontology. 24* (3-4), 93 – 96.

Seiglie, G. A., Haman, D. & Baker, M. B. (1993). *Parkiella* n. gen., a Late Cretaceous (sic) turrilinacid foraminifer from Gabon and Cameroon, West Africa. Micropaleontology, 39 (3), 288 – 291.

Semeniuk, T. A. (2000). Spatial variability in epiphytic foraminifera from micro- to regional scale. *Journ. Foram. Res.*, *30*, 99 - 109.

Sen-Gupta, B. K. (2002). *Systematics of modern foraminifera*. In:Sen Gupta, B.K. (Ed.), Modern Foraminifera. Kluwer, Dordrecht, 7 – 36.

Sen Gupta, B. K., Platon, E., Bernhard, J. M. & Aharon, P. (1997). Foraminiferal colonization of hydrocarbon-seep bacterial mats and underlying sediment, Gulf of Mexico slope. *Journ. Foram. Res.*, *27* (4), 292 - 300.

Sen-Gupta, B. K. & Machain-Castillo, M. L. (1993). Benthic Foraminifera in Oxygen pure habitats. *Marine Micropal.*, *20*, 182 – 201.

Sen Gupta, B. K., Smith, L. E. & Machain-Castillo, M. L. (2009). Foraminifera of the Gulf of Mexico. *In Felder, D.L. and D.K. Camp (eds.), Gulf of Mexico–Origins, Waters, and Biota. Biodiversity. Texas A&M Press, College Station, Texas*, 87 – 129.

Septfontaine, M. (1984). Le Dogger des Prealpes medians suisses et francaises. Stratigraphie, evolution paleogeographique et paleotectonique. *Mem. Soc. Helvet. des Sci. nat.*, *97*, 1 – 121.

Septfontaine, M. (1985). Milleux de depots et Foraminiferes (Litoulides) de la plate-forme carbonate du Lias moyen au Maroc. *Rev. Micropaleontol.*, *28* (4), 581 – 591.

Septfontaine, M. (1988). Towards an evolutionary classification of Jurassic lituolids (Foraminifera) in carbonate platform environment. *Revue de Paléobiologie Special Volume*, *2*, *Benthos '86*, *Geneve*, *Mars*, *1988*, 229 - 256.

Septfontaine, M. & Matos, J. E. (1998). *Pseudodictyopsella jurassica* n. gen., n. sp., a new foraminifera from the early Middle Jurassic of the Musandam Peninsula, N. Oman mountains; sedimentological and stratigraphical context. *Rev. Micropaleontologie*, *41* (1), 71 – 87.

Seravin, L. N. (1980). Makrosystema zhgutikonoscev. [Macrosystem of Mastigophora]. *Proc. Zool. Inst. Russian Acad. Sci.*, *94*, 4 – 22. (In Russian).

Sergeeva, N. G., Anikeeva, O. V. & Gooday, A. J. (2010). Soft-shelled, monothalamous foraminifera from the oxic/anoxic interface (NW Black Sea). *Micropaleontology*, *56* (3-4), 393 – 407.

Severtzov, A. N. (1925). *Glavnye napravlenija evolucionnogo processa. Progress, regress i adaptacii.* [*Main directions of the evolutionary process. Progress, Regress and Adaptation*]. Izdanie Tovaristschestva AV Dushnov i K°, Moscow. (In Russian).

Severtzov, A. N. (1939). *Morphologicheskie zakonomernosti evolucii.* [*Morphological regularities of evolution*]. BS Matveev and SV Emelianov (ed-s) Izdatelstvo Academii Nauk SSSR, Moscow- Leningrad. (In Russian).

Shires, R., Gooday, A. & Jones, A. (1994). The morphology and Ecology of an abundant New Komokiacean Mudball (Komokiacea, Foraminiferida) from the Bathyal and abyssal NE Atlantic. *Journ. Foram. Res.*, *24* (4), 214 – 225.

Shuaibi, A. (2011). *Pessagnoites*, a new foraminifer genus from the Cretaceous of the Austin Chalk, northeast Texas.*Micropaleontology*, *57* (1), 87 – 91.

Siano, R; Montresor, M., Probert, I., Not, F. & De Varga, C. (2010). *Pelagodinium* gen. nov. and *P. béii* comb. nov., a dinoflagellate symbiont of planktonic foraminifera. *Protist. 161* (3), 385 – 399.

Sigal, J. (1952). Ordre des Foraminifera. *Traite Paleontol*. 1. Paris: 133 - 301.

Sirel, E. (1992). *Cideina*, a new foraminiferal genus from the Maastrichtian limestone of the Cide region (North Turkey). *Turkey, Min. Res. Expl. Bull., (foreign edition), Ankara, 112* (1991), p. 68.

Sirel, E. (1996). *Praearchaias*, a new soritid genus (Foraminiferida) and its Oligocene shallow-water foraminiferal assemblage from the Diyarbakir region (SE Turkey). *Geologica Romana, Roma, 32*, 167 - 181.

Sirel, E. (1998). *Foraminiferal description and biostratigraphy of the Paleocene-Lower Eocene shallow-water limestones and discussion on the Cretaceous-Tertiary boundary in Turkey.* Ankara:General Directorate of the Mineral Research and Exploration, Monography Series, 2, 50, 51.

Sirel, E. (1999). Four new genera (*Haymanella, Kayseriella, Elasigella and Orduella*) and one new species of *Hottingerina* from the Paleocene of Turkey. *Micropaleontology, 45* (2), 126 – 128.

Sirel, E. & Acar, S. (1993). *Malatyna*, a new foraminiferal genus from the Lutetian of the Malatya region.- *Geologica Croatica, Zagreb, 42*, (2), 181 – 188.

Sliter, W. V. (1965). Laboratory experiments on the life cycle and ecologic controls of *Rosalina globularis* d'Orbigny. *Journal of Paleontology, 12*, 210–215

Sliter, W. V. (1970). *Bolivina doniezi* Cushman et Wickenden in clone Culture. *Contr. Cushm. Found. Foram. Res., 21*, 87 – 99.

Sliter, W. V. (1971). Predation on benthic Foraminifera. *Journ. Foram. Res., 1* (1), 20 – 29.

Sliter, W. V. (1974). Test ultrastructure of some living benthic Foraminifers. *Lethaia, 7*, 5 - 16.

Sliter, W. V. (1975). Foraminiferal life and residue assemblages from Cretaceous slope deposits. *Geol. Soc. Americ. Bull., 86*, 897 – 906.

Snider, L. J., Bernett, B. R. & Hessler, R. R. (1984). The composition and distribution of meiofauna and nanobiota in a central Noth Pacific deep-sea area. *Deep-sea Res., 31*, 1225 – 1249.

Sokolov, B. S. (2011). Paleontologija dokembrija i akrochrony biosphernoy evolucii (k teorii rasshirjajustschejsja biosphaery). [Paleontoly of Pre-Cambrian and akrochrones of the evolution of the biosphere (to the theory of the widening biosphere)]. In: The rate of the evolution of the organic world and biostratigraphy]. *Materials LVII session of Paleontol. Soc., Rus. Acad. Sci, Paleontol. Soc., VSEGEI (All-Russian Geol. Inst. in hon. A.P.Karpinski), 6 – 9 Apr. 2009, St. Petersburg*: 3 - 11. (In Russian).

Sokolova, M. N. (2000). Feeding and trophic structure of the deep-sea macrobenthos. *Am. Zool.*, 22:723 – 733.

Solovieva, M. N. (1978). Integrativnye systemy foraminifer. [Integrative systems of Foraminifera]. *Voprosy micropal., 21*, 3 – 6. (In Russian).

Songzhu, G., Quinglai, F. & Weihong, H. (2007). The last Permian deep-water fauna: Latest Chaghsingian small foraminifers from southwestern Guangxi, South China. *Micropaleontology, 53* (4), 311 – 330.

Souter, P. B., Cedhagen, T. & Souter, D. W. (2002). Spatial distribution of the large foraminiferan *Discobotellina biperforata*, off the West Moreton Island, Queensland, Australia. *Journ. Foram. Res.*, *32*, 448 - 452.

Spero, H. J. (1988). Ultrastructural examination of chamber morphogenesis and biomineralization in the planktonic foraminifer *Orbulina universa*. *Marine Biol.*, *99*, 9 – 20.

Spindler, M. (1978 a). The development of the organic lining in *Heterostegina depressa* (Nummulitidae: Foraminifera*). Journ. Foram. Res.*, *8* (3), 258 - 261.

Spindler, M. (1978 b). Anatomy of the test of *Heterostegina depressa* (Foraminiferida). *Journ. Foram. Res.*, *8* (4), 319 - 331.

Spindler, M. (1979). Lunar periodicity of reproduction in the planktonic foraminifer *Hastigerina pelagica*. *Marine ecology. Progress Ser.* V *1* (1), 61 – 64.

Spindler, M. (1980). The pelagic gulfweed *Sargassum natans* as a habitat for the benthhic Foraminifera *Planorbulina acervalis* and *Rosalina globularis*. *Neues. Jahrb. Geol. Paleontol. Monatsh.*, *9*, 569 - 580.

Spindler, M., Anderson, O. R., Hemleben, C. & Be, A. W. H. (1978). Light and electron microscopic observations of gametogenesis in *Hastigerina pelagica* (Foraminifere). *Journ.Protozool.*, *25* (4), 427 - 433.

Spindler, M. & Hemleben, Ch. (1982). Formation and possible function of annulate lamelle in a planktonic Foraminifer. *Journ. Ultrasstructure Res.*, *81*, 341 - 350.

Spindler, M., Hemleben, Ch., Salomons, J. B. & Smit, L. P. (1984). Feeding behavior of some planktonic foraminifers in laboratory cultures. *Journ. Foram. Res.*, *14*, 237 – 249.

Spindler, M. & Roettger, R. (1973). Der Kammerbauvorgang der Grossforaminifere *Heterostegina depressa* (Nummulitidae). *Marine Biol.*, *18*, 146 – 159.

Starobogatov, Ja. & Mikhalevich, A. (1985). On the heterogeneyty of the Sarcodina. *Abstr. 7 th Intern. Congress Protozool. Nairobi. Kenia*, *189*, 100.

Stathoplos, L. & Turos, N. (1994). Proteins and DNA from modern planktonic Foraminifera. *Journ. Foram. Res.*, *24* (1), 49 - 59.

Stevens, C. H., Luken, M. D. & Nestell, M. K. (1991). The Upper Permian fusulinids *Reichelina* and *Parareichelina* in Northern California: evidence for long-distance tectonic transport. *In: Cooper, J.D., Stevens, C.H. (eds). Paleozoic Paleogeography of the Western United States, II, Pacific Section SEPM, 67*, 635 – 642.

Stouff, V., Lesourd, M. & Debenay, J. -P. (1999 a). Laboratory observations of asexual reproduction (schizogony) and ontogeny of *Ammonia tepida* with the comments on the cycle. *Journ. Foram. Res.*, *29* (1), 75 – 85.

Stouff, V., Geslin, E., Debenay, J. -P. & Lesourd, M. (1999 b). Origin of morphological abnormalities in Ammonia (Foraminifera), Studies in laboratory and natural environments. *Journ. Foram. Res.*, *29*, 152 – 170.

Subbotina, N. N., Voloshinova, N. A. & Azbel, A. Ja., (Ed-s) (1981). *Vvedenie v izuchenie foraminifer.Clssifikacia melkich foraminifer mezzo-kajnozoja. [Introduction to the foraminiferal study. Classification of the smaller Foraminifera of Mezo-Cenozoic*]. Vsesojuzn. nauchno-issled. geologo-razvedochn. (VNIGRI)Inst. Leningrad: 211 (In Russian).

Suhr, S. B., Alexander, S. P., Gooday, A. J., David, W., Pond, D. W. & Bowser, S. S. (2008). Trophic modes of large Antarctic foraminifera: roles of carnivory, omnivory, and detritivory: *Marine Ecology Progress Series*, *371*, 155–164.

Sukhov, E.E. (2012). Novyj permskij rod *Petschorica* I novye vidy rodov *Petschorica, Crespinitella, Earlandia, Trochammina* i *Ammovertella* (Foraminifera) petschorskoy provincii. [New Permian genus *Petschorica* and new species of the genre *Petschorica, Crespinitella, Earlandia, Trochammina* and *Ammovertella* (Foraminifera) of the Pechora's province]. *Trudy XV Vserossijskogo mikropaleontol Sovestschanija, Gelendzhik, Sept. 2012*, 90 – 92. (In Russian).

Svavarsson, S., Gudmundsson, G. & Brattegard, T. (1993). Feeding by assellote isopods (Crustacea) on Foraminifers (Protozoa) in the deep sea. *Deep-Sea Research, Part I - Oceangraphic Research Papers, 40* (6), 1225 - 1239.

Tasli, K., Altiner, D., Koc, H. & Eren M. (2008). Benthic foraminiferal biostratigraphy of the Jurassic platform carbonate succession in the Bolkar Mountains (Southern Turkey). *Micropaleontology, 54* (5), 425 – 444.

Tasli, K. & Altiner, D. (2010). *Pseudomarssonella* Redmond 1965 (Foraminifera) and related taxa from the Jurassic platform carbonate succession of the Central Taurides, S Turkey: its phylogenetic relation with the subfamily Paleopfenderininae. *Bulletin de la Societe Geologique de France, 181*, (5), 443-451.

Temirbekova, U. T. & Antonova, Z. A. (1985). *Triasovye I jurskie miliolidy Severnogo Kavkaza. [Triassic and Jurassic miliolids of the North Caucasis*]. Publ. Inst. Geol. Acad. Sci. USSR (Dagestan Departm.), 1 – 104.

Tendal, O. S. (1979). Aspects of the biology of Komokiacea and Xenophyophoria. *Sarsia, 64*, 13 – 17.

Tendal ,O. S. & Hessler, R. R. (1977). An introduction to the biology and systematics of Komokiacea (Textulariana, Foraminiferida). *Scandinavian Sci. Press.*, 165 - 194.

Thalmann, H. E. (1960). *An index to the genera and species of foraminifera 1890 – 1950.* California George Vanderbilt Foundation. Stanford Univ. 393.

Todd, R. (1965). A new *Rosalina* (Foraminifera) parasitic on a bivalve. *Deep Sea Res., 12*, 831 - 837.

Toksvad, T. & Hansen, H. J. (1983). A study of calcareous cement in agglutinated Foraminifera. *In:* Verdenius, J.G., Van Hinte, J.E., Fortuin, A.R. (eds), *Proceedings of the First Workshop on Arenaceous Foraminifera 7-9 September 1981. Continental Shelf Institute, Trondheim, Publication, 108*, 159 - 169.

Toll-Riera, M., Castresana, J. & Alba, M. M. (2008). Accelerated evolution of Genes of Recent Origin. *Evolutionary Biology from Concept to Application*, Part I. Modelization of Evolution. P. Pontarotti (ed.), Springer-Verlag Berlin Heidelberg, 45 – 61.

Towe, K. M. (1971). Lamellar wall construction in planktonic foraminifera. In: *Proc. Second Planktonic Conference. 2*, 1213 – 1218.

Towe, K. M. & Cifelli, R. (1967). Wall ultrastructure in the calcareous Foraminifera: Crystallographic aspects and a model for calcification. *Journ. Paleontol.*, 41 (3), 742 - 762.

Towe, K. M. & Wright, R. C. (1963). Ultramicrostructure of some selected foraminiferal tests. *Micropaleontology, 9*, 171 - 195.

Towe, K. M., Berthold, W. U. & Appleman, D. E. (1977). The cristallography of *Patellina corrugate* Williamson: a-axis preffered orientation. *Journ. Foram. Res., 7* (1), 58 - 62.

Travis, J. L. & Bowser, S. S. (1986). A new model of Reticulopodial Motility and shape: Evidence for a Microtubule - Based Motor and an Actin Skeleton. *Cell Motility and the Cytoskeleton., 6*, 2 - 14.

Travis, J. L., Welnhofer, E. A. & Orokos, D. D. (2002). Autonomous reorganization of foraminiferan reticulopodia. *Journ.Foram.Res*, *32*, 425 - 433.

Triantaphyllou, M. V., Dimiza, M. D., Koukousioura, O. & Hallock, P. (2012). Observations on the life cycle of the symbiont-bearing foraminifer *Amphistegina lobifera* Larsen, an invasive species in coastal ecosystems of the Aegean sea (Greece, E. Mediterranean). *Jour.Foram.l Res.*, *42*, 143-150, doi:10.2113/gsjfr.42.2.143.

Tyszka, J. (1997). *Miliammina gerochi* n.sp. - a Middle Jurassic rzehakinid (Foraminiferida) from quasi-anaerobic biofacies. *Annales Societatis Geologorum Poloniae*, *67*, 355 - 364

Tyszka, J. (2001). *Spiroplectinata*, key benthic foraminifer genus for palaeoceanographic reconstruction of the Albian Lower Saxony Basin. *Palaeogeography*, *Palaeoslimatology*, *Palaeoecology*, *174*, 199 – 220.

Tyszka, J. (2006). Taxonomy of Albian Gavelinellidae (Foraminifera) from the Lower Saxony Basin, Germany. *Palaeontology, 49* (6), 1303 – 1334.

Tyszka, J. & Kaminski, M. A. (1995). Factors controlling the distribution of agglutinated Foraminifera in Aalenia-Bajocian dysoxic facies (Pieniny Klippen Belt, Poland). In: Kaminski M.A., Geroch S., Gasinski M.A.(Eds). Proc. Fourth Intern. Workshop Aggl. Foram., Krakow, Poland, Sept. 12 – 19, 1993. *Grzybowski Found. Sp. Publ.*, *3*, 271 - 291.

Ueno, K. (1992). Permian foraminifera from the Takakurayama group of the southern Abukuma Mountains, northeast Japan. *Pal. Soc. Japan, Trans. Proc., Tokyo, n.s., 168*, 1285 – 1287.

Ujiie, H. (1990). Bathyal benthic foraminifera in a piston core from East of the Miyako Islands, Ryukyu Island arc. *Univ. Ryukyus, Coll. Sci., Bull., Okinawa, Japan, 49*, p. 36.

Uspenskaya, A. V. (1982). Novye dannye o zhiznennych cyclach I biologii miksosporidij (Protista). [New data on the life cycles and biology of Mixosporidia (Protista)]. *Doklady Acad. Sci. USSR*, *262*, 503 – 507. (In Russian).

Vachard, D. (1991). An assemblage with *Yabeina* (fusulinid foraminifer) from the Midian (Upper Permian) of Whangaroa area (Orua Bay, New Zealand). *Rev. Micropaleontol.*, *34*, 201 – 230.

Vachard, D. (1994). Foraminiferes et moravamminides du Givetien et du Frasnien du domaine Ligerien (Massif Armoricain, France). *Palaeontographica*, *A231* (1–3), 1 – 92.

Vachard, D., Martini, R., Rettori, R. & Zaninetti, L. (1994). Nouvelle classification des foraminiferes Endothyroides du Trias. *Geobios*, *27* (5), 543 – 557.

Vachard, D. & Fourcade, E. (1997). Foraminiferes et algues du Permien du Guatemala. *Geobios*, *30*, 745 – 784.

Vachard, D. & Krainer, K. (2001 a). Smaller foraminifers of the upper Carboniferous Auerning group, Carnic Alps (Austria/Italy). *Riv. Italiana Paleontiol. di Stratigr.*, *107* (2), 147 – 168.

Vachard, D., Hauser, M., Martini, R., Zaninetti, L., Matter, A. & Peters, R. (2002). Middle Permian (Midian) foraminiferal assemblages from the Batain Plain (Eastern Oman), their significance to paleogeography. *Journ. Foram. Res.*, *32*, 155 – 172.

Vachard, D., Rettori, R., Angiolini, L. & Checconi, A. (2008). *Glomomidiella* gen. n. (Foraminifera, Miliolata, Neodiscidae), a new genus from the late Guadalupian-Lopingian of Hydra Island (Greece). *Riv. Italiana Paleontol. e Stratigrafia*, *114* (3), 349 – 361.

Vachard, D. & Cozar, P. (2010). An attempt of classification of the Palaeozoic incertae sedis Algospongia. *Rev. Espanol. Micropaleotol.*, *42* (2), 129 – 241.

Vachard, D., Pille, L. & Gaillot, J. (2010). Palaeozoic Foraminifera: Systematics, palaeoecology and responses to global changes. *Revue de Micropaléontologie*, *53* (4), 209 - 254

Van Ginkel, A. C. (2010). Systematics of the Eostaffellidae (Late Paleozoic Foraminifera). *Cushm. Found. Spec. Publ.*, *42*, 130.

Vdovenko, M. V., Rauser-Chernousova, D. M., Reitlinger, E. A. & Sabirov, A. A. (1993). *Spravochnik po sistematike melkich foraminifer Paleozoya (za iskluchеniem endotiroidey i permskikh mnogokamernykh lagenoidey.* [*A reference-book on the systematics of Paleozoic Foraminifera (excluding endothyrids and Permian multichambered lagenids*]. "Nauka" Publishers, Moscow, 125 (in Russian)

Vecchio, E. & Hottinger, L. (2007). Agglutinated conical foraminifera from the Lower-Middle Eocene of the Trentinara Formation (southern Italy). *Facies*, *53*, 509 - 533.

Venec-Peyre, M. T. (1981). Les foraminiferes et la pollution: etude de la microfaune de la cale du Dourduff (embouchure de la riviere de Morlaix). *Cahiers de Biologie Marine*, 25 – 33.

Venec-Peyre, M. T. (1987). Boring foraminifera in French Polynesian coral reefs. *Coral Reefs*, *5*, 205 – 212.

Venec-Peyre, M. T. (2005). "*Les Planches inedites de Foraminiferes d'Alcide d'Orbigny*". À l'aube de la micropaléontologie. Muséum national d'Histoire naturelle, Paris: 302.

Verhallen, P. J. (1986). Morphology and function of the internal structures of non- costate *Bulimina. Proc. Koninklijke Nederlads Akademie van Wetenschappen*, *89*, 367 – 385.

Vespermann, J. (1998). *Carlsiella brunsviga* gen.nov, sp.nov. and *Cerobertinella subhercynica* sp.nov.: new foraminifers in the Upper Barremian of the eastern Lower Saxony Basin. *Cretaceous Research, 19* (5), 663 - 671.

Vicedo, V., Calonge, A. & Caus, E. (2011).Cenomanian rhapidioninids (Foraminiferida), architecture of the shell and stratigraphy. *Journ. Foram. Res.*, *41*, 41 - 52, doi:10.2113/gsjfr.41.1. 41.

Vicedo, V., Frijia, G., Parente, M. & Caus, E. (2011). The late Cretaceous genera Cuvillierinella, Cyclopseudedomia, and Rhapydionina (Rhapydionodae, Foraminiferida) in shallow-water carbonates of Pylos (Peloponnese, Greece). *Journ.Foram.Res.*, *41*, 167 - 181, doi:10.2113/gsjfr.41.2.167.

Voronova, M. N. (1976). O popyploidii somaticheskich yader foraminifery *Cibicides lobatulus*. [On the polyploidy of the somatic nuclei of the foraminifer *Cibicides Lobatulus*]. *Cytology*, *18* (4), 509 – 512. (In Russian).

Voronova, M. N. (1978 a). Jadernyj dualism gamontov foraminifery *Cibicides lobatlus*. [Nuclear dualism of the gamonts of the foraminifera *Cibicides lobatulus*]. *Cytology*, *20* (8), 859 – 867. (In Russian).

Voronova, M. N. (1978 b). Mejos u foraminifery *Cibicides lobatulus*. [Meiosis in the foraminifera *Cibicides lobatulus*]. *Cytology*, *20* (11), 1328 – 1331. (In Russian).

Voronova, M. N. & Selivanova, G. V. (1976). Isledovanie generativnych jader foraminifery *Cibicides lobatulus* metodom cytophotometrii. [Investigation of the gennerative nuclei of the foraminifera *Cibicides lobatulus* with the cytophotometric method]. *Protozoology*, *1*, 144 – 150. (In Russian).

Voronova, M. N. (1979). Zhiznenyj cycl foraminifery *Cibicides lobatulus.* [The life cycle of the foraminifer *Cibicides lobatulus*]. Dr. thesis. Leningrad: 1 – 24. (In Russian).

Voronova, M. N. & Mikhalevich, V. I. (1985). Sovremenye predstavlenija o zhiznennych cyclach foraminifer (Recent concept of the foraminiferal life cycles). *Proc Zool Inst Acad Sci USSR*, *129*, 48 – 66. (In Russian).

Vuks, G. P. (1984). First findings of Geinitzinitas (Foraminifera) in the Permian of Transcaucasia. *Paleontologicheskiy Zhurnal*, *3*, 131 – 133.

Walker, D. A. (1971). Etching of the test surface of benthonic foraminifers due to ingestion by the gastropod *Littorina littorea* Linne. *Canadian Journ. Earth Sci.*, *8* (11), 1487 - 1491.

Wang, Y. & Ueno, K. (2009). A new fusulinoidean genus *Dilatofusulina* from the Lopingian (Upper Permian) of Southern Tibet, China. *Journ.Foram. Res.*, *39*, (1), 56 – 65.

Weber, F. W. (1965). Ueber die Paarung der Gamonten und der Kerndualismus der Foraminifere *Metarotaliella parva* Grell. Arch. *Protistenk.*, *108*, 217 - 270.

Wefer, G. & Berger, W. H. (1980). Stable isotopes in benthic Foraminifera: seasonal variation in large tropical species. *Science.*, *209* (4458), 803 - 805.

Weinberg, J. R. (1991). Rates of movement and sedimentary traces of deep-sea Foraminifera and mollusca in the laboratory. *Journ. Foram. Res.*, *21* (3), 213 - 217.

Weiner, S. & Erez J. (1984). Organic matrix of the shell of the foraminifer *Heterostegina depressa*. *Journ. Foram. Res.,14*. N 3. P. 206 - 212.

Weiner, A., Kurosawa, A., Aurahs, R., Darling, K. & Kucera, M. (2012). Global genetic diversity, species boundaries and their implications to morphotaxonomy in *Globigerinella siphonifera. The Micropaleontol Soc. Foraminifera and Nannofossil Groups Joint Meeting 2012*, Interdisciplinary advances in Foraminiferal and nannofossil Res., Pollock Halls, Univ.Edinburg, *47*.

Weinholz, Z. P. & Lutze, G. F. (1989). The *Stilostomella* extinction. In: Ruddiman, W.F., Sarnthein, M. et al., *Initial Reports of the Deep Sea Drilling Project.* Washington, D.C., U.S. Government Printing Office, *108*, 113 – 117.

Welnhofer, E. A. & Travis, J. L. (1996). In vivo microtubule dynamics during experimentally induced conservations between tubulin assembly states in *Allogromia laticollaris. Cell Motility and Cytoskeleton*, *34*, 81 – 94.

Welnhofer, E. A. & Travis, J. L. (2003). Functional integration of actin and microtubule cytoskeleton during reformation of *Allogromia* pseudopodia. Manuscript in preparation for *Journal of Eukaryotic Molecular Biology.*

Westall, F., Boni, L. & Guerzoni, E. (1995). The experimental silicification of microorganisms. *Paleontology*, *38* (3), 495 – 528.

Wetmore, R. A. (1988). Burrowing and sediment movement by benthic Foraminifera, as shown by time-lapse cinematography. *Rev. Paleobiol.*, *2*, 921 - 927.

Wetmore, K. L. (1999). Chamber formation in *Archaias angulatus. Journ. Foram. Res.*, *29*, (1), 69-74.

Whittaker, J. E. (1988). *Benthic Caenozoic foraminifera from Ecuador*. Natural History Mus., London, XI: p.

Williams, H. F. L. (1999). Foraminiferal distribution in tidal marshes bordering the strait of Juan de Duca: implications for paleosismicity studies. *Journ. Foram. Res.*, *29* (3) 196 - 208.

Williams, D. E., Hallock, P., Talge, H. K., Harney, J. N. & McRae, G. (1997). Responses of *Amphistegina gibbosa* populations in the Florida Keys (U.S.A.) to a multi-year stress event (1991-1996). *Journ. Foram.Res.*, *27* (4), 264 - 269.

Williamson D. F., Roettger, R., Schmaljohann, R. & Keigwin, L. (1981). Oxygen and, carbon isotopic fractination and algal symbiosis in the benthic foraminiferan *Heterostegina depressa*. *Palaeogeogr.*, *Palaeoclimatol.*, *Palaeoecol.* *33*, 231 - 251.

Winter, F. W. (1907). Zur Kenntnis der Thalamophoren. I. Untersuchung uber *Peneroplis pertusus* (Forskal). *Arch. Protistenkunde*, *10*, 1 – 113.

Wray, C. G., Lee, J. J. & De Salle, R. (1993). Extraction and enzymatic characterization of foraminiferal DNA. *Micropaleontology*, *39* (1), 69 - 74.

Wray, C. G., Langer, M. R., De Salle, R., Lee, J. J. & Lipps, J. H. (1995). Origin of the foraminifera. *Proceedings National Academy of Sciences*, *USA*, *92*, 141 - 145.

Yadrenkin, A. V. (2008). *Dagysina* – novyj rod Nodosariid (Foraminifera) iz verchnego triasa ostrova Kotelnyj, Novosibirskie ostrova. [*Dagysina* – a new nodosariid genus (Foraminifera) from the Upper Triassic of Koltel'ny Island, New Siberian Islands]. *Paleontologeskii Zhurnal*, *42*, (3), 221 – 223. (In Russian)

Yakovleva, S. P. (1979). K sisteme jurskich Lituolin. [On the system of the Jurassic Lituolins]. *Paleontol. Journ.*, *1*, 12 – 21. (In Russian).

Yalcin, H., Meric, E., Avsar, N, Tetiker, S., Barut, I. F., Yilmaz, S. & Dincer, F. (2008). Mineralogical and geochemical features of colored benthic foraminifers from Aegean and southwestern coasts of Turkey. *Micropaleontology*, *54* (3 - 4), 351 – 370.

Yanko, V. (1995). Benthic foraminifera as indicators of heavy nmetal pollution along Israeli Coasts (Cruise AVI-1, May, 1993). In: Yanko, V., Ed., *Avicenna (AVI CT92-0007), Benthic foraminifera as indicators of heavy metal pollution – a new kind of biological monitoring for the Mediterranean Task*, *Tel Aviv*, *5*, 58 – 152.

Yanko, V., Kronfeld, J. & Flexer, A. (1994). Response of Benthic Foraminifera to various Pollution Sources: Implications for Pollution Monitoring. *Journ. Foram. Res.*, *24* (1), 1 - 18.

Yanko, V., Ahmad, M. & Kaminski, M. (1998). Morphological deformities of benthic foraminiferal tests in response to pollution by heavy metals: Implication for pollution monitoring. *Journ. Foram. Res.*, *28* (3), 177 – 200.

Yanko, V., Arnold, A., J. & Parker., W. C. (1999). *Effects of marine pollution on benthic Foraminifera.* In: Sen Gupta, B.K. (ed), Modern Foraminifera. Kluwer Academic Publishers, New York, 217 – 235.

Yankovskaya A. I. & Mikhalevich, V. I. (1972). Foraminifery ozera Issyk-Kul I gruntovych vod Sredney Asii. [Foraminifera of the Issyk-Kul Lake and of the ground waters of the Middle Asia]. *Dokl. Acad. Sci. USSR.*, *205* (4), 1005 – 1008. (In Russian).

Zakrevskaya, E.Yu. (1993). Assiliny, Operculiny i Ranikotalii Kryma i ich biostratigrapficheskoe znachenie. [Assilins, operculins and ranikothaliins of Crimea and their biostratigraphic significance]. Nemkov G.I. (Ed). Russian Academy of Sciences, Gosudarstvennyj Geologicheskiy Musey in honour of V.I.Vernadski. Moscow, "Nauka", 112. (In Russian).

Zaninetti, L., Salvini-Bonnard, G. & Charollais, J. (1987). *Montsalevia*, n. gen. (Montsaleviidae, N. Fam., Foraminifere), dans le Cretace Inderieur (Beriasien Moyen-Valanginien) do Mont Saleve et du Jura Meridional (Haute-Savoie, France); note preliminaire. *Rev. Paleobiol.*, *6* (1), 165 – 168.

Zaninetti, L., Ciarapica, G., Cirilli, S. & Cadet, J. P. (1985). *Miliolechina stellata* n.gen., n. sp., et *Hirsutospirella pilosa*, n. gen., n. sp., (Foraminiferes), dans le Trias Superior (Norien) a facies recial des Dinarides. *Rev. Paleontolog.*, Geneve, *4*, 331 – 341.

Zaninetti, L., Ciarapica, G. & Decruez, D. (1987). Sur la subdivision des Involutinacea Butschli, 1880. *Rev. Paleobiol.*, *6* (1), 1 – 3.

Zapata, M. A. & Varela, M. S. (1981). Epibiosis de Turbelarios sobre *Buccella frigida* (Cushman)(Foraminifera, Protozoa). *Bol. Soc. Biologia Concepcion. Chile*, *52*, 245 - 246.

Zech, L. (1964). Zytochemische Messungen an den Zellkernen der Foraminiferen *Patellina corrugate* und *Rotaliella heterocariotica*. *Arch. Protistenk.*, *107*, 295 - 330.

Zhang, Y. C., Cheng, L. R. & Shen, S. Z. (2010). Late Guadalupian (Middle Permian) fusuline fauna from the Xiala Formation in Xainza County, Central Tibet: Implication of the rifting time of the Lhasa Block. *Journ. Paleontol.*, *84* (5), 955 – 973.

Zheng, S. & Zhaoxian, F. (2001). *Fauna Sinica, Phylum Granuloreticulosa, Class Foraminifera*. Science Press, Beijing, China: 788, 122 Plates.

Zmiri, A., Kahan, D., Hochstein, S. & Reiss, Z. (1974). Phototaxis and thermotaxis in some species of *Amphistegina* (Foraminifera). *Journ. Protozool.*, *21* (1), 133 - 139.

Zmiri, A., Kahan, D. & Reiss, Z. (1972). Aspects of the thermotactic response in Foraminifera. *Journ. Protozool. (Suppl.)*, *19*, 56.

Zucker, W. H. (1973). Fine structure of planktonic Foraminifera and their endosymbiotic algae. Ph.D. Thes., City Univ. of New York. 65.

About the Author

Dr. Alla-Valeria Mikhalevich
Zoological Institute Russian Academy of Sciences
Leading Research Officer, Laboratory of Protozoology
ul. Vs. Vishnevskogo 5, fl. 6, 197136
S. Petersburg, Russia
Tel. 7 (812) 346-32-53
mikha07@mail.ru

Professor Mikhalevich is a corresponding member of the European Academy of Sciences, Arts and Literature.

INDEX OF LATIN NAMES

A

B

C

D

E

F

G

H

I

M

N

Q

R

S

V

W

X

Z